ORDERED SETS

Advances in Mathematics

VOLUME 7

ORDERED SETS

By

EGBERT HARZHEIM
University of Düsseldorf, Germany

 Springer

Library of Congress Cataloging-in-Publication Data

A C.I.P. record for this book is available from the Library of Congress.

AMS Subject Classifications: 06-01, 06A05, 06A06, 06A07

ISBN 978-1-4419-3704-9

e-ISBN 978-0-387-24222-4

Printed on acid-free paper.

Printed in the United States of America.

9 8 7 6 5 4 3 2 1

springeronline.com

Contents

PREFACE

This book is written particularly for mathematics students and, of course, for mathematicians interested in set theory. Only some fundamental parts of naive set theory are presupposed, – not more than is treated in a textbook on set theory, even if this restricts us only to the most basic facts of this field. We have summarized all of this in Chapter 0 without longer discusssions and explanations, because there are several textbooks which can be consulted by the reader, e.g. Hrbacek/Jech [88], Kneebone/Rotman [99], Shen/Vereshchagin [159]. Besides this only elementary facts of analysis are used.

The theory of ordered sets can be divided into two parts, depending on whether the sets under consideration are finite or infinite. The first part is grounded mainly in combinatorics and graph theory and does not make essential use of set-theorical concepts, whereas the second part presupposes a knowledge of the fundamental notions of set theory, in particular of the system of ordinal and cardinal numbers. In this book we mainly deal with general infinite ordered sets. In this field the textbook literature is still very small. Therefore this book supplements the existing literature.

Chapter 1, and partially also Chapter 2, contains most of the basic notions of ordered sets. Chapter 3 treats some fundamental results on linearly ordered sets. Chapter 4 considers products of ordered sets, and Chapter 5 is more specialized and supplements results of Chapter 4 with respect to partially ordered sets. The remaining chapters 6 – 11 are generally independent of each other, with exception of Chapter 6 and the last three sections of Chapter 8.

In many cases ordered sets can be well illustrated by diagrams, and the reader is recommended to use them as visual tools of proof. Many considerations become more transparent if they are accompanied by drawings, which also provide a geometrical flavour. Nevertheless all proofs in this book are represented in great detail and independent of accompanying diagrams.

Ordered sets occur in many mathematical contexts, e.g. always if we have a structured system and study the nature of the set of all its substructures. In this set we have the natural order $\subseteq$ by set-inclusion. So, e.g. the set of all subgroups of a given group forms an ordered system (a lattice) with several special properties. The same holds for the system of all ideals of a ring or all subspaces of a topological space, to mention

x

only a few examples.

Moreover order theory is essentially equal to the theory of the relation $\subseteq$, for there is a fundamental theorem stating that every ordered set is isomorphic to a set of sets, which is ordered by $\subseteq$.

In particular, order theory has close connections with graph theory, for an ordered set can also be interpreted as a transitively directed graph. Topology also has a lot of cross connection to order theory, for an ordered set can be equipped in several ways with a topology which relates to the order. Among others the famous invariance theorems of the topology of the euclidean n - space $\mathbf{R}^n$ can be generalized to finite products of continuous linearly ordered sets.

Further, order theory overlaps in great part with the combinatorics of finite sets. This theory is treated in the books of Anderson [5] and Engel/Gronau [32]. In Chapter 9 we present some characteristic theorems of this field.

Herewith some remarks on the history of order theory and the relative literature.

Several concepts of order theory occurred early in the history of mathematics. Already the system $\mathbf{N}$ of the natural numbers is characterized by order properties, e.g. in the axiom system of Peano. Each element of $\mathbf{N}$ has an immediate successor, and each natural number n is an "iterated successor" of 1. If n is different from 1, it has also an immediate predecessor. The enlargement of $\mathbf{N}$ to the set $\mathbf{Z}$ of integers effects that the last property is possessed by all integers. Also the augmentation of $\mathbf{Z}$ by the rational numbers has an order-theoretical aspect insofar as it makes the set $\mathbf{Q}$ of rational numbers dense: Between each two of its elements there is another element of $\mathbf{Q}$. Finally the transition from $\mathbf{Q}$ to the set $\mathbf{R}$ of real numbers is a special case of an order-theoretical process, namely filling the gaps. This is the idea in Dedekind's construction of the reals. His method was carried on by MacNeille [116].

Another field, in which order-theoretical concepts are involved, is geometry. If we have a straight line L and have fixed two points a and b of L and declare a to be before (or less than) b, this generates a linear order on L. Here we have the prototype of the concept "linear order".

When Georg Cantor created set theory he also took into account sets with an order structure, in todays terminology he considered linearly ordered sets and among these, in particular, the important class of well-ordered sets. For a long time the interest of mathematics was restricted to linearly ordered sets; not until 1911 did C. S. Peirce [137] and E. Schröder [158] introduce the concept "partially ordered set". In the meantime it appeared that this more general concept has a much wider range of applications. In particular the occurring examples of partially ordered sets were mostly of a more special kind, namely lattices. Therefore there arose the field of lattice theory which began a rapid development after the appearance of Birkhoff's classic [9]. Despite the fact that lattices are also ordered sets, their theory is considered in their own framework. Therefore in this book it is minimally treated. As a standard reference we mention the book of G. Grätzer [60].

The theory of ordered sets was advanced substantially by Hausdorff who created many concepts which are fundamental to the theory. In his book "Grundzüge der Mengenlehre" [81] he compiled many results referencing this. A great progress in order theory was obtained in 1941 by the creation of the dimension theory of ordered sets by Dushnik/Miller [30], which is based on a theorem of Szpilrajn [167] of 1930, which states that every partial order on a set can be extended to a linear one. Also the important class of order-theoretical trees came into consideration in the years following 1935. Kurepa [104] had investigated this class of ordered sets in his thesis in1935, partly in connection with the Suslin problem of 1920, which also had stimulated the investigation of trees. In 1958 the book "Matematica del Orden" by N. Cuesta Dutari [17] was published, which seems to be the first dealing exclusively with general ordered sets. In 1982 Erné's book "Einführung in die Ordnungstheorie"appeared, and in 1986 the book "Theory of Relations" of R. Fraissé, with a new edition [45] in 2000. Rosenstein's book "Linear orderings" [154] considers a special class of ordered sets, in particular the countable linearly ordered sets. Fishburn's book [43] of 1985 is devoted to interval orders and interval graphs. In 1990 appeared "Introduction to Lattices and Order" of Davey and Priestley [18], with a second edition in 2002, in 1995 Trotter's survey article [172] in the "Handbook of Combinatorics", and just recently in 2003 the book "Ordered Sets" by B. S. Schröder [157].

In 1984 the journal ORDER was founded, which primarily publishes results on ordered sets. Much information on this field can also be found in the conference volumes [151], [152], [153], which were edited by I. Rival, and in [144], edited by M. Pouzet and D. Richard.

I thank Bruno Bosbach for his valuable information on the Latex

Code. I also which to thank my nephew Michael Harzheim who provided substantial help in the technical completion of my final typescript.

Finally some technical remarks for the reader: We use the abbreviation w.r.o.g. for "without restriction of generality", and iff for "if and only if". The symbol $\forall$ abbreviates "for all", and the symbol $\exists$ "there exists".

We have numbered the items (definitions, theorems, remarks) consecutively.

If e.g. we cite an item by 8.4.7 this means item 4.7 ($=$ the seventh item of section 8.4) of Chapter 8. If we cite something only by 4.7 then the item 4.7 of the same chapter is meant.

Many parts of the book contain material which is no longer needed in the subsequent parts, such as the Sections 2.4, 2.7, 4.6, 4.7, 5.3.

There is a close connection between the sections 6.1 and 8.7. And in 9.4 some formerly defined sets are considered from a combinatorial point of view. Apart from this the Chapters 5 - 11 are almost completely independent of each other. Each of these can be read to a great extent without referring to the other chapters.

Chapter 0
Fundamental notions of set theory

We presuppose that the reader is familiar with the fundamental notions and theorems of naive set theory, which are dealt with in almost every textbook on set theory. In this introductory chapter we mainly define our terminology and recall several basic facts which are often used in later chapters.

0.1 Sets and functions

Sets are usually denoted by capital latin letters, preferably with S, A, B, C, X, M, P. There are some standard notations in use: The set of positive integers $1,2,3,\ldots$ is denoted by $\mathbf{N}$, the set of non-negative integers $\{0,1,2,\ldots\}$ by $\mathbf{N}_0$ or by ω, the set of integers by $\mathbf{Z}$, the set of rational numbers by $\mathbf{Q}$, and the set of real numbers by $\mathbf{R}$. The last set is also called the *linear continuum*.

If $n \in \mathbf{N}$ the set of all n-tuples $(x_1, \ldots, x_n)$ of real numbers $x_1, \ldots, x_n$ is called the *euclidean n - space*, and we denote it by $\mathbf{R}^n$. For the *empty set*, which contains no element, the notation $\emptyset$ is used.

Sets are usually described in the form $S := \{x | P(x)\}$, where P is a given property, and where $P(x)$ means that x has the property P. Here we have a situation where *equality by definition* is applied. The item to be defined is at that side of the equality symbol $=$ where the colon appears.

If the elements of a set are again sets we prefer a notation by capital letters in fraktur: $\mathfrak{A}, \mathfrak{B}, \mathfrak{C}, \ldots$. The set of all subsets of a set S is denoted by $\mathfrak{P}(S)$, and it is also called the *power set* of S.

If A and B are sets such that each element of A is contained in B, we call A a *subset* of B and B a *superset* of A, and for this we write $A \subseteq B$ or $B \supseteq A$. If $A \subseteq B$ holds, but not $A = B$, we call A a *proper subset* of B, and B a *proper superset* of A, and for this we use the notation $A \subset B$ or $B \supset A$.

If A and B are sets, a *function* $f : A \to B$ assigns (or ascribes) to each element $x \in A$ an element $f(x) \in B$. For this we say: f maps A *into B*. Here A is called the *domain* of f or *range (of definition)* of f.

For $x \in A$ the element $f(x)$ is called the *f - image of x*, or the *value of f at x*, or the *image of x under f*. The set $\{f(x)|x \in A\}$ is said to be the *image set* of *f*. This is not necessarily identical with B. The term *mapping* is synonymous with function.

If $f : A \to A$ satisfies $f(x) = x$ for all $x \in A$, f is called the *identity mapping* of A. This is also denoted by id_A.

A function $f : A \to B$ is said to be *injective* if distinct elements $x, y \in A$ have distinct f - images $f(x), f(y)$, and f is called *surjective* if B is equal to the image set of $f : A \to B$. In this case we also say: f maps A *onto* B. The use of the term surjective of course presupposes that then the function is given as a triple f, A, B. If $f : A \to B$ is injective and surjective, it is called *bijective*. In this case there exists the *inverse function* $f^{-1} : B \to A$ of $f : A \to B$, which maps each $y \in B$ onto that $x \in A$, for which $f(x) = y$ holds.

If a function f maps a set A into itself, one can form the iterates f^n of f for $n \in \mathbf{N}$. Here $f^1 := f$, and if for an $n \in \mathbf{N}$ the function f^n is already defined, we put $f^{n+1}(x) := f(f^n(x))$ for $x \in A$. If $f : A \to A$ is bijective, one can define f^{-n} for $n \in \mathbf{N}$ inductively by $f^{-(n+1)}(x) := f^{-1}(f^{-n}(x))$ for $x \in A$.

A mapping $f : A \to B$ induces a mapping $f[\,] : \mathfrak{P}(A) \to \mathfrak{P}(B)$ of the power set of A into the power set of B by putting $f[T] := \{f(x)|x \in T\}$ for $T \subseteq A$. The distinction between $f(x)$ and $f[x]$ prevents misunderstanding, for it could happen that the same x is an element of A and also a subset of A. Despite this one can use also $f(T)$ instead of $f[T]$, if it is clear from the context, whether T is meant as an element or as a subset of A.

If $f : A \to B$ is a function and $T \subseteq A$ we call the mapping which assigns to each $t \in T$ the element $f(t)$, the *restriction* of f to T and denote it by $f \upharpoonright T$.

If $f : A \to B$ and $g : B \to C$ are functions, we define the *concatenation* (also *composition*) $g \circ f : A \to C$ (read this as g *after* f) by $(g \circ f)(x) := g(f(x))$ for $x \in A$.

In a set S each of its elements "occurs only once". There is also a concept *multiset*, which is defined for the situation where one wishes to allow multiple occurrences of the same element. E.g. it could make sense to quote the prime divisors of 12 in a collection 2,2,3, where 2 appears twice, — corresponding to the fact that $2 \cdot 2$ divides 12. A method to handle the possibility of multiple occurrence is given with the notion *family:*

Let I be a set, such that to each $i \in I$ an object a_i is assigned; then we call $(a_i)_{i \in I}$ or $(a_i | i \in I)$ a *family*, and I its *index set*. (And here it is possible that different indices i have the same object a_i.) The set which belongs to this family is nothing else than the set of the (different) a_i, — now without reference to I. In general, if we have given a set in a form $\{a_i | i \in I\}$, we assume that for different elements i the corresponding a_i are also different. Instead of "family" also the name "system" is frequently used.

0.2 Cardinalities and operations with sets

The fundamental notions of set theory concern the comparison of sets with respect to their "size":

A set A is called *equipotent* to a set B, if there exists a bijective function $F : A \to B$. In this situation we also say: A and B have the same *cardinality*. We have not defined here the concept of cardinality itself. But, following Frege, one can define the cardinality of a set A as the class of all sets which are equipotent to A. We denote it by $|A|$. Each set X which is equipotent to A is said to be a *realization* of the cardinality $|A|$.

If A and B are sets, and if there exists an injective mapping $f : A \to B$, we say that A has a cardinality *less than or equal to* that of B, in signs $|A| \leq |B|$. If $|A| \leq |B|$ holds, but at the same time $|A| \neq |B|$, we say that $|A|$ is *less than* $|B|$, in signs $|A| < |B|$. So we have $|A| \leq |B| \iff |A| < |B|$, or $|A| = |B|$. By the theorem of Cantor/Schröder/Bernstein the relation $\leq$ is antisymmetric: $|A| \leq |B|$ and $|B| \leq |A|$ implies $|A| = |B|$. Or in a formulation which does not mention cardinalities: If there is an injective mapping of A into B, and an injective mapping of B into A, then there also exists a bijective mapping of A onto B.

A *finite set* is empty or equipotent to a set $\{1, \ldots, n\}$ with $n \in \mathbf{N}$. For the cardinalities of the empty set resp. $\{1, \ldots, n\}$ the notations 0 resp. n are customary. A set which is equipotent to $\mathbf{N}$ is said to be *denumerable*, and if a set is finite or denumerable it is called *countable*.

The relations $\leq$ and $<$ for cardinals are transitive.

If I is a set, and S_i a set for $i \in I$, then the *union* $\cup\{S_i | i \in I\}$ of the sets $S_i, i \in I$, (resp. the *intersection* $\cap\{S_i | i \in I\}$ of the sets $S_i, i \in I$) is

4

the set of all elements which are contained in at least one (resp. all) of the sets S_i. Other notations for these concepts are $\cup_{i\in I} S_i$ resp. $\cap_{i\in I} S_i$.

The *cartesian product* $X\{S_i | i \in I\}$ (or $X_{i\in I} S_i$) is the set of all families $(x_i)_{i\in I}$ where $x_i \in S_i$ for $i \in I$.

If I is a finite set, e.g. $= \{1, \ldots, n\}$ for an $n \in \mathbf{N}$, then instead of the above notations one also writes $S_1 \cup \cdots \cup S_n$, $S_1 \cap \cdots \cap S_n$ and $S_1 \times \cdots \times S_n$.

If A and B are sets, $A \setminus B$ is the set of elements which are in A but not in B. (*Here it is not presupposed that B is a subset of A.*)

These definitions of unions and cartesian products lead in a very natural manner to addition and multiplication of cardinals: Let $(\,|S_i|, i \in I)$ be a family of cardinals (with realizations S_i for $i \in I$), then the sum $\sum_{i\in I} |S_i|$ is defined as the cardinality of the set $\cup_{i\in I} S_i'$, where for $i \in I$ we have $S_i' := \{(i, x) | x \in S_i\}$. By taking the pairs (i, x) instead of the x's themselves we have, so to speak, made the S_i's artificially disjoint. The product $\prod_{i\in I} |S_i|$ is the cardinality of the set $X_{i\in I} S_i$. (Here we need not make the factors artificially disjoint.)

If finally A, B are sets, we define the *power* A^B to be the set of all functions $f : B \to A$. And for the corresponding cardinal: $|A|^{|B|} := |A^B|$.

All these definitions for cardinals are independent of the choice of their realizations, as is immediately seen.

0.3 Well-ordered sets

Later the notion of linearly ordered sets is dealt with in great detail, so we presuppose it here. As a special case of this concept we now mention: A linearly ordered set is called *well-ordered*, if each non-empty subset of it has a first (or least) element.

If A and B are well-ordered sets, a mapping $f : A \to B$ is said to be $<$ - *preserving*, if $x < y$ in A entails $f(x) < f(y)$ in B. And two well-ordered sets A and B are called *isomorphic*, if there exists a surjective $<$ - preserving mapping of A onto B

From the theory of well-ordered sets we later need some elementary theorems:

3.1 Theorem. *If W is a well-ordered set with order $\leq$ and $f : W \to W$ a $<$ - preserving mapping, then $x \leq f(x)$ holds for all $x \in W$.*

3.2 Theorem. *If $f : W \to W$ is a $<$ - preserving mapping of a well-ordered set W onto itself, then f is the identity mapping of W.*

An *initial segment* of a well-ordered set W is a subset I of W, which satisfies: If $y \in I$ and $x \in W$ satisfies $x < y$, then also $x \in I$. Then the main theorem on well-ordered sets states:

3.3 Theorem. *If A and B are two well-ordered sets, then A is isomorphic to an initial segment of B, or B is isomorphic to an initial segment of A. And in both cases these initial segments are uniquely defined.*

So the well-ordered sets have nice properties, in particular it is easy to compare two of them by using the last theorem. Therefore it is of great interest under which conditions one can find a well-ordering for a given set. In this context the notion "choice function" plays an important role:

3.4 Definition. *A choice function f on a set S is a function which* ascribes to every non-empty subset $T \subseteq S$ an element $f(T) \in T$.

The *axiom of choice* then states that for every set S there exists a choice function. It is in short denoted by AC.

There is a variant of the axiom of choice which is frequently used, namely:

3.5 Theorem. AC *is equivalent to the following statement:*
If a set S is a union of disjoint non-empty subsets $S_i, i \in I$,   then there is a set $T \subseteq S$ which has exactly one element in common with each S_i, $i \in I$.

Now we can formulate the famous well-ordering theorem of Zermelo:

3.6 Theorem. *If we have a choice function for a set S, then S can be well-ordered.*

And as a consequence of this:

3.7 Theorem. *The axiom of choice implies that every set can be well-ordered.*

In the following we make use of the axiom of choice without mentioning this always. In some cases we have a more detailed look at it.

In well-ordered sets one can apply the *method of transfinite induction* which generalizes induction in **N**:

3.8 Theorem. *Let W be a well-ordered set and P a property which is possessed by certain elements of W. Suppose that for each $y \in W$ we have: If all elements $x < y$ have property P, then y also has it. Then* *it follows that all elements of W have property P. (Mind: If W is non-empty our assumption includes that the first element of W has property P.)*

0.4   Ordinals

We recall in short, without going into the details, the construction of the class of ordinal numbers by the method of von Neumann. Consider the sequence

$\emptyset$,$\{\emptyset\}$,$\{\emptyset, \{\emptyset\}\}$,$\{\emptyset, \{\emptyset\}, \{\emptyset, \{\emptyset\}\}\}$,...., where from every element x of the sequence we create its immediate successor by forming the set $x \cup \{x\}$. This step can be repeated transfinitely. The formal definition is:

A set M whose elements are again sets is called an *ordinal* if there holds: M, with the relation $\subseteq$, is well-ordered, and each $x \in M$ is the set of all elements of M which are $\subset x$.

This entails that the class On of all ordinals is well-ordered, and if α and β are ordinals, we have

$\alpha < \beta \Longleftrightarrow \alpha$ is a proper subset of $\beta \Longleftrightarrow \alpha \in \beta$.

The first infinite ordinal is denoted by ω or ω_0, and the finite ordinals by 0,1,2,.... So these symbols are used for ordinals and cardinals simultaneously.

An important property of the class On is:

4.1 Theorem. *Every well-ordered set is isomorphic to an initial segment of On, which is uniquely defined.*

So, if W is a well-ordered set and $f : W \to I$ an isomorphic mapping of W onto an initial segment I of On, then each $a \in W$ has a uniquely defined f - image ν in I, and we can append ν as an index to a, so that it appears as a_ν. In this manner we have represented W in the form $W = \{a_\nu | \nu \in I\}$ of a transfinite sequence. Now it makes sense to call a the νth *element* of W. Instead of "transfinite sequence" we also say "sequence".

Families of sets are often represented in a form $(S_\nu | \nu < \mu$), where the index set is an ordinal μ. We say that this family is an *ascending* (or *increasing*) *tower of sets* if $\kappa < \nu < \mu$ implies $S_\kappa \subseteq S_\nu$. If $S_\kappa \supseteq S_\nu$ for $\kappa < \nu < \mu$ holds, we have a *descending* or *decreasing tower of sets*. Replacing $\subseteq$ by $\subset$ resp. $\supseteq$ by $\supset$ we obtain the definitions *strictly ascending* (resp. *strictly descending*) *tower of sets*.

If λ is an ordinal $\neq 0$, for which the set of ordinals which are $< \lambda$ has no greatest element, then λ is called a *limit ordinal* or *limit number*. If this set has a greatest element κ, λ is said to be a *successor ordinal* or *successor number,* and then we also denote κ by $\lambda - 1$.

In On we have operations $+$ and $\cdot$, which can be defined by transfinite recursion:

If α is an ordinal we put $\alpha + 0 = \alpha$ and define $\alpha + 1$ to be the least ordinal which is $> \alpha$. If for an ordinal β we have already defined $\alpha + \beta$, we put $\alpha + (\beta + 1) = (\alpha + \beta) + 1$. And if λ is a limit ordinal such that $a + \xi$ is defined for all $\xi < \lambda$, we define $\alpha + \lambda$ to be the least ordinal which is greater than all $\alpha + \xi$ with $\xi < \lambda$. Then by induction $\alpha + \xi$ is defined for all $\xi \in$ On.

In a similar way the multiplication $\cdot$ is defined. If α is an ordinal, we put $\alpha \cdot 0 = 0$. And if for an ordinal β the product $\alpha \cdot \beta$ is already defined, we put $\alpha \cdot (\beta + 1) = (\alpha \cdot \beta) + \alpha$. If for a limit ordinal λ the product $\alpha \cdot \xi$ is defined for all $\xi < \lambda$, then we define $\alpha \cdot \lambda$ to be the least ordinal which is $> \alpha \cdot \xi$ for all $\xi < \lambda$. Now $\alpha \cdot \xi$ is defined for all $\xi \in$ On.

The idea of the multiplication is, intuitively speaking, that $\alpha \cdot \beta$ is an iterated sum $\alpha + \alpha + \alpha + \cdots$, where we have "β many" summands α.

Also powers of ordinals are defined, following the same pattern as in the definition of $+$ and $\cdot$.

If $\alpha \neq 0$ is an ordinal, we put $\alpha^0 := 1$. If for an ordinal β the power α^β is defined, we put $\alpha^{\beta+1} := \alpha^\beta \cdot \alpha$, and if λ is a limit ordinal for which α^ξ is defined for all $\xi < \lambda$, we define α^λ to be the least ordinal which is $> \alpha^\xi$ for all $\xi < \lambda$.

The last definition should not be confused with the definition of set exponentiation. We previously defined A^B when A and B are sets. Now ordinals α, β are also sets, say A and B, and then we would have two different definitions for α^β. So we agree that forming the exponentiation of ordinals is always understood in the sense of the last definition.

The operations $+$ and $\cdot$ are not commutative, and by transfinite induction one can verify that the following laws hold:
$$(\alpha+\beta)+\gamma = \alpha+(\beta+\gamma),\ (\alpha\cdot\beta)\cdot\gamma = \alpha\cdot(\beta\cdot\gamma),\ \alpha\cdot(\beta+\gamma) = (\alpha\cdot\beta)+(\alpha\cdot\gamma).$$

If $\alpha > \beta$ are ordinals there is exactly one ordinal γ which satisfies $\alpha = \beta + \gamma$.

Also families of functions $f_\nu : D_\nu \to E$ are often represented in a form $(f_\nu | \nu < \mu)$, where μ is an ordinal. If μ is a limit ordinal, and if for $\kappa < \nu < \mu$ we have $D_\kappa \subseteq D_\nu$ and $f_\nu \upharpoonright D_\kappa = f_\kappa$, we define the *limit mapping* $f : D \to E$ of the family as follows: $D = \cup\{D_\nu | \nu < \mu\}$, and for $x \in D$ we have $f(x) = f_\nu(x)$ for at least one ν, for which x is in D_ν. Then also $f(x) = f_\lambda(x)$ for all $\lambda < \mu$ which are $\geq$ the last mentioned ν.

If we have a set with a linear order represented in the form $\{x_\nu | \nu < \mu\}_<$, where μ is an ordinal, this shall indicate that for $\kappa < \nu < \mu$ we have $x_\kappa < x_\nu$.

0.5   The alephs

Ordinals are sets, and different ordinals can have the same cardinality. Then we define: The *number class* of an ordinal ν is the set of all ordinals which have the same cardinality as ν. Since this class is a well-ordered set it has a first element, which is said to be the *initial number* or *initial ordinal* of this class. The integers $0, 1, 2, \ldots$ are also initial numbers, — and each of these is the only element in its number class. The situation changes radically if we consider infinite ordinals:

Let μ be an infinite ordinal. Then the set of infinite initial ordinals which are $< \mu$ is well-ordered, and thus by 3.8 in a bijective and $<$ - preserving correspondence with an inital segment I of On. Then their set can be denoted by $\{\omega_\nu | \nu \in I\}$, where ω_ν is the initial ordinal which corresponds to the ordinal ν of I. The index ν, which is ascribed to the initial ordinal ω_ν, is independent of the choice of μ, and so we have an "enumeration" of all infinite initial ordinals in the form $\omega_\nu, \nu \in$ On, where for ordinals $\mu < \nu$ we have $\omega_\mu < \omega_\nu$. The least infinite ordinal is also the least infinite initial ordinal. It is denoted by ω_0, and also in short by ω.

The cardinality of ω_ν is denoted by $\aleph_\nu$. The sign $\aleph$ is called aleph, and the cardinalities $\aleph_\nu$ are called *alephs.*

With the use of the axiom of choice it follows that each cardinal $|S|$ is a non-negative integer or an aleph. For then S can by be well-ordered by 3.6, and thus it is by 3.8 equipotent to an initial segment of On, say to the set of ordinals $< \mu$, and then it is also equipotent to the initial number of the number class of μ.

The power set $\mathfrak{P}(S)$ of a set S has cardinality $2^{|S|}$. For there is a bijective mapping $f : \mathfrak{P}(S) \to \{0, 1\}^S$, namely the *characteristic function,* which ascribes to each subset $T \subseteq S$ the mapping which assigns to each $t \in T$ the number 1, and to each $t \notin T$ the number 0. By Cantor's theorem on the cardinality of the power set we have $|\mathfrak{P}(S)| > |S|$. Thus (by taking $S := \omega_\nu$) a consequence of AC is that $2^{\aleph_\nu}$ is $\geq \aleph_{\nu+1}$.

The aleph hypothesis states that for all ordinals ν we have $2^{\aleph_\nu} = \aleph_{\nu+1}$. The more general *generalized continuum hypothesis*, abbreviated GCH, states that strictly between an infinite cardinal a and 2^a there is no cardinal. One knows by work of Gödel and Cohen that the GCH is independent of the axioms of ZF (the axiom system of Zermelo–Fraenkel).

We mention some properties of the arithmetic of cardinals: We have several monotonicity laws: If a, b, a^*, b^* are cardinals with $a \leq a^*$ and $b \leq b^*$, then $a + b \leq a^* + b^*$, $a \cdot b \leq a^* \cdot b^*$ and $a^b \leq (a^*)^{b^*}$ hold.

Hessenberg's theorem states: $\aleph_\nu + \aleph_\nu = \aleph_\nu \cdot \aleph_\nu = \aleph_\nu$ *for all ordinals* ν. *From this follows, due to* AC, *that for cardinals* a, b, *of which at least one is infinite, the sum* $a + b$ *is the greatest of both. If in addition* a *and* b *are* $\neq 0$, *also* $a \cdot b$ *is the greatest of both.*

For exponentiation we have the law $(a^b)^c = a^{b \cdot c}$ for cardinals a, b, c. In particular this and Hessenberg's theorem implies Bernstein's equality which states: $2^{\aleph_\nu} = k^{\aleph_\nu} = (2^{\aleph_\nu})^{\aleph_\nu} = \aleph_{\nu+1}^{\aleph_\nu}$ *for every cardinal* k *satisfying* $2 \leq k \leq 2^{\aleph_\nu}$.

König's theorem states: *Let* I *be a non-empty index set, and let* a_i, b_i *be cardinals for* $i \in I$ *which satisfy* $a_i < b_i$ *for* $i \in I$. *Then the sum* $\sum\{a_i | i \in I\}$ *is* $< \prod\{b_i | i \in I\}$.

Let $\aleph_\alpha$ be an aleph, then there is a least ordinal γ, so that $\aleph_\alpha$ is a sum of $\aleph_\gamma$ many cardinals, which all are $< \aleph_\alpha$. This $\aleph_\gamma$ is called the *cofinality* of $\aleph_\alpha$. For this follows, using the theorem of König, that $\aleph_\alpha^{\aleph_\gamma} > \aleph_\alpha$ holds. And using GCH one can prove that $\aleph_\alpha^k = \aleph_\alpha$ holds for all cardinals k with $1 \leq k < \aleph_\gamma$.

Chapter 1
Fundamental notions

In this chapter we compile several notions which are essential for the theory of ordered sets. Most of them are well known to readers who are familiar with the fundamental concepts of set theory. First we review the most important notions on general relations.

1.1 Binary relations on a set

1.1 Definition. If S is a set and if R is a subset of $S \times S$, R is called a *binary relation on S*, in short a *relation on S*. Here S is called the *underlying set* or *carrier* or *ground set* or *base* of R.

If for $a, b \in S$ we have $(a, b) \in R$, we say: a is in relation R to b. We express this also by $a\,R\,b$.

If we write $x_1 R\, x_2 R \cdots R\, x_n$ for a natural number $n > 2$, this shall mean $x_i R\, x_{i+1}$ for $i = 1, \ldots, n-1$.

If $R' \subseteq R \subseteq S \times S$, then R' is called a *subrelation* of R. The empty set $\emptyset$ is the *empty* relation on every set S, and $S \times S$ is the *all-relation* on S.

(Of course, the concept relation could also be defined without referring to an underlying set S. Then a relation R is simply a set of pairs (a, b), and then such an R is a relation on every set S which contains all a and b for which $(a, b) \in R$.)

If R is a relation on S, then $R^c := (S \times S) \setminus R$ is called the *complementary* relation of R on S, and the set $\{(b, a) | (a, b) \in R\}$ is said to be the *inverse* relation of R, in short the *inverse* of R. It is denoted by R^{-1}. Of course one has $(R^c)^c = R$ and $(R^{-1})^{-1} = R$.

The relation $\{(x, x) | x \in S\}$ is called the *identity* relation or *diagonal* of S. It is denoted by $I(S)$ or id_S.

If R is a relation on S and T a subset of S, we define the *restriction* of R to T, in symbols $R \restriction T$, as the relation $R \cap (T \times T)$ on T.

1.2 Definition. If R_1 and R_2 are relations on a set S, we define the *product relation* $R_1 \circ R_2$ to be the relation R which is given by
$$R := \{(a, b) | a \in S, b \in S, \exists x \in S \mid a R_1 x \text{ and } x R_2 b\}.$$

[Remark: In some other texts this set is denoted by $R_2 \circ R_1$.] In the scope of this definition we introduce R^n, which should not be confused with the cartesian product, as follows:

If R is a relation on S, we put $R^2 := R \circ R$. If n is an integer >2 and R^{n-1} is already defined, we put $R^n := R \circ R^{n-1}$. The following is easily seen: For $a, b \in S$ we have aR^nb iff there exist elements $x_1, \ldots, x_{n-1} \in S$ such that $aRx_1R \cdots Rx_{n-1}Rb$ holds. The "powers" of a relation R commute, precisely:

1.3 Remark. *If $n, m \in \mathbf{N}$ and if R is a relation on S, we have $R^n \circ R^m = R^{n+m} = R^m \circ R^n$.*

For general relations R_1, R_2 on a set S usually $R_1 \circ R_2$ is different from $R_2 \circ R_1$, so that the operation $\circ$ of relational product is not commutative. But it is associative: If R_3 is also a relation on S, we have $(R_1 \circ R_2) \circ R_3 = R_1 \circ (R_2 \circ R_3)$.

1.2   Special properties of relations

We now discuss the most fundamental concepts for relations:

2.1 Definition. Let R be a relation on a set S. Then R is called *reflexive* iff $R \supseteq I(S)$, in other words, iff $(x, x) \in R$ for all $x \in S$, *irreflexive* iff $R \cap I(S) = \emptyset$,

symmetric, iff aRb entails bRa, (The same is expressed by $R \supseteq R^{-1}$, also by $R = R^{-1}$.)

antisymmetric, iff for $a, b \in S$ there holds: aRb and $bRa \implies a = b$. (The same is expressible in the form: For different elements $a, b \in S$ at most one of aRb and bRa can hold.),

transitive, if for $a, b, c \in S$ there holds: aRb and $bRc \implies aRc$. (This can also be formulated as $R \circ R \subseteq R$.)

2.2 Theorem and **Definition.** *Let R be a relation on a set S. Then we have:*

a) There exists a least relation $R_s \supseteq R$ on S, which is symmetric, namely $R_s = R \cup R^{-1}$. It is called the symmetric hull (or symmetric closure) of R.

b) There exists a least relation $R_t \supseteq R$ on S, which is transitive. It is called the transitive hull (or transitive closure) of R on S. We denote it by $\mathrm{TH}(R)$.

There holds $\mathrm{TH}(R) = \cup\{R^n|\ n \in \mathbf{N}\}$.

Proof. a) is trivial. b) The intersection of transitive relations on S is again transitive, also the all-relation $S \times S$. Then the intersection of all transitive relations on S which contain R is evidently the smallest transitive relation on S which contains R. Using induction on n one can see that $\mathrm{TH}(R)$ contains all relations R^n with $n \in \mathbf{N}$. Conversely $\cup\{R^n|n \in \mathbf{N}\}$ is a transitive relation on S which contains R, and therefore also $\mathrm{TH}(R) \subseteq \cup\{R^n|\ n \in \mathbf{N}\}$ holds.

1.3 The order relation and variants of it

By combining several notions of Definition 2.1 one obtains the notion of order relation and several variants of it.

3.1 Definition. Let R be a relation on a set S. Then R is called a *quasi-order* (or *pre-order*) if it is reflexive and transitive. If in addition to this R is also antisymmetric, R is called an *order relation* (or in short an *order*). Instead of *order* we also use *partial order* to emphasize the contrast to the following concept *linear order*. The pair (S, R) is called an *ordered set* or a *poset,* — in abbreviation of the name *partially ordered set*, which is also frequently used to emphasize the difference between partial and linear order, which we now define:

R is called a *linear order* (or *total order*) on S, if R is an order, and in addition the following holds:

(*) For every two elements $a \neq b$ of S either aRb or bRa holds.

(Both of these cannot hold since this would entail $a = b$ because of the antisymmetry of R.)

If a linearly ordered set S is finite, one can write its elements in their given order : $S = \{a_1, \ldots, a_n\}_<$ which means that $a_\nu < a_{\nu+1}$ for $\nu < n$. Similarly: If we have a linearly ordered set S, which is represented as $S := \{a_\nu|\nu < \mu\}$, where μ is an ordinal, $S := \{a_\nu|\nu < \mu\}_<$ shall mean, that for $\kappa < \nu < \mu$ we have $a_\kappa < a_\nu$, — so that the elements are indexed according to their order.

R is called a *strict order* if R is irreflexive and transitive. R is called a *strict linear order* if R is a strict order which satisfies (*). A strict order R is automatically antisymmetric: If a and b would be elements with aRb and bRa, the transitivity of R would entail aRa in contradiction to the irreflexivity of R.

14

The pair (S, R) is called a *quasi-ordered* (resp. *ordered, linearly ordered, strictly ordered, strictly linearly ordered*) *set*, if R is a quasi-order (resp. order, linear order, strict order, strict linear order) on S. In these cases one speaks also of the ordered (resp. ...) set S if it is clear which order (resp. ...) relation R on S is under consideration.

Several elementary facts about the above concepts are immediately clear. E.g. if R is a relation of one of the above introduced types, its inverse relation R^{-1} is this too. Further

R is a linear order on S $\iff$ the complementary relation R^c of R is a strict linear order on S.

Usually order relations and also quasi-orders R are denoted by the sign $\leq$, strict orders by $<$. And if we formulate "Let S be a poset" without mentioning a relation $\leq$, we always suppose that its order is denoted by $\leq$. If R is an order $\leq$ on a set S and if a and b are elements of S for which $a \leq b$ holds we describe this as: a is *less than or equal to* b, also: b is *greater than or equal to* a. The same means : a is *below* b, b is *above* a. For $a < b$ one says : a is *less* (or *smaller)* than b, or b is *greater* (or *larger*) than a. The negation of $a \leq b$ resp. $a < b$ is of course denoted by $a \nleq b$ resp. $a \nless b$.

In this context the following holds:

3.2 Theorem and **Definition**. *Let $\leq$ be a quasi-order on a set S. Then we define relations $<$ and $>$ on S as follows:*

For $a, b \in S$ we put $a < b \Leftrightarrow a \leq b$ and $b \nleq a$.

If $\leq$ is also an order relation, then we have $a < b \iff a \leq b$ and $a \neq b$. But for a quasi-order this is not generally valid ! Here we can have different elements a, b with $a \leq b$, but $a \nless b$, namely if different elements a, b satisfy $a \leq b$ and $b \leq a$.

If $\leq$ is an order on S, $<$ is a strict order on S. It is called the strict order belonging to $\leq$. Indeed: $a < b$ and $b < c$ entail $a \leq b \leq c$ and $a \leq c$; and $a = c$ cannot hold, — this would yield $b < a$ and $b \leq a$, and with $a \leq b$ also $a = b$ contradicting $a < b$.

If $\leq$ is a quasi-order on S, then $\geq$ denotes the inverse relation of $\leq$. (Hence: $a \geq b \iff b \leq a$.)

If a, b, c are elements of a poset which satify $a < b < c$, we say that b is between a and c. If only $a \leq b \leq c$ holds, b is said to be between a and c in the general sense.

Conversely to 3.2 there holds:

3.3 Theorem. *Let $<$ be a strict order on S. Then the relation $\leq$ $:= \; < \cup I(S)$ is an order on S.*

Proof. $\leq$ is reflexive by definition. Let now $a, b \in S$ and $a \leq b$ and $b \leq a$. If $a \neq b$ would hold, we would have $a < b$ and $b < a$, hence $a < a$ with contradiction to the irreflexivity of $<$. Thus $\leq$ is antisymmetic.

Let $a, b, c \in S$ and $a \leq b$ and $b \leq c$. If $a = b$ or $b = c$ holds, then $a \leq c$ is trivial. In the other case we have $a < b$ and $b < c$ and thus $a < c$ and $a \leq c$.

Linear orders can also be characterized in the following way:

3.4 Theorem *Let R be a reflexive and antisymmetric relation on a set S. Then the following two statements are equivalent:*
a) R is a linear order on S.
b) R and its complementary relation R^c are both transitive.

Proof. a) $\Rightarrow$ b) is trivial. Now, let b) be presupposed. If $a \neq b$ are elements of S either aRb or bRa must hold. Otherwise we would have aR^cb and bR^ca, hence aR^ca since R^c is transitive. But this contradicts aRa. And so R is a linear order.

In this context it follows that for a linear order R on S the inverse relation R^{-1} and the complementary relation R^c "nearly" coincide. They differ only in the diagonal relation $I(S)$:

3.5 Theorem *Let R be an order relation on S. Then the following two statements are equivalent:*
a) R is a linear ordering.
b) $R^c = R^{-1} \setminus I(S)$.

Proof. a) $\Rightarrow$ b) is trivial. Let now the assumption of b) be satisfied and let $a \neq b$ be elements of S. If neither aRb nor bRa would hold, we would have aR^cb and bR^ca and therefore, because of b), $aR^{-1}b$ and $bR^{-1}a$. This implies bRa and aRb, so that $a = b$ would follow in contradiction to our assumption. Thus R is a linear ordering.

3.6 Remark. If we have a reflexive and antisymmetric relation $\leq$ on a set P, from which we wish to show that it is also transitive, it suffices evidently to prove that the relation $<$, which is defined by $< \; := \; \leq \setminus \operatorname{id}_P$, is transitive. In signs: $a, b, c \in P$ and $a < b < c \Longrightarrow a < c$.

16

1.4 Examples

For every set S the relation id_S is trivially an order relation on S, and evidently the smallest one. It is a subrelation of every order relation on S, and nothing else than the equality relation $=$ over S. More interesting examples are:

4.1 Example. Let S be the set of all sequences $(a_n)_{n\in\mathbf{N}}$ of real numbers a_n. If $\mathfrak{a} = (a_n)_{n\in\mathbf{N}}$ and $\mathfrak{b} = (b_n)_{n\in\mathbf{N}}$ are elements of S we put $\mathfrak{a} \leq \mathfrak{b}$ if $\mathfrak{a}$ is a subsequence of $\mathfrak{b}$. Then $\leq$ is reflexive and transitive and thus a quasi-order. But $\leq$ is not antisymmetric and therefore no order. Indeed e.g. we have $0,1,0,1,\ldots \leq 1,0,1,0,\ldots \leq 0,1,0,1,\ldots$, but the last two sequences are different.

4.2 Example. Let S be a set, $\mathfrak{P}(S)$ the set of all subsets of S. For $T_1, T_2 \subseteq S$ we put $T_1 \leq T_2$ iff $T_1 \subseteq T_2$. Then $\leq$ is an order relation. It is linear iff $|S| \leq 1$. The corresponding relation $<$ is a strict order on $\mathfrak{P}(S)$.

4.3 Example. For $a, b \in \mathbf{N}$ we put $a \leq b$ iff $a|b$ (that means a divides b). This yields an order on $\mathbf{N}$, which is not linear. It is a subrelation of the usual linear order by magnitude of the naturals.

4.4 Example. For points (a,b), (c,d) of the two-dimensional euclidean space $\mathbf{R}^2$ we put $(a,b) \leq (c,d) \Leftrightarrow a \leq c$ and $b \leq d$, where on the right side $\leq$ has the usual meaning of $\leq$ in the reals. Then the $\leq$ of the left side yields an order on $\mathbf{R}^2$, which is not linear.

4.5 Example. Let S be a set of people. For $a, b \in S$ we put $a < b$ iff a is an ancestor of b. Then $<$ is a strict order on S (which normally is not linear).

A very important notion which has applications in all parts of mathematics is that of equivalence relation:

4.6 Definition. A relation R on a set S is called an *equivalence relation, if* R is reflexive, symmetric and transitive.

If R is an equivalence relation on S and $x \in S$ we call the set $\{y \in S | x \, R \, y\}$ the *equivalence class* of x (with respect to R).

An equivalence relation $\sim$ is also a quasi-order, but in general not an order relation. Indeed, if one of the equivalence classes of $\sim$ contains at least two elements a, b, we have $a \sim b$ and $b \sim a$ but $a \neq b$ so that $\sim$ is not antisymmetric.

The notion "equivalence relation" is closely related to that of partition:

4.7 Definition. Let S be a set, $\mathfrak{P}$ a set of non-empty subsets of S which satisfies the following two conditions:

1) $\cup\mathfrak{P} = S$.

2) Different sets of $\mathfrak{P}$ are disjoint.

Then $\mathfrak{P}$ is called a *partition* or *decomposition* of S.

The connection between the two concepts is described in the following theorem, which is easily verified:

4.8 Theorem *Let S be a set.*

a) If $\sim$ is an equivalence relation on S, we define $\mathfrak{P}(\sim)$ to be the set of all equivalence classes of $\sim$. Then $\mathfrak{P}(\sim)$ is a partition of S.

On the other hand we have:

b) If $\mathfrak{P}$ is a partition of S, we define a relation R on S by: If $x, y \in S$ we put $x \sim y$ iff x, y are in the same class of $\mathfrak{P}$. Then $\sim$ is an equivalence relation on S.

The notion of quasi-order is more general than that of order, but nevertheless it is closely related to it. The following reflection makes this evident:

4.9 Theorem and **Definition.** *Let $(S, \prec)$ be a quasi-ordered set. In S we introduce a relation $\sim$ as follows :*

For $a, b \in S$ we put $a \sim b \Leftrightarrow a \prec b$ and $b \prec a$.

Then $\sim$ is an equivalence relation on S.

If K_1, K_2 are equivalence classes of $\sim$, and if there holds $a_1 \prec a_2$ for some $a_1 \in K_1$ and some $a_2 \in K_2$, then there also holds $x_1 \prec x_2$ for all $x_1 \in K_1$ and all $x_2 \in K_2$, and then we put $K_1 \leq K_2$. The relation $\leq$, which is so defined, is an order relation in the set of equivalence classes of S. It is called the order $\leq$ which is generated by the quasi-order $\prec$.

Proof. It is immediately clear that $\sim$ is an equivalence relation. If now we have $a_1 \prec a_2$ for special elements $a_1 \in K_1$, $a_2 \in K_2$ and if $x_1 \in K_1$, $x_2 \in K_2$ are arbitrary, we have $x_1 \prec a_1 \prec a_2 \prec x_2$ and thus $x_1 \prec x_2$.

$\leq$ is trivially reflexive and transitive. Let now K_1 and K_2 be equivalence classes with $K_1 \leq K_2$ and $K_2 \leq K_1$. We choose arbitrary elements $a_1 \in K_1$ and $a_2 \in K_2$. Then there holds $a_1 \prec a_2$ and $a_2 \prec a_1$ and thus $a_1 \sim a_2$. This implies $K_1 = K_2$, and hence $\leq$ is antisymmetric.

The content of 4.9 can be interpreted as follows: If in a given quasi-ordered set each two elements, one superseding the other, are identified, we obtain a poset. On the other hand there holds: Every quasi-order $\prec$ can be obtained by starting from a poset $\mathfrak{A}$, in which every element K is replaced by a set M_K of elements, where the sets $M_K, K \in \mathfrak{A}$ are pairwise disjoint, and where the quasi-order $\prec$ is defined by

$a \prec b \Leftrightarrow a$ and b are in the same class $K \in \mathfrak{A}$, or $a \in M_{K_1}$ and $b \in M_{K_2}$ where $M_{K_1} \leq M_{K_2}$.

The above reflections make clear that the theory of quasi-ordered sets is entirely contained in the theory of posets.

4.10 Example. Let S be a set and $\mathcal{Z}(S)$ the set of all partitions of S. If $\mathfrak{Z}_1$ and $\mathfrak{Z}_2$ are in $\mathcal{Z}(S)$, we put $\mathfrak{Z}_1 \leq \mathfrak{Z}_2$ and call $\mathfrak{Z}_1$ *finer* than $\mathfrak{Z}_2$, iff there holds: Every class $Z_1 \in \mathfrak{Z}_1$ is $\subseteq$ a class $Z_2 \in \mathfrak{Z}_2$,— which, of course, is uniquely defined. It follows immediately that $\leq$ is an order in $\mathcal{Z}(S)$, — which is not linear, if S has at least three elements.

4.11 Example. Let $\mathfrak{J}$ be the set of all closed intervals of **R**. If $[a, b]$ and $[c, d]$ are in $\mathfrak{J}$, we put $[a, b] \leq [c, d]$, if $[a, b] = [c, d]$ or $b \leq c$ in **R**. Then this defines an order in $\mathfrak{J}$, which is not linear.

1.5 Special remarks

The content of this section is no longer needed in the rest of the book. We consider here systematically the relationship of possible combinations of properties which were introduced before. To this purpose we abbreviate the words for reflexive, irreflexive, symmetric, antisymmetric, transitive by their initials r, i, s, a, t.

A relation on a set S which is at the same time s and a, is evidently a subset of the diagonal $I(S)$. Further the properties r and i exclude themselves on non-empty sets S. So we need only study those combinations of at least two of the five properties r,i,s,a,t, in which at most one of r and i and at most one of s and a is contained. Let now R be a relation on S.

Case: R is r and t: Then R is a quasi-order.

Case: R is i and t: Then R is a strict order.

Case: R is s and t: If an element $x \in S$ is in relation R to an element $y \in S$ we also have xRx because xRy entails yRx and then xRx. Then we have $S = U \cup (S \backslash U)$, where U is the set of elements of S which are not in relation R to any element of S and $R \upharpoonright S \backslash U$ is an equivalence relation on $S \backslash U$.

Case: R is a and t: Then $R \cup I(S)$ is an order relation $\leq$ on S. Its corresponding strict order $<$ is then a subrelation of R. And so we have: A relation R on S with properties a and t is a subset of an order relation and a superset of its corresponding strict order.

Case: R is r and s and t: Then R is an equivalence relation.

Case: R is i and s and t: This combination is only possible for the trivial case $R = \emptyset$. For if there would exist elements $x, y \in S$ with xRy, we also would have yRx and then also xRx which contradicts i.

Case: R is r and a and t: Then R is an order relation.

Case: R is i and a and t. This is the same case as that where R is i and t. Indeed, if a relation R on S is i and t, it is also a. For if there exist elements $x \neq y$ in S satisfying xRy and yRx we would have xRx in contradiction to property i.

If we don't require transitivity from the relation R, we still can consider the four combinations i,s resp. r,s resp. i,a resp. r,a. They correspond to graph-theoretical concepts. So the first type i,s leads to the notion *graph without loops and multiple edges*.

1.6 Neighboring elements. Bounds

6.1 Definition. Let $(P, \leq)$ be a poset, $a, b \in P$. If $a < b$ holds, a is said to be a *predecessor* of b and b a *successor* of a.

If $a \leq b$ or $b \leq a$ holds, then a and b are said to be *comparable* (also: *a comparable* with b). If none of these holds, a and b are called *incomparable*, and this is denoted by $a \parallel b$.

A subset $T \subseteq P$ is called a *chain* (resp. an *antichain*), if every two different elements of T are comparable (resp. incomparable). The *length* of a finite chain is the same as its cardinality.

If $a < b$ holds and if there is no element z in P which satisfies $a < z < b$, then a is called a *lower* (or *left*) *neighbor* or *immediate predecessor* of b. We denote this by $a \lhd b$ and call $\lhd$ the *left-neighbor-relation* or *lower-neighbor-relation* to $\leq$. In this situation we also call b an *upper* (or *right*) *neighbor* or *immediate successor* of a. For this we use the sign $b \rhd a$. In each of the cases $a \lhd b$ and $b \lhd a$ we call a and b *neighbors* or *neighboring* or *adjacent elements*.

6.2 Example. In the set $\mathbf{Z}$ of integers, ordered in the usual sense, every element g has exactly one immediate predecessor and exactly one immediate successor, namely $g-1$ (resp. $g+1$). In the sets $\mathbf{Q}$ of rational numbers (resp. $\mathbf{R}$ of real numbers), ordered in the usual sense, no element has an immediate predecessor, and no element has an immediate successor.

Between an order relation $\leq$ and its corresponding (left neighbor)-relation $\lhd$ there holds the following reconstruction theorem:

6.3 Theorem. *Let $(P, \leq)$ be a poset, where P is finite. Then $\leq$ is uniquely determined by $\lhd$, hence $\leq$ reconstructible from $\lhd$. Indeed, $\leq$ is the transitive hull $\mathrm{TH}(\lhd \cup I(P))$ $(=: H)$.*

Proof. $H \subseteq \leq$ is trivial. Let now $a, b \in P$ and $a \leq b$. We have to show aHb. If $a = b$ holds this is trivial. So we assume $a < b$. Among the chains of P which have a as least and b as greatest element there must be one with a highest number of elements. Then this has a form $\{a = a_1, \ldots, a_{n+1} = b\}$, where $a_i < a_{i+1}$ for $i = 1, \ldots, n$. Here we also must have $a_1 \lhd \cdots \lhd a_{n+1}$. Indeed, if there would exist an element z with $a_i < z < a_{i+1}$ for some $i \in \{1, .., n\}$, this could be inserted into the previous chain contradicting the fact that this had already a maximal number of elements.

In the following we still characterize those relations which are the lower-neighbor-relation of an order relation $\leq$. First we define:

6.4 Definition. We call a relation R on a set S *totally intransitive,* if there holds: For every sequence $a_1, \ldots, a_{n+1}$ with a natural number $n > 1$ of elements of S, where $a_i R a_{i+1}$ holds for every $i = 1, \ldots, n$, we have $a_1 R^c a_{n+1}$ (the same: not $a_1 R a_{n+1}$) and $a_1 \neq a_{n+1}$.

A totally intransitive relation R on a set S is irreflexive. Indeed, if there would exist an element $a \in S$ with aRa, then it would follow (by putting $a = a_1 = a_2 = a_3, n = 2$) $a \neq a$.

6.5 Theorem. *Let R be a relation on a set S. Then the following two statements are equivalent:*

a) R is the relation $\lhd$ which belongs to some order $\leq$.

b) R is totally intransitive.

Proof. Suppose a) holds. And let a sequence $a_1 \lhd \cdots \lhd a_{n+1}$ with $n > 1$ be given. Then of course $a_1 <$ (and thus $\neq$) a_{n+1} holds and also not $a_1 \lhd a_{n+1}$ because there is an element a_2 with $a_1 < a_2 < a_{n+1}$.

Suppose now that b) holds. And let $\leq$ be the transitive hull $\mathrm{TH}(R\cup I(S))$. Then this $\leq$ is indeed an order relation: It is obviously reflexive and transitive. It is also antisymmetric: Let a, b be elements of S satisfying $a \leq b$ and $b \leq a$. We assume indirectly that $a \neq b$ holds. Then there must exist finite sequences $aR \cdots Rb$ and $bR \cdots Ra$. Together these would form a sequence $aR \cdots RbR \cdots Ra$ which contradicts b).

Now we shall prove that the left-neighbor-relation $\lhd$ to $\leq$ is R. To this purpose we assume that a, b are elements of S with $a \lhd b$. Then a fortiori $a < b$ holds, and according to the construction of $\leq$ there follows aRb or the existence of a finite sequence $a\ R \cdots R\ b$. But the last possibility cannot happen because in the last expression between the two R's no element can occur which is different from a and b. For this follows from $a \lhd b$ and $R \subseteq\ \leq$. So we have obtained $\lhd\subseteq R$.

Let now $a\ R\ b$ hold, and then also $a \leq b$. Then there is no $c \in S$ with $a < c < b$. Otherwise we would have a finite sequence $aR \cdots RcR \cdots Rb$. But since R is totally intransitive, we then would also have $a\ R^c\ b$ contradicting $a\ R\ b$. So $a \lhd b$ holds, and thus $R \subseteq\lhd$ and finally $R =\lhd$.

6.6 Definition. Let $(P, \leq)$ be a poset, $T \subseteq P$, $a \in P$. If $a \leq t$ for all $t \in T$ holds, a is called a *lower bound* of T. (It is not required, that a is in T.) Analogously u is called an *upper bound* of T if $u \geq t$ for all $t \in T$ holds. If here $\leq$ is replaced by $<$, resp. $\geq$ by $>$, a is said to be a *strict lower* (resp. a *strict upper*) *bound* of T.

Every element of P is a lower bound and at the same time an upper bound of the empty set $\emptyset$.

A subset $T \subseteq P$ is said to be *bounded from below* (resp. *from above*) if there exists a lower (resp. upper) bound of T in P. If T is bounded from below and from above, it is called *bounded*.

$(P, \leq)$ is called *directed from above (resp. below)*, if for every two different elements $a, b \in P$ there exists an upper (resp. lower) bound of the two-element set $\{a, b\}$. We call $(P, \leq)$ *directed*, if it is directed from below and from above.

The following statement is easily verified:

6.7 Theorem. *If a poset $(P, \leq)$ is directed from above, then there exists for every non-empty finite subset T of P an upper bound. And then every finite subset of P is bounded from above.*

Proof. Let $T = \{t_1, \ldots, t_n\}$ with $n \geq 3$, u_2 an upper bound of $\{t_1, t_2\}$, u_3 an upper bound of $\{u_2, t_3\}, \ldots,$ u_n an upper bound of

$\{u_{n-1}, t_n\}$. Then u_n is an upper bound of T.

6.8 Definition. Let $(P, \leq)$ be a poset with $P \neq \emptyset$, $T \subseteq P$. If for T there exists a lower bound $a \in P$ which satisfies $a \geq x$ for all lower bounds x of T, then a is called the *infimum* (or *greatest lower bound*) of T. It is denoted by $\inf T$.

If $\inf T$ exists it is indeed uniquely determined: If a_1 and a_2 are lower bounds of T which are $\geq$ all lower bounds of T, they must also fulfill $a_1 \leq a_2$ and $a_2 \leq a_1$, and thus they must be equal.

Dually to infimum the concept *supremum* (or *least upper bound*) of T is defined. It is denoted by $\sup T$.

In cases where the set S shall be indicated, relative to which $\inf T$ resp. $\sup T$ are constructed, we also use the notation $\inf_S T$ resp. $\sup_S T$.

So we have the following: A subset T of a poset P has not necessarily a lower resp. upper bound in P, and if it has one, this is usually not uniquely determined. And if $\inf T$ or $\sup T$ exist they need not belong to T. For the situation where this is yet the case we define: a is called the *least* or *first element* or *minimum* of T, if $a \in T$ and $a \leq t$ for all $t \in T$. We express this also as $a = \min T$. Of course, then also $\min T = \inf T$ holds. Dually one defines the concept *greatest* or *last element* or *maximum* of T. This is denoted by $\max T$.

Now we pay attention to the special case where T is empty: If $T = \emptyset$, every $a \in P$ is a lower and at the same time an upper bound of T. And then we have:

$\inf \emptyset$ exists iff $(P, \leq)$ has a greatest element $\max P$, which then is $= \inf \emptyset$, and dually

$\sup \emptyset$ exists iff $(P, \leq)$ has a least element $\min P$, which then is $= \sup \emptyset$.

$(P, \leq)$ is called inf-*complete* (resp. sup-*complete*), if for every subset T of P (also to $T = \emptyset$) the element $\inf T$ (resp. $\sup T$) exists. Instead of inf-complete resp. sup-complete we can also say *complete* or *completely ordered*, because each of these two properties entails the other as we prove now:

6.9 Theorem. *Let $(P, \leq)$ be an* inf-*complete poset and $T \subseteq P$. Then there exists also* sup *T, and this is $= \inf U$, where U is the set of upper bounds of T. So $(P, \leq)$ is also* sup-*complete.*

Proof. P has a greatest element $g = \inf \emptyset$ because it is inf-complete. Then g is an upper bound of T, and so the set U of all upper bounds of T is non-empty. According to our assumption inf U exists. We shall prove inf $U = $ sup T.

First we have $t \leq u$ for every $t \in T$ and every $u \in U$. And so every $t \in T$ is a lower bound of U, hence $t \leq $ inf U (which is the greatest lower bound of U). So inf U is an upper bound of T, and then also the least upper bound of T, that means inf $U = $ sup T.

Usually the values sup T and inf T depend on the "surrounding" set P. They are ascribed to T from outside. Contrary to this the values max T and min T are determined internally. In this context we introduce another concept (minimal element), which should not be confused with the concept "least element":

6.10 Definition. An element a of a poset $(P, \leq)$ is said to be *minimal* (resp. *maximal*), if no element x of P exists which satisfies $x < a$ (resp. $x > a$). We denote the set of all minimal (resp. maximal) elements of P by $\mathrm{Mi}(P)$ (resp. $\mathrm{Ma}(P)$).

6.11 Remark. If a poset has a least element, then this is also a minimal element, analogously the greatest element is also a maximal element. In a linearly ordered set the notions "minimal element" and "least element" evidently coincide, also "maximal" and "greatest". But this is usually not the case:

6.12 Example. Consider a set of five elements a, b, c, d, e, whose order $\leq$ is given by $a < c < e$, $b < c < d$, $a \parallel b$, $d \parallel e$. See Figure 1.

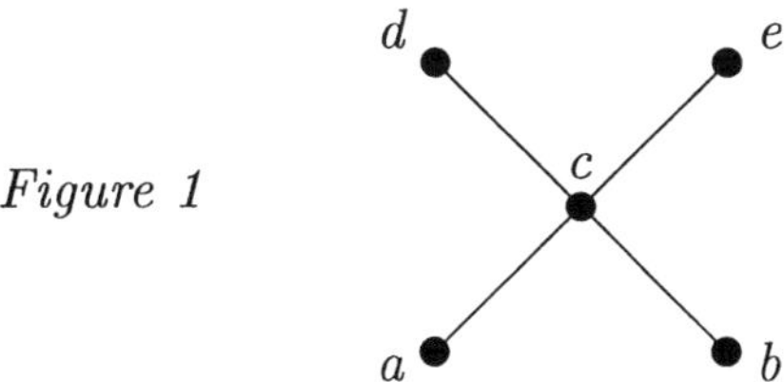

Figure 1

In this set a and b are minimal elements, d and e are maximal elements. This set has neither a greatest nor a least element.

6.13 Example. Even if a poset contains exactly one minimal element, this need not be the least element: Let P be the following poset: P contains all integers and in addition to this exactly one element x,which shall be incomparable with all other elements of P whilst the elements of $\mathbf{Z}$ are ordered in the usual way. This defines an order relation on P, relative to which x is the only minimal element and at the same time the only maximal element, but x is neither least nor greatest element of P.

We supplement Definition 6.6 by:

6.14 Definition. A non-empty poset $(L, \leq)$ is called a *lattice* if for every two-element subset $\{a, b\}$ of L there exist $\inf\{a, b\}$ and $\sup\{a, b\}$.

If $\inf T$ (and with it $\sup T$) exists for every subset T of L, then L is called a *complete lattice*. So "complete lattice" is the same as "completely ordered set".

1.7 Diagram representation of finite posets

The concept "poset" is rather intuitive and offers good possibilities for illustrations. To this purpose several notions have been introduced:

7.1 Definition. Let $(P, \leq)$ be a finite poset, a an element of P. The *height* $h(a)$ (of a in P), (also denoted by $\mathrm{Ht}(a)$) is the greatest non-negative integer h, so that there exists a chain $\{a_0, \ldots, a_h = a\}$, where $a_0 < \cdots < a_h$ (the same: $a_0 \lhd \cdots \lhd a_h$). Informally speaking, $h(a)$ is the greatest number for which a chain of P with $h(a) + 1$ elements exists which ends in a.

For $n \in \mathbf{N}_0$ let L_n denote the set of all elements of P which have height n. It is called the *n-level* or *height-n-set* of P. If $P \neq \emptyset$ we call the maximum of all numbers $h(a) + 1$, $a \in P$, the *height* of P and denote it by $h(P)$.

So $h(P)$ is the number of levels L_n of P which are non-empty. And $h(P) = m + 1$, if $m := \max\{h(a) | a \in P\}$.

The system of the levels of a poset P is illustrated in the following statement:

7.2 Theorem. *Let $(P, \leq)$ be a finite poset.*
1) For non-negative integers $i, j < h(P)$, $i \leq j, x \in L_i$, $y \in L_j$, we have $x \leq y$ or $x \parallel y$. ($y < x$ is impossible.)
2) For every non-negative integer $n < h(P)\}$ the level L_n is an antichain. The levels L_n are pairwise disjoint and form a partition of P in antichains (the so-called level-antichains).

Proof. 1) Every chain among those chains which have $y \in L_j$ as last element and a maximal number of elements has $j + 1$ elements. If now we would have $y < x$, one could extend these chains by adjoining x and so obtain a chain of cardinality $j + 2 > i + 1$ which has x as last element. But this contradicts $x \in L_i$. Then 2) follows from 1) by putting $i = j$ in 1).

The next theorem is also very informative. Together with 7.2 it shows how the levels L_n partition the set P in a layered manner:

7.3 Theorem. *Let h be the height of a non-empty finite poset $(P, \leq)$. For $i < h$ then there holds: L_i is the set of minimal elements of $P \backslash \cup\{L_\nu | \nu < i\}$. In particular, L_0 is the set of all minimal elements of P.*

Proof. Let $a \in L_i$. If a would not be minimal in the set $\cup\{L_\nu | \nu \geq i\}$, then this set would contain an element b with $b < a$, and we would have $b \in L_\nu$ for an index $\nu \geq i$. Hence there would exist a chain of cardinality $\nu + 1$ with b as last element and then also a chain of cardinality $\nu + 2$ with a as last element. This would yield $h(a) \geq \nu + 1 > i$, contradicting $a \in L_i$.

7.4 Theorem. *Let a, b be elements of a finite poset $(P, \leq)$ with $a \lessdot b$. Then we have $h(b) \geq h(a) + 1$.*

Proof. Every chain of cardinality $h(a) + 1$ which ends in a yields a chain of cardinality $h(a) + 2$ $(\leq h(b) + 1)$ which ends in b by attaching the element b.

7.5 Remark. By the way, in 7.4 one cannot conclude that $h(b) = h(a) + 1$ holds. If e.g. P has four elements a, b, c, d, where $a < b < c$ and $d < c$ hold and where d is incomparable with a and b, then $h(c) = 2$, $h(d) = 0$, but $d \lessdot c$ holds. See Figure 2.

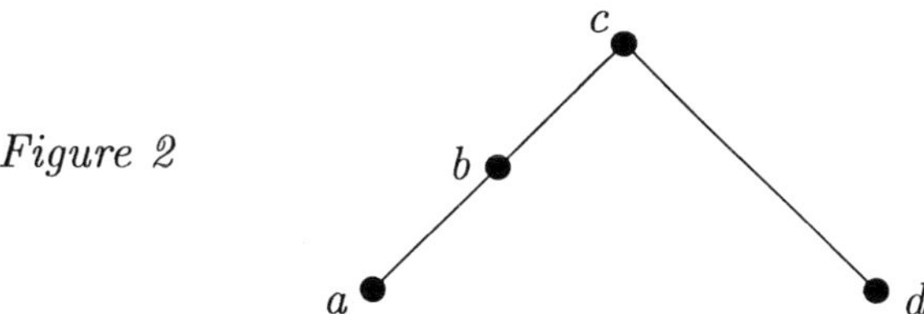

Figure 2

26

We now describe more details in connection with 7.3 with the aim to obtain a good illustration for finite posets.

7.6 Theorem. *Let the assumptions of 7.3 be fulfilled and $a \in L_n$ for a non-negative integer $n < h$. Let further $\{a_0, \dots, a_n\}_<$ be a longest chain among the chains which have $a = a_n$ as last element. Then there holds $a_\nu \in L_\nu$ for $\nu = 0, \dots, n$.*

Proof. The sets $L_0, L_1, \dots$ are antichains and therefore the elements $a_0, \dots, a_n$ are all in different levels. Because of 7.4 we have $0 \leq h(a_0) < \cdots < h(a_n) = n$. This is only possible if $h(a_\nu) = \nu$ for $\nu = 0, \dots, n$.

Now we can illustrate very well the finite posets by diagrams: Let $h \geq 1$ be the height of a finite poset $(P, \leq)$. We draw h parallel horizontal lines. On the lowest of these we put those elements of P which have height 0, — that means the minimal elements of P. On the next higher line we place the elements of height 1 and so on. On the highest line we place the elements of P which have height $h-1$, and here the construction finishes, since P has (only) height h.

If $h \geq 1$ and $i \in \{0, \dots, h - 1\}$, $x \in L_i$, $y \in L_j$, $j > i$ and $x \lhd y$ hold, we link x with y by a line segment. From the diagram which so arises one can realize all relations $x \lhd y$. According to 6.3 the order $\leq$ is then determined by $\lhd$. The levels can easily be recognized, — all of them are antichains. Also one has a good insight into how the chains of P look. They correspond to paths, which can be traveled along the line segments of the diagram, where one starts with an element and moves upwards only in ascending steps. See e.g. Figure 3.

Figure 3

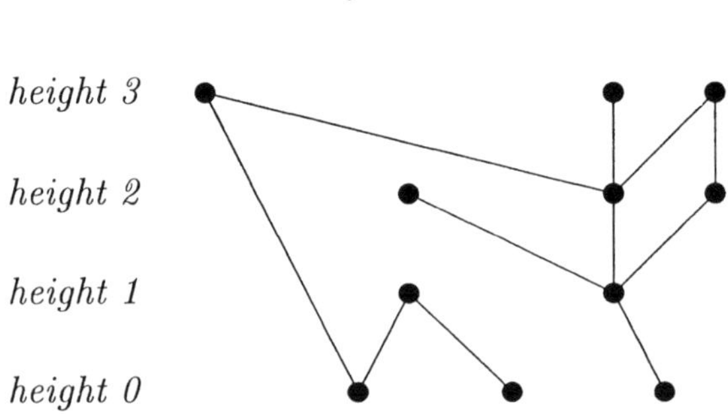

From the last consideration we can extract a useful formula:

7.7 Theorem. *Let m,n be natural numbers, P a finite poset with $|P| \geq m \cdot n + 1$. Then there exists a chain with $m + 1$ elements or an antichain of $n + 1$ elements. (Namely $h(P) \geq m + 1$ holds, or one of the $|L_\nu|$ is $\geq n + 1$.)*

Proof. Let h be the height of P (we have $h \geq 1$) and $L_0, \ldots, L_{h-1}$ the levels of P. Then $\sum_{\nu=0}^{h-1} |L_\nu| = |P| \geq m \cdot n + 1$. Then, of course, the number h of summands is $\geq m + 1$ or at least one summand $|L_\nu|$ is $\geq n + 1$. In the case where $h \geq m + 1$ we have the existence of a chain with $h \geq m + 1$ elements. In the second case we are done because each L_ν is an antichain.

With 7.7 we immediately obtain a theorem of Erdös and Szekeres [34]:

7.8 Theorem. *Let a set P of $m \cdot n + 1$ points $p_i = (a_i, b_i)$ of the euclidean plain $\mathbf{R}^2$ be given. Then there are $m+1$ points $p_1, \ldots, p_{m+1}$ with $a_1 \leq \cdots \leq a_{m+1}$ and $b_1 \leq \cdots \leq b_{m+1}$ or $n + 1$ points $p_1, \ldots, p_{n+1}$ with $a_1 \leq \cdots \leq a_{n+1}$ and $b_1 \geq \cdots \geq b_{n+1}$.*

Proof. We introduce an order relation in P as follows: $(a, b) \leq (c, d) \iff a \leq c$ and $b \leq d$ (see Example 4.4). Then 7.7 implies our statement.

By the way, using 7.2 we easily obtain a partial statement of a theorem of Szpilrajn [167]:

7.9 Theorem. *Every order $\leq$ on a finite set P is a subset of a linear order on this set. In other words: Every order relation on a finite set is extendible to a linear order.*

Proof. Let $(P, \leq)$ be a non-empty finite poset. We consider the levels $L_0, \ldots, L_{h-1}$ of P where h is the height of P. Then we take an arbitrary linear order in each of these levels, and if i, j are elements of $\{0, \ldots, h-1\}$ with $i < j$ we put $x < y$ for all $x \in L_i$ and all $y \in L_j$. Together we so obtain a linear order in P, which extends the original order $\leq$.

The observations of 7.4 and 7.5 motivate the following notion:

7.10 Definition. Let $(P, \leq)$ be a poset in which all bounded chains are finite. Then a mapping $r : P \to \mathbf{Z}$ is called a *rank function*, if for all $x, y \in P$ with $x \lhd y$ there holds $r(y) = r(x) + 1$. Then $r(x)$ is called the *rank* of x. For $n \in \mathbf{Z}$ we denote the set of elements of P which have the rank n by $r_n(P)$. It is called the nth *rank* of $(P, \leq, r)$.

If P is finite, and if the least number of $\{r(x)|x \in P\}$ is 0, then $\max\{r(x)|x \in P\}$ is called the *rank* of P.

The concepts "height" and "rank" have several common features, but must not be confused. The height function is defined for every finite poset. But generally finite posets have no rank function:

7.11 Example. Let P be the poset of 7.5. Then P has a rank function r, for which $r(a) = 0$, $r(b) = r(d) = 1$, $r(c) = 2$. But the height of d is 0. For the other three elements of P the height and the rank are equal.

In analogy to 7.2 we have

7.12 Theorem. *Let $(P, \leq)$ be a poset with a rank function r. Then all ranks of P are antichains. For integers $m \neq n$ the ranks $r_m(P)$ and $r_n(P)$ are disjoint. So also the ranks of P form a partition of P in antichains.*

Proof. The conclusion is the same as in 7.2: If a, b are different elements with the same rank, they cannot be comparable. If e.g. we would have $a < b$ there would exist a sequence $a = a_0 \lhd \cdots \lhd a_k = b$ for some natural number k. And here every a_i, $i = 1, \ldots, k$, would have the rank $r(a_{i-1}) + 1$, so that $r(a) < r(b)$ would follow with contradiction.

The height function of a finite poset is of course uniquely defined. The question arises, whether this also holds for the rank function if there exists one for a given poset. First we introduce a new concept:

7.13 Definition. A poset $(P, \leq)$ is said to be *connected*, if for every two elements $a \neq b$ of P there exist finitely many elements $a = a_0, \ldots,$ $a_n = b$, such that a_i is comparable with a_{i+1} for $i = 0, \ldots, n - 1$.

Then we can prove:

7.14 Theorem. *Let $(P, \leq)$ be a connected poset with a rank function r and and e a fixed element of P. Then the rank function r is uniquely determined by the value $r(e)$.*

Proof. All immediate successors of e must have rank $r(e) + 1$, their immediate successors must have rank $r(e) + 2$ and so on. The immediate predecessors of e must have rank $r(e) - 1$ and so on. Because of the connectedness of P every element of P can be reached in this way, and its rank is uniquely defined.

1.8 Special subsets of posets. Closure operators

Now we introduce several concepts for subsets of posets which suggest themselves:

8.1 Definition. Let $(P, \leq)$ be a poset, $T \subseteq P$. We call T an *initial segment* of P, if there holds:

$a \in T$, $x \in P$ and $x < a \implies x \in T$. (Other denotations are *lower set, down set* or *decreasing set*.) We denote the set of initial segments of P by $\mathfrak{I}(P)$.

Dually T is called a *final segment* of P, if there holds

$a \in T$, $x \in P$ and $x > a \implies x \in T$.

T is called *convex* or a *segment* of P, if there holds

$a, b \in T$, $z \in P$ and $a \leq z \leq b \implies z \in T$.

A chain C of P is said to be *saturated*, if every chain $(K, \leq \restriction K)$ of P which contains C has C as a segment. So if one adds to C an element $e \in P \setminus C$, then $C \cup \{e\}$ can only be a chain if e is $<$ resp. $>$ all elements of C.

If $\mathfrak{S}$ is a set of pairwise disjoint non-empty segments of a linearly ordered set $(L, \leq)$, we have a natural linear order in $\mathfrak{S}$, which is given by: If $S_1, S_2 \in \mathfrak{S}$ we put $S_1 < S_2 \iff s_1 < s_2$ for all $s_1 \in S_1$ and all $s_2 \in S_2$. It is clear that this $<$, augmented by the identity relation $\mathrm{id}_\mathfrak{S}$, is a linear order on $\mathfrak{S}$. We call it the *natural order* of $\mathfrak{S}$ (and use the same sign $\leq$ for it as that of L).

It follows rather immediately:

8.2 Theorem. *If $\mathfrak{A}$ is a non-empty set of initial segments of a poset $(P, \leq)$, then $\cup \mathfrak{A}$ and $\cap \mathfrak{A}$ are again initial segments of P. The system of initial segments of a poset is thus fully union-compatible and fully intersection-compatible. Then this system is also a topology on P, the topology of initial segments of P.*

Dually we obtain an analogous statement for the system of final segments.

For general segments we only obtain:

8.3 Theorem. *The intersection of a set of segments of a poset is again a segment of it.*

From 8.2 we obtain:

8.4 Theorem and **Definition.** *Let T be a subset of a poset $(P, \leq)$. Then there exists a smallest initial segment of P which contains T. It is called the initial segment generated by T. It is the intersection of all initial segments of P which contain T. We denote it by $IS(T)$. Dually the concept final segment generated by T is defined and denoted by $FS(T)$.*

The convex hull (or convex closure) $Conv(T)$ is defined as the intersection of all convex subsets of P which contain T. It is evidently the smallest convex subset of P which contains T.

Also $IS(T)$ and $FS(T)$ are convex sets, and there holds $ConvT = IS(T) \cap FS(T)$.

Proof. We show $IS(T) \cap FS(T) \subseteq Conv(T)$. The rest is trivial. Let $x \in IS(T) \cap FS(T)$. Then there exist elements $a, e \in T$ with $x \leq a$ and $e \leq x$, and $e \leq x \leq a$ shows that $x \in Conv(T)$ holds.

Now we define another important notion of order theory. It was introduced by Kuratowski:

8.5 Definition. Let $(P, \leq)$ be a poset, $c : P \to P$ be a mapping of P into itself. Then c is called a *closure operator* if the following three conditions are fulfilled:

1) $c(x) \geq x$ for every $x \in P$. (Extensionality)
2) For $a, b \in P$ with $a \leq b$ there follows $c(a) \leq c(b)$. (Isotonicity)
3) For every $x \in P$ we have $c(c(x)) = c(x)$. (Idempotency)

The element $c(x)$ is called the *c-closure* of x. If $c(x) = x$ holds, we call x a *fixed element* of c. The set of all fixed elements of c is denoted by Fix(c).

Then one can immediately verify:

8.6 Theorem. *The mappings IS, FS and Conv of the power set $\mathfrak{P}(S)$ of a poset $(S, \leq)$ into $\mathfrak{P}(S)$ are closure operators with respect to the (usual) order $\subseteq$. The initial segments, final segments, resp convex subsets are their corresponding fixed elements.*

We present two further examples of closure operators : If $\mathfrak{S}$ is the set of all subsets of $\mathbf{R}^2$ and $T \in \mathfrak{S}$, then let $c(T)$ be the least closed subset (in the standard topology of $\mathbf{R}^2$) which contains T. Then c is a closure operator on $\mathfrak{S}$.

If we define $c(T)$ as the least convex subset of $\mathbf{R}^2$ which contains T, then again c is a closure operator on $\mathfrak{S}$.

8.7 Example. Every antichain of a poset is convex. This does not hold for every chain, even if this chain is a maximal one: In the four-element set $\{a, b, c, d\}$ where $a < b < c$, $a < d < c$ and $b \parallel d$ (see Figure 4) we have: $\{a, b, c\}$ is a maximal chain, but it does not contain all elements between a and c because d is missing.

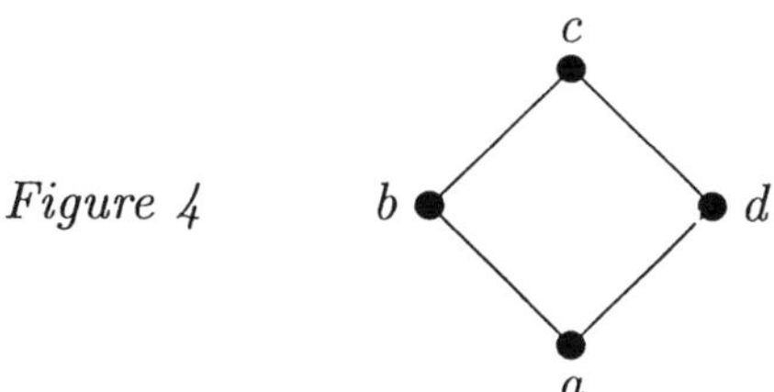

Figure 4

8.8 Theorem and **Definition.** *Let a be an element of a poset $(P, \leq)$ and $Inc(a)$ the set of all elements of P which are incomparable with a. Then $Inc(a)$ is convex.*

Proof. Let $x, y \in Inc(a)$ and $x \leq z \leq y$ for an element $z \in P$. Then $a \leq z$ cannot hold, otherwise we would also have $a \leq y$ contradicting $y \parallel a$. Further $z \leq a$ cannot hold, because this would entail $x \ (\leq z) \leq a$ contradicting $x \parallel a$. Therefore $z \in Inc(a)$ follows.

We introduce some more concepts:

8.9 Definition. Let $(P, \leq)$ be a poset, $a \in P$. Then we call the set $(P < a) := \{x \in P | x < a\}$ the *open initial segment of a in P*, and the set $(P \leq a); = \{x \in P | x \leq a\} =: (a]$ the *closed initial segment* (or *principal ideal*) *of a in P*. Analogously we define $(P > a)$ and $(P \geq a) = [a)$ and call it the *open* (resp. *closed*) *final segment* or *principal filter* of a. Evidently the sets so defined are indeed initial (resp. final) segments of $(P, \leq)$. Further $(P \parallel a)$ is the set of those elements of P which are incomparable with a. For subsets A, B of P we put $A < B \Longleftrightarrow a < b$ for all $a \in A$ and all $b \in B$. Further $A > B$, $A \leq B$, $A \geq B$ are defined analogously.

Let $a, b \in P$ and $a < b$. Then we call

$[a, b] := \{x \in P | a \leq x \leq b\}$ the *closed interval with ends a, b of* $(P, \leq)$,

$(a, b] := \{x \in P | a < x \leq b\}$ the *left open, right closed interval with ends a, b of* $(P, \leq)$.

$[a,b) := \{x \in P|\ a \leq x < b\}$ the *left closed, right open interval with ends a,b of* $(P, \leq)$.

$(a,b) := \{x \in P|\ a < x < b\}$ the *open interval with ends a,b of* $(P, \leq)$. The intervals $(a,b]$, $[a,b)$, (a,b) are also denoted by $]a,b]$, $[a,b[$, $]a,b[$ respectively.

Every set of one of these four types is called an *interval* of $(P, \leq)$. Every interval is of course the intersection of an (open or closed) initial segment of an element $b \in P$ and an (open or closed) final segment of an element $a \in P$. Every interval is a fortiori a segment, but not conversely. If e.g. $(P, \leq)$ is the set of rational numbers which are $> \sqrt{2}$ and $< \sqrt{3}$, then this set is a segment, but not an interval of the set $\mathbf{Q}$ of rational numbers, equipped with their usual order. (But some authors use the word "interval" in the wider sense of "segment".)

If $a = b$ we put $[a,b] := \{a\}$, and if we only know that a and b are comparable elements of P, then $[a,b]$ and $[b,a]$ denote the same subset of P. The same holds for (a,b) and (b,a). But if we formulate: Let $[a,b]$ resp. (a,b) be an interval, we always assume that $a < b$ holds.

To be an initial segment is transitive in the following sense:

8.10 Theorem. *Let A be an initial segment of a poset $(P, \leq)$ and let B be an initial segment of A (with the induced order), then B is also an initial segment of $(P, \leq)$. Analogous statements hold, if we replace "initial segment" by "final segment" or by "segment".*

8.11 Example. If S is an arbitrary set. Then the set of all finite subsets of S is an initial segment of the power set $(\mathfrak{P}(S), \subseteq)$.

Initial and final segments can be complementary:

8.12 Theorem and **Definition.** *Let A be an initial segment of a poset $(P, \leq)$. Then $P \backslash A := E$ is a final segment of $(P, \leq)$. It is called the final segment complementary to A.*

Proof. Let $e \in E$, $x \in P$, $e \leq x$. If x would not be in E, we would have $x \in A$ and further $e \in A$ in contradiction to $e \in E$.

8.13 Example. In every poset $(P, \leq)$ the sets $\emptyset$ and P are initial and final segments at the same time. But they need not be the only subsets of P, which are both. If e.g. P is a union of two disjoint non-empty chains C_1, C_2, where every element of C_1 is incomparable with every element of C_2, then C_1 and C_2 are both, an initial and a final segment. Moreover C_1 is the initial segment which is complementary to the final segment C_2. Also C_1 is the final segment which is complementary to the initial segment C_2.

Now we study the set $\mathrm{Fix}(c)$ of a closure operator c in some more detail:

8.14 Theorem. *The set* $\mathrm{Fix}(c)$ *of a closure operator* $c : P \longrightarrow P$ *on a poset* P *is the image set* $c[P]$.

Proof. If $y \in c[P]$ holds, y is $= c(x)$ for some $x \in P$ and then, due to 3) from 8.5 we have $c(y) = c(c(x)) = c(x) = y$. On the other hand, if $x \in \mathrm{Fix}(c)$, we have $x = c(x) \in c[P]$.

The set $\mathrm{Fix}(c)$ of a closure operator c is inf-closed, precisely:

8.15 Theorem (Ward [176]). *Let* c *be a closure operator on a poset* P, *and* T *a subset of* $F := \mathrm{Fix}(c)$ *for which* $i := \inf_P T$ *exists. Then* i *also belongs to* F.

Proof. We have $i \leq x \forall x \in T$, and then by the isotonicity $c(i) \leq c(x) = x \forall x \in T$. So $c(i)$ is a lower bound of T and hence $c(i) \leq i$. Due to the extensionality also $i \leq c(i)$ holds and thus $i = c(i) \in F$.

8.16 Remark. The statement which replaces inf by sup in the last theorem is not true, as the following example exhibits:
Let $\mathfrak{S}$ be the set of all subsets of $\mathbf{R}$, ordered by inclusion. And for $T \in \mathfrak{S}$ let $c(T)$ be the least closed subset of $\mathbf{R}$ which contains T. Further, let F be the set of all closed intervals $[0, \frac{1}{2} + \frac{1}{4} + \cdots + \frac{1}{2^n}]$ with $n \geq 2$ of $\mathbf{R}$. Then F has a supremum in $\mathbf{R}$, namely the half-open interval $[0, 1)$, but this does not belong to F.

But if now we strengthen the assumption of 8.15 we obtain the following stronger statement:

8.17 Theorem (Ward [176]). *Let* L *be a complete lattice,* $c : L \to L$ *a closure operator. Then* $F := \mathrm{Fix}(c)$ *is again a complete lattice, and for every subset* T *of* F *there holds*

$$(1)\ \inf_F T = \inf_L T \quad \text{and} \quad (2)\ \sup_F T = c(\sup_L T).$$

Proof. Let T be a subset of F. Then 8.15 yields $\inf_L T \in F$, and so already $\inf_F T = \inf_L T$ follows.

By (1) the set U of upper bounds of T in F satisfies $\sup_F T = \inf_F U = \inf_L U$. So for all subsets T of F there exist $\inf_F T$ and $\sup_F T$, and so F is complete.

We still have to prove (2). We have $u := c(\sup_L T) \geq \sup_L T$. Thus u is an upper bound of T in L and by 8.14 an element of F. Now let $y \in F$ be an arbitrary upper bound of T. Then $y \geq \sup_L T$ and $c(y) \geq c(\sup_L T) = u$. Now y is ($= c(y)$ because of $y \in F$ and thus) $\geq u$. Therefore u is the least upper bound $\sup_F T$ of T in F.

1.9 Order-isomorphic mappings. Order types

Everytime a mathematical structure is under consideration, those mappings of the carrier sets are interesting which preserve the structure fully or partially. In this context it becomes possible to compare mathematical structures and to establish relationships between them. Applying a suitable notion of structural equality, one can to some extent transfer the discussion from the carrier sets to structure types. For our purposes especially those mappings are interesting which preserve the order of a given poset. First we define in full generality:

9.1. Definition Let S_1, S_2 be sets, and let R_i be a relation on S_i for $i = 1, 2$. Then the relational system (S_1, R_1) is said to be *isomorphic* to the relational system (S_2, R_2), if there exists a bijection $f : S_1 \to S_2$ with the property: For arbitrary $a, b \in S_1$ there holds
$$a \, R_1 \, b \Leftrightarrow f(a) \, R_2 \, f(b).$$
If this holds, then f is called an *isomorphism* or an *isomorphic mapping* of (S_1, R_1) onto (S_2, R_2).

If here $S_1 = S_2$ and $R_1 = R_2$ holds, f is called an *automorphism* of (S_1, R_1).

Isomorphic relational systems have the same structure, they only differ in the carrier sets.

One can immediately see that the property to be isomorphic yields an equivalence relation in every set of relational systems.

If in the above text we replace "isomorphic" by "homomorphic", "bijection" by "surjection", $\Leftrightarrow$ by $\Rightarrow$, "isomorphism" by "homomorphism", one obtains the concepts which correspond to the situation where the structure of the relational system (S_1, R_1) is only partially preserved, namely "in one direction".

Generally homomorphisms are not reversible, but isomorphisms $f : S_1 \to S_2$ are, and one can easily see that then $f^{-1} : S_2 \to S_1$ is an isomorphism from (S_2, R_2) onto (S_1, R_1).

Now we address the special situation where R_1 and R_2 are order relations.

9.2 Definition. Let $(P_1, \leq)$ and $(P_2, \leq)$ be posets. Then $(P_1, \leq)$ is said to be *order-isomorphic*, in short *isomorphic*, or *similar* to $(P_2, \leq)$, if there exists a bijection $f : P_1 \to P_2$ with the property: For all $a, b \in P_1$ there holds $a \leq b \Leftrightarrow f(a) \leq f(b)$. In this case we write $(P_1, \leq) \simeq (P_2, \leq)$, also in short $P_1 \simeq P_2$, and call f a *similar* or *order-isomorphic*, in short *isomorphic, mapping* or an *(order-) isomorphism* of $(P_1, \leq)$ onto $(P_2, \leq)$.

A poset $(P, \leq)$ is said to be *symmetric* if there exists an isomorphism of $(P, \leq)$ onto $(P, \geq)$. The sets $\mathbf{Z}$, $\mathbf{Q}$, $\mathbf{R}$ are evidently symmetric, $\mathbf{N}$ is not.

The weaker homomorphism-concept is here described as follows: Let $f : P_1 \to P_2$ be a mapping. Then

f is called $\leq$ - *preserving* (or *isotone, increasing, ascending*) if for arbitrary elements $a, b \in P_1$ there holds

$$a \leq b \Longrightarrow f(a) \leq f(b).$$

f is said to be $<$ - *preserving* (or *strictly isotone, strictly increasing, strictly ascending*), if for arbitrary elements $a, b \in P_1$ there holds

$$a < b \Longrightarrow f(a) < f(b).$$

f is called $\leq$ - *reversing* (or *anti-isotone, decreasing, descending*),if for arbitrary elements $a, b \in P_1$ we have

$$a \leq b \Longrightarrow f(a) \geq f(b).$$

f is said to be $<$ - *reversing* (or *strictly anti-isotone, strictly decreasing, strictly descending*), if for arbitrary elements $a, b \in P_1$ we have

$$a < b \Longrightarrow f(a) > f(b).$$

f is called $\|$- *preserving* if $a \| b \Longrightarrow f(a) \| f(b)$.

Isomorphic mappings preserve of course "everything", $\leq$, $<$, and $\|$. On the other hand there follows:

9.3 Theorem. *Let P_1 and P_2 be posets, and let $f : P_1 \to P_2$ be a surjective mapping, which preserves $<$ and $\|$. Then f is an isomorphic mapping.*

Proof. For $a \neq b$ of P_1 there holds $a < b$, $b < a$ or $a \| b$. From this follows the injectivity of f because f preserves $<$ and $\|$. Let now

$a, b \in P_1$ and $a \le b$. Then evidently $f(a) \le f(b)$ holds. On the other hand, if $f(a) \le f(b)$, we cannot have $a \| b$ or $b < a$ since otherwise $f(a) \| f(b)$ or $f(b) < f(a)$ would follow, and so we must have $a \le b$.

9.4 Example. A strictly isotone mapping $f : P_1 \to P_2$ need not be injective; only its restrictions on chains must be injective. If e.g. P_1 is the three-element set $\{a, b, c\}$, which is ordered by $a < b$, $a < c$, $b \| c$, and if P_2 is the two-element chain $\{d, e\}$ in which $d < e$ holds (see Figure 5), then we consider the mapping $f : P_1 \to P_2$ which maps a onto d, and b and c onto e. Then f is strictly isotone, but not injective.

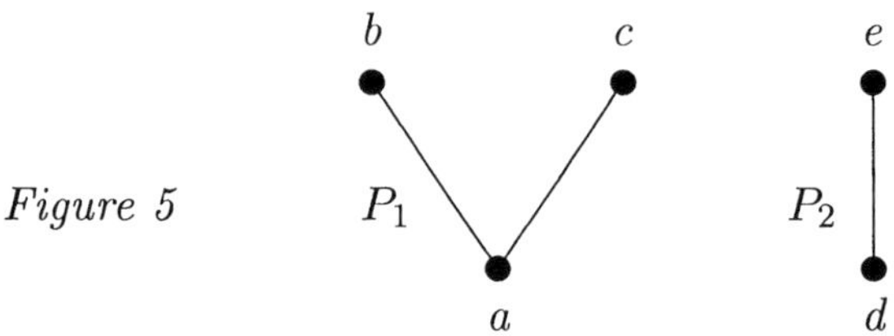

Figure 5

The relation to be order-isomorphic is an equivalence relation, and this makes it possible to subdivide the class of all posets in disjoint subclasses of posets which are pairwise isomorphic, the so-called order types:

9.5 Definition. The class of all posets which are isomorphic to a given poset $(P, \le)$ is called the *order type* of $(P, \le)$. We denote it by $\mathrm{tp}(P, \le)$, also in short with $\mathrm{tp}(P)$, if it is clear which order in P is under consideration. The poset $(P, \le)$ is said to be a *realization* of this order type. So we have for posets A, B the correspondence: $\mathrm{tp}(A) = \mathrm{tp}(B) \iff A \simeq B$. Order types are preferably denoted by small greek letters.

A well-ordered set is isomorphic to a uniquely determined ordinal number, and therefore two well-ordered sets are isomorphic, iff they are isomorphic to the same ordinal τ. For this reason one can call τ the order type of these well-ordered sets. Then an ordinal τ, seen as a linearly ordered set, has as order type again τ.

Since isomorphic mappings preserve all essential things, isomorphic sets have the same structural properties. E.g. if $f : P \to T$ is an isomorphic mapping of the posets P and T, we have statements like these: If A is an initial segment of P, $f[A]$ is one of T. If m is a minimal (resp. the least) element of P, $f(m)$ is a minimal (resp. the least) element in T, and so on.

Order types can be compared in a natural way by the following definition which was introduced by Fraissé [44]:

9.6 Definition. If α is an order type, we define α^* as follows: We choose a poset $(P, \leq)$ which has type α. Then α^* is the type of $(P, \geq)$. We call α^* the *reverse* order type of α.

An order type α is said to be *symmetric*, if $\alpha = \alpha^*$. The same: α is symmetric if at least one (the same: each) realization $(P, \leq)$ of α is symmetric.

Let α and β be order types. We say: α is *embeddable* in β, and denote this with $\alpha \leq \beta$, if the following holds: If $(A, \leq)$ is a poset of type α and $(B, \leq)$ a poset of type β, then there exists an isomorphic mapping of A onto a subset of B. For this situation we also say: A is *embeddable in B*, and α is a *subtype* of β. (It is easily seen that this definition does not depend on the choice of the sets A and B, which realize the types α and β.)

If τ is an order type, a poset P is said to be τ- *free* or *free from* τ, if P has no subset which , with the induced order, has order type τ.

We write $\alpha < \beta$ and call α *less* than β, if $\alpha \leq \beta$, but not $\beta \leq \alpha$ holds. We say: α and β *have the same content*, if $\alpha \leq \beta$ and $\beta \leq \alpha$ hold. For this we write $\alpha \ (=) \ \beta$. For the relations $\leq$, $<$, and $(=)$ we have the following statements which are easily verifiable:

9.7 Theorem. *Let T be an arbitrary set of order types. Then the relation $\leq$ (on T) is reflexive and transitive, hence a quasi-order. The relation $<$ (on T) is irreflexive and transitive, hence a strict order. The relation $(=)$ is an equivalence relation in T. It partitions T in classes of order types, where all posets which have order types in the same class are pairwise embeddable in each other.*

We remark that the above defined relation $\leq$ is not antisymmetric. E.g. a closed interval A of $\mathbf{R}$ and an open interval B of $\mathbf{R}$ have different order types α resp. β, but we evidently have $\alpha \leq \beta$ and $\beta \leq \alpha$. And here $\alpha \leq \beta$ does not imply $\alpha = \beta$ or $\alpha < \beta$.

But if α and β are (not only order types, but also) ordinals, the usual relation $\leq$ for ordinals coincides with the above defined $\leq$, and then again we have antisymmetry.

38

With the concepts of the above definitions there result possibilities to classify the system of posets by forbidden subtypes. Here we compile several notions in this context. (They will be treated later in more detail.)

9.8 Definition. If k is a cardinal number, let a_k denote the order type of a k-element set X, which is *totally unordered*, that means its order-relation contains only the pairs (x, x) with $x \in X$.

The order types of the sets of natural numbers, resp. negative integers, resp. rational numbers, resp. real numbers (all equipped with their natural order) are denoted by ω, resp. ω^*, resp. η, resp. λ.

Let $(P, \leq)$ be a poset. Then we have:

$(P, \leq)$ is linearly ordered $\Leftrightarrow$ P has no two-element antichain, the same: no subset of type a_2.

We now define several other concepts using forbidden subtypes:

$(P, \leq)$ is *well-ordered* $\Leftrightarrow$ P has no subset of type a_2 and no subset of type ω^*.

$(P, \leq)$ is *well-founded* $\Leftrightarrow$ P has no subset of type ω^*.

$(P, \leq)$ is *well-quasi-ordered* $\Leftrightarrow$ P has no subset of type ω^* and no infinite antichain.

$(P, \leq)$ is a *tree* $\Leftrightarrow$ P has a least element, and P has no subset of type ω^* and no subset which is a Λ-*hook*, that is an ordered set of three elements a, b, c, where a and b are $< c$, and where a and b are incomparable.

$(P, \leq)$ is *scattered* if P has no subset of type η.

Now we can formulate a very informative statement which expresses the fact that, roughly speaking, the whole theory of posets is completely contained in the theory of the subset relation $\subseteq$.

9.9 Theorem. *Let $(P, \leq)$ be a poset. Then $(P, \leq)$ is isomorphic to a subset of the power set $\mathfrak{P}(P)$, which is equipped with the order $\subseteq$. More detailed: $(P, \leq)$ is isomorphic to the set of all initial segments $(a]$ $:= \{x \in P \mid x \leq a\}$ with $a \in P$. The mapping $f : P \longrightarrow \{(a] \mid a \in P\}$, which satisfies $f(a) = (a]$, is an isomorphism.*

Proof. Let $a, b \in P$. If $a < b$ holds, then $(a]$ is a proper subset of $(b]$ because of $b \notin (a]$. Thus f is strictly isotone. If $a \| b$ holds, then also $(P \leq a)$ and $(P \leq b)$ are incomparable. Indeed, if w.r.o.g. we would have $(P \leq a) \subseteq (P \leq b)$, this would entail $a \leq b$, which is impossible. So f preserves also $\|$ and is therefore an order-isomorphism.

The statement of 9.9 can be sharpened insofar as f also preserves infima:

9.10 Theorem (MacNeille [116]). *Under the assumption of 9.9 there holds: If for a set T with $\emptyset \notin T \subseteq P$ we have the existence of $\inf T$, then also $\inf f[T]$ exists in $(\mathfrak{P}(P), \subseteq)$, and there holds $\inf f[T] = f(\inf T)$.*

Proof. Let T be a non-empty subset of P, for which $\inf_P T$ exists. Since f is an isomorphic mapping we then have $f(\inf_P T) \leq f(t)$ for all $t \in T$, hence:

(1) $f(\inf_P T)$ is a lower bound of $f[T]$.

Let now $L \in \mathfrak{P}(P)$ be a lower bound of $f[T]$. Then $L \subseteq f(t) = (P \leq t)$ holds for all $t \in T$ and then further $L \subseteq D := \cap\{(P \leq t) | t \in T\}$. Now there holds:

(2) $D \subseteq f(\inf_P T)$.

Indeed we have: $x \in D \implies x \leq t \; \forall t \in T \implies x \leq \inf_P T \implies x \in (P \leq \inf_P T) = f(\inf_P T)$.

So we have obtained $L \ (\subseteq D) \subseteq f(\inf_P T)$. Then, because of (1), $f(\inf_P T)$ is the greatest lower bound of $f[T]$ in $\mathfrak{P}(P)$, which means $= \inf f[T]$.

Here the question arises whether in 9.10 one can replace inf by sup. But then the statement needs no longer to be true. A counter-example can easily be found using the following theorem:

9.11 Theorem. *Let P be a poset. For $x \in P$ we put $f(x) := (P \leq x) \ (\in \mathfrak{P}(P))$. Let T be a subset of P for which $\sup T$ exists.*

Then $\sup f[T] = f(\sup T)$ holds iff T has a greatest element $\max T$ $(= \sup T)$. (Here, of course, $\sup f[T]$ is formed in $\mathfrak{P}(P)$.)

Proof. $\sup f[T] = f(\sup T) \iff \cup\{(P \leq t) | t \in T\} = (P \leq \sup T) \iff IS(T) = (P \leq \sup T)$, where $IS(T)$ is the initial segment of P which is generated by T *(see 8.4)*. We always have $IS(T) \subseteq (P \leq \sup T)$, and so the equation $IS(T) = (P \leq \sup T)$ is equivalent to $IS(T) \supseteq (P \leq \sup T)$. This is equivalent to $\sup T \in IS(T)$. This now means that $\sup T$ is $\leq$ an element of T which implies that it is equal to this element. And then $\sup T$ is the greatest element of T.

1.10 Cuts. The Dedekind–MacNeille completion

Since completely ordered sets have nice properties it is desirable to embed arbitrary posets P in completely ordered posets. A simple possibility to attain this was shown in 9.9. But here $\mathfrak{P}(P)$ has a higher cardinality than P, and also the dimension (which is defined later) of $\mathfrak{P}(P)$ usually exceeds that of P. So it suggests looking for a complete superset of P, which arises from P by adjoining a "minimal" set of new elements.

This aim was reached by a construction of MacNeille [116], which generalizes a method of Dedekind, namely that by which he introduced the irrational numbers. The idea behind this was the following: After the rational numbers have been constructed one can define the irrational numbers as initial segments of the linearly ordered set of rational numbers, so that each irrational number x is the set of all rationals $\leq x$. The effect of this construction is, concerning the order-theoretical aspect, that the linearly ordered set $\mathbf{Q}$ of rational numbers is embedded into the set $\mathbf{R}$ of real numbers as a dense subset. MacNeille then obtained a generalization of this for general posets. First we introduce some concepts:

10.1 Theorem and **Definition.** *Let $(P, \leq)$ be a poset, $T \subseteq P$. Then we denote the set of lower bounds of T by $L(T)$, analogously the set of upper bounds of T by $U(T)$.*

Then $L(T)$ is an initial segment of P, and $U(T)$ a final segment of P. Every element of $L(T)$ is $\leq$ every element of $U(T)$.

The set $P \setminus L(T)$ is a final segment and $P \setminus U(T)$ an initial segment of P.

Further we define $C(T) := L(U(T))$ and call this set the cut generated by T in P.

The set of all cuts $C(T)$, $T \subseteq P$, ordered by inclusion, is called the Dedekind–MacNeille completion of P. We denote it by $DM(P)$ and equip it with the order $\subseteq$ of set inclusion. So $DM(P)$ is an ordered subset of $\mathfrak{P}(P)$.

The set $C(T)$ contains the initial segment $IS(T)$ because every element of $IS(T)$ is a lower bound of $U(T)$, hence in $L(U(T))$. But $IS(T)$ can be a proper subset of $C(T)$: Consider the poset of three elements a, b, c, where $a < c$, $b < c$, $a \parallel b$. For its subset $T := \{a, b\}$

there holds $U(T) = \{c\}$, $L(U(T)) = \{a, b, c\}$, but $IS(T) = \{a, b\}$.

For one-element sets T we yet have, as can easily be seen:

10.2 Theorem. *For every element a of a poset $(P, \leq)$ there holds $C(\{a\}) = L(U(\{a\})) = (a]$.*

In accordance with 9.2 we have: A mapping $f : \mathfrak{P}(S) \longrightarrow \mathfrak{P}(S)$ of the power set of a set S into itself is *isotone*, if $T_1 \subseteq T_2 \subseteq S$ entails $f(T_1) \subseteq f(T_2)$. It is *anti-isotone*, if $T_1 \subseteq T_2 \subseteq S$ implies $f(T_1) \supseteq f(T_2)$.

Now there follows:

10.3 Theorem a) *The mappings L and U of $\mathfrak{P}(P)$ into $\mathfrak{P}(P)$, where P is a poset, are both anti-isotone.*

b) *Their composition $C := L \circ U$ (and analogously $U \circ L$) is isotone and extensional.*

c) *C is a closure operator in $\mathfrak{P}(P)$.*

d) *For every $T \subseteq P$ the set $C(T)$ is the least element of the Dedekind–MacNeille completion $DM(P)$ which contains T.*

Proof. a) is trivial, and b) follows by twofold application of a), and since $L(U(T)) \supseteq T$ holds for all $T \subseteq P$. Indeed, every element of T is of course a lower bound of the set of all upper bounds of T.

c) We still have to prove that C is idempotent: $T \subseteq P \Longrightarrow ULU(T) \supseteq U(T)$ since $U \circ L$ is extensional. From this follows, since L is anti-isotone, $LULU(T) \subseteq LU(T)$, the same $CC(T) \subseteq C(T)$. Because of the extensionality of C we also have $CC(T) \supseteq C(T)$. Together we obtain $C(C(T) = C(T)$, and thus C is a closure operator.

d) Let M be an element of $DM(P)$ with $M \supseteq T$. Then $M = C(X)$ for some $X \subseteq P$. From $M = C(X) \supseteq T$ there follows $C(X) = CC(X) \supseteq C(T)$ since C is isotone.

Another elementary fact is:

10.4 Theorem. *The Dedekind–MacNeille completion of a linearly ordered set S is again linearly ordered.*

Proof. Let $A, B \subseteq S$. Then $C(A) = L(U(A))$ and $C(B) = L(U(B))$ are initial segments of S, and then one is a subset of the other.

10.5 Definition. A poset P is called *dense*, if for each two elements $a, b \in P$ with $a < b$ there is an element $c \in P$ satisfying $a < c < b$.

A subset $S \subseteq P$ is said to be *dense in P*, if for each two elements $a, b \in P$ with $a < b$ there is an element $s \in S$ satisfying $a \leq s \leq b$.

42

Note: We can have the situation that a subset S of a poset P is dense in P, but not dense.

The importance of the Dedekind–MacNeille completion of a poset S stems mainly from the fact that it is a complete lattice L, into which S can be embedded as a subset, which is dense in L. First we prove:

10.6 Theorem. *Let P be a poset, then its Dedekind–MacNeille completion $DM(P)$ is a complete lattice. It contains among others all principal ideals $(a]$ with $a \in P$.*

Proof. The power set $\mathfrak{P}(P)$ of P with the order $\subseteq$ is a complete lattice, and the mapping C is a closure operator on it with $DM(P) = \{C(T)|T \subseteq P\}$ as the set of its fixed points. By 8.17 this is also a complete lattice. The rest is clear.

In this connection MacNeille [116] proved that every poset P is embeddable into a complete lattice (e.g. its Dedekind–MacNeille completion $DM(P)$), whereby the embedding is compatible with the construction of inf and sup. Precisely he proved:

10.7 Theorem. *Let P be a poset, $D := DM(P)$ its Dedekind–MacNeille completion, $\varphi : P \longrightarrow D$ the mapping which ascribes to every $x \in P$ its initial segment $(x] \subseteq P$, then there holds: $\varphi[P]$ is isomorphic to P, and further:*

If for a subset $T \subseteq P$ the infimum $\inf_P T$ resp. supremum $\sup_P T$ exists, then we have

(1) $\inf_D \varphi[T] = \varphi(\inf_P T)$ *resp.*

(2) $\sup_D \varphi[T] = \varphi(\sup_P T)$.

Proof. The first part of the statement is already contained in 9.9. We put $u := \inf_P T$, if this exists. Then $(u] = \cap\{(t]|t \in T\}$, for both sides of this equation contain exactly all lower bounds of T in P. Therefore $\varphi(u)$ $(= (u]) = \cap\{(t]|t \in T\}$ is in D, and then $\varphi(u)$ is the greatest lower bound $\inf_D \varphi[T]$ of $\varphi[T] = \{(t]|t \in T\}$.

Assume now that $v := \sup_P T$ exists. The left side of (2) is the least set of D, which contains all $\varphi(t) = (t]$ with $t \in T$, i.e. which contains the initial segment $A = \cup\{(t]|t \in T\}$ of P, which is generated by T. So by 10.3 d) the left side of (2) is exactly $= C(A) = L(U(A))$.

The right side of (2) is $= (v]$, where $v = \sup_P T$ is also $= \sup_P A$, for T and A have the same set of upper bounds. The statement (2) is therefore equivalent to $L(U(A)) = (\sup_P A]$, and this holds because both sides contain exactly all those elements of P which are $\leq$ every upper bound of A.

Next we define:

10.8 Definition [156]. A subset D of a poset P is called σ- *dense* (resp. δ- *dense*) in P if every $x \in P$ is $= \sup M$ (resp. $= \inf M$) for some subset $M \subseteq D$.

10.9 Theorem. (J. Schmidt [156], Banaschewski [8]). *Let P be a poset. Then P is embeddable into the complete lattice $DM(P)$, in which its image set P^* is σ- dense and δ- dense.*

Proof. Let $\varphi : P \longrightarrow DM(P)$ be the mapping which ascribes to every $x \in P$ its initial segment $(x]$ in P. Then the image set $P^* := \varphi[P]$ is isomorphic to P *(see 9.9)*.

Let now X be an element of $DM(P)$. Then we have $X = L(U(T))$ for a set $T \subseteq P$. Now $X = \cup\{(x] | x \in L(U(T))\}$ is the supremum of the subset $\{(x] | x \in L(U(T))\}$ of P^*.

On the other hand $X = L(U(T))$ is $= \cap\{(x] | x \in U(T)\}$ since both sets consist of all lower bounds of $U(T)$ in P. And therefore $X = \inf\{(x] | x \in U(T)\}$ is the infimum of a subset of P^*.

An immediate consequence of 10.9 and 10.4 is:

10.9' Theorem. *Every ordered (resp. linearly ordered) set P has an ordered (resp. linearly ordered) superset, which is a complete lattice, and in which P is σ- dense and δ- dense.*

The structure of the Dedekind–MacNeille completion of a poset can also be characterized by the following isomorphism theorem (J. Schmidt [156], Banaschewski [8]), which supplements 10.9:

10.10 Theorem. *Let V be a complete lattice and S a subset of V which is σ- dense and δ- dense in V. Then V is isomorphic to the Dedekind–MacNeille completion $DM(S)$ of S.*

Proof. If t is an element or subset of V we denote the set of all upper (resp. lower) bounds of t in S by $U_S(t)$ (resp. $L_S(T)$). To every $x \in V$ we ascribe the set $\varphi(x) := L_S(U_S(x)) \in DM(S)$. First we show:

(1) $L_S(U_S(x)) = (S \leq x)$.

For $x \in S$ this is trivial. So let $x \in V \backslash S$. An element of $(S \leq x)$ is $\leq$ every element of $(S \geq x)$. And thus it is also in the left side of (1).

Let now a be an element of $L_S(U_S(x))$. Then $a \in S$, and a is $\leq$ every element of $(S \geq x)$. Then also $a \leq \inf_V(S \geq x) = x$ holds, and thus $a \in (S \leq x)$, and (1) is proved.

Now the mapping φ is an isomorphism. For, let x,y be elements of V. Then there holds :

(2) $x < y \iff (S \leq x) \subset (S \leq y)$.

Since $x = \sup_V(S \leq x)$ and $y = \sup_V(S \leq y)$ holds by assumption, then $x < y$ entails that $(S \leq x)$ and $(S \leq y)$ must be different, and the right side of (2) follows.

On the other hand, if $(S \leq x) \subset (S \leq y)$ holds, we have for the suprema (in V) x and y of both sides $x \leq y$. Indeed, x is the least upper bound in V of $(S \leq x)$, and y is an upper bound in V of $((S \leq y)$ and then also of) $(S \leq x)$. Now $x < y$ must hold since $x = y$ would entail $(S \leq x) = (S \leq y)$. So (2) is proved.

From (2) one obtains also that, if x and y are incomparable, the images $(S \leq x)$ and $(S \leq y)$ are also incomparable, and this together with (2) implies that the mapping $\varphi : V \to DM(S)$ is injective. It is also surjective. For every element of $DM(S)$ is of the form $L_S(U_S(T))$ where T is a subset of S. In V there exists $u := \inf U_S(T)$. Then $L_S(U_S(T)) = L_S(U_S(u)) = \varphi(u)$.

The next theorem makes a statement, which can roughly be interpreted as: The Dedekind–MacNeille completion of a poset is a smallest completely ordered superset of it. For every completely ordered superset of a poset S has a subset which is isomorphic to $DM(S)$:

10.11 Theorem. *Let K be a complete ordered set and $S \subseteq K$. Then the Dedekind–MacNeille completion $DM(S)$ is isomorphic to a subset of K.*

More detailed: For every $T \subseteq S$ we put $h_T := \inf_K U_S(T)$, where $U_S(T)$ is the set of upper bounds of T in S. Then the set $C^*(S) := \{h_T | T \subseteq S\}$, equipped with the order induced from K, is isomorphic to $DM(S)$.

Proof. Let T be a subset of S. Then for the cut $C_S(T)$ of T in S there holds:

(1) $C_S(T) := L_S(U_S(T)) = (S \leq h_T)$.

Further we have for sets $A, B \subseteq S$:

(2) $C_S(A) \subseteq C_S(B) \iff h_A \leq h_B$ (in K).

To show this, let $h_A \leq h_B$. Then every element of $C_S(A) = L_S(U_S(A))$ is a lower bound of $U_S(A)$ and thus $\leq \inf_K U_S(A) = h_A \leq h_B = \inf_K U_S(B)$. And then every lower bound of $U_S(A)$ is also a lower bound of $U_S(B)$, and so it is in $L_S(U_S(B)) = C_S(B)$.

Assume now that $C_S(A) \subseteq C_S(B)$ holds. By (1) we then have $A \subseteq C_S(A) = (S \leq h_A) \subseteq C_S(B) = (S \leq h_B)$, which means $A \subseteq (S \leq h_B)$. Therefore h_B is an upper bound of A and a lower bound of $U_S(B)$. Thus all elements of $U_S(B)$ are upper bounds of A, and then $U_S(B) \subseteq U_S(A)$ implies $h_B = \inf_K U_S(B) \geq \inf_K U_S(A) = h_A$.

From (2) our assertion now follows immediately since (2) implies:

(3) $C_S(A) \subset C_S(B) \iff h_A < h_B$.

Indeed, $C_S(A) \subset C_S(B)$ implies $h_A \leq h_B$, and if here also $h_A = h_B$ would hold, we would have $h_B \leq h_A$ and by (2) $C_S(B) \subseteq C_S(A)$, a contradiction. Analogously the right side of (3) implies the left side of (2) and also of (3), since $C_S(A) = C_S(B)$ would by (2) imply $h_B \leq h_A$ with contradiction. Finally (3) implies that, if $\|$ means incomparability, $C_S(A) \| C_S(B) \iff h_A \| h_B$.

As a corollary to 10.11 we obtain:

10.12 Theorem. *Let T be a subset of a poset P. Then for their Dedekind–MacNeille completions follows: $DM(T)$ is isomorphic to a subset of $DM(S)$.*

Proof. By 10.7 S is isomorphic to a subset of the completely ordered set $DM(S)$. Then T is isomorphic to a subset $T^* \subseteq DM(S)$. With 10.11, — by putting $DM(S)$ for K and T^* for S —, it follows that $DM(T^*)$, and with it $DM(T)$, is isomorphic to a subset of $DM(S)$.

10.13 Remark. Non-isomorphic posets can have isomorphic Dedekind–MacNeille completions. E.g. the linearly ordered sets of the rational (resp. real) numbers both have completions which are isomorphic to the linearly ordered interval $[0, 1]$.

10.14 Example. In the following we present an illustration of the Dedekind–MacNeille completion for a poset P which is a subset of $\mathbf{R}^2$ and equipped with the restriction of the order of $\mathbf{R}^2$, which was defined in 4.4. In Figure 6 the set P contains the points marked with $\circ$ and those marked with $\bullet$. Consider the set T, which is indicated by the three points, whose set is encircled: Then the set U of upper bounds of T consists of all points of P which are situated in the right upper quadrant

46

Q_1 of $h_T = \inf U$ in $\mathbf{R}^2$. Now the Dedekind–MacNeille completion of T contains all points of P which are $\leq h_T$, i.e. which are situated in the (left lower) quadrant Q_2 of h_T. Besides the points of T, which are marked with •, it contains also the points marked with ∘

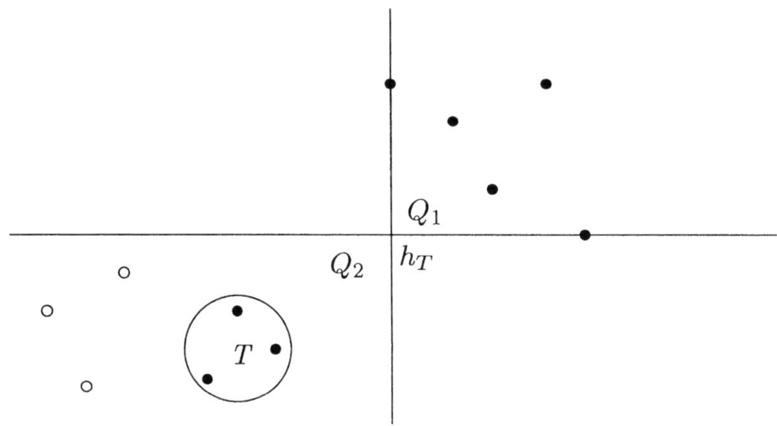

Figure 6

Finally we shall make some additional remarks on cuts in linearly ordered sets. One great difference between the linearly ordered sets and the (only partially) ordered sets is the following. If I is an initial segment and F its complementary final segment, then in the linearly ordered case we have $a < b$ for every $a \in I$, $b \in F$. In the partially ordered case this usually is not valid. See e.g. 8.13.

For these reasons the concept of cut is for linearly ordered sets mostly defined in another way than that which was used in 10.1. Namely in this definition a *cut* in a linearly ordered set S is a pair (A, B), where A is an initial and B the corresponding complementary final segment of S. And this cut is called *proper,* if A and B are both non-empty. Of course, if a cut in a linearly ordered set is defined in one of these senses, it determines also the concept in the other sense.

For the proper cuts in linearly ordered sets we introduce a rough classification, which will be intensively refined later. This is obtained by indicating of the components A, B of a cut (A, B), whether A has a greatest element or not, and whether B has a least element or not:

10.15 Definition. Let $(S, \leq)$ be a linearly ordered set. A proper cut (A, B) in S is said to be of type

$(1, 1)$, if A has a last and B a first element,

$(\infty, 1)$, if A has no last and B a first element,

$(1, \infty)$, if A has a last and B no first element,

(∞, ∞) if A has no last and B no first element.

A cut of type $(1, 1)$ is also called a *jump* in S. A proper cut of type (∞, ∞) is called a *gap* in S. If all proper cuts of S have one of the types $(\infty, 1)$ or $(1, \infty)$, then S is said to be *continuously ordered* or in short *continuous*.

10.16 Example. A linearly ordered set is dense iff it has no jumps. If we take the sets $\mathbf{Z}, \mathbf{Q}, \mathbf{R}$ of integers, rationals, reals with their usual order $\leq$ we have: In $(\mathbf{Z}, \leq)$ all proper cuts are of type $(1, 1)$. In $(\mathbf{Q}, \leq)$ there occur cuts of the types $(1, \infty)$ and $(\infty, 1)$ and (∞, ∞). In $(\mathbf{R}, \leq)$ all proper cuts are of type $(1, \infty)$ resp. $(\infty, 1)$. Thus the set of real numbers is continuously ordered. The construction of the real number system was exactly so performed that this resulted.

1.11 The duality principle of order theory

An *order-theoretical statement* is a statement S, in which besides logical symbols and variables only the signs $\leq$ and $\geq$ occur. If then we exchange in S each occurrence of $\leq$ by an occurrence of $\geq$, we obtain the *dual statement* $D(S)$ of S. Evidently $D(D(S)) = S$. And so the property to be dual is symmetric.

If an order-theoretical statement S is a consequence of a set $\mathfrak{A}$ of order-theoretical statements, and if we exchange each statement of $\mathfrak{A}$ by its dual, thus obtaining a set $D(\mathfrak{A})$, and if we do so also in each sentence of the proof, which derives S from $\mathfrak{A}$, then we obtain a proof of $D(S)$ from $D(\mathfrak{A})$. This procedure is called the *dualization* of the first proof.

Dual concepts are e.g. lower and upper bound, inf and sup, min and max.

The duality principle of order theory implies the duality principle of lattice theory, which (in consequence of the exchange of inf and sup) implies the exchange of $\wedge$ by $\vee$ and conversely.

Chapter 2
General relations between posets and their chains and antichains

In this chapter we treat several structure theorems for posets, for which no special assumptions are required. In later chapters we then consider special classes of posets, e.g. trees, partially well-ordered sets and so on. In particular we consider the relationships between general posets and chains.

2.1   Components of a poset

1.1 Definition. Let S be a set with a relation R. We call the relational system (S, R) *reducible*, if one can split S into two non-empty disjoint subsets A, B (so that $S = A \cup B$), such that no element of A is in relation R to an element of B, and no element of B is in relation R to an element of A. If (S, R) is not reducible, it is called *irreducible* or *R-connected.* Under these assumptions there follows:

1.2 Theorem and **Definition.** *Each element $x \in S$ is contained in a greatest subset $C \subseteq S$ such that $(C, R \upharpoonright C)$ is irreducible. This set C is uniquely defined and called the R-component of x. Then the set of all R-components is a partition of S.*

Proof. Let $x \in S$ be given. We put $X_0 := \{x\}$. Then we define X_1 to be the set of all elements $y \in S$ for which $x \; R \; y$ or $y \; R \; x$ holds. If X_n is already defined for a natural number n, we define X_{n+1} to be the set of all $z \in S$ for which $y \; R \; z$ or $z \; R \; y$ holds for at least one element $y \in X_n$. Then evidently $\cup_{n=0}^{\infty} X_n$ is the greatest irreducible subset of S which contains x.

By the way, the above definitions are in accordance with the graph-theoretical concepts "connectivity component" and "connected graph". For us the most important case is that in which R is an order relation $\leq$.

2.2   Maximal principles of order theory

In the theory of posets there are several principles which are equivalent to the axiom of choice, and which are often applied in different parts of mathematics, e.g. in the proof that every vector space has a basis, or that every field can be extended to an algebraically closed field, or that every filter can be extended to an ultrafilter. In this section we discuss the relationship between these principles. First we introduce (resp. recall) some definitions:

2.1 Definition.   A *choice function* in a set S is a mapping of $\mathfrak{P}(S)\backslash\{\emptyset\}$ into S, where for every non-empty set $T \subseteq S$ there holds $f(T) \in T$.

A set $\mathfrak{F}$ of sets is said to be of *finite character*, if the following two conditions hold:

1) If $T \in \mathfrak{F}$, then also every finite subset of T belongs to $\mathfrak{F}$. And conversely:

2) If every finite subset of a set T belongs to $\mathfrak{F}$, then also T belongs to $\mathfrak{F}$.

The axiom of choice states the following: For every set there exists a choice function. Then the following theorems, which can be proved using the axiom of choice, are equivalent for a given set S :

2.2 Theorem.   *There exists a choice function in S.*

2.3 Theorem.   *S can be well-ordered.*

2.4 Theorem (Hausdorff's maximal principle [81, p. 140]).
If an order is given in S, then for every chain C of S there exists a maximal chain of S which contains C.

2.5 Theorem (Zorn's lemma [182]) *Let $(S, \leq)$ be a poset for which there holds: To every non-empty chain C of S there exists an upper bound in S. Then S has a maximal element.*

Proof. $2.2 \Rightarrow 2.3$ is the content of the well-ordering theorem of Zermelo. A proof of it is contained in every textbook of set theory, and so we omit a proof here.

$2.3 \Rightarrow 2.4$.   We assume that $S\backslash C$ is infinite since the finite case is trivial. Because of 2.3 we can find a representation $S\backslash C = \{a_\nu | \nu < \lambda\}$, where λ is an ordinal number and moreover a limit ordinal. We put $C_0 := C$. Then there holds for $\nu = 0$:

(*) $C \subseteq C_\nu$, and C_ν is a maximal chain in $C \cup \{a_\iota | \iota < \nu\}$, and $C_{\nu_1} \subseteq C_{\nu_2}$ for $\nu_1 < \nu_2 < \nu$.

Assume now that (*) holds for all $\nu < \mu$, where μ is an ordinal $< \lambda$. If μ is a successor ordinal $\kappa + 1$ we define $C_\mu := C_\kappa \cup \{a_\kappa\}$ if this set is a chain, and $C_\mu := C_\kappa$ otherwise. Then (*) also holds for $\nu = \mu$.

If μ is a limit ordinal we put $C_\mu := \cup\{C_\nu | \nu < \mu\}$. Then (*) is also valid for μ instead of ν. Indeed, if C_μ would not be a maximal chain of $C \cup \{a_\iota | \iota < \mu\}$, there would exist an element $a_\tau \notin C_\mu$ with $\tau < \mu$ such that $C_\mu \cup \{a_\tau\}$ is a chain. But then $C_{\tau+1}$ had to be the set $C_\tau \cup \{a_\tau\}$, which is not the case.

By transfinite induction it follows that (*) also holds for $\nu = \lambda$, and then C_λ is a maximal chain in S which contains C.

2.4 $\Rightarrow$ 2.5. Let M be a maximal chain of S. According to the assumption of 2.5, M has an upper bound u in S. Now u must belong to M since M is a maximal chain, and thus u is the greatest element of M. But then u must also be a maximal element of S, for if there would exist an element $a \in S$ with $a > u$, then the set $M \cup \{a\}$ would be a chain of S, which is a strict superset of M in contradiction to the maximality of M.

2.5 $\Rightarrow$ 2.3. Let $\mathfrak{W}$ be the set of all pairs (T, R), where $T \subseteq S$ holds and where $R = \leq\restriction T$ is a well-ordering of T. For pairs $W_1, W_2 \in \mathfrak{W}$ we put $W_1 \leq W_2$ if the carrier set of W_1 is an initial segment of W_2. Then it is clear that this relation $\leq$ is an ordering in $\mathfrak{W}$. Let $\mathfrak{C}$ be a chain in $\mathfrak{W}$. Then the union U of the carrier sets of the sets $C \in \mathfrak{C}$, taken with the union of the orders of the $C \in \mathfrak{C}$, is an upper bound of $\mathfrak{C}$ in $\mathfrak{W}$. Now the assumptions of Zorn's lemma are satisfied, and so $\mathfrak{W}$ has a maximal element W. Its carrier set T must now contain all elements of S since otherwise T could be enlarged by an element of $S \backslash T$ to a well-ordered set which is strictly greater than W, contradicting the maximality of W.

Since 2.3 $\Rightarrow$ 2.2 is trivial we now have a ring-proof for the equivalence of 2.2 and 2.5.

Another theorem which is equivalent to the axiom of choice is the following:

2.6 Theorem of Teichmüller [168]. *Hausdorff's maximal principle (Every chain of a poset is a subset of a maximal chain) is equivalent to: If $\mathfrak{F}$ is a set family of finite character, then it has a maximal element.*

Proof. We assume Hausdorff's maximal principle. Let $\mathfrak{F}$ be a family of finite character. Then there exists a maximal chain $\mathfrak{M}$ in the poset $(\mathfrak{F}, \subseteq)$. The union $\cup\mathfrak{M}$ belongs to $\mathfrak{F}$ because everyone of its finite subsets is already a subset of a set $M \in \mathfrak{M}$ and thus in $\mathfrak{F}$. Now $\cup\mathfrak{M}$ is an upper bound of $\mathfrak{M}$ and thus an element of $\mathfrak{M}$ because $\mathfrak{M}$ is a maximal chain. So finally $\cup\mathfrak{M}$ is the greatest element of $\mathfrak{M}$ and then also a maximal element of $\mathfrak{F}$, since otherwise $\mathfrak{M}$ would not be a maximal chain of $\mathfrak{F}$.

Assume now that every family of finite character has a maximal element, and that C is a chain of a poset $(P, \leq)$ Let $\mathfrak{F}$ be the set of all sets $T \subseteq P\backslash C$, for which $T \cup C$ is a chain of $(P, \leq)$. Clearly $\mathfrak{F}$ is of finite character, and so there exists a maximal element M in $\mathfrak{F}$. Then $M \cup C$ is a maximal C containing chain of P.

2.7 Remark. The set of all chains of a poset is evidently of finite character. The same holds for the set of all antichains of a poset. So, using one of the before mentioned principles, e.g.Teichmüller's theorem, we immediately obtain that every antichain of a poset is contained in a maximal antichain. The question arises whether the latter statement entails the axiom of choice. In this context we have the following statement:

2.8 Theorem of Kurepa [110]. *The axiom of choice is equivalent to the validity of the following two conditions:*

a) *Every set can be linearly ordered. And*

b) *Every antichain of a poset is a subset of a maximal antichain of this set.*

Proof. Because of 2.3 and 2.7 it is clear that the axiom of choice implies a) and b). Assume now that a) and b) hold, and that $\mathfrak{F}$ is a family of disjoint non-empty sets. According to a well-known theorem of set theory it suffices now to find a function f which assigns to every $X \in \mathfrak{F}$ an element $f(X) \in X$.

First we take a linear order L in the set $S := \cup\mathfrak{F}$. Then we define an ordering $P \subseteq L$ on S by $P := \cup\{L \restriction X | X \in \mathfrak{F}\}$. So P is an order-relation on S which is a union of linear orderings on disjoint sets X. Let now A be a maximal antichain of (S, P). Clearly A must have exactly one element from every $X \in \mathfrak{F}$. We denote this by $f(X)$, and then f fulfills our requirement.

2.3   Linear Extensions of posets

An application of the maximum principle yields the important extension theorem of Szpilrajn, according to which every order is a subset of a linear order on the same set. The following main part of its proof does not use the axiom of choice:

3.1 Theorem. *Let $(P, \leq)$ be a poset, a and b two incomparable elements of P. Then there exists an order-relation $\leq_*$ on P which contains $\leq$ and in which $a \leq_* b$ holds.*

In more detail there holds: The transitive closure H of the relation $\leq \cup\{(a, b)\}$ is an order on P. It contains exactly all pairs of $\leq$ and all pairs (x, y) for which $x \leq a$ and $b \leq y$ holds.

Proof. An order-relation $\leq_*$which shall contain $\leq$ and (a, b) must also contain the transitive closure of $\leq \cup\{(a, b)\}$. Hence we put for $x, y \in P$:

(1) $x \leq_* y \Leftrightarrow x \leq y$ or $(x \leq a$ and $b \leq y)$.

For this we now have the following compatibility property:

(2) If $x \leq a$ and $b \leq y$, then $x \geq y$ cannot hold.

For otherwise we would have $b \leq y \leq x \leq a$ and thus $b \leq a$, but a and b are incomparable. And so our definition of $x \leq_* y$ does not contradict a relation $x > y$ which might have been valid already.

We verify that $\leq_*$ is an order relation, the rest is then clear. It is reflexive because $\leq$ is already reflexive. Let now $x \leq_* y$ and $y \leq_* x$. If here also $x \leq y$ and $y \leq x$ hold, we are done. We consider the three other cases.

$x \ (\leq_* \setminus \leq) \ y$ and $y \leq x \implies b \leq y \leq x \leq a$ in contradiction to $a \parallel b$,

$x \leq y$ and $y \ (\leq_* \setminus \leq) \ x \implies b \leq x \leq y \leq a$ in contradiction to $a \parallel b$,

$x \ (\leq_* \setminus \leq) \ y$ and $y \ (\leq_* \setminus \leq) \ x \implies b \leq x \leq a$ in contradiction to $a \parallel b$.

$\leq_*$ is also transitive: Let $x \leq_* y$ and $y \leq_* z$. For (x, y) and (y, z) of $\leq$ also $(x, z) \in \leq$ holds. If $(x, y) \in \leq$ and $(y, z) \in \leq_* \setminus \leq$, then we have $y \leq a$ and $b \leq z$, hence $x \leq y \leq a$ and thus $x \leq_* z$. If $(x, y) \in \leq_* \setminus \leq$ and $(y, z) \in \leq$, then analogously we have $x \leq a$ and $b \leq y \leq z$ and thus $x \leq_* z$. Finally the case where (x, y) and (y, z) both belong to $\leq_* \setminus \leq$ cannot occur. For then we would have $y \leq a$ and $b \leq y$ and thus $b \leq a$ contradicting $a \parallel b$.

Now there follows in full generality:

3.2 Theorem (of Szpilrajn [167],1930) *Using the axiom of choice there holds: Let $(P, \leq)$ be a poset. Then there exists a linear order $\leq_*$*

on P which contains $\leq$, a so-called linear extension of $\leq$. In particular: If a and b are incomparable elements, this linear order can be chosen in such a way that $a \leq_* b$ holds.

Proof. Let $\mathfrak{A}$ be the set of all order-relations on P, which contain $\leq$ and the pair (a, b). Because of 3.1 $\mathfrak{A}$ is non-empty. If we order $\mathfrak{A}$ by inclusion, it satisfies the assumptions of Zorn's lemma: For every chain $\{R_i | i \in I\}$ of $\mathfrak{A}$ the set $\cup\{R_i | i \in I\}$ is an upper bound. Thus $\mathfrak{A}$ has a maximal element $\leq_*$. This must be a linear order on P, because otherwise there would exist a strictly greater order in $\mathfrak{A}$ because of 3.1.

We can sharpen the statement of 3.2 as follows:

3.3 Theorem. *Let $(P, \leq)$ be a poset, A an antichain of P. If we choose an arbitrary linear order $\leq_A$ on A, then there exists a linear order on P which contains $\leq$ and $\leq_A$.*

Proof. Let X (resp. Y) be the set of all elements $x \in P$ (resp. $y \in P$) for which there exists an element $a \in A$ with $x < a$ (resp. $a < y$). Then for every $y \in Y$ and every $x \in X$ the relation $y \leq x$ cannot hold. In particular then X and Y are disjoint. Indeed, if there would exist a pair $(y, x) \in Y \times X$ with $y \leq x$, then there would exist elements $a_1, a_2 \in A$ with $x < a_1$ and $a_2 < y$. This has as consequence $a_2 < y \leq x < a_1$, and then a_1 and a_2 would be comparable, a contradiction.

First we consider the case where A is a maximal antichain. Then every element of $P \backslash A$ is comparable with at least one element of A and thus in X or in Y. Let then $\leq_X$ (resp. $\leq_Y$) be a linear extension of the order (induced from $\leq$) of X (resp. Y). Then there exists a linear extension $\leq_L$ of $\leq$, of which X is an initial segment, A a segment, Y a final segment, where $x < a < y$ holds for every $x \in X$, $a \in A$, $y \in Y$, and where the restrictions of $\leq_L$ onto X, A, Y are $\leq_X, \leq_A, \leq_Y$ respectively.

If finally the antichain A is not maximal, there exists by 2.7 a maximal antichain $A^* \supseteq A$. Of course there exists a linear extension $\leq^*$ of $\leq_A$ to A^*, and then, according to the first part of the proof, we obtain a linear extension of $\leq \cup \leq^*$ which then also is a linear extension of $\leq \cup \leq_A$ to P.

2.4 The linear kernel of a poset

4.1 Definition. Let $(P, \leq)$ be a poset. An element $a \in P$ shall be called *fully comparable* or a *kernel element,* if a is comparable with all elements of P. The set of all kernel elements of P shall be called the *linear kernel* of $(P, \leq)$. We denote it by $LK(P)$.

Using the axiom of choice we can characterize the linear kernel also in the following way:

4.2 Theorem. *The linear kernel of a poset $(P, \leq)$ is the intersection of all maximal chains of P.*

Proof. That every maximal chain of a poset $(P, \leq)$ contains $LK(P)$ is trivial. Assume now that an element $a \in P$ is contained in all maximal chains of P, and that x is an arbitrary element of P. According to the maximal principle there exists a maximal chain M of P which contains x; M must also contain a, and so a and x are comparable. Since x was arbitrary, a is in $LK(P)$.

The linear kernel gives us some useful information on the structure of a poset. To this purpose we introduce a kind of class:

4.3 Definition. Let $(P, \leq)$ be a poset and L its linear kernel. We call two elements $a, b \in P \backslash L$ *equivalent,* in signs $a \sim b$, if $(L < a) = (L < b)$ and $(L > a) = (L > b)$ hold. Roughly speaking: a and b are equivalent if they are in the same position to L. Then it is clear that $\sim$ is indeed an equivalence relation on $P \backslash L$. Figure 7 illustrates the situation. The equivalence classes are encircled.

Figure 7 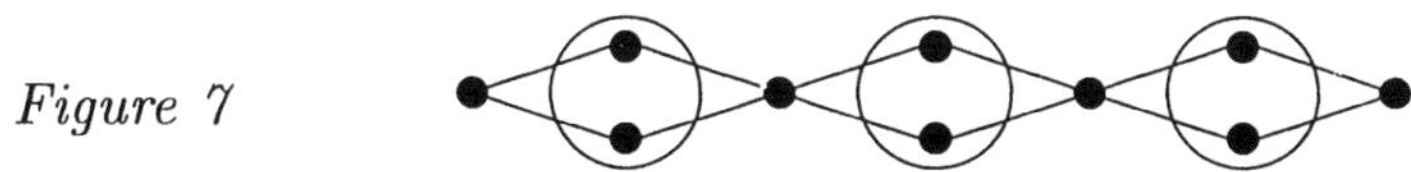

If e.g L is empty, then all elements of P belong to the same class, and this is the whole set P. With the assumptions of Definition 4.3 there follows rather immediately:

4.4 Theorem. *Every $\sim -$class of $P \backslash L$ is a convex subset of $(P, \leq)$.*

Proof. If a, b are elements of an $\sim -$ class C with $a < b$ and if x is an element of P satisfying $a < x < b$, then there holds $(L < x) \subseteq$

$(L < b) = (L < a)$ and $(L < a) \subseteq (L < x)$ and thus $(L < x) = (L < a)$. Similarly we have $(L > x) = (L > a)$ and so $x \sim a$.

4.5 Theorem and **Definition.** *Let the assumptions of 4.3 be fulfilled and A and B two different $\sim -$classes of $P\backslash L$. Then there holds $a < b$ for every $a \in A$ and every $b \in B$, or $b < a$ for every $b \in B$ and every $a \in A$.*

In the first case we put $A < B$, in the second $B < A$. This now defines a linear order in the set of equivalence classes.

Proof. Let $a \in A$, $b \in B$. Since a and b are not equivalent we have $(L < a) \neq (L < b)$ or $(L > a) \neq (L > b)$. Suppose the first inequality. The other case is symmetric to this. Then we assume w.r.o.g. that there exists an element $x \in (L < b)\backslash(L < a)$. This entails $a < x < b$. For x is in L and thus comparable with a, but not $< a$ and also $\neq a$ because of $a \notin L$. Now all elements of A (resp. B) are in the same position to L, and therefore $a^* < x < b^*$ holds for all elements $a^* \in A$, $b^* \in B$.

4.6 Theorem. *Let L be the linear kernel of a poset P. An $\sim -$ class C of $P\backslash L$ cannot be a chain of P. In particular: Every $\sim -$ class has more than one element.*

Proof. Because of 4.5 all elements of C are comparable with all elements of $P\backslash C$. If now C would be a chain, then all elements of C would be comparable with all elements of P, and we would have $C \subseteq L$, but in fact C is disjoint to L.

Finally we mention the following observation:

4.7 Theorem. *Every antichain of a poset $(P, \leq)$ is contained in exactly one $\sim -$ class.*

Proof. Let a, b be elements of an antichain A of $(P, \leq)$. If $a \in K_a$, $b \in K_b$, where K_a and K_b would be different $\sim -$ classes of $P\backslash LK(P)$, we have $K_a < K_b$ or $K_b < K_a$ (in the sense of 4.5). From 4.5 then there would follow $a < b$ or $b < a$ which is impossible.

2.5 Dilworth's theorems

When we investigate the structure of a given poset we are mainly interested in the properties of its chains and antichains. In this context there holds a fundamental theorem of Dilworth, which is one of the most important of order theory. First we introduce several concepts:

5.1 Definition. Let $(P, \leq)$ be a poset. The supremum of the cardinals of the antichains of P is called the *width* of P. It is denoted by $w(P)$.

It is clear that for finite posets P there always exists an antichain of P whose cardinality is $w(P)$. For infinite ordered sets this is not always valid. If e.g. the set P of all pairs of natural numbers is ordered by: $(a, b) \leq (c, d) \iff a \leq c$ and $b \leq d$, then every antichain of P is finite, and for every natural number n there exists an antichain with n elements, so that P has the width $\aleph_0$.

We define the *chain-covering-number* $CCN(P)$ to be the least cardinal number k, such that P is a union of k chains of P. For finite P this number exists trivially, for infinite P the existence follows from the axiom of choice (according to which every non-empty set of cardinals has a least one).

It is immediately clear that for a finite poset $(P, \leq)$ the CCN must be at least as great as the width w of P. For if A is an antichain of w elements, then in a covering of P by chains each of these chains can contain at most one element of A, and thus at least w chains are needed to cover P. Conversely now Dilworth's theorem states that if w is finite, w chains are also sufficient to cover P. First we formulate it only for finite sets:

5.2 Theorem (Dilworth [24],1950). *Let $(P, \leq)$ be a finite poset, w its width. Then P is a union of w chains.*

Proof (by Perles [139]). If P has only one element, the statement is clear. Let now n be a natural number, so that our assertion is already proved for all posets with at most n elements, and let P be a poset with $n + 1$ elements. Let A be an antichain of P with w elements. We define (recall definition 1.8.4) $B := IS(A)$ and $R := FS(A)$. (The idea of this setting is to divide the set P into two parts, one "below" A and the other "above" A, which overlap in the set A. If both have at most n elements our induction hypothesis can work.) First we see
$$B \cup R = P.$$

Indeed, every element $x \in P$ is $\leq$ an element of A or $\geq$ an element of A. In the other case $A \cup \{x\}$ would still be an antichain, but A was already maximal. Next we prove
$$B \cap R = A.$$

$B \cap R \supseteq A$ is trivial. Let now $x \in B \cap R$. Then from $x \in R$ we obtain the existence of an $a_1 \in A$ with $a_1 \leq x$, and from $x \in B$ the existence of an $a_2 \in A$ with $x \leq a_2$. From $a_1 \leq x \leq a_2$ there follows $a_1 = a_2$ because A is an antichain. And also $x(= a_1) \in A$.

Now we distinguish two cases:

Case 1. There exists a w - element antichain A, so that the sets B and R each contain at most n elements.

Then, according to our induction hypothesis, B and also R is a union of w chains. Hereby each of the chains covering B ends in an element of A, and each of the chains covering R begins with an element of A. Then we can link the w chains ending in elements of A with the w chains which begin in elements of A. Together we obtain a system of w chains which cover P.

Case 2. For every w - element antichain A one of the sets B and R contains $n+1$ elements, — and is hence $= P$.

If then for a given antichain A we have $B = P$, then every element of P is $\leq$ an element of A, and then we have $A = Ma(P)$ (the set of maximal elements of P). Indeed, every element a of A is maximal: If there would exist an $x \in P$ with $a < x$, there would also exist an element $a^* \in A$ with $x \leq a^*$. Then $a < a^*$ would follow, contradicting the antichain-property of A. If $R = P$ we obtain analogously $A = Mi(P)$.

So we have : In Case 2 every w - element antichain of P is one of the sets $Mi(P)$, $Ma(P)$, (possibly both).

If now M is an arbitrary maximal chain of P, then M contains of course a minimal and a maximal element of P (which could also be equal). At any rate M contains of every w - element antichain of P an element. The set $P\backslash M$ has then only the width $w-1$ and at most n elements. According to the induction hypothesis it is the union of $w-1$ chains. These, together with M, are w chains, the union of which is P. And so our theorem is proved by induction.

5.3 Remark. By the way: Theorem 1.7.7 (and with it also Theorem 1.7.8 of Erdös/Szekeres) is also an immediate consequence of Dilworth's theorem: If there P would not contain an antichain of size $n+1$, then the width w of P would be $\leq n$, and P would be the union of w disjoint chains. If none of them would contain at least $m+1$ elements, we would have $|P| \leq n \cdot m$, a contradiction.

Using the axiom of choice we now can generalize 5.2 to infinite posets: For this purpose we shall make use of a graph-theoretical theorem of De Bruijn/Erdös. First we introduce several graph-theoretical concepts:

5.4 Definition. If S is a set, we denote the set of all two-element subsets of S by $[S]^2$. Let V be a set and $\mathfrak{E}$ a subset of $[V]^2$. Then $(V, \mathfrak{E})$ is called a *graph,* the elements of V are called *vertices,* those of $\mathfrak{E}$ *edges* of the graph, V is called its *vertex set,* $\mathfrak{E}$ its *edge set.* If $\{a, b\} \in \mathfrak{E}$ (resp. $\notin \mathfrak{E}$) we say: a and b are *linked* (resp. *unlinked*).

Let k be a natural number. A subset $V' \subseteq V$ is called k - *colorable,* if there exists a mapping $f : V' \to \{1, \ldots, k\}$ such that for every $\{a, b\} \in \mathfrak{E} \cap [V']^2$ we have $f(a) \neq f(b)$.

5.5 Theorem (De Bruijn/Erdös [20],1950). *Let k be a natural number and $(V, \mathfrak{E})$ a graph with the following property: Every finite subset $F \subseteq V$ is k - colorable (in the graph $(F, \mathfrak{E} \cap [F]^2)$). Then also the whole graph $(V, \mathfrak{E})$ is k - colorable.*

Proof (of Posa). We consider the set $\mathfrak{K}$ of all edge-sets $\mathfrak{T}$ which satisfy $\mathfrak{E} \subseteq \mathfrak{T} \subseteq [V]^2$, and have the property that every finite subset $F \subseteq V$ is k - colorable in $(F, \mathfrak{T} \cap [F]^2)$. It can easily be verified that $(\mathfrak{K}, \subseteq)$ fulfills the assumptions of Zorn's lemma, and so $\mathfrak{K}$ has a maximal element $\mathfrak{M}$. (Now $(V, \mathfrak{M})$ is a so-called *extremal* graph with respect to the property of being finitely k - colorable.) For the rest of the proof the axiom of choice is no longer needed.

For $a, b \in V$ we put $a \sim b \iff \{a, b\} \notin \mathfrak{M}$. The relation $\sim$ is trivially reflexive and symmetric. But it is also transitive: Let $a \sim b$ and $b \sim c$. Then $\{a, b\} \notin \mathfrak{M}$ holds, and since $\mathfrak{M}$ is maximal, there exists a finite set $\mathfrak{M}_1 \subseteq \mathfrak{M}$ such that the graph $(V, \mathfrak{M}_1 \cup \{a, b\})$ is not k-colorable. Analogously $b \sim c$ yields: There exists a finite subset $\mathfrak{M}_2 \subseteq \mathfrak{M}$ such that $(V, \mathfrak{M}_2 \cup \{b, c\})$ is not k-colorable. Then in every mapping $f : V \longrightarrow \{1, \ldots, k\}$, in which for all $\{x, y\} \in \mathfrak{M}_1 \cup \mathfrak{M}_2$, there holds $f(x) \neq f(y)$, a and b must have the same value. This also holds for b and c and thus also for a and c. Finally, this means that $(V, \mathfrak{M}_1 \cup \mathfrak{M}_2 \cup \{\{a, c\}\})$ is not k - colorable, and this implies $\{a, c\} \notin \mathfrak{M}$, so that $a \sim c$ holds, and $\sim$ is an equivalence relation.

Now V is partitioned into the equivalence classes of $\sim$. All elements of the same class are unlinked, and every element of an equivalence class is linked with every element of every other equivalence class. Then there are at most k classes $C_1, \ldots, C_j$, where j is a natural number $\leq k$. Finally we map all elements of C_1 to 1, ..., all elements of C_j to j, and obtain so a j - coloring of $(V, \mathfrak{M})$ and a fortiori of $(V, \mathfrak{E})$. Of course it is also a k - coloring.

Now we can prove Dilworth's theorem in its general form (following [68]. p.187):

5.6 Theorem (Dilworth [24]): *Let $(P, \leq)$ be a poset with width w, where w is a natural number. Then P is a union of w chains.*

Proof. Let $\|$ denote the relation to be incomparable in P. Every finite subset $F \subseteq P$ has width $\leq w$ and is thus a union of at most w disjoint chains because of 5.2. Now we ascribe to all elements of the same chain a color $\in \{1, \ldots, w\}$, so that elements of different chains obtain different colors. Evidently this yields a w - coloring of the graph $(F, \|)$. According to 5.5 then also $(P, \|)$ has a w - coloring. In this all elements with the same color must form a chain, and then P is a union of at most w chains.

Dilworth's theorem dealt with posets of finite width. The question arises of how the situation changes if we drop the finiteness-condition. The next theorem shows that then the situation is entirely different. First we mention:

5.7 Lemma. *Let ω_α be an initial ordinal and P the set of all pairs (μ, ν) with $\mu, \nu \in \omega_\alpha$. If (μ, ν) and (κ, λ) are in P we put $(\mu, \nu) \leq (\kappa, \lambda) \iff \mu \leq \kappa$ and $\nu \leq \lambda$. This defines an order relation on P. Then this order has a linear extension $\leq_L$ which has the order type ω_α.*

Proof. For $\nu < \omega_\alpha$ let P_ν be the set of all pairs (μ, κ) with $\max\{\mu, \kappa\} = \nu$, and let $\leq_\nu$ be a well-ordering of P_ν which extends the order which P_ν has as restriction of the order of P. To this purpose we take in the set A of elements (ν, ξ) with $\xi < \nu$ the order $\leq$, further in the set B of elements (ξ, ν) with $\xi \leq \nu$ the order $\leq$ and put $x \leq_\nu y$ for $x \in A$, $y \in B$.

Then we define a well-ordering $\leq_L$ of P as follows: For $x \in P_\nu$ and $y \in P_\mu$ with $\nu < \mu$ we put $x \leq_L y$. If x and y are in the same set P_ν we put $x \leq_L y \iff x \leq_\nu y$. Then one can easily verify that $\leq_L$ is a well-ordering of P which extends the order $\leq$. It has an order type $\leq \omega_\alpha$ because every principal ideal of $(P, \leq_L)$ has a cardinality $< \aleph_\alpha$. Indeed, every $x \in P$ is in some P_ν with $\nu < \omega_\alpha$, and then $|(P \leq_L x)|$ is $\leq \sum\{2|\mu| \mid \mu \leq \nu\} < \aleph_\alpha$.

5.8 Theorem (Wolk [178], Perles [138]). *Let P be the set of 5.7. For every natural number n there exists an n-element antichain of P. But P has no infinite antichain. Then $(P, \leq)$ has width $\aleph_0$. Every covering*

of P by chains contains $\aleph_\alpha$ chains, and so the chain-covering-number
$CCN(P)$ of P is $\aleph_\alpha$.

Proof. If n is a natural number, the pairs $(\nu, n-\nu)$, $\nu = 0, \ldots, n-1$, form an n - element antichain.

We assume now indirectly that P has an infinite antichain A. Then all pairs $(\mu, \nu) \in A$ must have different first components and different second components. We arrange the elements of A according to the size of the first components: $A = \{(\mu_\iota, \nu_\iota) | \iota < \tau\}$; here τ is an ordinal $\geq \omega$ (= the first infinite ordinal) and $\mu_\iota < \mu_\kappa$ for $\iota < \kappa < \tau$. Then the set of second components $\{\nu_\iota | \iota < \tau\}$ must satisfy $\nu_\iota > \nu_\kappa$ for $\iota < \kappa < \tau$. This yields a contradiction because in a well-ordered set there is no strictly descending infinite sequence.

We now make the indirect assumption that $\mathfrak{C} = \{C_\nu | \nu < \sigma\}$ is a covering of S with fewer than $\aleph_\alpha$ chains C_ν, so that $\sigma < \omega_\alpha$. First we assume that $\aleph_\alpha$ is a regular cardinal. For the ordinals $\mu < \omega_\alpha$ we put $M_\mu := \{(\mu, \xi) | \xi < \omega_\alpha\}$. Now $M_\mu = \cup_{\nu < \sigma}(M_\mu \cap C_\nu)$ is the union of $|\sigma| < \aleph_\alpha$ summands, and then one of these, say $M_\mu \cap C_{\nu(\mu)}$, must contain $\aleph_\alpha$ elements because the regular cardinal $\aleph_\alpha$ cannot be the sum of fewer than $\aleph_\alpha$ cardinals, which all are $< \aleph_\alpha$.

For every chain C_ν there is at most one μ so that $|C_\nu \cap M_\mu| = \aleph_\alpha$. For if there would be ordinals $\sigma < \tau < \omega_\alpha$ with $|C_\nu \cap M_\sigma| = |C_\nu \cap M_\tau| = \aleph_\alpha$, then C_ν would have a subset of type ω_α in common with each of the sets M_σ and M_τ. Because of $\sigma < \tau$, then C_ν would also contain a subset of type $\omega_\alpha + \omega_\alpha$. But this is impossible because, due to 5.7, every chain of P has an order type $\leq \omega_\alpha$.

So the mapping which ascribes to every $\mu < \omega_\alpha$ the ordinal $\nu(\mu)$ had to be injective. But this is impossible because of $\aleph_\alpha > |\sigma|$. And so the theorem is proved for the regular case.

If finally $\aleph_\alpha$ is a singular cardinal, α is a limit ordinal. Let then ρ be an ordinal $< \alpha$, which is not a limit ordinal. Then $\aleph_\rho$ is a regular cardinal. Now we consider the set $P_\rho := \{(\mu, \nu) | \mu, \nu < \omega_\rho\}$. It has the chain-covering-number $\aleph_\rho$ because of the first part of this proof. Every covering of P with chains encompasses in a natural way a covering of P_ρ. This must contain at least $\aleph_\rho$ chains, and then also the covering of P must contain at least $\aleph_\rho$ chains. Since this holds for every non-limit ordinal $\rho < \alpha$ there now follows $CCN(P) = \aleph_\alpha$.

Finally we mention without proof a generalization of Dilworth's Theorem. Oellrich and Steffens [130] proved:

5.9 Theorem. *For every countable poset without infinite chains there exists a (so-called Dilworth-) partition $(C_i, i \in I)$ of P into pairwise disjoint chains and an antichain A such that $C_i \cap A \neq \emptyset$ for every $i \in I$.*

2.6 The lattice of antichains of a poset

First we introduce (resp. recall) some concepts of lattice theory:

6.1 Definition. A poset $(P, \leq)$ is called a *lattice,* if for every two elements $a, b \in P$ there exists a greatest lower bound $\inf(a, b)$ and a lowest upper bound $\sup(a, b)$ in P. These are denoted by $a \wedge b$ resp. $a \vee b$. A lattice $(P, \leq)$ is called *distributive,* if for arbitrary elements a, b, c of P there holds $a \wedge (b \vee c) = (a \wedge b) \vee (a \wedge c)$.

An example for distributive lattices is furnished by the following theorem which can easily be verified:

6.2 Theorem *Let $\mathfrak{F}$ be a family of sets, which contains with any two sets A and B also their intersection $A \cap B$ and their union $A \cup B$. Then $(\mathfrak{F}, \subseteq)$ is a distributive lattice. Its operations $\wedge$ and $\vee$ coincide with $\cap$ and $\cup$ respectively.*

A special case of this theorem is:

6.3 Theorem. *Let $\mathfrak{I}$ be the set of all initial segments of a finite poset $(P, \leq)$. Then $(\mathfrak{I}, \subseteq)$ is a distributive lattice.*

Proof. In 1.8.2 we already mentioned that the intersection and the union of initial segments of a poset are again initial segments of it.

In finite posets we now have an interesting correspondence between initial segments and antichains:

6.4 Theorem. *Let $(P, \leq)$ be a poset, $\mathfrak{I}$ its set of initial segments, $\mathfrak{A}$ its set of antichains. Then the mapping $f : \mathfrak{A} \longrightarrow \mathfrak{I}$, which is defined by $f(A) := IS(A)$ (= the initial segment of A in P) for $A \in \mathfrak{A}$, is an injection of $\mathfrak{A}$ into $\mathfrak{I}$.*

Proof. If $A_1, A_2 \in \mathfrak{A}$ and $f(A_1) = f(A_2)$, then every $x_1 \in A_1$ is $\leq$ an element $y \in A_2$, and this y is $\leq$ an element $x_2 \in A_1$. Now $x_1 \leq y \leq x_2$ entails $x_1 = x_2$ since x_1, x_2 are in the antichain A_1. Then also $x_1(= y) \in A_2$. Thus we have obtained $A_1 \subseteq A_2$; analogously there follows $A_2 \subseteq A_1$, and then $A_1 = A_2$.

If in 6.4 the set P is finite, one can prove more, namely:

6.5 Theorem. *Let $(P, \leq)$ be a finite poset, $\mathfrak{I}$, $\mathfrak{F}$, and f as in Theorem 6.4. Then $f : \mathfrak{A} \to \mathfrak{I}$ is a bijection of $\mathfrak{A}$ onto $\mathfrak{I}$.*

Proof. Because of 6.4 we only have to prove that f is surjective. Let I be an initial segment of P. Let then A be the set $\mathrm{Ma}(I)$ of maximal elements of I. Then A is an antichain and $I = IS(A) = f(A)$, because every element of I is $\leq$ a maximal element of I, since P is finite; and every element of $IS(A)$ is $\leq$ an element of $A \subseteq I$ and thus in I.

If in 6.5 we drop the condition that P is finite, the statement no longer holds: If e.g. P is the set of rational numbers with their usual ordering, then $\mathfrak{A}$ is the set of one-element subsets of P, and thus $\mathfrak{A}$ is countable, whereas the set $\mathfrak{I}$ is equipotent with the set of real numbers with their natural order, and therefore of a higher cardinality.

Now we introduce in harmony with 6.5 an ordering in the set of antichains:

6.6 Definition. Let $(P, \leq)$ be a poset, $\mathfrak{A}$ the set of its antichains. For $A, B \in \mathfrak{A}$ we put
$A \leq B \iff IS(A) \subseteq IS(B)$. ($\iff$ Every element of A is $\leq$ an element of B). (This $\leq$ should not be confused with $\subseteq$.)

Then, as an immediate consequence of 6.5, 6.3 and the last definition we obtain:

6.7 Theorem. *Under the assumptions of 6.5 the mapping $f : \mathfrak{A} \to \mathfrak{I}$ is an order-isomorphism. And then $(\mathfrak{A}, \leq)$ is a distributive lattice.*

Proof. By 6.2, $\mathfrak{I}$ is a distributive lattice (and then also $\mathfrak{A}$, which is order-isomorphic to it).

We also mention two elementary properties:

6.8 Theorem. *Let $(P, \leq)$ be a poset, A and B subsets of P.*
a) *Then $IS(A \cup B) = IS(A) \cup IS(B)$.*
b) *If P is finite and if X is any subset of P, then $Ma(IS(X)) = Ma(X)$.*

Proof. a) is trivial. b) If a is a maximal element of $IS(X)$, then a must be an element of X, and then it is a fortiori a maximal element of X. Let now b be a maximal element of X. If b would not be maximal in $IS(X)$, there would exist an element $c \in IS(X)$ with $b < c$ and further an element $d \in X$ with $c \leq d$. This would entail $b < d$, which is impossible since b is maximal in X.

Now we can describe the structure of the lattice of 6.7 in more detail:

6.9 Theorem. *Let $(P, \leq)$ be a finite poset. Then the set $\mathfrak{A}$ of its antichains with the order $\leq$ of 6.6 is a distributive lattice whose operations $\wedge$ and $\vee$ satisfy the following relations: For $A, B \in \mathfrak{A}$ we have*

$$A \wedge B = Ma(IS(A) \cap IS(B)) \ \text{and} \ A \vee B = Ma(A \cup B).$$

Proof. For the mapping f of 6.4 we have $f(A) = IS(A)$ and $f(B) = IS(B)$. Since f is an isomorphism, $f(A \wedge B) = IS(A) \cap IS(B)$. But also $Ma(IS(A) \cap IS(B))$ is mapped by f onto $IS(A) \cap IS(B)$. Therefore, because of the injectivity of f, $A \wedge B = Ma(IS(A) \cap IS(B))$.

On the other hand $f(A \vee B) = IS(A) \cup IS(B) = IS(A \cup B)$ by 6.8 a): Also $Ma(A \cup B)$ is mapped by f onto $IS(A \cup B)$. Since f is injective, then $A \vee B = Ma(A \vee B)$.

For order theory and combinatorics specially the sublattice of the maximal-number-antichains of a finite poset has nice properties:

6.10 Theorem. *Let $(P, \leq)$ be a finite non-empty poset of width w. Then, as we have already proved, P is a union of w disjoint chains $C_1, \ldots, C_w$. If now A is a maximal-number-antichain, then A has exactly one element a_i in common with every chain $C_i, i = 1, \ldots, w$.*

If then A, B are two w-element antichains, $A = \{a_1, \ldots, a_w\}$ and $B = \{b_1, \ldots, b_w\}$, where a_i and $b_i \in C_i$ for $i = 1, \ldots, w$, then there holds:

(1) $A \leq B \iff a_i \leq b_i$ for $i = 1, \ldots, w$.

Proof. Assume that $A \leq B$ holds, that means $IS(A) \subseteq IS(B)$. Let i be a fixed number of $\{1, \ldots, w\}$. Then we have $a_i \ (\in IS(A) \subseteq IS(B)) \leq$ an element $b_j \in B$. For $j = i$ we are done. In the case $j \neq i$ we conclude: a_i and b_i are comparable because they belong to the chain C_i. If now $b_i < a_i$ would hold, this would yield $b_i < a_i \leq b_j$ and thus $b_i < b_j$, contradicting the fact that B is an antichain. So there remains only the possibility $a_i \leq b_i$, and the direction $\Rightarrow$ of (1) is proved. The direction $\Longleftarrow$ is trivial.

6.11 Theorem. *Let the assumptions of 6.10 be satisfied. We put $s_i := \max\{a_i, b_i\}$ and $l_i := \min\{a_i, b_i\}$ for $i = 1, \ldots, w$. Then $\{s_1, \ldots, s_w\}$ and $\{l_1, \ldots, l_w\}$ are w-element (hence maximal-number-) antichains and there holds $A \vee B = \{s_1, \ldots, s_w\}$ and $A \wedge B = \{l_1, \ldots, l_w\}$.*

Proof. If $\{s_1, \ldots, s_w\}$ would not be an antichain, then there would exist two indices $i \neq j$ in $\{1, \ldots, w\}$ with $s_i < s_j$. Then we consider the four possible cases:

Case 1. $s_i = a_i, s_j = a_j$. This would yield $a_i < a_j$, which is impossible.

Case 2. $s_i = b_i, s_j = b_j$. This would yield $b_i < b_j$, which is impossible.

Case 3. $s_i = a_i, s_j = b_j$. This entails $b_i \leq s_i < s_j = b_j$, a contradiction.

Case 4. $s_i = b_i, s_j = a_j$. This entails $a_i \leq s_i < s_j = a_j$, a contradiction.

Together with 6.10 there follows $A \vee B = \{s_1, \ldots, s_w\}$. The rest is obtained analogously or via the duality principle.

We can bring 6.11 to a formally more general form:

6.12 Theorem. *If $(P, \leq)$ is a finite poset of width w, then, however P is represented as the union of w disjoint chains $C_1, \ldots, C_w$, the greatest w-element antichain of P is $\{a_1, \ldots, a_w\}$, where $a_i = \max C_i$ for $i = 1, \ldots, w$. Analogously the least w-element antichain of P is $\{b_1, \ldots, b_w\}$, where $b_i = \min C_i$ for $i = 1, \ldots, w$.*

We now can comprehend some of the last reflections:

6.13 Theorem (Dilworth [25]). *The maximal-number-antichains of a finite poset constitute a distributive lattice with respect to the order of 6.6. If A, B are maximal-number antichains, then $A \vee B = Ma(A \cup B)$ and $A \wedge B = Mi(A \cup B)$.*

As we have seen in 6.9 the first of these relations is valid also in the lattice of all antichains. The second, $A \wedge B = Mi(A \cup B)$, does not always hold in this more general case.

Using the last results one can also obtain proofs of Dilworth's theorem 5.2 (see e.g. [145], [77], [49]):

6.14 Remark. Let Dilworth's theorem 5.2 already be proved for all posets with at most n elements, where n is a natural number, and let $(P, \leq)$ be a poset with $n + 1$ elements and a a maximal element of P. Let w be the width of the set $P' := P \backslash \{a\}$ and $A := \{a_1, \ldots, a_w\}$ the greatest w-element antichain of P'. Then P' is a union of w disjoint chains $C_1, \ldots, C_w$ where a_i is the greatest element of C_i for $i = 1, \ldots, w$. If $A \cup \{a\}$ is an antichain, then P has width $w + 1$ and is the union of the $w + 1$ chains $C_1, \ldots, C_w, \{a\}$. Otherwise a is comparable with an element $a_i \in A$, and then also $a > a_i$ must hold since a is maximal. Then P is the union of the w chains C_j, $j \neq i$, and $C_i \cup \{a\}$, and has width w. Thus the theorem follows by induction on $|P|$.

2.7 The ordered set of initial segments of a poset

In many situations the set of initial segments of a poset P gives some insight into the structure of P, e.g. in 1.9.9 and in connection with the Dedekind–MacNeille completion. A related notion is defined in the following:

7.1 Definition. Let P be a poset. And let $\mathfrak{W}$ be the set of all well-ordered subsets of P. Then we introduce a partial order in $\mathfrak{W}$ by:

If $a = \{a_\nu | \nu < \alpha\}$ and $b = \{b_\nu | \nu < \beta\}$ are well-ordered subsets of P with index sets α (resp.β) which are ordinals, and where for $\mu < \nu < \alpha$ we have $a_\mu < a_\nu$, and similarly for b, we put
$$a \leq b \iff \alpha \leq \beta \text{ and } a_\nu = b_\nu \text{ for } \nu < \alpha.$$

For this we also say: a is an *initial segment* of b. It is immediately clear that this relation $\leq$ is indeed a (partial) order in $\mathfrak{W}$.

In this context there holds a theorem of Kurepa [105], [109]:

7.2 Theorem. *Let $(P, \leq)$ be a poset, $\mathfrak{W}$ the set of all well-ordered subsets of P. We equip $\mathfrak{W}$ with the order $\leq$ of 7.1*
Then there is no $<$ - preserving mapping of $\mathfrak{W}$ into P.

Proof. Assume the contrary, that $f : \mathfrak{W} \to P$ is a $<$ - preserving mapping. Then $\emptyset \in \mathfrak{W}$ has an f - image $f(\emptyset) =: a_1 \in P$. And $\emptyset$ is a proper initial segment of the well-ordered set $\{a_1\}$ which is formed by (the single element) a_1. Since f is $<$ - preserving we further have $a_1 = f(\emptyset) < f(\{a_1\}) := a_2$.

In general, suppose that we have for an ordinal λ already defined a well-ordered subset $\{a_\nu | \nu < \lambda\}$ of P, and where for every $\nu < \lambda$ we have $a_\nu = f(\{a_\kappa | \kappa < \nu\})$. Then we put $a_\lambda := f(\{a_\nu | \nu < \lambda\})$ and obtain $\{a_\nu | \nu < \mu\} < \{a_\nu | \nu < \lambda\}$ for every $\mu < \lambda$. And this entails $a_\mu = f(\{a_\nu | \nu < \mu\}) < f(\{a_\nu | \nu < \lambda\}) = a_\lambda$. So we have prolonged the well-ordered set $\{a_\nu | \nu < \lambda\}$ by an element $a_\lambda \in P$. This process could be extended over all ordinals so that for every ordinal λ a subset of P of type λ would exist, which is impossible.

As a corollary to 7.2 we obtain:

7.3 Theorem. *Let $(P, \leq)$ be a poset, $\mathfrak{I}$ the set of all initial segments of P, and $\mathfrak{I}_w$ the set of those initial segments I of P which are generated by well-ordered subsets of P, i.e. for which $I = IS(W)$ holds for a well-ordered subset W of P. We equip $\mathfrak{I}$ and $\mathfrak{I}_w$ with the inclusion oder $\subseteq$.*

Then there is no $<$ - preserving mapping $f : \mathfrak{I}_w \to P$ and a fortiori no $<$ - preserving mapping of $\mathfrak{I}$ into P.

Proof. The mapping g, which maps every well-ordered subset $W \subseteq P$ onto $IS(W)$, is $<$ - preserving. Indeed, if W_1, W_2 are in the set $\mathfrak{W}$ of all well-ordered subsets of P, and if W_1 is a proper initial segment of W_2, then there exists an element $x \in W_2 \backslash W_1$. This x is in $IS(W_2) \backslash IS(W_1)$, and so $IS(W_1)$ is a proper subset of $IS(W_2)$.

If now, contrary to our assertion, a $<$ - preserving mapping $f : \mathfrak{I}_w \to P$ would exist, then $f \circ g$ would be a $<$ - preserving mapping of $\mathfrak{W}$ into P, which contradicts 7.2.

A part of 7.3 was also proved by Dilworth and Gleason in [26], where they also proved a counterpart. Their proof is totally different from the above proof. It applies an interesting method which has some common features with Cantor's proof, where he showed that the set of all subsets of a set has a higher cardinality than this set.

7.4 Theorem (Dilworth/Gleason [26]). *Let $\mathfrak{I}(P)$ be the set of all initial segments of a poset P. We order it by $\subseteq$. Let $f : \mathfrak{I}(P) \to P$ be injective. Then neither f nor $f^{-1} : f(\mathfrak{I}(P)) \to \mathfrak{I}(P)$ is $<$ - preserving.*

Proof. a) Assume indirectly that f is $<$ - preserving. Let then $\mathfrak{I}_0$ be the set of all initial segments I with $f(I) \in I$, and $A := \cap \mathfrak{I}_0$. For $I \in \mathfrak{I}_0$ we have $A \subseteq I$ and, because f is supposed to be $<$ - preserving, $f(A) \leq f(I) \in I$, so that $f(A) \in I$. This holds for all $I \in \mathfrak{I}_0$, and so $f(A) \in \cap \mathfrak{I}_0 = A$. This entails $A \in \mathfrak{I}_0$.

Now we define $B := \{x \in P | x < f(A)\}$. For $x \in B$ we then have $x < f(A) \leq f(I) \in I$ for all $I \in \mathfrak{I}_0$, and thus $x \in I$ for all $I \in \mathfrak{I}_0$, so that $x \in A$, and so $B \subseteq A$ follows. Here also $B \subset A$ holds because of $f(A) \in A \backslash B$. ($f(A) \in B$ would entail the contradiction $f(A) < f(A)$.) This yields $f(B) < f(A)$, and by definition of B, $f(B) \in B$, further $B \in \mathfrak{I}_0$ and $A \subseteq B$. This contradicts $B \subset A$, and therefore f is not $<$ - preserving.

b) Assume that f^{-1} is $<$ - preserving. Let then $\mathfrak{I}_1$ be the set of all initial segments I of P with $f(I) \notin I$ and let

$C := \{x \in P | x \leq f(I) \text{ for some } I \in \mathfrak{I}_1\}$.

C is an initial segment of P. If we would have $f(C) \notin C$ then $C \in \mathfrak{I}_1$ would follow. This and the trivial fact $f(C) \leq f(C)$ entails $f(C) \in C$ because of the definition of C. Due to this contradiction $f(C) \notin C$ cannot hold.

So assume $f(C) \in C$. Then, by definition of C, there exists an $I \in \mathfrak{I}_1$ with $f(C) \leq f(I)$. Since we assumed f^{-1} to be $<$ - preserving this entails $C \subseteq I$. This and $f(I) \in C$ yields $f(I) \in I$, contrary to the definition of $\mathfrak{I}_1$.

7.5 Remark. The second part of the last theorem is a generalization of Cantor's theorem, according to which the set of all subsets of a given set S has a higher cardinality than S. To see this we equip S with the smallest possible order on S, namely with $\{(x, x) | x \in S\}$. Then every subset of S is an initial segment of S, so that $\mathfrak{I}(S) = \mathfrak{P}(S)$ holds. If now $f : \mathfrak{P}(S) \to S$ would be an injective mapping, then $f^{-1} : f(\mathfrak{I}(S)) \to \mathfrak{I}(S)$ would be $<$ - preserving, — this is trivial since in S there are no elements a, b with $a < b$. And this would contradict 7.4.

In the next theorems, which were established by Bonnet and Pouzet [11], relations between the set of initial segments of a poset and the linear extensions of its order are studied:

7.6 Theorem. *Let $(S, \leq)$ be a linearly ordered set. Then the set $\mathfrak{I}(S)$ of initial segments of S, ordered by $\subseteq$, is a maximal chain of the ordered power set $(\mathfrak{P}(S), \subseteq)$.*

Proof. Let $X \subseteq S$ be comparable with all elements of $\mathfrak{I}(S)$. We have to show: $X \in \mathfrak{I}(S)$. Let x be a fixed element of X. By assumption then X is comparable with the initial segment (and principal ideal) $(x]$ of S. And so

a) $X \subseteq (x]$ or b) $(x] \subseteq X$ holds.

Let $y \in S$ be an element with $y < x$. Then $(y] \cap X = X$ or $= (y]$ since X is comparable with all initial segments of S. But $(y] \cap X = X$ cannot hold because of $x \notin (y]$ and $x \in X$. So we have $(y] \subseteq X$. In the case a) we therefore have $y \in X$ for all $y \leq x$. This, of course, also holds in case b). Then X is also an initial segment of S, hence $\in \mathfrak{I}(S)$.

The next theorem states, roughly speaking: The smaller the order relation on a set S, the more elements has $\mathfrak{I}(S, \leq)$.

7.7 Theorem. *Let P be a set with two order relations $\leq_1$ and $\leq_2$,* *where $\leq_1$ is a subrelation of $\leq_2$. Then for the corresponding sets of initial segments $\mathfrak{I}_1 := \mathfrak{I}(P, \leq_1)$ and $\mathfrak{I}_2 := \mathfrak{I}(P, \leq_2)$ we have $\mathfrak{I}_2 \subseteq \mathfrak{I}_1$.*

Proof. Let $X \in \mathfrak{I}_2, x \in X$. If an element $y \in P$ satisfies $y <_1 x$ we have $y <_2 x$ and $y \in X$. Then X is also in $\mathfrak{I}_1$.

7.8 Theorem. *Let $(P, \leq)$ be a poset, $\leq_L$ a linear extension of $\leq$. Then the set $\mathfrak{I}(P, \leq_L)$ of initial segments of $(P, \leq_L)$ is a maximal chain of $\mathfrak{I}(P, \leq)$.*

Proof. $\mathfrak{I}(P, \leq_L) \subseteq \mathfrak{I}(P, \leq)$ holds by 7.7. Now $\mathfrak{I}(P, \leq_L)$ is by 7.6 a maximal chain of $(\mathfrak{P}(P), \subseteq)$, and a fortiori of $\mathfrak{I}(P, \leq)$.

7.9 Theorem. *Let $(P, \leq)$ be a poset, $\mathfrak{M}$ a maximal chain of $\mathfrak{I}(P, \leq)$. Then $\mathfrak{M}$ is of the form $\mathfrak{I}(P, \leq_L)$, where $\leq_L$ is a linear extension of $\leq$. And here $\leq_L$ is unique.*

Proof. Let x, y be different elements of P. Then there exists a segment $M \in \mathfrak{M}$, which separates x and y, i. e.: M contains exactly one element of $\{x, y\}$. Indeed, if this would be false, all elements of $\mathfrak{M}$ would contain both elements x, y or none of them. But then $\mathfrak{M}$ would be a chain with an initial segment $\mathfrak{A}$, which contains only elements which contain neither x nor y, and with a corresponding final segment $\mathfrak{B}$, whose elements all contain x and y. But then $\mathfrak{M}$ would not be maximal, because it would be possible to enlarge the chain $\mathfrak{M}$ by an initial segment of $(P, \leq)$, namely by the initial segment $\cup \mathfrak{A} \cup (x]$, if we have $x < y$, resp. by $\cup \mathfrak{A} \cup (y]$, if $y < x$ or $y \parallel x$ holds. In both cases the initial segment contains exactly one element of x, y.

We put $x <_L y$ iff there exists an $M \in \mathfrak{M}$ with $x \in M, y \notin M$.

According to what we mentioned before, then always $x <_L y$ or $y <_L x$ holds, and not both, since otherwise there would follow the existence of initial segments $A, B \in \mathfrak{M}$ with $x \in A, y \notin A, y \in B, x \notin B$. In the case $A \subseteq B$ there would follow $x \in (A \subseteq) B$, and in the case $B \subseteq A$ we would have $y \in (B \subseteq) A$, both with contradiction. So it turned out that $\leq_L$ is antisymmetric.

It is also transitive: If $x <_L y$ and $y <_L z$ hold for elements $x, y, z \in P$, we have initial segments $A, B \in \mathfrak{M}$ with $x \in A, y \notin A, y \in B, z \notin B$. Then $A \subset B$ holds and $x \in B$, so that $x <_L z$ follows.

$\leq_L$ extends $\leq$. For let a, b be elements of P with $a < b$. Then there exists an a, b separating initial segment $M \in \mathfrak{M}$. Then only $a \in M$ and $b \notin M$ is possible, and so $a <_L b$ holds.

Different orders $\leq_L$ would define different sets $\mathfrak{I}(P, \leq_L)$, and therefore the above $\leq_L$ is uniquely defined.

7.10 Example. If a set S has only the least order, namely the relation $=$, then $\mathfrak{I}(S, =)$ is the set $\mathfrak{P}(S)$ of all subsets of S.

Chapter 3
Linearly ordered sets

Among the posets first the linearly ordered ones have been investigated in more detail, in particular by Hausdorff ([81],[82],[83]), who established most of the notions and theorems of this chapter.

3.1 Cofinality

The structure of a poset is largely characterized by its chains and antichains, especially its maximal ones. A linearly ordered set has only one maximal chain, namely itself. In order to describe and analyze the structure of a poset we take into consideration its well- and its inversely well-ordered subsets. To proceed in this manner makes sense because one has a very good oversight of the class of well-ordered (resp. inversely well-ordered) sets. Indeed, the order types of the well-ordered sets are nothing else than the ordinal numbers. In this connection the notion of cofinality has turned out to be of fundamental interest.

1.1 Definition. We recall some notions of set theory: An ordinal number α is the set of all ordinal numbers β that are $< \alpha$.

For ordinals α, β we have $\beta < \alpha \Leftrightarrow \beta \in \alpha \Leftrightarrow \beta$ is a proper subset of α.

So we always consider an ordinal α as the set of ordinals $< \alpha$ which is well-ordered by the relation $\in$.

For a set M we denote by $\omega(|M|)$ the least ordinal number which has cardinality $|M|$.

If M is a well-ordered set, $\mathrm{ord}(M)$ denotes the (uniquely defined) ordinal to which it is isomorphic.

1.2 Definition. Let P be a poset and $T \subseteq P$. Then we call T *cofinal* in P, if there holds: For every $x \in P$ there exists a $t \in T$ with $x \leq t$. (The same: $IS(T) = P$.)

Dually T is called *coinitial* in P, if for every $x \in P$ there exists a $t \in T$ with $t \leq x$. (The same: $FS(T) = P$.)

Since these concepts are dual it suffices to study only one of them, say cofinality.

It should be remarked that Hausdorff's definition of cofinality in [81] differs from the above one in the cases where P is only ordered, but not linearly ordered.

1.3 Remark. If a poset P has a greatest element x, then $\{x\}$ is cofinal in P. The whole set P is of course cofinal and also coinitial in P. If a set T is cofinal in P, every set S with $T \subseteq S \subseteq P$ is also cofinal in P.

So it is interesting to find such subsets of P which have "few" elements.

A trivial fact is the transitivity of the property to be cofinal:

1.4 Theorem. *If X is cofinal in the poset $(P, \leq)$, and if Y is cofinal in the poset $(X, \leq)$ with the order induced by $(P, \leq)$, then Y is also cofinal in $(P, \leq)$.*

In the following we restrict our considerations on linearly ordered sets. First we shall show that in every linearly ordered set S there exist well-ordered subsets which are cofinal in S. And then we are mainly concerned with cofinality in well-ordered sets.

1.5 Theorem. *Let $(S, \leq)$ be a linearly ordered set. Then there exists a well-ordered subset $W \subseteq S$ which is cofinal in S and which satisfies $\mathrm{ord}(W, \leq) \leq \omega(|S|)$.*

Proof. If S is finite or if S has a last element the statement is trivial. In the other case $|S|$ is an aleph and thus $= |\omega(|S|)|$. Then there exists a well-ordering $\prec$ of S so that $\omega(|S|)$ is the order type of the well-ordered set $(S, \prec)$. And then S is representable as $S = \{a_\nu | \nu < \omega(|S|)\}$, where for $\nu_1 < \nu_2 < \omega(|S|)$ we have $a_{\nu_1} \prec a_{\nu_2}$. Now we put
$$W := \{a_\nu | \nu < \omega(|S|) \text{ and } a_\kappa < a_\nu \text{ for all } \kappa < \nu\}.$$
Then the orders $\leq$ and $\prec$ coincide on W, and therefore $(W, \leq)$ is isomorphic to a subset of $\omega(|S|)$, so that we have $\mathrm{ord}(W, \leq) \leq \mathrm{ord}(\omega(|S|) = \omega(|S|)$.
Now W is cofinal in S. Indeed, in the other case there would exist a least ordinal μ such that a_μ is $>$ all elements of W. But then also $a_\mu > a_\nu$ would hold for all $\nu < \mu$, and then a_μ had to be in W, a contradiction.

We supplement Definition 1.2 as follows:

1.6 Definition. Let α be an ordinal and $(S, \leq)$ a linearly ordered set. Then α is said to be *cofinal* in $(S, \leq)$, if there holds: S has a cofinal subset which is isomorphic to α.

If β is an ordinal, then (consequently) α is called *cofinal* in β, if the well-ordered set of ordinals $< \beta$ has a cofinal subset which is isomorphic to α. (By the way: $\emptyset$ is cofinal and coinitial in $\emptyset$.)

Among the ordinals which are cofinal in a linearly ordered set $(S, \leq)$ is a least one. This is called the *cofinality number* of $(S, \leq)$. It is denoted by $\mathrm{cf}(S, \leq)$. Analogously we define $\beta^* := \mathrm{coin}(S, \leq)$, if β is the least ordinal for which S has a coinitial subset which is isomorphic to β^*, and for this we say: β^* is coinitial in S.

So we have: For an ordinal α the cofinality number $\mathrm{cf}(\alpha)$ is the least ordinal γ so that γ is cofinal in the well-ordered set of ordinals $< \alpha$. Therefore we always have $\mathrm{cf}(\alpha) \leq \alpha$. The two possibilities which are contained in this inequality give rise to the following definition:

1.7 Definition. An ordinal α is called *regular*, if $\alpha = \mathrm{cf}(\alpha)$. It is said to be *singular*, if $\alpha > \mathrm{cf}(\alpha)$ holds. We put $\mathrm{cf}(\emptyset) = 0$.

1.8 Example. The ordinals 0 and 1 are evidently regular. All natural numbers $n > 1$ are singular; their cofinality number is 1. The initial ordinal ω_0 (which is the order type of the set of natural numbers, ordered by magnitude) is regular, because every cofinal subset of ω_0 has again ω_0 as corresponding order type.

The ordinals $\omega_0 + 1$, $\omega_0 + 2, \ldots$, $\omega_0 + n$ with $n \in \mathbf{N}$ are singular, because the one-element set of their last element is cofinal in them. The least ordinal $> \omega_0$ which is regular is then ω_1; this follows later from a much more general theorem.

Of course it is now desirable to have a characterization of the regular ordinals. The next theorem narrows the scope, in which one has to seek them, significantly:

1.9 Theorem. *A regular ordinal is 0 or 1 or an infinite initial number ω_α.*

Proof. According to 1.8 a regular ordinal ζ which is $\neq 0$ and $\neq 1$ is infinite. Then there exists exactly one initial ordinal ω_α such that $\omega_\alpha \leq \zeta < \omega_{\alpha+1}$. (Here the index α is determined by $|\zeta| = \aleph_\alpha$.) According to 1.5 now there exists a well-ordered cofinal subset W of ζ with ord $W \leq \omega(|\zeta|) = \omega_\alpha$. Thus we have $\mathrm{cf}(\zeta) \leq$ ord $W \leq \omega_\alpha \leq \zeta$. Since ζ is regular, there follows $\mathrm{cf}(\zeta) = \zeta$ and then also $\zeta = \omega_\alpha$.

Now we can sharpen 1.5 substantially:

1.10 Theorem. *Let $(S, \leq)$ be a linearly ordered set. Then* $\mathrm{cf}(S, \leq)$ *is 0 or 1 or a regular initial number. And there is no other regular ordinal, which is also cofinal in S.*

Proof. Let W be a set which is cofinal in S and whose corresponding ordinal is $\alpha := \mathrm{cf}(S, \leq)$. If α wouldn't be regular, we would have $\alpha > \mathrm{cf}(\alpha)$. Then there would exist a set T which is cofinal in W, and then also in S, with $\mathrm{ord}\ T = \mathrm{cf}(\alpha) < \alpha$ contradicting the fact that α was minimal.

If now α' would be another regular ordinal which is cofinal in S, we would have $\alpha < \alpha'$, because α was the least ordinal which is cofinal in S. And there would exist two cofinal subsets A (resp. A') with corresponding ordinals α (resp. α'). Then we map every $x \in A$ onto the least element of A' which is $> x$ and denote it by $f(x)$. The set $f[A]$ is cofinal in A' since A is cofinal in S. Then there follows $\mathrm{ord}\ f[A] \geq \mathrm{cf}\ A' = \mathrm{cf}\ \alpha' = \alpha'$, because α' is regular. For the corresponding cardinal numbers we now obtain $|\alpha| = |A| \geq |f[A]| \geq |\alpha'|$ and from this $\alpha \geq \alpha'$ since α' is an initial ordinal, contradicting $\alpha < \alpha'$.

Supplementing 1.9 we now prove:

1.11 Theorem. *Every initial ordinal ω_β whose index β is not a limit ordinal is regular.*

Proof. ω_0 is regular. If now β is neither 0 nor a limit ordinal, there exists an ordinal α with $\beta = \alpha + 1$. Then $\omega_{\alpha+1}$ is regular. Indeed, let T be a cofinal subset of $\omega_{\alpha+1}$. Then there holds
$$S := \omega_{\alpha+1} = \cup\{(S < \tau) | \tau \in T\} \text{ and further}$$

(1) $\aleph_{\alpha+1} = |\omega_{\alpha+1}| = |\cup\{(S < \tau)|\tau \in T\}| \leq \sum_{\tau \in T} |(S < \tau)| = \sum_{\tau \in T} |\tau|$.

Because of $\tau < \omega_{\alpha+1}$ we have $|\tau| < \aleph_{\alpha+1}$, which means $|\tau| \leq \aleph_\alpha$ for $\tau \in T$, and then (1) yields $\aleph_{\alpha+1} \leq \sum_{\tau \in T} \aleph_\alpha = \aleph_\alpha \cdot |T|$. If now we would have $|T| \leq \aleph_\alpha$, there would follow $\aleph_{\alpha+1} \leq \aleph_\alpha \cdot \aleph_\alpha$, a contradiction. Therefore we have $|T| = \aleph_{\alpha+1}$ and thus $\mathrm{ord}\ T \geq \omega_{\alpha+1}$. Of course also $\mathrm{ord}\ T$ *is* $\leq \omega_{\alpha+1}$ and hence $= \omega_{\alpha+1}$. So we have $\mathrm{cf}\ \omega_{\alpha+1} = \omega_{\alpha+1}$, so that $\omega_\beta = \omega_{\alpha+1}$ is regular.

An initial ordinal ω_λ can therefore only then be singular if λ is a limit ordinal. In addition to this an initial ordinal ω_λ, whose index λ is a limit ordinal, can only be regular if the following very constricting condition is satisfied (and this is only a necessary condition):

1.12 Theorem and **Definition.** *Let λ be a limit ordinal and ω_λ regular, then we have $\lambda = \omega_\lambda$. (If ω_λ satisfies $\lambda = \omega_\lambda$ it is called exorbitant.)*

Proof. In the linearly ordered set ω_λ the subset of the ω_ν, $\nu < \lambda$, is cofinal: Let an ordinal $\zeta < \omega_\lambda$ be given. Then $|\zeta| < \aleph_\lambda$ holds, so that $|\zeta|$ is finite or an $\aleph_\alpha$ with $\alpha < \lambda$. Since λ is a limit number there exists an ordinal β with $\alpha < \beta < \lambda$. Then there follows $|\zeta| \le \aleph_\alpha < \aleph_\beta$ and $\zeta < \omega_\beta$. So $\{\omega_\nu | \nu < \lambda\}$ is a cofinal subset of ω_λ with corresponding ordinal number λ, and we obtain $\omega_\lambda = \mathrm{cf}(\omega_\lambda) \le \lambda$. Since also $\lambda \le \omega_\lambda$ holds, we finally have $\lambda = \omega_\lambda$.

Hausdorff has shown that such exorbitant numbers exist ([81]). But the last theorem does not establish that exorbitant numbers are always regular.

The importance of the notion of cofinality becomes evident by the following simple, but informative statements:

1.13 Theorem. *Let $(S, \le)$ be a linearly ordered set and $\alpha = \mathrm{cf}(S)$. If a set $T \subseteq S$ has cardinality $|T| < |\alpha|$, then there exists a strict upper bound of T in S.*

Proof. T cannot be cofinal in S. Otherwise also $\mathrm{cf}(T)$ would be cofinal in S. Because of $|\mathrm{cf}(T)| \le |T| < |\alpha|$ this would yield $\mathrm{cf}(S) \le \mathrm{cf}(T) < \alpha$, a contradiction.

In particular there follows:

1.14 Theorem. *If ω_α is an initial ordinal and $\omega_\gamma = \mathrm{cf}(\omega_\alpha)$, $T \subseteq \omega_\alpha$ and $|T| < \aleph_\gamma$, then there exists an ordinal $\beta < \omega_\alpha$ such that $\zeta < \beta$ holds for all $\zeta \in T$.*

If especially ω_α is regular (hence $= \omega_\gamma$), then for every $T \subseteq \omega_\alpha$ with $|T| < \aleph_\alpha$ there exists a strict upper bound β of T with $\beta < \omega_\alpha$.

Next we prove a theorem which gives an informative illustration how the concept of cofinality can be applied in proving theorems. This was already formulated by Cantor. A proof of it was found in the Nachlass of Hausdorff.

1.15 Theorem. *Let $(S, \le)$ be a linearly ordered set, which has no subset of type ω_1 and no subset of type ω_1^* (in other words: all well-ordered and all inversely well-ordered subsets of S are countable), but* *for every ordinal $\nu < \omega_1$ S shall contain a subset of type ν. Then S has also a subset of type η (of **Q**).*

Proof. For elements $a, b \in S$ we put $a \sim b$ iff there exists an ordinal $< \omega_1$, which is not embeddable in $[a, b]$. Then $\sim$ is reflexive and symmetric by definition. We also prove its transitivity:

Let $a \sim b$ and $b \sim c$ for elements a, b, c of S and w.r.o.g. $a < b$. Also we assume $b < c$; all other cases are trivial. Let α_1 and α_2 be ordinals $< \omega_1$ such that α_1 is not embeddable in $[a, b]$ and α_2 is not embeddable in $[b, c]$. Then there holds $\alpha_1 + \alpha_2 < \omega_1$ because ω_1 is additively indecomposable. We put $\beta := \alpha_1 + \alpha_2 + 1$. Then β is not embeddable in $[a, c]$. Otherwise $[a, c]$ would have a subset B with $\mathrm{ord}(B) = \beta$, and considering $B = (B \cap [a, b]) \cup (B \cap [b, c])$ there must hold $\mathrm{ord}(B \cap [a, b]) \geq \alpha_1$ or $\mathrm{ord}(B \cap [b, c]) \geq \alpha_2$, otherwise $\mathrm{ord}(B) \leq \alpha_1 + \alpha_2 < \beta$ would follow and with this a contradiction. Thus we have obtained $a \sim c$, and $\sim$ is indeed an equivalence relation. Let $\mathfrak{K}$ be the set of its equivalence classes. Evidently they are segments of S and then the set $\mathfrak{K}$ has a natural linear order which is induced by the order of S. We prove:

(1) For every class $K \in \mathfrak{K}$ there exists an ordinal $< \omega_1$, which is not embeddable in K.

Let a fixed $K \in \mathfrak{K}$ be given. We choose an element $d \in K$. Then we first obtain:

(2) There exists an ordinal $\gamma < \omega_1$, which is not embeddable in $(K \geq d)$.

If K has a last element e, then (2) is trivial because of $d \sim e$. In the other case there exists a regular initial number which is cofinal in $(K \geq d)$, and this can only be ω_0, because ω_1 is not embeddable in S. Let then $\{d_\nu | \nu < \omega_0\}$ be a subset which is cofinal in $(K \geq d)$, where $d = d_0$ and $d_{\nu_1} < d_{\nu_2}$ for $\nu_1 < \nu_2$ holds. For every interval $[d_\nu, d_{\nu+1}]$ of S there exists an ordinal $\zeta_\nu < \omega_1$, which is not embeddable in this interval. The supremum $\sup\{\zeta_\nu | \nu < \omega_0\}$ is an ordinal $\sigma < \omega_0$. Then $(K \geq d)$ has no subset T with $\mathrm{ord}\, T = \sigma \cdot \omega_1 + 1 =: \gamma$, and (2) holds.

In a similar way, but not quite analogous, there follows:

(3) There exists an ordinal $\delta < \omega_1$, which is not embeddable in $(K \leq d)$.

If $(K \leq d)$ has a first element, then this is clear. In the other case this set has a coinitial subset $\{k_\nu | \nu < \omega_0\}$, where $k_0 = d$ holds and $k_{\nu_1} < k_{\nu_2} \iff \nu_1 > \nu_2$. Let ζ_ν be an ordinal $< \omega_1$ which is not embeddable in $[k_{\nu+1}, k_\nu]$. We put $\sigma := \sup\{\zeta_\nu | \nu < \omega_0\}$. Then we have $\sigma < \omega_1$ because of 1.14. Now there follows: For every well-ordered subset $T \subseteq (K \leq d)$ there holds ord $(T) < \sigma \cdot \omega_0 =: \delta$ because T meets only finitely many of the intervals $[k_{\nu+1}, k_\nu]$; otherwise T would have an infinite descending chain, contradicting its well-ordering. Since $\delta < \omega_1$ holds (3) is proved.

From (2) and (3) now (1) follows in an analogous way as in the proof of the transitivity of $\sim$.

(4) The linearly ordered set $\mathfrak{K}$ of classes is dense.

Indeed, if $\mathfrak{K}$ would contain two classes $K_1 < K_2$, such that there is no class between them, then we choose elements $a_1 \in A_1$, $a_2 \in A_2$. Since a_1 and a_2 are not in relation $\sim$, there exists for every $\nu < \omega_1$ a subset $T \subseteq [a_1, a_2]$ with $\mathrm{ord}(T) > \nu + \nu$, and then K_1 or K_2 has a subset U with $\mathrm{ord}(U) > \nu$. Since this holds for all $\nu < \omega_1$ one of the sets K_1, K_2 would yield a contradiction to (1).

(5) $\mathfrak{K}$ has more than one class.

This follows directly from (1) and from the assumption that for every $\nu < \omega_1$ there is a subset $V \subseteq S$ with ord $V = \nu$.

From (4) and (5) we finally obtain that $\mathfrak{K}$ contains a subset of type η, and this, of course, then also holds for the set S.

3.2   Characters

In this section we introduce the element- and gap-characters, which were introduced by Hausdorff ([81], Chapter VI). Using these one can make much more detailed statements about the "environment" of an element (resp. a gap) than using only the types of Definition 1.10.15. Therefore these notions are also important in the topology of ordered sets.

2.1 Definition. Let S be a linearly ordered set. If A and B are non-empty subsets of S, they are called *neighboring* (in S), if $A < B$ or $B < A$ holds and if there is no element $z \in S$ which lies between A and B, i. e. which satisfies $A < z < B$ or $B < z < A$. We then also say: Neighboring sets cannot be separated by an element.

2.2 Example. In $\mathbf{Q}$ the sets $\{x \in \mathbf{Q} | x < \sqrt{2}\}$ and $\{x \in \mathbf{Q} \,| x > \sqrt{2}\}$ are neighboring, in $\mathbf{R}$ they are not.

In $\mathbf{R}$ the sets A, B are neighboring, where $A := \{\sqrt{2}\}$ and where B is the set of all rational numbers of the form $\frac{a}{b} > \sqrt{2}$ where b *is a* power of 2 and a an integer. Here B is not a segment of $\mathbf{R}$; but this is also not required in the definition.

2.3 Definition. Let (A, B) be a cut in a linearly ordered set $(S, \leq)$. Then, according to 1.10, there exists exactly one regular ordinal α which is cofinal in A, and there exists exactly one regular ordinal β, such that its inverse β^* is coinitial in B. Then the pair (α, β^*) is called the *character* of the cut (A, B), and the latter is said to be an (α, β^*) - *cut* of S. We have $\alpha = \mathrm{cf}(A, \leq)$ and $\beta^* = \mathrm{coin}(B, \leq)$.

The cut (A, B) is called *proper*, if $A \neq \emptyset \neq B$. Then α and β are both > 0.

In the case where (A, B) is a gap, (α, β^*) is said to be a *gap-character*. Since in this case A has no last and B has no first element, α and β are regular initial numbers. Hence gap-characters have a form $(\omega_\xi, \omega_\eta^*)$.

We call $\mathrm{cf}(S)$ the *end-character* of S. And if γ is the regular ordinal for which S has a coinitial subset of type γ^*, then γ^* is said to be the *initial-character* of S.

Evidently (A, B) has the character $(1,1)$ iff (A, B) is a jump.

If we are interested in studying the environment of an element a of a linearly ordered set S, we would do well to partition S into the three subsequent segments $(S < a), \{a\}$, and $(S > a)$. Then we define the *element-character* of A to be the pair (α, β^*), where α is the regular ordinal which is cofinal in $(S < a)$, and where β is the regular ordinal for which β^* is coinitial in $(S > a)$. If there holds $\alpha = \omega_\xi$, then a is also said to be an ω_ξ - *element* or an ω_ξ - limit. If $\beta = \omega_\eta$, then a is also called an ω_η^* - *element* or an ω_η^* - limit.

If $a \in S$ has the character (α, β^*), then a is said to be an (α, β^*) - *element* or an (α, β^*) - limit. A gap with character (α, β^*) is called an (α, β^*) - *gap*.

If in the character (α, β^*) of an element (resp. a gap) the numbers α and β are equal, then the element (resp. gap) is said to be *symmetric*. A cut (A, B) is called *symmetric* if in its character (α, β^*) the numbers α and β are equal. A symmetric cut in a non-empty linearly ordered set is then a jump or a gap of a type $(\omega_\xi, \omega_\xi^*)$.

A symmetric element has either an immediate predecessor and an immediate successor or none of them.

2.4 Example. In $\mathbf{Q}$ all element-characters and all gap-characters are $= (\omega_0, \omega_0^*)$. In $\mathbf{R}$ also all element-characters are $= (\omega_0, \omega_0^*)$, gap-characters don't exist here. So $\mathbf{Q}$ and $\mathbf{R}$ only contain symmetric elements, and $\mathbf{Q}$ has only symmetric gaps.

Applying Hausdorff's maximal principle one can prove the existence of symmetric elements, resp. gaps:

2.5 Theorem. (Hausdorff [81], p.142) *Let S be a linearly ordered set of cardinality $\aleph_\alpha$. And let $x < y$ be two elements of S. Then there exist in the interval $[x, y]$ two neighboring elements or a symmetric element of a character $(\omega_\xi, \omega_\xi^*)$ or a symmetric gap of a character $(\omega_\xi, \omega_\xi^*)$, where ω_ξ is a regular initial ordinal $\leq \omega_\alpha$.*

Proof. Let P be the set of all pairs (a, b) with $a, b \in S$ and $a < b$. We introduce a relation $\prec$ in P by:

For (a, b) and $(a', b') \in P$ we put $(a, b) \prec (a', b') \Longleftrightarrow a < a' < b' < b$. (So $\prec$ is a subrelation of $\supset$.) It is immediately clear that this yields a strict order on P. Augmenting it by the diagonal on P we obtain an order $\preceq$ on P.

Let $\mathfrak{C}$ be a maximal chain in $(P, \preceq)$, which contains (x, y). There exists a regular ordinal which is cofinal in $\mathfrak{C}$. If this is 1, then $\mathfrak{C}$ has a last element (a, b), and then a and b are neighboring, or there is exactly one element of S between them, because in the other case it would have been possible to extend $\mathfrak{C}$, but this was maximal. Then in both cases we have the existence of two neighboring elements in $[x, y]$.

If the regular ordinal which is cofinal in $\mathfrak{C}$ is an initial ordinal ω_ξ, then $\mathfrak{C}$ has a cofinal subset $\{(a_\nu, b_\nu) | \nu < \omega_\xi\}$. Here the a_ν, $\nu < \omega_\xi$, constitute a subset A of S of type ω_ξ, and the b_ν, $\nu < \omega_\xi$, a subset B of type ω_ξ^*. We have $a_\mu < b_\nu$ for all $\mu, \nu < \omega_\xi$ and thus $A < B$. Between A and B is at most one element of S, - otherwise $\mathfrak{C}$ could again be enlarged. If there is no element between A and B, then the pair $(\mathrm{IS}(A), \mathrm{FS}(B))$ is a gap in S of character $(\omega_\xi, \omega_\xi^*)$. If there is exactly one element between A and B, then this has that character.

3.3 η_α - sets

Hausdorff began to study universally ordered sets. In this connection he came across the following important concept:

3.1 Definition. A linearly ordered set S is called an η_α - *set*, if there holds: S has no non-empty neighboring subsets which both have a cardinality $< \aleph_\alpha$, and every cofinal and every coinitial subset of S has a cardinality $\geq \aleph_\alpha$. (The last condition implies that $S \neq \emptyset$ holds.)

3.2 Example. Every non-empty linearly ordered set that is dense and has no first and no last element is an η_0 - set. In particular $\mathbf{Q}$ and $\mathbf{R}$ and also the set of irrational numbers are η_0 - sets.

The following statement is trivial:

3.3 Remark. Every η_α - set is also an η_β - set for all ordinals $\beta < \alpha$.

One can characterize η_α - sets also in the following way:

3.4 Theorem. *A linearly ordered set S is an η_α - set iff the following conditions are satisfied:*
1) Every element-character of S is of the form $(\omega_\xi, \omega_\eta^*)$, where ξ and η are ordinals $\geq \alpha$.
2) Every gap-character of S is of the form $(\omega_\xi, \omega_\eta^*)$ where ξ or η are $\geq \alpha$.
3) $\mathrm{cf}(S) \geq \omega_\alpha$ and $\mathrm{coin}(S)$ is the inverse of an ordinal $\geq \omega_\alpha$.

Proof. An η_α - set S has the above three properties: If an element $a \in S$ has the character $(\omega_\xi, \omega_\eta^*)$, then the one-element set $\{a\}$ is neighboring with a subset $A \subseteq (S < a)$ of type ω_ξ which then has cardinality $\aleph_\xi$, which must be $\geq \aleph_\alpha$ because of 3.1. So we have $\xi \geq \alpha$. Similarly there follows $\eta \geq \alpha$ and also 2). The condition 3) is fulfilled by definition.

Let now S have the above three properties, and let $A < B$ be non-empty neighboring subsets of S. Then there exists a subset A' which is cofinal in A and whose order type is a regular ordinal ρ, and a subset B' which is coinitial in B whose order type is the inverse of a regular ordinal τ. If $\rho = 1$ holds, then A' has only one element and then there must hold $\tau \geq \omega_\alpha$ and thus $|B| \geq \aleph_\alpha$. If $\tau = 1$ there follows analogously $\rho \geq \omega_\alpha$ and $|A| \geq \aleph_\alpha$ because of 1). If ρ and τ are both > 1, then they are initial numbers ω_ξ resp. ω_η. Now $(\omega_\xi, \omega_\eta^*)$ forms a gap-character, and then 2) yields $|A| \geq |A'| = \aleph_\xi \geq \aleph_\alpha$ or $|B| \geq |B'| = \aleph_\eta \geq \aleph_\alpha$, and then S is an $\eta_\alpha-$ set.

There follows, as was also already observed by Hausdorff:

3.5 Theorem. *If ω_α is a singular initial number and S an η_α - set, then S is also an $\eta_{\alpha+1}$ - set.*

Proof. According to 3.4 all element-characters of S are of the form $(\omega_\xi, \omega_\eta^*)$ where ξ and η are $\geq \alpha$. But ξ and η must be different from α because ω_α is singular, and then they are also $\geq \alpha + 1$. Analogously all gap-characters of S must have a form $(\omega_\xi, \omega_\eta^*)$ where ξ or η is $\geq \alpha + 1$. Finally $\mathrm{cf}(S)$ must be $\geq \omega_\alpha$ and at the same time $\neq \omega_\alpha$. Therefore it must be $\geq \omega_{\alpha+1}$. Analogously the rest of 3) follows.

Due to 3.5, in studying η_α - sets we can restrict ourselves to the case where ω_α is regular.

In η_α - sets we have the following extension theorem:

3.6 Theorem. *Let ω_α be a regular initial number and S a linearly ordered set which has no subset of type ω_α and no subset of type ω_α^*. Further, let T be a subset of S, E an η_α - set, and $f_T : T \longrightarrow f_T[T] \subseteq E$ an order-isomorphic mapping. Then f_T is extendable to an order-isomorphic mapping $f : S \longrightarrow f[S] \subseteq E$.*

Proof. We take a well-ordering for the set $A := S \setminus T$. Then we can represent this set in the form $\{a_\nu | \nu < \lambda\}$ for some ordinal λ. Since f_T is isomorphic, the subset $f_T[(T < a_0)]$ is situated before the set $f_T[(T > a_0)]$ in the set E. But these two sets are not neighboring, for otherwise there would exist a set K which is cofinal in $f_T[(T < a_0)]$ and for which $\mathrm{cf}(K) < \omega_\alpha$ holds, and further a set L which is coinitial in $f_T[(T > a_0)]$ and for which $\mathrm{coin}(L)$ is the inverse of an ordinal $< \omega_\alpha$. Thus there exists an element b_0 between $f_T[(T < a_0)]$ and $f_T[(T > a_0)]$, and we can choose this as f-image of a_0.

If now μ is an ordinal with $\mu \leq \lambda$, and if we have a tower of $<$ - preserving mappings $f_\tau : T \cup \{a_\nu | \nu < \tau\} \longrightarrow E$ for all $\tau < \mu$, then we define a mapping $f_\mu : T \cup \{a_\nu | \nu < \mu\} \longrightarrow E$ as follows: If μ is a limit ordinal we define f_μ to be the limit mapping of the $f_\nu, \nu < \mu$, (which extends all $f_\tau, \tau < \mu$). Then f_μ is again $<$ - preserving.

If μ is a successor number $\nu + 1$ we construct (in analogy to f_0) a $<$ - preserving mapping $f_\mu : T \cup \{a_\nu | \nu \leq \mu\} \longrightarrow E$ which extends the function f_ν. Applying transfinite induction we so construct an order-preserving mapping $f : S \longrightarrow E$.

As a special case of the last theorem, by taking $T = \emptyset$, we obtain an embedding theorem of Hausdorff (see [173]):

3.7 Theorem. *Let ω_α be a regular initial ordinal, S a linearly ordered set, which has no subset of type ω_α and no subset of type ω_α^*. Then S is embeddable into every η_α - set.*

Later we shall prove a generalization of the last theorem by exchanging one of the two occurences of ω_α by $\omega_{\alpha+1}$. (See 6.2.10)

We shall generalize the last theorem and prepare this with:

3.8 Lemma. *Let ω_α be an initial ordinal, τ an ordinal $< \omega_\gamma = \mathrm{cf}(\omega_\alpha)$, $S = \cup_{\nu < \tau} S_\nu$ a poset, where all summands S_ν, $\nu < \tau$, are free from ω_α. Then also S is free from ω_α. And symmetrically for ω_α^*.*

Proof. We assume indirectly that $T \subseteq S$ is a subset of type ω_α. Then in view of $T = \cup_{\nu < \tau}(T \cap S_\nu)$ we obtain $\aleph_\alpha = |T| \leq \sum_{\nu < \tau} |T \cap S_\nu|$. In this sum every summand is $< \aleph_\alpha$, and there are only $|\tau| < \aleph_\gamma$ many summands. This gives a contradiction to the regularity of $\aleph_\alpha$. The case concerning ω_α^* runs symmetrically.

Now we can generalize 3.7:

3.9 Theorem. *Let ω_α be regular, $S = \cup\{S_\nu | \nu < \omega_\alpha\}$ a linearly ordered set, where every S_ν is free from ω_α and ω_α^*. Then S is embeddable into every η_α - set.*

Proof. Let E be an η_α - set. We define $T_\nu := \cup_{\mu < \nu} S_\mu$ for $\nu < \omega_\alpha$. According to 3.8 then also T_ν is free from ω_α and ω_α^*. And the sets T_ν, $\nu < \omega_\alpha$, form a tower of sets which increases with ν.

If now for every $\nu < \omega_\alpha$ one has a $<$-preserving mapping $f_\nu : T_\nu \longrightarrow E$ (which exists according to 3.7), then because of 3.6 there also exists a $<$ - preserving mapping $f_{\nu+1} : T_{\nu+1} \longrightarrow E$, which extends f_ν. (To see this exchange S by $T_{\nu+1}$, T by T_ν in 3.6.) Now we apply transfinite induction. If τ is an ordinal $< \omega_\alpha$, and if for every $\nu < \tau$ already a $<$ - preserving mapping $f_\nu : T_\nu \longrightarrow E$ is defined so that the T_ν form a tower which increases with ν, and that the mappings f_ν also form a tower of mappings, we conclude as follows:

If τ is a successor number $\nu + 1$, we construct $f_{\nu+1}$ in the same way as before. If τ is a limit number — so that $T_\tau = \cup_{\nu < \tau} T_\nu$ — we define f_τ to be the limit mapping which extends all f_ν, $\nu < \tau$. Then finitely f_{ω_α} is defined and furnishes the embedding of S into E.

The property to be an η_α - set is inherited by the subsets which are dense in an η_α - set:

3.10 Theorem. *Let E be an η_α - set and $D \subseteq E$ a subset which is dense in E. Then D is also an η_α - set.*

Proof. If A, B are two subsets of D which are neighboring in D and which satisfy $A < B$, then there exists at most one element in E which lies between them, for if there would be two, say $x < y$, then there would also exist an element $z \in D$ between them, and this would satisfy $A < z < B$, contradicting our assumption. If now there is no element of E between A and B, then A and B define a gap in E, and then A has a well- or B an inversely well-ordered subset of cardinality $\geq \aleph_\alpha$, because E is an η_α - set. If there is exactly one element of E between A and B, then there follows $|A| \geq \aleph_\alpha \leq |B|$. Finally every subset of D that is cofinal in D is also cofinal in E, analogously for coinitial, and thus all three conditions of 3.4 are satisfied.

The next theorem follows immediately with 3.4:

3.11 Theorem. *Each open interval (a, b) of an η_α - set S is again an η_α - set. And the initial (resp. final) segments $(S < a)$, $(S > a)$ are also η_α - sets.*

Chapter 4
Products of orders

In mathematics one often has a situation where one creates a new structure which is composed of several structures which are already given. E.g. in algebra one studies direct products of groups, rings and so on, or in topology one investigates product spaces and quotient spaces. In order theory these concepts have analogous definitions. In particular we introduce among others sum orders, product orders, with the order of cardinal product or in special cases with a lexicographic order.

4.1 Construction of new orders from systems of given posets

1.1 Definition. Let $(I, \prec)$ be a poset, and let $(P_i, \leq_i)$ be posets for $i \in I$ with pairwise disjoint carrier sets P_i. Then we define the *ordered sum* $\sum_{i \in I}(P_i, \leq_i)$ of the posets $(P_i, \leq_i)$ over the ordered argument $(I, \prec)$ as the poset $(V, \leq)$, where $V := \cup_{i \in I} P_i$ and where $\leq$ is now defined by:

For $a, b \in V$ we put $a \leq b \iff a$ and b are in the same summand P_i, and there holds $a \leq_i b$, or

$a \in P_i$ and $b \in P_j$ with $i \neq j$ and $i \prec j$.

We call the so defined $\leq$ the *sum-order* of the $\leq_i$ over the ordered argument $(I, \prec)$. It is easily seen that this $\leq$ is indeed an order.

If I is an arbitrary set (with no order) and if for every $i \in I$ a poset $(P_i, \leq_i)$ is given we define also $\sum_{i \in I}(P_i, \leq_i)$ to be the ordered sum of the P_i, $i \in I$, over the set I, after we have equipped I with the diagonal order $\prec$. (Here $\prec$ is the set of all pairs (x,x) with $x \in I$). We call this poset the *ordered sum* of the posets $(P_i, \leq_i)$ over the argument I.

Then the following statement is immediately clear, and that was the sense of the definition:

1.2 Remark. In both situations of 1.1 every order $\leq_i$ is the restriction of $\leq$ to the set P_i.

We mention a special case of 1.1:

1.3 Theorem. *If $(I, \prec)$ is a linearly ordered set, and if the $(S_i, \leq_i)$, $i \in I$, are linearly ordered sets with pairwise disjoint carrier sets S_i, then $\sum_{i \in I}(S_i, \leq_i)$ is a linearly ordered set.*

We also wish to define a sum of order-types. To this purpose we represent the types by posets of the corresponding type and apply Definition 1.1. In this connection we make use of the usual procedure to make set families pairwise disjoint. If $(S_i)_{i \in I}$ is a family of sets, these need not be disjoint. But we can obtain such a situation by defining $S_i' := \{(i,x) | x \in S_i\}$ for $i \in I$. Then we can define:

1.4 Definition. Let $(I, \prec)$ be a poset, and let τ_i, $i \in I$, be order types. Then the *sum* $\sum_{i \in I} \tau_i$ of the order types τ_i over the ordered argument $(I, \prec)$ is defined to be the order-type τ of the following poset V: We take a realization $(P_i, \leq_i)$ of τ_i for every $i \in I$, and this so that the P_i, $i \in I$, are pairwise disjoint. Then τ is the order-type of the ordered sum $\sum_{i \in I} (P_i, \leq_i)$.

One can easily verify that this definition is independent from the choice of the realizations P_i. Similar to 1.3 we have:

1.5 Theorem. *If $(I, \prec)$ is a linearly ordered set, and if the τ_i, $i \in I$, are linear order types, then $\sum_{i \in I} \tau_i$ is also a linear order type.*

Next we introduce a product of posets. The idea of product order is that it holds exactly then, when it holds "coordinatewise". This leads to:

1.6 Definition. Let I be a set, and let $(P_i, \leq_i)$, $i \in I$, be posets. Then we define the *ordered product* or *direct product* of the $(P_i, \leq_i)$, $i \in I$, to be the poset $(\mathsf{X}_{i \in I} P_i, \leq)$, where the cartesian product of the sets P_i over the argument I is equipped with the order $\leq$ which is defined by: For two families $(a_i)_{i \in I}$ and $(b_i)_{i \in I} \in \mathsf{X}_{i \in I} P_i$ we put

$(a_i)_{i \in I} \leq (b_i)_{i \in I} \iff a_i \leq_i b_i$ for all $i \in I$.

$\leq$ is said to be the *product order* of the $\leq_i$, $i \in I$. It is evident that it is an order.

If the sets $(P_i, \leq_i)$, $i \in I$, are all the same poset $(P, \leq)$, we also write $(P, \leq)^I$ instead of $\mathsf{X}_{i \in I}(P_i, \leq_i)$.

By the way, another possibility to define a product order would be to put $(a_i) \leq (b_i) \iff a_i = b_i$ for all $i \in I$ or $a_i < b_i$ for all $i \in I$. Again one can verify that this yields an order in the direct product $\mathsf{X}_{i \in I} P_i$. It is a subrelation of the previously defined product order (and usually a proper subset of it).

1.7 Example. A simple example for 1.6 is the n-dimensional euclidean space $\mathbf{R}^n$. In its product order $\leq$ two of its points are related by
$(a_1, \ldots, a_n) \leq (b_1, \ldots, b_n) \iff a_i \leq b_i$ for $i = 1, \ldots, n$.

If p, q are points of the euclidean plane $\mathbf{R}^2$, then $p \leq q$ holds iff q is situated in the right upper quadrant of p.

In 1.6 the set I was not equipped with an order-structure. But if we have a well-ordering on I, then there arises the possibility to introduce a linear order into $X_{i \in I} P_i$ in a natural way by relaying onto the order of I. This leads to the concept of lexicographic order:

1.8 Theorem and **Definition.** *Let $(I, \prec)$ be a well-ordered set, and let $(P_i, \leq_i)$ be posets. Then the lexicographic product $L_{i \in I}(P_i, \leq_i)$ of the $(P_i, \leq_i)$, $i \in I$, is defined to be the poset $(X_{i \in I} P_i, \leq)$, where in the cartesian product of the sets P_i the order relation $\leq$ is introduced (in another way than in 1.6):*

For $(a_i)_{i \in I} \neq (b_i)_{i \in I}$ from $X_{i \in I} P_i$ let j be the least index $i \in I$, for which $a_j \neq b_j$. (Its existence is guaranteed by the well-ordering of the index set I.) And then we put $(a_i) < (b_i) \iff a_j <_j b_j$. (This definition implies that, if a_j and b_j are incomparable in $\leq_j$, the families (a_i) and (b_i) are incomparable in $\leq$.)

And further we put $(a_i) \leq (b_i) \iff (a_i) < (b_i)$ or $(a_i) = (b_i)$. The relation $\leq$ which is so defined is called the lexicographic order or the order by first differences of $X_{i \in I} P_i$.

If this order is considered in comparison with the product order, which was defined in 1.6, we denote it by $\leq_L$, and then the sign $\leq_P$ shall be used to denote the product order.

The product of order types $\tau_i, i \in I$, is the type of $L_{i \in I}(P_i, \leq_i)$, where $(P_i, \leq_i)$ has type τ_i.

Proof. We have to verify that the lexicographic order $\leq_L$ is indeed an order relation. The definition immediately yields that it is reflexive and antisymmetric. Further $<_L$ (and with it also $\leq_L$) is also transitive: Let $(a_i) < (b_i)$ and $(b_i) < (c_i)$ hold for elements of $X_{i \in I} P_i$. Let then i_1 be the first element $i \in I$ with $a_i \neq b_i$, and then also $a_i <_i b_i$, and let i_2 be the least element $i \in I$ with $b_i \neq c_i$, and then also $b_i < c_i$. We put $i^* := \min\{i_1, i_2\}$. If now $i^* = i_1$ holds, we have $a_{i^*} < b_{i^*} \leq c_i^*$. For $i^* = i_2$ we obtain $a_{i^*} \leq b_{i^*} < c_{i^*}$. So we have in any case $a_{i^*} < c_{i^*}$, but $a_i = b_i = c_i$ for all $i < i^*$ of I, and $(a_i) < (c_i)$ follows.

1.9 Remark and **Definition.** The definitions 1.6 and 1.8 are often applied in the special case, where all $(P_i, \leq_i)$ are the same poset $(P, \leq)$. In this case the cartesian product $X_{i \in I} P_i$ obtains the simple form P^I, and one has the product order $\leq_P$ on P^I. If in addition to this I is also well-ordered we also have the lexicographic order $\leq_L$ on P^I.

If in particular I and P are ordinals (and insofar well-ordered sets) τ, resp. μ, we write $\mu((\tau))$ instead of μ^τ in order to avoid a collision with the definition of the exponentiation of ordinal numbers. So we can say:

$\mu((\tau))$ is the set of all functions $f : \tau \longrightarrow \mu$. Practically $\mu((\tau))$ is the set of all transfinite sequences of length τ, containing ordinals $< \mu$. In this set we then have the product order and the lexicographic order as well. If we don't mention details we always presuppose that the lexicographic order is meant by $\leq$.

Also if S is a set and τ an ordinal we define $S((\tau))$ to be the set of all mappings from τ into S, — the same: the set of all sequences of length τ with elements in S. If here S is linearly ordered, we again have a linear order in $S((\tau))$ by first differences.

The following statement is immediately clear:

1.10 Theorem. *If $(I, \prec)$ is a well-ordered set and if $(P_i, \leq_i)$ is a poset for $i \in I$, then the product order $\leq_P$ of $X := X_{i \in I} P_i$ is a subrelation of the lexicographic order $\leq_L$ of X.*

Now one could also define product operations for order types in analogy to 1.6, 1.8 and 1.9. But we don't need them in the sequel.

The property to be linearly ordered is transferred from the factors onto their lexicographic product, as can immediately be seen:

1.11 Theorem. *Let I be a well-ordered set and let the $(S_i, \leq_i)$, $i \in I$, be linearly ordered sets. Then the lexicographic product $L_{i \in I}(S_i, \leq_i)$ is also linearly ordered.*

Contrary to 1.11 the property to be linearly ordered is usually not transferred from the factors to their ordered product, as can easily be seen. Further the property to be well-ordered is in general not transferred from the factors to their lexicographic product:

1.12 Example. Let I be the set of non-negative integers with their usual order. For every $i \in I$ let S_i be the two-element set $\{0, 1\}$ which is well-ordered by $0 < 1$. Then the lexicographic product $2((\omega_0))$ is not well-ordered. This set is the set of all sequences of length ω_0 of numbers 0 and 1. For $n \in \mathbf{N}$ let f_n be the sequence which has a digit 1 at the

n^{th} position and otherwise only zeros. Then the set $\{f_n | n \in \mathbf{N}\}$ has no first element in the lexicographic order.

But we obtain a positive result if we sharpen the assumption so that I is finite:

1.13 Theorem. *Let I be a finite well-ordered set, and for $i \in I$ let S_i be a well-ordered set. Then the lexicographic product $\mathrm{L}_{i \in I} S_i$ is well-ordered.*

Proof. W.r.o.g we assume that $I = \{1, \ldots, n\}$ (with the natural order). Let T be a non-empty subset of $\mathrm{X}_{i \in I} S_i$. The set T_1 of all first components of the sequences of T has a first element k_1 because of the well-ordering of S_1. Let then E_1 be the set of all those sequences of T that have k_1 as first component. The set E_2 of all elements of S_2, that are the second component of an element of E_1 has a first element k_2 because of the well-ordering of S_2. Analogously we construct all k_i for $i = 1, \ldots, n$. The sequence $(k_1, \ldots, k_n)$ is then evidently the least element of T.

Between the inclusion order $\subseteq$ of a power set $\mathfrak{P}(S)$ of a set S and the lexicographic product $\{0,1\}^S$ there holds a simple, but very informative relation:

1.14 Theorem. *Let S be an arbitrary set, and let the set $\{0,1\}$ be ordered by $0 < 1$. Then the poset $(\mathfrak{P}(S), \subseteq)$ is isomorphic to the ordered product $\{0,1\}^S$. The mapping f which ascribes to every $T \subseteq S$ its characteristic function $\phi_T \in \{0,1\}^S$ yields an isomorphism between them. Here $\phi_T(x) = 1$ for $x \in T$, and $\phi_T(x) = 0$ for $x \in S \setminus T$.*

Proof. It is trivial that f is bijective. Further we have $T_1 \subseteq T_2 \subseteq S \iff$ For the x that satisfy $\phi_{T_1} = 1$ also $\phi_{T_2} = 1$ holds, and this is equivalent to $\phi_{T_1} \leq \phi_{T_2}$.

By means of 1.14 one can give another proof for the theorem of Szpilrajn (2.3.2). The idea of it goes back to Novotný [128]. First we prove a lemma:

1.15 Lemma. *Let S be a set. Then the order $\subseteq$ of $\mathfrak{P}(S)$ can be extended to a linear order $\leq$ of $\mathfrak{P}(S)$.*

Proof. The mapping f of 1.14 yields an isomorphism of $(\mathfrak{P}(S), \subseteq)$ onto $(\{0,1\}^S, \leq)$, where $\leq$ is the product order $\leq_P$ .This is a subrelation of the lexicographic ordering $\leq_L$ which is a linear order. If now A, B

$\in \mathfrak{P}(S)$ we put $A \leq B \Longleftrightarrow f(A) \leq_L f(B)$. Evidently $\leq$ is a linear order on $\mathfrak{P}(S)$ (because $\leq_L$ is one) which extends $\subseteq$ (because $\leq_L$ extends $\leq_P$).

1.16 Remark. Now we obtain a new proof for the theorem of Szpilrajn: $(S, \leq)$ is isomorphic to a subset $\mathfrak{T}$ of $\mathfrak{P}(S)$, with the order $\subseteq$, say by a mapping f. Because of 1.15 the order $\subseteq$ of $\mathfrak{T}$ can be extended to a linear order $\leq_T$ of $\mathfrak{T}$. Then S can be linearly ordered by: For s_1, $s_2 \in S$ we put $s_1 \leq_e s_2 \Longleftrightarrow f(s_1) \leq_T f(s_2)$. And $\leq_e$ is an extension of $\leq$.

Hausdorff had introduced some more general concepts, initial and final orders, which we now describe. Here arbitrary linearly ordered arguments are considered.

1.17 Definition. Let $(I, \leq)$ be a linearly ordered set and $(S_i, \leq_i)$ a linearly ordered set for $i \in I$. If $(a_i)_{i \in I}$ and $(b_i)_{i \in I}$ are elements of the cartesian product $X := \mathrm{X}_{i \in I} S_i$ we put $(a_i) < (b_i)$ if there exists an initial segment A of I such that $a_i \leq b_i$ holds for all $i \in A$, and $a_i < b_i$ for at least one $i \in A$. Then one verifies that this $<$ is a strict order. Enlarged by the diagonal relation it becomes an order $\leq_I$ on X, which is called the *initial order* on X. The product order of X is a subrelation of $\leq_I$. And it can easily be seen that the lexicographic order is a special case of this initial order $\leq_I$.

A variant of this definition is obtained, if we put $(a_i) < (b_i)$ if there exists a non-empty initial segment A of I such that $a_i < b_i$ for all $i \in A$. Enlarged by the diagonal of S we obtain again an order on S. But this is no longer a generalization of the lexicographic order.

Dually to these notions one can define the *final order* of X by replacing in the definitions "initial segment" by "final segment". Following the terminology of Hausdorff, we denote by final order the order $\leq$ which is given by $(a_i) < (b_i)$ if $a_i < b_i$ for all i of a non-empty final segment A of I , which is then enlarged by the diagonal of S.

In mathematics a property is called *productive* if there holds: If all factors of a cartesian product have this property the product also has this. In our case the property to be complete is productive:

1.18 Theorem. *Let the sets* $(S_i, \leq_i)$, $i \in I$, *be completely ordered sets. Then their ordered product* $\mathrm{X}_{i \in I}(S_i, \leq_i)$ *is also completely ordered.*

Proof. Let T be a non-empty subset of the ordered product, T_i the set of all i-components of the elements of T, $a_i := \inf T_i$ for $i \in I$. Then evidently $(a_i)_{i \in I}$ is the infimum of T in the ordered product. And so this is also completely ordered.

4.2 Order properties of lexicographic products

In lexicographic products one always has as argument a well-ordered set. W.r.o.g. we than can assume that this is an ordinal number λ. This is the set of all ordinal numbers $< \lambda$, and the order of λ is the usual order in the class of the ordinals. In the following λ will be the length of certain transfinite sequences. And in most cases (but not everytime) λ is also a limit ordinal. In particular we are interested in the well- and inversely well-ordered subsets of lexicographic products, and what kind of cuts these sets possess. First we define (resp. recall) some concepts:

2.1 Definition. Let P be a poset and τ an order type. Then P is said to be τ - *free*, if P has no subset of type τ.

If $(a_\nu)_{\nu < \mu}$ is a (transfinite) sequence a, which means μ an ordinal and the a_ν arbitrary elements, then a_ν is called the ν - *component* of a. For $\rho \leq \mu$ the sequence $(a_\nu)_{\nu < \rho}$ is called an *initial segment* of $(a_\nu)_{\nu < \mu}$, more precisely the ρ - *segment* of it. For this we also say: $(a_\nu)_{\nu < \mu}$ *begins* with $(a_\nu)_{\nu < \rho}$. If $\rho < \mu$ holds, $(a_\nu)_{\nu < \rho}$ is called a *proper* initial segment. Then we have the following fundamental statement:

2.2 Theorem. *Let ω_α be a regular initial ordinal and λ an ordinal with $0 < \lambda < \omega_\alpha$. For $\nu < \lambda$ let $(S_\nu, \leq_\nu)$ be a linearly ordered set. We consider the cartesian product $X := \mathsf{X}_{\nu < \lambda} S_\nu$ with the lexicographic order $\leq_L$. Then there hold:*
   a) If all $(S_\nu, \leq_\nu)$, $\nu < \lambda$, are ω_α - free, then X is also ω_α - free.
   b) If all $(S_\nu, \leq_\nu)$, $\nu < \lambda$, are ω_α^* - free, then X is also ω_α^* - free.

Proof. We assume that the assumption of a) holds and assume indirectly that X has a well-ordered subset W of type ω_α, say $W = \{w_\nu | \nu < \omega_\alpha\}$, where the order of the w_ν corresponds to the order of their indices ν. If there exists an ordinal $\tau < \omega_\alpha$ such that all w_ν with $\tau \leq \nu < \omega_\alpha$ begin with a sequence F, we say: *Nearly all w_ν begin with F.*

Let K_0 be the set of all 0-components of the elements of W. Then there follows:

(1) K_0 has a greatest element a_0.

For otherwise K_0 would be cofinal in W, which has type ω_α, and then also K_0, and a fortiori S_0, had to contain a subset of type ω_α, which yields a contradiction. So we have obtained:

For $\mu = 1$ the following statement (*) holds:

(*) μ is an ordinal $\leq \lambda$, and there exists a sequence $(a_\nu)_{\nu<\mu}$ with $a_\nu \in S_\nu$ for $\nu < \mu$, such that nearly all w_ν begin with $(a_\nu)_{\nu<\mu}$.

Let now τ be an ordinal $\leq \lambda$ such that (*) holds for all μ with $0 < \mu < \tau$.

Case 1. τ is a successor ordinal $\rho + 1$.

Then, according to our induction hypothesis, there exists a sequence $(a_\nu)_{\nu<\rho}$ with $a_\nu \in S_\nu$ for $\nu < \rho$, such that nearly all w_ν begin with it. Then, quite analogous to the existence of a_0, there follows the existence of an element $a_\rho \in S_\rho$ such that nearly all w_ν have a_ρ as ρ - component. And then also nearly all w_ν begin with $(a_\nu)_{\nu<\tau}$. So (*) also holds for τ (instead of μ).

Case 2. τ is a limit ordinal.

Then for very $\mu < \tau$ there exists an ordinal $\nu_\mu < \omega_\alpha$ and a sequence $(a_\nu)_{\nu<\mu}$ such that all w_ν with $\nu_\mu \leq \nu < \omega_\alpha$ begin with $(a_\nu)_{\nu<\mu}$. Since ω_α is regular and because of $|\tau| < \aleph_\alpha$ the supremum σ of $\{\nu_\mu | \mu < \tau\}$ is still $< \omega_\alpha$. And then all w_ν with $\sigma \leq \nu < \omega_\alpha$ begin with $(a_\nu)_{\nu<\tau}$. Again (*) holds for τ (instead of μ).

Using transfinite induction now there follows that (*) also holds for λ (instead of μ). But this yields a contradiction, because then nearly all w_ν must begin with a sequence $(a_\nu)_{\nu<\lambda}$. And that means that they must be equal which gives a contradiction.

For the proof of b) we can use a). Therefore we consider the sets S_ν with the inverse orders $\geq_\nu$ of the $\leq_\nu$ and consider the set X, but now with the lexicographic order $\leq_I$ with respect to the orders $\geq_\nu$. This $\leq_I$ is evidently the inverse order of $\leq_L$. If the $(S_\nu, \leq_\nu)$ are ω_α^* - free, then the $(S_\nu, \geq_\nu)$ are ω_α - free. According to a) now $(X, \leq_I)$ is ω_α - free, and thus $(X, \leq_L)$ is ω_α^* - free.

If in 2.2 all $(S_\nu, \leq_\nu)$ are the same linearly ordered set $(S, \leq)$ we obtain as a corollary to 2.2:

2.3 Theorem. *Let ω_α be a regular initial ordinal, $\lambda < \omega_\alpha$ an ordinal, and S a linearly ordered set. Then there holds:*

a) If S is ω_α - free, then the lexicographic product $S((\lambda))$ is also ω_α - free.

b) If S is ω_α^* - free, then $S((\lambda))$ is also ω_α^* - free.

2.4 Remark. The Theorems 2.2 and 2.3 are no longer valid if we don't require that ω_α is regular: Let S be a linearly ordered set of order type $\sum\{\mathrm{tp}(\omega_\nu)|\nu \in (\mathbf{Z} \leq 0)\}$ $(= \cdots + \omega_1 + \omega_0)$. In 2.2 we take $\lambda = \alpha := \omega_0$. Then $X := S((\omega_0))$ has a subset of type ω_{ω_0} $(= \omega_0 + \omega_1 + \cdots + \omega_\nu + \cdots |\nu < \omega_0)$. Indeed, in X we have subsets $X_\nu = \{\nu\} \times \omega_\nu \times \{1\} \times \{1\} \times \cdots$ for $\nu < \omega_0$, and for $\mu < \nu < \omega_0$ each element of X_μ is $<$ each element of X_ν so that the union $\cup\{X_\nu|\nu < \omega_0\}$ is a subset of X of type ω_{ω_0}. But S has no subset of type ω_{ω_0} since every well-ordered subset of S intersects only finitely many of the summands of one of the types ω_ν.

Now we consider the special case where λ and S are ordinals:

2.5 Theorem. *Let μ and λ be ordinals. Then there holds:*
a) *For every well-ordered set $T \subseteq \mu((\lambda))$ we have* $\mathrm{cf}(T) \leq \sup\{\mu, \lambda\}$.
b) *For every inversely well-ordered subset $S \subseteq \mu((\lambda))$ there holds* coin $S \leq \lambda^*$.

Proof. If μ or λ is 0, this is trivial. So let both be $\neq 0$.
a) If T has a last element, we have $\mathrm{cf}(T) = 1$, and the statement a) is trivial, also if T is empty. In the other case $\mathrm{cf}(T)$ is a regular ordinal ω_α, and $\mu((\lambda))$ is not ω_α - free. Then μ or λ is $\geq \omega_\alpha$ because of 2.3 a), and this yields $\sup\{\mu, \lambda\} \geq \omega_\alpha = \mathrm{cf}(T)$.
b) Let S be an inversely well-ordered subset of $\mu((\lambda))$. If S is empty or has a first element, then our statement is clear. In the other case we have coin $S = \omega_\beta^*$, where ω_β is regular. Then $\mu((\lambda))$ is not ω_β^* - free. According to 2.3 b) now μ is not ω_β^* - free or $\lambda \geq \omega_\beta$. But every well-ordered set, and then also μ, is already ω_0^* - free. So λ must be $\geq \omega_\beta$ and thus $\omega_\beta^* = $ coin $S \leq \lambda^*$.

From 2.3 we can derive, and this now also for singular ω_α :

2.6 Theorem. *Let ω_α be an arbitrary initial ordinal, $n \in \mathbf{N}$, $\lambda < \omega_\alpha$. Then there holds:*
a) $n((\lambda))$ *has no subset of type ω_α and no subset of type ω_α^*.*
b) $n((\omega_\alpha))$ *has no subset of type $\omega_{\alpha+1}$ and no subset of type $\omega_{\alpha+1}^*$.*

Proof. If ω_α is regular, a) follows immediately from 2.3. If now ω_α is singular, and then α a limit ordinal, then from $\lambda < \omega_\alpha$ there also follows $\lambda < \omega_\beta$ for an ordinal $\beta < \alpha$, where β is a successor number so that ω_β is regular. Because of 2.3, then $n((\lambda))$ has no subsets of one of the types $\omega_\beta, \omega_\beta^*$, a fortiori no subset of one of the types $\omega_\alpha, \omega_\alpha^*$.

b) is now a consequence of a) by replacing λ by ω_α, and ω_α by $\omega_{\alpha+1}$.

Until now we investigated which ordinals (resp. their inverses) do not occur as subtypes of linearly ordered sets. In the other direction we have a trivial statement:

2.7 Theorem. *Let μ be a limit ordinal, λ an ordinal > 0. Then $\mu((\lambda))$ has a cofinal subset of type μ.*

Proof. Every set $\{a_\nu | \nu < \mu\}$ of elements $a_\nu \in \mu((\lambda))$, where a_ν has ν as 0-component, is cofinal in $\mu((\lambda))$.

Many posets, e.g. **Z,Q** and **R,** have the property to be symmetric in the following sense:

2.8 Definition. A linear order $\leq$ on a set S is called *symmetric* and $(S, \leq)$ a *symmetrically ordered set*, if there exists a bijective mapping $f : S \longrightarrow S$ such that for $x, y \in S$ we have $x \leq y \Longrightarrow f(x) \geq f(y)$. Then f is said to be a *symmetry* of $(S, \leq)$.

An order type is said to be *symmetric* if it has a symmetric realization.

The symmetry of the factors of a lexicographic product of linearly ordered sets is transferred also to their product:

2.9 Theorem. *Let λ be an ordinal, and $(S_\nu, \leq_\nu)$ a symmetrically linearly ordered set for every $\nu < \lambda$. Then also the lexicographic product $X := \mathsf{X}_{\nu<\lambda} S_\nu$ is symmetrically ordered.*

Proof. For every $\nu < \lambda$ there exists a symmetry $g_\nu : S_\nu \longrightarrow S_\nu$. For $a = (a_\nu)_{\nu<\lambda} \in X$ we put $f(a) := (g_\nu(a_\nu))_{\nu<\lambda}$. If now $b = (b_\nu)_{\nu<\lambda} \in X$, we have: $a \leq b \Longleftrightarrow a_\nu \leq b_\nu$ for $\nu < \lambda \Longleftrightarrow g_\nu(b_\nu) \leq g_\nu(a_\nu)$ for $\nu < \lambda \Longleftrightarrow f(b) \leq f(a)$, and thus we have obtained a symmetry f of $(X, \leq)$ because f is also bijective.

Our next investigations concern the existence of infima and suprema of subsets of lexicographic products. Together with this we can describe which lexicographic products are dense (resp. continuous).

2.10 Theorem. *Let λ be an ordinal, and for $\nu < \lambda$ let S_ν be a linearly ordered set without gaps, which has a first element e_ν and a last element z_ν, so that S_ν is completely ordered. Then for every non-empty subset T of the lexicographic product $X = \mathsf{X}_{\nu<\lambda} S_\nu$ there exist $\inf T$ and $\sup T$. So the set X has no gaps, but X has a first and a last element, and thus it is a complete lattice.*

Proof. Let T be a non-empty subset of X. Let $a_0 \in S_0$ be the infimum of the set of 0-components of the elements of T. We make the following induction hypothesis: Let τ be an ordinal $< \lambda$, for which already elements $a_\nu \in S_\nu$ have been defined for all $\nu < \tau$. Let T_τ be the set of all sequences of T that have $(a_\nu)_{\nu<\tau}$ as τ - segment. If $T_\tau = \emptyset$ we define $a_\tau := z_\tau$. If $T_\tau \neq \emptyset$ we define a_τ to be the infimum of the set of τ - components of the elements of T_τ. Recursively we have so defined a sequence $a = (a_\nu)_{\nu<\lambda} \in X$. It will follow that $a = \inf T$. First there holds:

(1) a is a lower bound of T.

If there would exist a $t = (t_\nu)_{\nu<\lambda} \in T$ with $t < a$, let τ be the first index where the components t_τ and a_τ are different, i. e. where $t_\tau < a_\tau$ holds. Then t has $(a_\nu)_{\nu<\tau}$ as τ - segment. But then an element $a_\tau \leq t_\tau$ had to be defined, which gives a contradiction.

Also a is the greatest lower bound of T. For assume indirectly that there exists an element $b = (b_\nu)_{\nu<\lambda} \in X$ with $a < b \leq T$. Let σ be the first ordinal for which $a_\sigma < b_\sigma$. If there is no element in T which begins with $(a_\nu)_{\nu<\sigma}$, then by definition $a_\sigma = z_\sigma \geq b_\sigma$ would follow with contradiction. And if an element of T begins with $(a_\nu)_{\nu<\sigma}$, then a_σ is the infimum of the set of σ - components of the elements of T_σ, and b_σ must be $\leq$ all σ - components of the elements of T_σ, hence $b_\sigma \leq a_\sigma$ with contradiction.

The first element of X is $(e_\nu)_{\nu<\lambda}$ and the last $(z_\nu)_{\nu<\lambda}$. And so X is inf-complete and complete. If finally (A, B) is a proper cut in X, it is no gap because there exists $\inf B$, and this is the least element of B or the greatest of A.

Ordinals are linearly ordered sets without gaps, and so 2.10 yields the following statement:

2.11 Theorem. *Let μ be a successor ordinal and λ an arbitrary ordinal. Then $\mu((\lambda))$ has no gaps. If also λ is > 0, then $\mu((\lambda))$ has a first and a last element and is therefore a complete lattice.*

We supplement 2.11 by:

2.12 Theorem. *Let μ be a limit ordinal and $\lambda \leq \omega$. Then $\mu((\lambda))$ has no gaps. It is no complete lattice because it has no greatest element. But every non-empty subset has an infimum.*

Proof. If $\lambda < \omega$ holds, then $\mu((\lambda))$ is well-ordered according to 1.13. Assume now $\lambda = \omega$, and let T be a non-empty subset of $\mu((\omega))$.

We define elements $a_0, a_1, \ldots, a_\nu, \ldots$ for $\nu < \omega$ as in the proof of 2.10. Then evidently $(a_\nu)_{\nu<\omega} = \inf T$.

If now (A, B) is a proper cut in $\mu((\omega))$, then there exists $\inf B$, and it is the first element of B or the last element of A, so that (A, B) is no gap.

We cannot sharpen 2.12, so that also each non-empty subset T of $\mu((\omega))$ has a minimum. E.g. the set T of all sequences $\in \mu((\omega))$, in which exactly one component is $= 1$, and all others $= 0$, is a counter-example.

For the sets $\mu((\lambda))$ there is only one case left which is not yet settled by 2.11 and 2.12, namely:

2.13 Theorem. *Let μ be a limit ordinal and λ an ordinal $> \omega$. Then $\mu((\lambda))$ has gaps.*

Proof. Let $a = (a_\nu | \nu < \omega)$ be an ω - sequence of ordinals $a_\nu < \mu$. And let A be the set of all sequences $x = (x_\nu | \nu < \lambda)$, where $x_\nu = a_\nu$ for $\nu < \omega$, $x_\omega < \mu$, $x_\nu = 0$ for $\omega < \nu < \lambda$. Then A has no last element since μ is a limit ordinal. The element $m = (m_\nu | \nu < \lambda)$, where $m_\nu = a_\nu$ for $\nu < \omega, m_\nu = 0$ for $\omega \leq \nu < \lambda$, is $= \min A$.

Let now $b = (b_\nu | \nu < \lambda)$ be an upper bound of A, and thus $> x$ for all $x \in A$ since μ has no greatest element. Then we also have $b > m$. Let δ be the least ordinal with $b_\delta > m_\delta$. Now $\delta \geq \omega$ is impossible. So $\delta < \omega$ holds, and then b cannot be a least upper bound of A. The cut formed by the initial segment $IS(A)$ and the final segment of upper bounds of A is thus a gap.

Finally we investigate what can be said about the existence of neighboring elements in the sets $\mu((\nu))$.

2.14 Theorem. *Let μ and λ be ordinals > 0.*

1) If μ and λ are limit numbers, then $\mu((\lambda))$ is dense. There are no neighboring elements.

2) If μ is a limit number, λ a successor number $\rho + 1$, then there exist in $\mu((\lambda))$ neighboring elements. Two sequences $a = (a_\nu)_{\nu<\lambda}$ and $b = (b_\nu)_{\nu<\lambda}$ of $\mu((\lambda))$ with $a < b$ are neighboring, iff $a_\nu = b_\nu$ for all $\nu < \rho$, and for their last components a_ρ, b_ρ there holds $b_\rho = a_\rho + 1$.

3) If μ is a successor number $\rho + 1$, λ arbitrary, then there exist neighboring elements in $\mu((\lambda))$. Two elements $(a_\nu)_{\nu<\lambda} < (b_\nu)_{\nu<\lambda}$ of $\mu((\lambda))$ are neighboring iff for the first index τ where they differ there *holds: $b_\tau = a_\tau + 1$, and $a_\nu = \rho$, $b_\nu = 0$ for all ν with $\tau < \nu < \lambda$. (Such ordinals ν of course don't exist if $\lambda = \tau + 1$ holds.)*

Proof. 1) If $a = (a_\nu)_{\nu<\lambda} < (b_\nu)_{\nu<\lambda} = b$ holds for two elements $a, b \in \mu((\lambda))$, let τ be the index where a and b differ for the first time, so that $a_\tau < b_\tau$ holds. Every $z \in \mu((\lambda))$, that has $(a_\nu | \nu \leq \tau)$ as initial segment, and as $(\tau+1)$ - component the ordinal $a_{\tau+1} + 1$, is then strictly between a and b. And thus $\mu((\lambda))$ is dense.

2) Assume that $a < b$ are neighboring elements, and let τ be the index where a and b differ for the first time, hence $a_\tau < b_\tau$. If also $a_\tau + 1 < b_\tau$ would hold, then a and b would evidently not be neighboring. So we assume $a_\tau + 1 = b_\tau$. If now $\tau + 1 < \lambda$ would hold, then one could find as in 1) an element z strictly between a and b. Therefore we have $\tau + 1 = \lambda$ and thus $\tau = \rho$. So the condition of 2) is necessary. Clearly it is also sufficient.

3) is immediately clear.

Now we can give an answer to the question, for which ordinals μ, λ the set $\mu((\lambda))$ is continuously ordered, i. e. without steps and gaps:

2.15 Theorem. *$\mu((\lambda))$ is continuously ordered iff μ is a limit number and $\lambda = \omega_0$.*

Proof. $\mu((\lambda))$ can only be continuous, if it is dense, and then μ and λ have to be limit numbers due to 2.14. Because of 2.13, λ must be ω_0. So the condition is necessary, and 2.14 and 2.12 show that it is also sufficient.

4.3 Universally ordered sets and the sets $\mathbf{H}_\alpha$ of normal type η_α

In this section we study $\aleph_\alpha$ - universally ordered sets and in particular the sets of type h_α (which is the same as Hausdorff's normal type η_α). These sets have many nice properties, and they are important to bring a good oversight into the system of linearly ordered sets.

3.1 Definition. A linearly ordered set $(S, \leq)$ is said to be $\aleph_\alpha$ - *universal*, if there holds: Every linearly ordered set M of cardinality $\leq \aleph_\alpha$ is isomorphic to a subset of S (which is equipped with the induced order), the same: M is embeddable in S.

The statement of the following example can easily be verified, but it follows also later in a more general context:

3.2 Example. The set $\mathbf{Q}$ of rational numbers with their usual order is an $\aleph_0$ - universal set of cardinality $\aleph_0$.

Of course it is an easy thing to construct a linearly ordered set which is $\aleph_\alpha$ - universal, if α is a given ordinal. But if we seek such a set which has the minimal possible cardinality the situation changes. Using the GCH we shall see that there always exists an $\aleph_\alpha$ - universal set of cardinality $\aleph_\alpha$. For the theory of universally ordered sets the following sets H_α are of fundamental importance. They were introduced by Sierpinski in 1949 ([164]) subsequently to previous work of Hausdorff ([81], Chapter VI):

3.3 Definition. A *dyadic sequence* is a sequence $(a_\nu | \nu < \mu)$, where μ is an ordinal, and where each a_ν is 0 or 1. Let α be an ordinal. Then we define H_α to be the set of all dyadic sequences $\varphi = (\varphi_\nu | \nu < \omega_\alpha) \in 2((\omega_\alpha))$ that have a last digit 1. So for every $\varphi \in H_\alpha$ there exists an ordinal (which depends on φ) $\mu < \omega_\alpha$ with $\varphi_\mu = 1$ and $\varphi_\nu = 0$ for $\mu < \nu < \omega_\alpha$. We always consider H_α as equipped with the lexicographic order of first differences.

It can easily be verified that H_0 is isomorphic to the set $\mathbf{Q}$ with its usual order, and so H_0 is $\aleph_0$ - universal. This makes the conjecture plausible that generally H_α is $\aleph_\alpha$ - universal. This is indeed the case. It was proved by Sierpinski [164] for regular cardinals $\aleph_\alpha$ and by Gillman [53] also for singular cardinals $\aleph_\alpha$. In 1958 Mendelson [118] gave a very short and elegant proof which we now present:

3.4 Theorem. *For every ordinal α the set H_α is $\aleph_\alpha$ - universal.*

Proof. Let $(S, \leq)$ be a linearly ordered set with $|S| = \aleph_\alpha$. Then we can represent S in the form $S = \{u_\nu | \nu < \omega_\alpha\}$, where for $\nu < \mu < \omega_\alpha$ we have $u_\nu \neq u_\mu$.

Now we construct a $<$ - preserving mapping $f : S \longrightarrow H_\alpha$ recursively. The image sequence $f(u_\mu)$ of u_μ shall be denoted by $a_\mu = (a_{\mu\nu} | \nu < \omega_\alpha)$. Let $f(u_0)$ be the sequence $(1, 0, 0, \ldots)$ of length ω_α, which has the digit 1 at the first place and at the other components only zeros. Let now μ be an ordinal $< \omega_\alpha$, and for all $\nu < \mu$ let $f(u_\nu)$ already be defined. Then $f(u_\mu)$ is defined as the sequence $a_\mu = (a_{\mu\nu} | \nu < \omega_\alpha)$, for which $a_{\mu\mu} = 1$, $a_{\mu\nu} = 0$ for $\mu < \nu < \omega_\alpha$, and which for $\nu < \mu$ satisfies

(1) $a_{\mu\nu} = 0 \iff u_\mu < u_\nu$, and (then of course)

(1') $a_{\mu\nu} = 1 \iff u_\nu < u_\mu$.

Evidently the a_μ, $\mu < \omega_\alpha$, are elements of H_α. We now verify, that f maps the set S isomorphically onto the set $A := \{a_\mu | \mu < \omega_\alpha\}$.

Let $\nu < \mu < \omega_\alpha$. Then we distinguish two cases:

Case 1. $u_\nu < u_\mu$.

Here we have $a_{\mu\nu} = 1$ because of (1'), and further

(2) If $a_{\nu\kappa} = 1$ holds for a $\kappa < \nu$, then also $a_{\mu\kappa} = 1$.

Indeed, from the assumption and (1') there now follows $u_\kappa < u_\nu$ ($< u_\mu$), and then again with (1') $a_{\mu\kappa} = 1$.

Then (2) has as a consequence, that at the place, where a_ν and a_μ differ for the first time, the corresponding component of a_ν must be 0 and that of a_μ must be 1, so that $a_\nu < a_\mu$ follows.

Case 2. $u_\mu < u_\nu$.

Then (1) yields $a_{\mu\nu} = 0 < a_{\nu\nu} = 1$. If further we have $a_{\mu\kappa} = 1$ for a $\kappa < \nu$, then (1') yields $u_\kappa < u_\mu$ ($< u_\nu$), hence $u_\kappa < u_\nu$, and again because of (1') this results in $a_{\nu\kappa} = 1$. Because of $a_{\mu\nu} = 0 < 1 = a_{\nu\nu}$ the least ordinal κ for which $a_{\mu\kappa} \neq a_{\nu\kappa}$ holds is $\leq \nu$, and then we have $a_{\mu\kappa} < a_{\nu\kappa}$. So finally $a_\mu < a_\nu$ follows.

Next we wish to determine the cardinality of the sets H_α. But before we introduce the cardinals $\mathfrak{k}_\alpha$, which often occur in set theory (partly in other notation):

3.5 Definition. Let α be an ordinal. We put $\mathfrak{k}_0 := \aleph_0$. For successor numbers $\alpha + 1$ we put $\mathfrak{k}_{\alpha+1} := 2^{\aleph_\alpha}$. And if α is a limit ordinal we define $\mathfrak{k}_\alpha := \sum_{\nu < \alpha} 2^{\aleph_\nu}$. Then we have

3.6 Theorem. 1) *For every ordinal α there holds $\aleph_\alpha \leq \mathfrak{k}_\alpha$.*

2) *If we assume the GCH, there holds $\aleph_\alpha = \mathfrak{k}_\alpha$ for all ordinals α.*

3) *Also for the non-limit-ordinals $\alpha > 0$ we have $\mathfrak{k}_\alpha = \sum_{\nu < \alpha} 2^{\aleph_\nu}$.*

Proof. 1) For $\alpha = 0$, 1) holds by definition. Let now α be an ordinal > 0, and 1) already be proved for all ordinals $\nu < \alpha$. If α is no limit ordinal and thus of the form $\alpha = \beta + 1$, then there follows:

(1) $\aleph_\alpha = \aleph_{\beta+1} \leq 2^{\aleph_\beta} = \mathfrak{k}_{\beta+1} = \mathfrak{k}_\alpha$.

If α is a limit number, then we have:

(2) $\aleph_\alpha = \sum_{\nu < \alpha} \aleph_\nu \leq \sum_{\nu < \alpha} \aleph_{\nu+1} \leq \sum_{\nu < \alpha} 2^{\aleph_\nu} = \mathfrak{k}_\alpha$.

So 1) is also valid for α, and then by transfinite induction it holds in general.

2) Applying the GCH the signs $\leq$ in (1) and (2) can be sharpened to $=$, and 2) follows.

3) Let $\alpha = \beta+1$. Then $\mathfrak{k}_\alpha = \mathfrak{k}_{\beta+1} = 2^{\aleph_\beta} = \sum_{\nu \leq \beta} 2^{\aleph_\nu} \leq |\beta+1| \cdot 2^{\aleph_\beta} = 2^{\aleph_\beta}$.

3.7 Theorem. *For every ordinal α we have $|H_\alpha| = \mathfrak{k}_\alpha$. Using the* GCH *there follows $|H_\alpha| = \aleph_\alpha$.*

Proof. First we define: If H is a set of (transfinite) sequences $s = (s_\nu)_{\nu < \lambda}$ of zeros and ones, where λ is an ordinal and where κ is an ordinal $\leq \lambda$, we define $(H|\kappa)$ to be the set of all those sequences of H, in which all digits 1 occur at positions $< \kappa$.

Then we have $H_0 = \cup_{n \in \mathbb{N}}(H_0|n)$. Then $|(H_0|n)| \leq 2^n$ and $|H_0| \leq \sum_{n \in \mathbb{N}} 2^n = \aleph_0$. Since also $|H_0| \geq \aleph_0$ holds the statement is valid for $\alpha = 0$.

Let now α be an ordinal > 0 and the statement already be proved for all ordinals $\beta < \alpha$. Then $H_\alpha = \cup\{(H_\alpha|\nu)|\nu < \omega_\alpha\}$.

If α is a successor number $\beta + 1$, then for every $\nu < \omega_{\beta+1}$ we have $|(H_\alpha|\nu)| \leq 2^{|\nu|} \leq 2^{\aleph_\beta}$, hence $|H_{\beta+1}| \leq \aleph_{\beta+1} \cdot 2^{\aleph_\beta} = 2^{\aleph_\beta} = \mathfrak{k}_{\beta+1}$. Also $|H_{\beta+1}| \geq \mathfrak{k}_{\beta+1}$ holds because $H_{\beta+1}$ has a subset which is equipotent to $2((\omega_\beta))$, where the last set has cardinality $2^{\aleph_\beta}$.

Let now α be a limit ordinal. Then $H_\alpha = \cup_{\nu < \alpha}(H|\omega_\nu)$. So there follows $|H_\alpha| \leq \sum_{\nu < \alpha} |H_\nu| = \sum_{\nu < \alpha} \mathfrak{k}_\nu = \mathfrak{k}_\alpha$, further $|H_\alpha| \geq |H_\nu| = \mathfrak{k}_\nu$ for all $\nu < \alpha$, and then also $|H_\alpha| \geq \mathfrak{k}_\alpha$ $(= \sum_{\nu < \alpha} \mathfrak{k}_\nu)$.

In the rest of this section α always denotes an ordinal and $\omega_\gamma = \mathrm{cf}(\omega_\alpha)$.

Now we shall prove that H_α is an η_γ - set, and thus an η_α - set if ω_α is regular. First we show:

3.8 Theorem. *Let a be an element of H_α. Then $(H_\alpha < a)$ has a cofinal subset of type ω_α, and $(H_\alpha > a)$ has a coinitial subset of type ω_α^*.*

Thus every element of H_α has the element-character $(\omega_\gamma, \omega_\gamma^*)$. And so H_α has only symmetric elements.

Proof. Let $a := (a_\nu)_{\nu < \omega_\alpha}$. There is a τ with $a_\tau = 1$, and $a_\nu = 0$ for $\tau < \nu < \omega_\alpha$. For the ordinals μ with $\tau < \mu < \omega_\alpha$ let f_μ be that sequence of H_α, which has $(a_\nu)_{\nu < \tau}$ as initial segment, and which satisfies $f_\mu(\tau) = 0$, $f_\mu(\nu) = 1$ for $\tau < \nu \leq \mu$, and $f_\mu(\nu) = 0$ for $\mu < \nu < \omega_\alpha$. The set $\{f_\mu|\tau < \mu < \omega_\alpha\}$ then has the order type ω_α and is cofinal in $(H_\alpha < a)$.

On the other hand we define for the ordinals μ with $\tau < \mu < \omega_\alpha$ a sequence $g_\mu \in H_\alpha$ as follows: g_μ has $(a_\nu)_{\nu \leq \tau}$ as initial segment, $g_\mu(\mu) = 1$ and $g_\mu(\nu) = 0$ for the ordinals $\nu \neq \mu$ that satisfy $\tau < \nu < \omega_\alpha$. Then $\{g_\mu | \tau < \mu < \omega_\alpha\}$ has the order type ω_α^* and is coinitial in $(H_\alpha > a)$.

For the gaps of H_α we have only a "one-sided" analog to 3.8:

3.9 Theorem. *To every gap (A, B) of H_α there exists a subset of type ω_α which is cofinal in A, or a subset of type ω_α^* which is coinitial in B.*

Every gap of H_α has a character $(\omega_\gamma, \omega_\beta^*)$ or $(\omega_\beta, \omega_\gamma^*)$, where ω_β is a regular ordinal $\leq \omega_\alpha$.

Proof. Let (A, B) be a gap of H_α. We consider A as a subset of $2((\omega_\alpha))$. Then according to 2.11, A has a supremum $s = (s_\nu)_{\nu < \omega_\alpha}$ in $2((\omega_\alpha))$. This cannot belong to A or to B, because in the first case s would be the greatest element of A, in the second the least element of B. Both is impossible because (A, B) is a gap. Thus we have $s \in 2((\omega_\alpha)) \setminus H_\alpha$. Of course s has $\aleph_\alpha$ zeros or $\aleph_\alpha$ ones, precisely:

There exists a subset $T \subseteq \omega_\alpha$ of type ω_α, so that $s_\nu = 0$ for $\nu \in T$, or there exists a subset $U \subseteq \omega_\alpha$ of type ω_α, so that $s_\nu = 1$ for $\nu \in U$. In the first case we define for every $\tau \in T$ a sequence $f_\tau \in H_\alpha$ by:

(*) f_τ has $(s_\nu)_{\nu < \tau}$ as initial segment, $f_\tau(\tau) = 1$, and $f_\tau(\nu) = 0$ for $\tau < \nu < \omega_\alpha$.

Then $\{f_\tau | \tau \in T\}$ is a subset of H_α of type ω_α^* which is coinitial in $B = (H_\alpha > s)$.

In the second case we also define $f_\tau \in H_\alpha$ by (*) for the ordinals $\tau \in U$. Now $\{f_\tau | \tau \in U\}$ is a subset of type ω_α which is cofinal in $A = (H_\alpha < s)$.

The rest now follows from the fact that $2((\omega_\alpha))$, and then a fortiori H_α, has by 2.6 no subset of type $\omega_{\alpha+1}$ and no subset of type $\omega_{\alpha+1}^*$.

In addition to 3.9 there still follows in the same manner as before, that H_α has a coinitial subset of type ω_α^* and a cofinal subset of type ω_α. So finally we can resume:

3.10 Theorem. *H_α is an η_γ - set. If ω_α is regular, then H_α is an η_α - set.*

3.11 Definition. We denote the order type of H_α by h_α. If ω_α is regular, we shall see in the following that h_α is nothing else but what Hausdorff called the *normal type η_α*.

The types h_α have many features in common with the order type of rational numbers, because they are the generalizations of this to higher cardinalities. Now we establish a characterization theorem for the types h_α with regular ω_α, which has no connection to the special realizing sets H_α.

3.12 Theorem [64] **(Characterization Theorem).** *Let ω_α be regular and S a linearly ordered set. Then S has the order-type h_α exactly if the following two conditions are satisfied:*

1) S is an η_α - set.

2) S is a union of $\aleph_\alpha$ many subsets S_ν, $\nu < \omega_\alpha$, where every S_ν is free from ω_α and ω_α^*.

Proof. H_α is an η_α - set, and we have $H_\alpha = \cup\{T_\mu | \mu < \omega_\alpha\}$, where T_μ is the set of all sequences $(a_\nu | \nu < \omega_\alpha\}$ of H_α, in which the last digit 1 is already in the segment $(a_\nu)_{\nu<\mu}$. Then we have tp $T_\mu \leq$ tp $2((\mu))$, and thus T_μ has no subset of type ω_α and no subset of type ω_α^* because of 2.6, and the "only if" part is proved.

Let now S be a linearly ordered set which has the above properties 1) and 2). Then we have to construct an isomorphic mapping of S onto H_α. We do this by applying a method of alternating mappings. First S_0 is embeddable in H_α by 3.3.7 by a mapping $f_0 : S_0 \longrightarrow f[S_0] \subseteq H_\alpha$. According to 3.3.6 then the inverse mapping $f_0^{-1} : f[S_0] \longrightarrow S_0$ can be extended to an isomorphic mapping of $T_0 \cup f[S_0]$ onto a subset of S. And then we also have an isomorphic mapping $F_0 : A_0 \longrightarrow B_0$, where A_0 is a subset of S which contains S_0, and where B_0 is a subset of H_α which contains T_0, and where A_0 and B_0 have no subset of type ω_α and no subset of type ω_α^*.

Now we make the following induction hypothesis: Let τ be an ordinal $< \omega_\alpha$, and let for every $\nu < \tau$ an isomorphic mapping $F_\nu : A_\nu \longrightarrow B_\nu$ be given, where for $\mu < \nu < \tau$ the mapping F_ν extends F_μ such that there holds:

$(*)$ $A_\nu \supseteq S_\nu$ and $B_\nu \supseteq T_\nu$, where all A_ν and all B_ν with $\nu < \omega_\alpha$ are ω_α - free and ω_α^* - free.

If τ is a successor number $\sigma + 1$, then one can construct, in analogy to the construction of F_0, an isomorphic mapping $F_\tau : A_\tau \longrightarrow B_\tau$, which satisfies $(*)$ for $\nu = \tau$ and which extends F_σ.

If τ is a limit ordinal, let f_τ be the limit-mapping of the F_ν, $\nu < \tau$. It maps $\cup_{\nu<\tau} A_\nu$ isomorphically onto $\cup_{\nu<\tau} B_\nu$. These two sets are free from ω_α and ω_α^* because of 3.3.8, and so one can extend f_τ by a twofold application of 3.3.6 to an isomorphic mapping $F_\tau : A_\tau \longrightarrow B_\tau$, where the following holds:

$$S \supseteq A_\tau \supseteq \cup\{A_\nu | \nu < \tau\} \cup S_\tau \text{ and } H_\alpha \supseteq B_\tau \supseteq \cup\{B_\nu | \nu < \tau\} \cup T_\tau,$$

where A_τ and B_τ are free from ω_α and ω_α^*.

Then for $\mu < \nu < \omega_\alpha$ the mapping F_ν extends F_μ, and the limit mapping of the F_ν, $\nu < \omega_\alpha$, is an isomorphic mapping of S onto H_α.

The characterization theorem has a lot of applications. There holds the following minimal property, whose substance is due to Hausdorff [81], p. 183:

3.13 Theorem. *Every η_α - set contains a subset of type h_α.*

Proof. Let E be an η_α - set. If ω_α is regular, then H_α is a union of $\aleph_\alpha$ many subsets which are free from ω_α and from ω_α^*, and thus H_α is by 3.3.9 embeddable in E. So H_α is isomorphic to a subset $T \subseteq E$ of type h_α.

If ω_α is singular, then E is also an $\eta_{\alpha+1}$ - set because of 3.3.5. Then it has a subset of type $h_{\alpha+1}$ according to the already proved regular case. This yields a fortiori our statement.

3.14 Corollary. *An η_α - set has a cardinality $\geq \mathfrak{k}_\alpha$. If here ω_α is singular, then it has a cardinality $\geq \mathfrak{k}_{\alpha+1}$.*

For a regular ordinal ω_α there exists an η_α - set of cardinality $\aleph_\alpha$ iff $\mathfrak{k}_\alpha = \aleph_\alpha$.

Hausdorff ([81], p.182) proved:

3.15 Theorem. *If two η_α - sets both have the cardinality $\aleph_\alpha$, then they are isomorphic. (Indeed, each of them is isomorphic to H_α.)*

Proof. Let E be an η_α - set of cardinality $\aleph_\alpha$. Then E is a union of $\aleph_\alpha$ one-element sets, - which of course are free from ω_α and ω_α^*. Then E is isomorphic to H_α because of 3.12.

3.16 Remark. 3.15 contains the following statement: If $\mathfrak{k}_\alpha = \aleph_\alpha$ holds, then every two η_α - sets of cardinality $\mathfrak{k}_\alpha$ are isomorphic.

Here the assumption that $\mathfrak{k}_\alpha = \aleph_\alpha$ holds, is not superfluous: Gillman [52] proved: If $2^{\aleph_\alpha} \neq \aleph_{\alpha+1}$ holds, then there exist two non-isomorphic $\eta_{\alpha+1}$ - sets of cardinality $2^{\aleph_\alpha}$.

And further: If for ordinals δ, β we have $\delta \geq \beta + 2$ and $\aleph_\delta \geq 2^{\aleph_\beta}$, then there exist two non-isomorphic $\eta_{\beta+1}$ - sets of cardinality $\aleph_\delta$.

104

Concerning the existence of η_α - sets of cardinality $\aleph_\alpha$, Gillman [52] proved: If α is a limit number, then there exists an η_α - set of cardinality $\aleph_\alpha$ iff $\aleph_\alpha$ is regular and $2^{\aleph_\beta} \leq \aleph_\alpha$ is valid for every $\beta < \alpha$. The latter holds for the so-called *strongly inaccessable* numbers.

The characterization theorem gives very short proofs in some cases:

3.17 Theorem. *If ω_α is regular and E a linearly ordered set of type h_α, and if T is a subset of E, which is again an η_α - set, then T has the order type h_α.*

Proof. T satisfies the conditions of 3.12 a fortiori because H_α does this.

The last theorem immediately yields the next one, which is related to a result of Padmavally [135], p. 67:

3.18 Theorem. *If ω_α is regular and E a linearly ordered set of type h_α, and if $D \subseteq E$ is dense in E, then D has also the type h_α.*

Proof. According to 3.3.10 D is an η_α - set, and so 3.17 proves the statement.

3.19 Theorem. *Let ω_α be a regular ordinal. Then every open interval (a,b) of H_α has again the order type h_α. And for every $a \in H_\alpha$ the sets $(H_\alpha < a)$ and $(H_\alpha > a)$ have the order type h_α.*

Proof. Let $a < b$ be elements of H_α. Then the open interval (a,b) of H_α and the sets $(H_\alpha < a)$ and $(H_\alpha > a)$ are $\eta_\alpha-$ sets by 3.3.11 and the union of $\aleph_\alpha$ many subsets which are free from ω_α and ω_α^*. Thus 3.12 proves our assertion.

In 4.6.3′ the first part of the last theorem is generalized to arbitrary initial ordinals ω_α. Also the next theorem will be generalized later to arbitrary ω_α in 4.4.5.

3.20 Theorem. *For regular ω_α there holds $h_\alpha \cdot h_\alpha = h_\alpha$.*

Proof. We realize the type $h_\alpha \cdot h_\alpha$ by the lexicographic product $H_\alpha \times H_\alpha$. We prove first that the latter is an η_α - set. Let A and B be non-empty subsets of $H_\alpha \times H_\alpha$ with $|A| < \aleph_\alpha > |B|$, which satisfy $a < b$ for all $a \in A$ and all $b \in B$. We assume indirectly that A and B are neighboring.

Case 1. There exists an element $z \in H_\alpha$ such that A and B contain elements which have z as first component. We put $A' := \{x \in$

$H_\alpha | (z, x) \in A\}$ and $B' := \{y \in H_\alpha | (z, y) \in B\}$. Then A' and B' are neighboring in H_α because A and B are neighboring in $H_\alpha \times H_\alpha$. But this and $|A'| \leq |A| < \aleph_\alpha > |B| \geq |B'|$ contradict the fact that H_α is an η_α- set.

Case 2. There is no element z which satisfies the assumption of Case 1. Then we put $A'' := \{a \in H_\alpha | \exists x \in H_\alpha \text{ with } (a, x) \in A\}$ and $B'' := \{b \in H_\alpha | \exists x \in H_\alpha \text{ with } (b, x) \in B\}$. By assumption we have $A'' \cap B'' = \emptyset$ and then $a < b$ for all $a \in A''$ and all $b \in B''$. Now A'' and B'' must be neighboring in H_α since A and B are neighboring in $H_\alpha \times H_\alpha$. This leads to a contradiction as in Case 1.

If C is cofinal (resp. coinitial) in $H_\alpha \times H_\alpha$ then the set of its first components is cofinal (resp. coinitial) in H_α and thus of a cardinality $\geq \aleph_\alpha$. The latter then also holds for the set C.

We now define the sets which were introduced by Hausdorff as carrier sets for his normal types η_α :

3.21 Definition. $3((\omega_\alpha))$ is by 1.9 the set of all transfinite sequences of length ω_α of digits 0,1,2. We order it lexicographically. Let then H'_α be the subset of those sequences $(a_\nu | \nu < \omega_\alpha) \in 3((\omega_\alpha))$ for which there exists an ordinal $\mu < \omega_\alpha$ such that $a_\nu = 1$ for all ν with $\mu \leq \nu < \omega_\alpha$.

Then there holds:

3.22 Theorem. *For regular ω_α the set H'_α has the order type h_α.*

Proof. Quite analogously to the proof of 3.10 it follows that H'_α is an η_α - set. And it is a union of $\aleph_\alpha$ subsets which are free from ω_α and ω_α^*; this follows from 2.3.

With the last theorem we immediately see that h_α is a symmetric order type (see 2.8):

3.23 Theorem. *For every ordinal α the lexicographically ordered set H'_α is symmetric.*

Proof. We define a bijective mapping $f : H'_\alpha \longrightarrow H'_\alpha$ by: If $(a_\nu | \nu < \omega_\alpha)$ is an element of H'_α we put $f(a_\nu | \nu < \omega_\alpha) := (2 - a_\nu | \nu < \omega_\alpha)$. This mapping interchanges 0 with 2, and we have for $x, y \in H'_\alpha$: $x \leq y \iff f(x) \geq f(y)$.

The next theorem is related to a result of Padmavally [134], p.260. In 4.6 it will be generalized on arbitrary ω_α :

3.24 Theorem. *Let ω_α be a regular initial number, S a linearly ordered set, and $S = \cup\{S_\nu | \nu < \omega_\alpha\}$, where every S_ν is embeddable in H_α. Then also S is embeddable in H_α.*

Proof. By 3.12 every S_ν is a fortiori representable in the form $S_\nu = \cup\{S_{\nu\mu} | \mu < \omega_\alpha\}$, where each $S_{\nu\mu}$ is free from ω_α and ω_α^* .Then $S = \cup_{\nu<\omega_\alpha}(\cup_{\mu<\omega_\alpha}S_{\nu\mu}) = \cup_{\tau<\omega_\alpha}(\cup\{S_{\nu\mu} | \nu < \tau, \mu < \tau\}$. Here every summand $\cup S_{\nu\mu}$ is a union of $|\tau|^2 < \aleph_\alpha$ sets which are free from ω_α and ω_α^*. Therefore itself is also free from ω_α and ω_α^* by 3.3.8. According to 3.3.9 then S is embeddable in every η_α - set and thus also in H_α.

In 1.9.6 we had introduced the relations $\leq$ and $<$ for order types. The next two theorems describe properties of the types h_α.

3.25 Theorem. *Let E be an η_α - set, where ω_α is regular. Then there holds $h_\alpha = \mathrm{tp}(E)$ or $h_\alpha < \mathrm{tp}(E)$.*

Proof. If E is isomorphic to a subset T of H_α, then according to 3.17, T and with it E, has the type h_α.

If E is not isomorphic to a subset of H_α we obtain, because of $h_\alpha \leq \mathrm{tp}\, E$, $h_\alpha < \mathrm{tp}(E)$.

3.26 Theorem. *Let ω_α be a regular ordinal, $T \subseteq H_\alpha$ and $\mathrm{tp}(T) < h_\alpha$. Then $\mathrm{tp}(H_\alpha \setminus T) = h_\alpha$.*

Proof. No interval of H_α is a subset of T because of 3.19. Then $H_\alpha \setminus T$ is dense in H_α, and therefore it has the type h_α because of 3.18.

In 3.1.17 we had introduced the final order in a cartesian product of linearly ordered sets. We now present an application of this concept for sequences of real numbers. If $(a_n)_{n\in\mathbf{N}}$ and $(b_n)_{n\in\mathbf{N}}$ are converging sequences of real numbers with limits a resp. b, then we have the following elementary fact:

If $a < b$ holds, then there exists an index $n_0 \in \mathbf{N}$ such that $a_n < b_n$ holds for all $n \geq n_0$.

This simple observation gives rise to the following definition:

3.27 Definition. Let S be the set of all sequences $(a_n)_{n\in\mathbf{N}}$ of real numbers. For (a_n) and $(b_n) \in S$ we put $(a_n) < (b_n) \iff \exists n_0 \in \mathbf{N}$ such that $a_n < b_n$ for all $n \geq n_0$.

Then $<$ is a transitive relation, and if we put $(a_n) \leq (b_n) \iff (a_n) < (b_n)$ or $(a_n) = (b_n)$, then $\leq$ is an order in S, which of course is not linear.

In this context Hausdorff ([81], p.191) proved:

3.28 Theorem. *Let A be a countable subset of S. Then there exists an element in S which is $<$ all elements of A. And symmetrically: There exists an element in S which is $>$ all elements of A.*

Let A, B be countable subsets of S, such that $a < b$ holds for all $a \in A$, $b \in B$, then there exists an element $x \in S$ with $a < x < b$ for all $a \in A$, $b \in B$.

Proof. Let $A = \{a_n | n \in \mathbf{N}\}$ be a countable subset of S, where $a_n = (a_{ni})_{i \in \mathbf{N}}$ for $n \in \mathbf{N}$. We choose $x_1 < a_{11}$, $x_2 < \min\{a_{12}, a_{22}\}, \ldots,$ $x_n < \min\{a_{1n}, \ldots, a_{nn}\}, \ldots$. Then $x := (x_n)_{n \in \mathbf{N}}$ is evidently $< a$ for every $a \in A$.

Let now $B := \{b_n | n \in \mathbf{N}\}$ be another countable subset of S such that $a < b$ holds for all $a \in A$, $b \in B$, and $b_n = (b_{ni})_{i \in \mathbf{N}}$ for $n \in \mathbf{N}$. We say:

a_k is less than b_l *from i on,* if $a_{k\nu} < b_{l\nu}$ for all $\nu \geq i$.

Now for every pair $k, l \in \mathbf{N}$ there exists a least number $i(k,l)$ such that a_k is less than b_l from $i(k,l)$ on. For every $n \in \mathbf{N}$ we put

$i_n := \max\{i(k,l) | k, l \leq n\}$.

Then we have $a_k < b_l$ from i_n on for all $k, l \leq n$. Then there also exists a strictly increasing sequence $\lambda_1 < \lambda_2 < \cdots < \lambda_n < \cdots$ of natural numbers such that for every $n \in \mathbf{N}$ there holds:

$a_k < b_l$ from λ_n on for all $k, l \leq n$.

Every $j \in \mathbf{N}$ that is $\geq \lambda_1$ satisfies $\lambda_n \leq j < \lambda_{n+1}$ for   exactly one $n \in \mathbf{N}$. Then we choose an element $x_j \in \mathbf{R}$ satisfying $\max\{a_{kj} | k \leq n\} < x_j < \min\{b_{kj} | k \leq n\}$.

Let x be the sequence $(x_j)_{j \in \mathbf{N}}$. Then we have $a < x < b$ for all $a \in A$ and $b \in B$.

Now Hausdorff's observation ([81],p.192) follows:

3.29 Theorem. *Let M be a maximal chain of the set S of all sequences of real numbers, equipped with the final order. Then M is an η_1 - set.*

M has cardinality $2^{\aleph_0} = |\mathbf{R}|$. If the Continuum Hypothesis is assumed, then M has the order type h_1 (= normal type η_1).

Proof. The first statement follows immediately from 3.28. And then by 3.13 we have $\aleph_1 \leq |M| \leq |\mathbf{R}|^{\aleph_0} = 2^{\aleph_0} = \aleph_1$. If CH is assumed, then $\aleph_1 = \aleph_1$ holds, and the rest follows from 3.15.

Hausdorff ([81], p.193) had mentioned that several other order relations for sets of sequences of real numbers can be defined (and were inves-

108

tigated in the literature), which have some resemblance with the notion which we have treated. In particular the following order for sequences with non-vanishing elements was studied: $(a_n)_{n\in\mathbf{N}} < (b_n)_{n\in\mathbf{N}} \iff \lim \frac{a_n}{b_n} = 0$. The variations of Definition 3.27 have a similar behaviour as the order which we studied, and so we don't treat them in more detail.

4.4 Generalizations to the case of a singular ω_α

In the last section we established several theorems for the sets H_α (and types h_α) where ω_α was regular. We now generalize most of them to the case where ω_α is singular, which implies that α is a limit ordinal.

4.1 Lemma. *Let S be a linearly ordered set of type h_α, where ω_α is regular. Let further $S^* \subseteq S$ be a set which is free from ω_α and from ω_α^*. Let (A, B) be a cut (see 1.10.15) in the set S^* (where S^* is equipped with the order induced from S). Then the set T of all $x \in S$, that satisfy $A < x < B$, has order type h_α.*

Proof. For the set A there holds: A is empty, or A has a last element, or A has a cofinal subset whose type is an ordinal $< \omega_\alpha$. For B the corresponding properties hold. In every possible case it follows that A and B are not neighboring in the η_α - set S. Thus T is not empty. Now A and T are neighboring in S, and therefore T must be coinitial with ω_α^*. Analogously T is cofinal with ω_α. Since T is also a segment of S, it follows easily that T is an η_α - set, and then it has the type h_α because of 3.17.

4.2 Definition. If M is a set of transfinite sequences $(s_\nu)_{\nu<\tau}$ of length τ where the components s_ν are 0 or 1, then for $\mu < \tau$ we define the subset $M[\mu]$ to be the set of all those $s \in M$ that have a last digit 1, and this at an s_ν with $\nu < \mu$.

Now we can supplement our characterization theorem 3.12 by

4.3 Theorem. *Let ω_α be a singular initial ordinal. Then a linearly ordered set S has the order type h_α iff S is representable in the form $S = \cup_{\nu<\alpha} S_{\nu+1}$, where $S_{\nu+1}$ has the order type $h_{\nu+1}$, and where there holds $S_{\mu+1} \subseteq S_{\nu+1}$ for $\mu < \nu < \alpha$.*

Proof. α is a limit ordinal, because ω_α is singular, and then we have $H_\alpha = \cup_{\nu<\alpha} H_\alpha[\omega_{\nu+1}]$. Here $H_\alpha[\omega_{\nu+1}]$ has the type $h_{\nu+1}$, and the summands increase with ν. The above condition is thus necessary.

Let now S satisfy the given condition. Then we first define $S_\lambda := \cup_{\nu<\lambda} S_{\nu+1}$ for the limit ordinals λ which are $< \alpha$. Then there exists an isomorphic mapping of S_1 onto $H_\alpha[\omega_1]$ because both sets have the type h_1. Applying Lemma 4.1 this mapping can be extended to an isomorphic mapping of $S_2 \supseteq S_1$ onto $H_\alpha[\omega_2] \supseteq H_\alpha[\omega_1]$. Indeed, the sets S_1 and $H_\alpha[\omega_1]$ are free from ω_2 and ω_2^*, and S_2 and $H_\alpha[\omega_2]$ are sets of type h_2.

If in general τ is an ordinal with $1 < \tau < \alpha$ and if for every ν with $0 < \nu < \tau$ an isomorphic mapping $\varphi_\nu : S_\nu \longrightarrow H_\alpha[\omega_\nu]$ is already defined such that for $\nu > \mu > 0$ the mapping φ_ν extends the mapping φ_μ, we define φ_τ inductively as follows: If τ is a limit ordinal then φ_τ shall be the limit of the mappings φ_ν, $0 < \nu < \tau$. Then φ_τ is an isomorphic mapping of S_τ onto $H_\alpha[\omega_\tau]$. If τ is a successor number, then again using Lemma 4.1 we extend the mapping $\varphi_{\tau-1}$ to an isomorphic mapping of S_τ onto $H_\alpha[\omega_\tau]$. By transfinite induction we so obtain an isomorphic mapping of S onto H_α as the limit of the mappings $\varphi_\nu, \nu < \alpha$.

For later use we establish the following lemma, which can be proved in a very similar way as 4.3 using 4.1:

4.4 Lemma. *If ω_α is singular, $S = \cup_{\nu<\alpha} S_\nu$ a linearly ordered set, where every summand S_ν, $\nu < \alpha$, is free from $\omega_{\nu+1}$ and $\omega_{\nu+1}^*$, then S is embeddable in H_α.*

We can now generalize 3.20 to the following

4.5 Theorem. $h_\alpha \cdot h_\alpha = h_\alpha$ *for every ordinal α.*

Proof. Let ω_α be singular, the regular case was already settled in 3.20. For every ordinal $\nu \leq \alpha$ let P_ν be the set of all pairs of elements of $H_\alpha[\omega_\nu]$ (see Definition 4.2!), ordered according to the principle of first differences. Then $P_\alpha = \cup_{\nu<\alpha} P_{\nu+1}$, tp $P_\alpha = h_\alpha \cdot h_\alpha$, and $\mathrm{tp}(P_{\nu+1}) = h_{\nu+1} \cdot h_{\nu+1} = h_{\nu+1}$ because of 3.20. The sets $P_{\nu+1}$ increase with ν, and so Theorem 4.3 yields our statement.

Next we generalize Theorem 3.24:

4.6 Theorem. *Let $S = \cup\{S_\nu | \nu < \omega_\alpha\}$ be a linearly ordered set, where every $S_\nu, \nu < \omega_\alpha$, is embeddable in H_α. Then also S is embeddable in H_α.*

Proof. The case of a regular ω_α was already settled in 3.24. Let therefore ω_α be singular and thus α a limit ordinal. Then we have $\omega_\alpha = \cup_{\mu<\alpha}\omega_\mu$. Since S_ν is embeddable in $H_\alpha = \cup_{\mu<\alpha}H_\alpha[\omega_\mu]$, we have a representation $S_\nu = \cup_{\mu<\alpha}S_{\nu\mu}$, where $S_{\nu\mu}$ is embeddable in $H_\alpha[\omega_\mu]$ and thus free from $\omega_{\mu+1}$ and $\omega^*_{\mu+1}$. For $\kappa < \mu$ we put $F_\kappa := \cup\{S_{\nu\mu}|\nu < \kappa$ and $\mu < \omega_\kappa\}$. Then F_κ is a union of $\aleph_\kappa$ sets which are free from $\omega_{\kappa+1}$ and $\omega^*_{\kappa+1}$. Then also F_κ is free from $\omega_{\kappa+1}$ and $\omega^*_{\kappa+1}$. Indeed, a well- or inversely well-ordered subset of F_κ of cardinality $\aleph_{\kappa+1}$ would have an intersection of the same cardinality with one of the $\aleph_\kappa$ summands. Then finally $S = \cup_{\kappa<\alpha}F_\kappa$ is embeddable in H_α by Lemma 4.4.

The next theorem is a special case of Theorem 6.3, which we present later.

4.7 Theorem. *Let $a < b$ be elements of H_α. Then the open interval (a, b) has a subset of type h_α, and moreover a segment of type h_α.*

Proof. For regular ω_α this is already contained in 3.19. The set H_α is dense, and so there exists a $z \in H_\alpha$ with $a < z < b$. For $\nu < \omega_\alpha$ let a_ν, b_ν, z_ν be the ν - component of a, b, z respectively, and let δ be the first ordinal, for which the δ - components a_δ and b_δ differ, so that $a_\delta = 0$ and $b_\delta = 1$. If $z_\delta = 0$, then, due to $a < z$, there is a $\mu > \delta$ with $a_\mu = 0$ and $z_\mu = 1$. Then all sequences $(x_\nu|\nu < \omega_\alpha)$ of H_α, for which $x_\nu = z_\nu$ for $\nu \leq \mu$ holds, are in (a, b).

If $z(\delta) = 1$, there is a $\mu > \delta$ where $z_\mu = 0$ and $b_\mu = 1$. Then all sequences $(x_\nu|\nu < \omega_\alpha)$ of H_α, for which $x_\nu = z_\nu$ holds for $\nu \leq \mu$, are in (a, b). In both cases we have a dyadic sequence s of length $\mu + 1 < \omega_\alpha$, such that all prolongations of s to a sequence $\in H_\alpha$ are in (a, b). If now we take the set of all sequences that arise from s by adhering to s the sequences of H_α we obtain a subset of (a, b), which is isomorphic to H_α.

4.5 The method of succesively adjoining cuts

In this section we deal with a principle of how one can obtain from a given linear order type a greater one by "adjoining cuts". Dedekind had already applied a similar method in order to give an exact construction of the irrational numbers. The more general method which is treated here was applied by Cuesta Dutari [16], [17]. Subsequently the author rediscovered his method and some of his results and extended some of them. Later the method of successively adjoining cuts was also used by Conway [14] in the context of the construction of surreal numbers. For details see e.g. the book of Norman Alling [4].

Our construction principle yields interesting insights into the hierarchy of order types. In particular we obtain a new characterization of the order types h_α, namely one which does not make use of representations by lexicographic orderings.

We recall the former definition (1.10.15) of a cut (A, B) of a linearly ordered set $(T, \leq)$ and extend it:

5.1 Definition. Let $(T, \leq)$ be a linearly ordered set. A *cut* of T is a pair (A, B), where A is an initial segment of T and where $B = T \backslash A$. Here A and B, also T, are allowed to be empty ! We denote the set of all cuts of T by $S(T)$.

If T and $S(T)$ are disjoint we now introduce a linear order in the set $T \cup S(T)$ as follows : In this set the elements of T shall have the same order to each other as in $(T, \leq)$. If (A, B) and (C, D) are cuts of T, we put $(A, B) \leq (C, D) \iff A \subseteq C$. If $t \in T$ and $(A, B) \in S(T)$ then we put $t < A, B) \iff t \in A$, and $(A, B) < t \iff t \in B$. It can easily be verified that the relation $\leq$ which is so defined is a linear order. $(T, \leq)$ is an ordered subset of it.

The following is easily seen: Every element x of T has an immediate predecessor and an immediate successor in $T \cup S(T)$, which are not in T, namely the cuts $(T < x, T \geq x)$ and $(T \leq x, T > x)$.

5.2 Definition. If T_1 and T_2 are subsets of a linearly ordered set, we call T_1 *strictly dense* (resp. *dense*) in T_2, if for every two elements $a < b$ of T_2 there exists an element $c \in T_1$ with $a < c < b$ (resp. $a \leq c \leq b$).

Then we verify:

5.3 Lemma. *If T is linearly ordered, then T is strictly dense in $S(T)$, and $S(T)$ is strictly dense in T. Then further each of the sets T and $S(T)$ is dense in $T \cup S(T)$.*

Proof. Let (A, B) and (C, D) be cuts of T with $(A, B) < (C, D)$. Then $C \backslash A \neq \emptyset$, and there exists an element $x \in C \backslash A$ which now fulfills $(A, B) < x < (C, D)$.

If $a < b$ are elements of T, the cut $(\{x \in T | x \leq a\}, \{x \in T | x > a\})$ is strictly between a and b. The rest is evident.

An easy consequence of Lemma 5.3 is:

112

5.4 Lemma. *If T is linearly ordered and if $T \cup S(T)$ has a subset L of order type λ (resp.λ^*), where λ is a limit ordinal, then already T has a subset L' of type λ (resp. λ^*), which generates the same initial (resp. final) segment of T as L does.*

Proof. Let $A = \{a_\nu | \nu < \lambda\}$ be a well-ordered subset of $T \cup S(T)$ with $a_\nu < a_\mu$ for $\nu < \mu < \lambda$. Because of 5.3 there now exist elements $t_\nu \in T$ with $a_\nu \leq t_\nu \leq a_{\nu+1}$. The set of all t_ν, $\nu < \lambda$, then evidently contains a well-ordered set of type λ which fulfills our statement. The case of λ^* is of course symmetric to that of λ^*.

The next lemma stands in analogy to the completeness theorem of Dedekind, which states — roughly speaking — that if one adjoins to a linearly ordered set the cuts that are gaps, one obtains a linearly ordered set without gaps.

5.5 Lemma. *If T is linearly ordered, then the set $T \cup S(T)$ has a first element $(\emptyset, T)$ and a last element $(T, \emptyset)$ and has no gaps. For this one also says: It is order-theoretically closed.*

Proof. We assume indirectly that there exists a gap (A, B) in $T \cup S(T)$. Then A has a well-ordered cofinal subset A' of order type λ_1, and B has a coinitial inversely well-ordered subset B' of order type λ_2, where λ_1 and λ_2 are limit ordinals. According to 5.4 there now exist subsets A'' and B'' of T of types λ_1 resp. λ_2^* such that A'' and B'' are neighboring in T. So they define a cut of T which is a gap. This cut had to be an element of $S(T)$. It would be greater than all elements of A and smaller than all elements of B. This contradicts the fact that (A, B) is a cut in $T \cup S(T)$. The rest is clear.

Now we derive a relation between the order types of T and $S(T)$, which can be proved using a method of Padmavally [133]:

5.6 Lemma. *Let T be a linearly ordered set. Then there holds* $\text{tp}(T) < \text{tp}(S(T))$, *and then of course also* $\text{tp}(T) < \text{tp}(T \cup S(T))$.

Proof. The relation $\text{tp}(T) \leq \text{tp}(S(T))$ is trivial: If we map $t(\in T) \longrightarrow (A, B)$, where $A = \{x \in T | x \leq t\}$ and $B = T \backslash A$, this yields an isomorphic mapping of T in $S(T)$.

We now assume indirectly that there exists a $<$ - preserving mapping $f : S(T) \longrightarrow T$. Let a_0 be the first element (this is $(\emptyset, T)$) of $S(T)$. It fulfills evidently $a_0 < f(a_0)$. Let then a_1 be the first element of $S(T)$ which is $> f(a_0)$ (if such an element still exists). In general we proceed as follows: Let $a_\nu < f(a_\nu)$ hold for an ordinal ν. Then there exists the least cut in T which is $> f(a_\nu)$, namely $(T \leq f(a_\nu), T > f(a_\nu)) =: a_{\nu+1}$. We now have $a_\nu < a_{\nu+1}$, and since f is $<$ - preserving, $f(a_\nu) < f(a_{\nu+1})$. This entails that $f(a_{\nu+1})$ is in the second component of the cut $a_{\nu+1}$ and thus $> a_{\nu+1}$. So we can resume:

From $a_\nu < f(a_\nu)$ there follows $a_{\nu+1} < f(a_{\nu+1})$.

If for a limit ordinal λ we have already defined elements $a_\nu \in S(T)$ for all $\nu < \lambda$, such that $a_\nu < f(a_\nu)$ holds for all $\nu > \lambda$, and such that the a_ν increase with ν, we define a_λ to be the cut (A, B), where A is the initial segment of T which is generated by the elements $f(a_\nu)$, $\nu < \lambda$, and $B = S(T) \backslash A$. Then a_λ is greater than all elements $f(a_\nu)$, $\nu < \lambda$, and then also greater than all a_ν, $\nu < \lambda$. By transfinite induction we could so construct for every ordinal ν an element $a_\nu \in S(T)$, such that the a_ν strictly increase with ν. This is impossible of course, and thus our statement is proved.

Now we introduce the method of successively adjoining cuts:

5.7 Definition. Let T_0 be the empty set $\emptyset$. In this set, which we consider as linearly ordered, exists exactly one cut, namely $(\emptyset, \emptyset)$. We denote the set which contains this pair and nothing else by T_1, so that $T_1 = T_0 \cup S(T_0)$ holds. Then we define $T_2 := T_1 \cup S(T_1)$ and so on. For instance, T_1 has one element, T_2 has three, in general for $n \in \mathbf{N}$ the set T_{n+1} has $2 \cdot |T_n| + 1$ elements.

Generally our construction runs as follows: Let τ be an ordinal such that for all $\nu < \tau$ a linearly ordered set T_ν is already defined so that for $\mu < \nu < \tau$ the set T_μ is an ordered subset of T_ν. (The sets T_ν, $\nu < \tau$, thus form a tower of linearly ordered sets.) If now τ is a successor number, so that $\tau - 1$ exists, we put $T_\tau := T_{\tau-1} \cup S(T_{\tau-1})$ and order it according to 5.1. If τ is a limit number, we put $T_\tau := \cup_{\nu < \tau} T_\nu$. And we order T_τ in the obvious way: We put $a < b$ for elements of T_τ iff $a < b$ holds in at least one T_ν with $\nu < \tau$. Instead of T_ν we also write $T(\nu)$.

The above definition generalizes at the same time the construction of von Neumann of the system of ordinal numbers and the construction of the Dedekind cut.

Now e.g. $T(\omega_0)$ has the order type of the set of rational numbers, ordered by magnitude. In general we have the following theorem of Cuesta Dutari [16]:

5.8 Theorem. $T(\omega_\alpha)$ *is an* η_α - *set, if* ω_α *is regular.*

Proof. Let ω_α be regular, and let A and B be neighboring subsets of $T(\omega_\alpha)$. We assume indirectly that both have a cardinality $< \aleph_\alpha$. Then also $|A \cup B| < \aleph_\alpha$ holds. We have $T(\omega_\alpha) = \cup\{T_\nu | \nu < \omega_\alpha\}$, and then the set of indices ν, for which T_ν has elements in common with $A \cup B$ cannot be cofinal with the regular ω_α. So there exists an index $\beta < \omega_\alpha$ for which $A \cup B \subseteq T_\beta$ holds. But then our construction would furnish an element $z \in T_{\beta+1} \subseteq T(\omega_\alpha)$, which is greater than all elements of A and smaller than all elements of B, but this contradicts the assumption that A and B are neighboring. In a similar way it follows that $T(\omega_\alpha)$ is cofinal with ω_α and coinitial with ω_α^*.

5.9 Definition. We denote the order type of T_ν by t_ν or by $t(\nu)$. According to 5.6 we then have $t_\mu < t_\nu$ for $\mu < \nu$.

The statement of 5.8 can be sharpened essentially to the following:

5.10 Theorem ([64], Satz 14). *For every ordinal α the set $T(\omega_\alpha)$ has the type h_α.*

Proof. Let first ω_α be regular, and let ν be an ordinal $< \omega_\alpha$. We prove:

(1) T_ν has no subset of type ω_α (and analogously no subset of type ω_α^*).

If (1) would be false there would exist a least ordinal λ for which T_λ has a subset of type ω_α. According to 5.4 then λ must be a limit number. If now W is a well-ordered subset of T_λ of cardinality $\aleph_\alpha$, then we have $W = \cup_{\kappa < \lambda}(W \cap T_\kappa)$. Now $|W| = \aleph_\alpha$ is regular and $|\lambda| < \aleph_\alpha$, and so at least one of the summands $W \cap T_\kappa$ with $\kappa < \lambda$ must have the cardinality $\aleph_\alpha$. But this contradicts the minimal property of λ, and (1) is proved. Then $T(\omega_\alpha)$ is the union of the $\aleph_\alpha$ many subsets T_ν, $\nu < \omega_\alpha$, which are all free from ω_α and ω_α^*, and then 5.8 and the characterization theorem 3.12 yield that $T(\omega_\alpha)$ has the type h_α.

If ω_α is singular our assertion follows immediately from 4.3, for then $T(\omega_\alpha) = \cup_{\nu < \alpha} T(\omega_{\nu+1})$, where the $T(\omega_{\nu+1})$ have the type $h_{\nu+1}$.

4.6 Special properties of the sets T_λ for indecomposable λ.

The sets T_λ with an indecomposable λ have some remarkable properties. We recall this concept:

6.1 Definition. An ordinal λ is said to be *indecomposable,* if in every representation $\lambda = \alpha + \beta$ as sum of ordinals α, β there follows $\alpha = \lambda$ or $\beta = \lambda$.

A well-known theorem of set theory states that the indecomposable ordinals are exactly the powers ω^τ, where τ is an ordinal. In particular the initial ordinals ω_ν are indecomposable.

Further we define: A linearly ordered set S is said to be *homogeneous,* if for each two elements $a, b \in S$ there is an isomorphism of S onto S, which maps a onto b.

If S is a linearly ordered set, $T \subseteq S$, (A, B) a cut in T, x an element of S, then we say: x is *within* the cut (A, B) iff $A < x < B$ holds.

For every $x \in T_\lambda$ there exists a smallest index $\delta \le \lambda$ for which $x \in T_\delta$. Then δ is called the *degree* of x.

Now there holds:

6.2 Lemma. *Let λ be an indecomposable ordinal, and let $x < y$ be elements of T_λ, t an element of $T_\lambda \cap (x, y)$ of least degree δ. Then δ is a successor number, and t is uniquely defined.*

Proof. T_λ is dense, and so there exists an element $t \in T_\lambda \cap (x, y)$, and then also one with lowest degree δ. If δ would be a limit ordinal we would have $T_\delta = \cup_{\nu < \delta} T_\nu$, and then t would already be in a T_ν with $\nu < \tau$, a contradiction.

If now u would be another element of $T_\delta \cap (x, y)$, say w.r.o.g. with $u < t$, there would exist by 5.3 an element $z \in T_{\delta-1}$ with $u < z < t$. Its degree is $\le \sigma - 1 < \delta$, and this is a contradiction to the minimal property of t.

6.3 Theorem. *Let λ be an indecomposable ordinal. Then every open interval of T_λ is isomorphic to T_λ.*

In general we have the following: If S_1 and S_2 are segments of T_λ that are coinitial with the same ω_β^* and cofinal with the same ω_γ, then S_1 and S_2 are isomorphic.

Proof. First we prove as the main part: If $a < b$ are elements of T_λ, then the open interval (a, b) of T_λ has again the type t_λ. To this purpose we construct an isomorphic mapping f_λ of T_λ onto (a, b). Let α be the least ordinal such that a and b are both in T_α. We map the one-element set T_1 into (a, b), where $T_1^* := f[T_1]$ contains only the (by 6.2 uniquely defined) element of least degree $\delta + 1$ which is in (a, b). This degree is

116

indeed a successor number. And we have $\delta + 1 \leq \alpha + 1$ because a and b are in T_α, so that there exist elements of $T_{\alpha+1}$ between them.

Now for $\nu = 1$ the following induction hypothesis is satisfied:

(1) $f_\nu : T_\nu \to T_\nu^*$ is an isomorphic mapping with $(a,b) \cap T_{\delta+\nu} \subseteq T_\nu^* \subseteq (a,b) \cap T_{\alpha+\nu+1}$.

Suppose now that (1) is valid for a fixed ordinal $\nu < \lambda$. Then we define $f_{\nu+1}(x) := f_\nu(x)$ for all $x \in T_\nu$.

If $x \in T_{\nu+1} \backslash T_\nu$ holds, then x is a cut $(A, B) \in T_{\nu+1}$ of T_ν. If here A and B are non-empty, then there is exactly one element x^* of $(a,b) \cap T_{\delta+\nu+1}$ between $A^* := f_\nu[A]$ and $B^* := f_\nu[B]$. For if there would be two, there would by 5.3 exist an element of $(a,b) \cap T_{\delta+\nu}$ between them, and this contradicts the fact that $A^* \cup B^* \supseteq f_\nu[T_\nu] = T_\nu^* \supseteq (a,b) \cap T_{\delta+\nu}$. Then we put $f_{\nu+1}(x) := x^*$.

If $x = (A, B)$ is an improper cut, where A or B is empty, we have to discuss the cuts $(\emptyset, T_\nu)$ and $(T_\nu, \emptyset)$. These cases are symmetric, and so we restrict ourselves to the first case. Here there is an element x^* of $T_{\alpha+\nu+2}$ between a and T_ν^* because of (1) and $a \in T_\alpha \subseteq T_{\alpha+\nu+1}$. If there is also such an element which is in $T_{\delta+\nu+1}$ (and then this is uniquely determined by 6.2) we choose this as x^*. In any case we map x onto x^* by $f_{\nu+1}$. We treat $(T_{\nu+1}, \emptyset)$ analogously. So we obtain an isomorphic mapping $f_{\nu+1} : T_{\nu+1} \to T_{\nu+1}^* := f_{\nu+1}[T_{\nu+1}]$ which satisfies (1) for $\nu + 1$.

If μ is a limit ordinal $\leq \lambda$ such that (1) holds for all $\nu < \mu$, we define $f_\mu : T_\mu := \cup\{T_\nu | \nu < \mu\} \to \cup\{T_\nu^* | \nu < \mu\} =: T_\mu^*$ to be the limit mapping of the $f_\nu, \nu < \mu$. Then f_μ satisfies (1) for μ, and so (1) holds for all $\nu < \lambda$. The limit mapping f_λ of the $f_\nu, \nu < \lambda$, is then an isomorphism of T_λ onto (a,b).

The rest now follows easily: Let S_1 and S_2 be segments as introduced in the statement. Then we choose elements $d_0 \in S_1$ and $e_0 \in S_2$ and further a sequence $d_0 < d_1 < \cdots < d_\tau < \cdots | \tau < \omega_\gamma$ which is cofinal in S_1 and a sequence $e_0 < e_1 < \cdots < e_\tau < \cdots | \tau < \omega_\gamma$ which is cofinal in S_2. According to that what we have proved already $[d_\tau, d_{\tau+1}]$ is isomorphic to $[e_\tau, e_{\tau+1}]$ for every $\tau < \omega_\gamma$. Together this yields an isomorphic mapping of $\{x \in S_1 | x \geq d_0\}$ onto $\{x \in S_2 | x \geq e_0\}$. The counterpart of this follows symmetrically, and therefore S_1 is isomorphic to S_2.

If s_ν is the smallest and g_ν the greatest element of $T_{\nu+1}$ for $\nu < \lambda$, then $\{s_\nu | \nu < \lambda\}$ is coinitial in T_λ, and $\{g_\nu | \nu < \lambda\}$ is cofinal in T_λ. Thus T_λ and all open intervals of T_λ are cofinal with λ and coinitial with λ^*, and so everything is proved.

By 5.10 we have $H_\alpha \simeq T(\omega_\alpha)$, and so 6.3 implies the corollary, which generalizes 3.19 to arbitrary initial ordinals ω_α:

6.3′ Corollary. *For every ordinal α holds: Each open interval of H_α is isomorphic to H_α.*

Our last theorem implies that T_λ is a homogeneous chain if λ is indecomposable :

6.4 Theorem. *Let λ be indecomposable. Then for every two elements a,b of T_λ there exists an isomorphic mapping of T_λ onto itself (a so-called automorphism) which maps a onto b.*

Proof. By 6.3 the initial segments $(T_\lambda < a)$ and $(T_\lambda < b)$ are isomorphic, say by a mapping φ, also $(T_\lambda > a)$ and $(T_\lambda > b)$ are isomorphic, say by a mapping ψ. Then the mapping which maps a onto b and the other elements x of T_λ onto $\varphi(x)$ (resp. $\psi(x)$) fulfills our assertion.

The question arises whether the statement of 6.3 can be generalized to more general ordinals. But this is not the case. To see this we first mention:

6.5 Lemma. *If α and β are ordinals, and if C is a cut in T_α, then the set of all elements of $T_{\alpha+\beta}$ that are within the cut C, has the type t_β.*

The proof follows immediately from the construction of the sets T_ν.

6.6 Remark. Let λ be a decomposable limit ordinal.Then there exist open intervals of T_λ which are not isomorphic. Indeed, there exists a representation $\lambda = \alpha + \beta$, where α and β are ordinals between 0 and λ. Then also there holds $\lambda = (\alpha + 1) + \beta$ since the limit number β satisfies $1 + \beta = \beta$. In $T_{\alpha+1}$ there exist neighboring elements x, y and because of 6.5 the set of elements of $T_{\alpha+\beta}$, that are between x and y, is an open interval of T_λ of type t_β. On the other hand there exist open intervals of T_λ of type t_λ. If e.g. a is the least and b the greatest element of $T_2 \setminus T_1$, then the open interval (a, b) of T_λ evidently has the type t_λ.

If λ is a successor ordinal ≥ 2 there trivially exist open intervals of T_λ which are not isomorphic. For then there exist neighboring elements in T_λ, and the open interval with these endpoints is empty. And also there exist non-empty open intervals.

Next we study the idempotency property of certain types t_λ. To this purpose we define:

6.7 Definition. For every ordinal α we denote $T_\alpha \times T_\alpha$ by $P(\alpha)$. We consider $P(\alpha)$ as linearly ordered by first differences. An easy consequence is then:

6.8 Lemma. $P(\alpha)$ *is isomorphic to a subset of* $T_{\alpha+\alpha}$.

Proof. $P(\alpha)$ has the order type $t_\alpha \cdot t_\alpha$. If now we have a cut in T_α, then there exists because of 6.5 a subset M of elements of $T_{\alpha+\alpha}$, which lie within this cut, such that M has order type t_α. If then $a \in T_\alpha$, let $s(a)$ denote the cut (A, B) in T_α, where A contains all elements $\leq a$. Let $S(a)$ be a subset of $T_{\alpha+\alpha}$ of type t_α , whose elements lie within the cut $s(a)$. Then the set $S = \cup\{S(a)|a \in T_\alpha\}$ has the type $t_\alpha \cdot t_\alpha = \mathrm{tp}(P(\alpha))$, and so we have an isomorphic mapping of $P(\alpha)$ onto $S \subseteq T_{\alpha+\alpha}$.

We shall prove that $t_\lambda \cdot t_\lambda = t_\lambda$ holds for indecomposable ordinals λ and prepare this with some lemmas.

6.9 Lemma. *Let* α *and* β *be ordinals,* $C = (A, B)$ *a cut in* $P(\alpha)$, M *the set of all elements of* $P(\alpha + \beta)$ *that lie within* C. *Then* M *has an order type* $\geq t_\beta$.

Proof. Case 1. There exists an element $x \in T_\alpha$ such that A and B have elements which have x as first component.

Then $\{y|(x, y) \in A\}$ and $\{y|(x, y) \in B\}$ are neighboring in T_α, and so they define a cut C_α in T_α. By 6.5 the set U of elements of $T_{\alpha+\beta}$ that lie within C_α has the order type t_β. Then the elements of $\{x\} \times U$ lie within the cut C, and we have $\{x\} \times U \subseteq P(\alpha + \beta)$ so that $M \supset \{x\} \times U$ holds. The last set has order type $\mathrm{tp}\ U = t_\beta$, and so Case 1 is settled.

Case 2. No element of T_α is the first component of an element of A and of an element of B. Then the set A_1 of the first components of the elements of A and the set B_1 of the first components of the elements of B are disjoint and constitute a cut $C_1 = (A_1, B_1)$ in T_α. By 6.5 the set V of all elements of $T_{\alpha+\beta}$ that lie within C_1 has order type t_β. Then all elements of $(V \times T_{\alpha+\beta})$ belong to $P(\alpha + \beta)$, and then also to M because they lie within C. The order type of their set is $\geq \mathrm{tp} V = t_\beta$.

6.10 Lemma. *Let* α *and* β *be ordinals with* $\omega^\nu \leq \alpha, \beta < \omega^{\nu+1} (= \omega^\nu \cdot \omega)$. *Then we can choose natural numbers* $n, m \geq 2$ *satisfying*

(1) $\alpha \cdot (n - 1) \geq \beta \cdot 2$, *and*

(2) $\alpha \cdot n \leq \beta \cdot (m - 1)$.

a) *Let $f : T_\alpha \longrightarrow P(\beta)$ be a $<$ - preserving mapping. Then there exists a set $S_\alpha \subseteq T_{\alpha \cdot n}$ and a $<$ - preserving surjective mapping $f^* : S_\alpha \to P(\beta)$ which extends f.*

b) *There exists a $<$ - preserving mapping $f' : T_{\alpha \cdot n} \longrightarrow P(\beta \cdot m)$, which extends f^*, and then also f.*

Proof. First one can choose n so great that (1) holds because of $\alpha \geq \omega^\nu$ and $\beta \cdot 2 < \omega^{\nu+1}$. Then in (2) m can be chosen so great that (2) holds because of $\alpha \cdot n < \omega^{\nu+1}$ and $\beta \geq \omega^\nu$.

a) Let $C = (C_1, C_2)$ be a cut in T_α. The set $M(C)$ of elements of $T_{\alpha \cdot n}$ that lie within C has the order type $t(\alpha \cdot (n-1))$. For we have $T(\alpha \cdot n) = T(\alpha + \alpha \cdot (n-1))$, and then this follows from 6.5 by putting $\beta = \alpha \cdot (n-1)$.

Let further $C' := (f(C_1), f(C_2))$ be the corresponding cut in $f(T_\alpha) \subseteq P(\beta)$. Then the set M' of elements of $P(\beta)$ that lie within the cut C' has (trivially) an order type $\leq$ tp $P(\beta)$ which is $\leq t_{\beta+\beta}$ by 6.8. Now we can define, due to (1), an isomorphic mapping between a subset of M and the set M'. We perform this action with all cuts C of T_α and combine this with f. Then there results a mapping f^* which satisfies a). Here S_α is $T_\alpha \cup \cup \{M(C) | C$ is a cut of $T_\alpha\}$.

b) Let $D = (D_1, D_2)$ be a cut in $S_\alpha \subseteq T_{\alpha \cdot n}$ and $D' := (f^*(D_1), f^*(D_2))$ the corresponding cut in $f^*(S_\alpha) \subseteq P(\beta)$. The set E of elements of $T_{\alpha \cdot n}$ that lie within D has trivially an order type $\leq t_{\alpha \cdot n}$. And the set E' of elements of $P(\beta \cdot m)$ that lie within D' has (according to 6.9 by taking there $\beta \cdot (m-1)$ for β and β for α) an order type $\geq t_{\beta \cdot (m-1)}$. And so by (2) there exists an isomorphic mapping f_E^* of E onto a subset of E'. Then f^*, together with the mappings f_E^*, furnishes a mapping f' which satisfies b).

In a special form of 6.10, namely by replacing α by $\omega^\nu \cdot n_i$, β by $\omega^\nu \cdot m_i$, f by f_i, and f' by f_{i+1} we obtain:

$6.10'$. Lemma. *Suppose we have a $<$ - preserving mapping $f_i : T(\omega^\nu \cdot n_i) \to P(\omega^\nu \cdot m_i)$, where n_i and m_i are natural numbers. Then there exists a $<$ - preserving mapping $f_{i+1} : T(\omega^\nu \cdot n_{i+1}) \to P(\omega^\nu \cdot m_{i+1})$, where n_{i+1} and m_{i+1} are natural numbers with $n_{i+1} > n_i$ and $m_{i+1} > m_i$, which extends f_i, and which satisfies $f_{i+1}(T(\omega^\nu \cdot n_{i+1})) \supseteq P(\omega^\nu \cdot m_i)$.*

6.11 Theorem [64]. $t_\lambda \cdot t_\lambda = t_\lambda$ *holds for every indecomposable ordinal λ, which means for the ordinals $\lambda = \omega^\nu$ where ν is an ordinal.*

Proof. For $\nu = 0$ we have $\omega^0 = 1$, and $t_0 = 1 = 1 \cdot 1$. In the case $\nu = 1$ we have $\lambda = \omega^1 = \omega_0$ and we must show that $T(\omega_0) \times T(\omega_0)$ is isomorphic to $T(\omega_0)$. But these sets have order types $h_0 \cdot h_0$ and h_0 which are equal by 4.5. By the way: Every set $T(\omega_\alpha)$, where α is an arbitrary initial ordinal has by 5.10 the order type h_α, and for this there holds $h_\alpha \cdot h_\alpha = h_\alpha$ by 4.5.

Now we assume that the theorem is already proved for a fixed ordinal ν, so that we have an isomorphic mapping $f_0 : T(\omega^\nu) \longrightarrow P(\omega^\nu)$. Then for $i = 0$, $n_0 = 1 = m_0$ the following condition is satisfied:

(1) $f_i : T(\omega^\nu \cdot n_i) \longrightarrow P(\omega^\nu \cdot m_i)$ is a $<$ - preserving mapping.

Let now i be a fixed natural number for which (1) holds. Then by $6.10'$ there exists a $<$ - preserving mapping $f_{i+1} : T(\omega^\nu \cdot n_{i+1}) \to P(\omega^\nu \cdot m_{i+1})$ with $n_{i+1} > n_i$ and $m_{i+1} > m_i$ which extends f_i and which satisfies:

(2) $f_{i+1}(T(\omega^\nu \cdot n_{i+1})) \supseteq P(\omega^\nu \cdot m_i)$.

By induction on i we can so define a sequence of $<$ - preserving mappings $f_i : T(\omega^\nu \cdot n_i) \to P(\omega^\nu \cdot m_i)$ for all $i \in \mathbf{N}$, where the $n_i, i \in \mathbf{N}$, and the $m_i, i \in \mathbf{N}$, form strictly ascending sequences of natural numbers, and where for $i < j$ the function f_j is an extension of f_i with
$f_j(T(\omega^\nu \cdot n_{i+1})) \supseteq P(\omega^\nu \cdot m_i)$.

The limit mapping of the f_i, $i \in \mathbf{N}$, (which extends all of these), maps $\cup_{i \in \mathbf{N}} T(\omega^\nu \cdot n_i) = T(\omega^\nu \cdot \omega) = T(\omega^{\nu+1})$ isomorphically onto $\cup_{i \in \mathbf{N}} P(\omega^\nu \cdot m_i) = P(\omega^{\nu+1})$. Therefore our statement also holds for $\nu + 1$.

If now μ is a limit number such that for all ordinals $\nu < \mu$ the sets $T(\omega^\nu)$ and $P(\omega^\nu)$ have the same order type, so that there exists an isomorphic mapping $f_\nu : T(\omega^\nu) \longrightarrow P(\omega^\nu)$, the limit mapping which extends all f_ν with $\nu < \mu$ is an isomorphism of $T(\omega^\mu)$ onto $P(\omega^\mu)$.

The property which was expressed in 6.11 is restricted to the indecomposable ordinals. Namely there holds:

6.12 Theorem [64]. *Let δ be a decomposable ordinal. Then there holds $t(\delta) \cdot t(\delta) > t(\delta)$.*

Proof. For finite ordinals δ the assertion is trivial. So we assume $\delta > \omega_0$. The decomposable δ has a representation $\delta = \alpha + \beta$, where α and β are ordinals $< \delta$. In the set $P(\delta)$ (see Definition 6.7), whose order type is $t(\delta) \cdot t(\delta)$, exists a system $\mathfrak{I}_P$ of disjoint segments, such that each of them contains a subset of type $t(\delta)$, and such that also $\mathfrak{I}_P$ has the order type $t(\delta)$ with respect to the natural order of $\mathfrak{I}_P$, which is given by

$I_1 \leq I_2$ for $I_1,\, I_2 \in \mathfrak{I}_P \iff x \leq y$ for all $x \in I_1$ and all $y \in I_2$.

Now it suffices to prove:

(1) In $T(\delta)$ there is no such system of disjoint segments.

Let $\mathfrak{I}_T$ be a system of disjoint segments of $T(\delta)$ such that each of them contains a subset of type $t(\delta)$. Then we have:

(2) Every $S \in \mathfrak{I}_T$ contains an element of $T(\alpha)$.

For otherwise there would exist a segment $S \in \mathfrak{I}_T$ such that all elements of S would lie within the same cut of $T(\alpha)$. But then S would not contain a subset of type $t(\delta)$ because of 6.5, which states that the set of all elements of $T(\delta)$ that lie within a cut of $T(\alpha)$ has the order type $t(\beta) < t(\delta)$. So (2) holds.

Now all sets $I' := I \cap T(\alpha)$, $I \in \mathfrak{I}_T$, are non-empty, and so to the set of segments $I \in \mathfrak{I}_T$ there corresponds a set $\mathfrak{S}_T$ of non-empty disjoint segments I' of $T(\alpha)$ which has the (induced) order type $\mathrm{tp}\,\mathfrak{S}_T = \mathrm{tp}\,\mathfrak{I}_T$.

So finally by 5.6 we have $\mathrm{tp}\,\mathfrak{I}_T = \mathrm{tp}\,\mathfrak{S}_T \leq \mathrm{tp}\,T(\alpha) < \mathrm{tp}\,T(\alpha+1) \leq \mathrm{tp}\,T(\delta)$, so that $\mathrm{tp}\,\mathfrak{I}_T < \mathrm{tp}\,T(\delta)$ follows, and (1) is proved.

6.13 Definition. We call an order type τ *interval-homogeneous*, if all non-empty open intervals of a realization of τ are isomorphic.

We call an order type τ *unattainable* (by summation), iff there holds: If A and B_i, $i \in A$, are linearly ordered sets with order types $\mathrm{tp}\,A < \tau$ and $\mathrm{tp}\,B_i < \tau$ for every $i \in A$, then also the ordered sum $\sum_{i \in A} \mathrm{tp}\,B_i$ is $< \tau$.

This definition entails that for unattainable order types τ there holds: If α and β are order types $< \tau$, then also $\alpha \cdot \beta < \tau$ holds. A sufficient condition for being unattainable is presented in:

6.14 Theorem. *Let τ be an interval-homogeneous and idempotent linear order type. Then τ is unattainable.*

Proof. Let A and B_i, $i \in A$, be linearly ordered sets with types $\mathrm{tp}(A) < \tau$ and $\mathrm{tp}(B_i) < \tau$ for all $i \in A$. Then there holds $\sum_{i \in A} \mathrm{tp}\,B_i \leq \sum_{x \in T} \tau_x$, where T is a set of type τ and every $\tau_x = \tau$. The latter sum is $\tau \cdot \tau = \tau$, so that $\sum_{i \in A} \mathrm{tp}\,B_i \leq \tau$ holds.

We now assume indirectly that $\tau \leq \sum_{i \in A} \mathrm{tp}\,B_i$ holds. Then there exists a $<$ - preserving mapping f of T into the ordered sum $\sum_{i \in A} B_i$, which is the set of all pairs (a, b) with $a \in A$, $b \in B_a$. Let A' be the set of all elements a that occur as the first component of such a pair (a, b) which is ascribed by f to an element of T. Because of $\tau > \mathrm{tp}A \geq \mathrm{tp}A'$ there exist two elements $t' < t''$ in T, whose f - images have the same first component, say a'. Then a' is also the first component of the f - images of all $t \in T$ that fulfill $t' < t < t''$. This set $\{t \in T | t' < t < t''\} =: T^*$ has again the type τ because of the interval-homogenity of τ. Due to the injectivity of f the second components of the images of the elements of T^* must be different and their set, which is contained in $B_{a'}$, must have the type τ. But this contradicts $\mathrm{tp}\, B_{a'} < \tau$.

An immediate consequence of the last theorem, 6.3 and 6.11 is now:

6.15 Theorem. *If λ is an indecomposable ordinal, $t(\lambda)$ is an unattainable order type.*

4.7   Relations between the order types of lexicographic products

As we have seen in the previous sections the ordered sets $\mu((\nu))$, where μ and ν are ordinals, and certain subsets of them like the sets H_α are rather important for the theory. They are comparatively easy to describe because the ordinals μ, ν which determine them, are well-ordered sets. They give a good oversight over the structure of the class of linearly ordered sets. In this section we investigate, following [71], how the sets $\mu((\nu))$ and the types h_α are related.

7.1 Definition. If T is a linearly ordered set and σ an ordinal, let $T((\sigma))$ denote the lexicographically ordered set of all transfinite sequences of length σ of elements of T. Its order type shall be denoted by $\tau^{(\sigma)}$, where τ is the order type of T. The property to be an η_α - set is productive in the following sense:

7.2 Theorem. *Let τ be an ordinal > 0, and let E_ν, $\nu < \tau$, be η_α - sets. Then the lexicographic product $L := \mathrm{L}_{\nu<\tau} E_\nu$ is again an η_α - set.*

Proof. Let (A, B) be a proper cut in L. We denote the set of the 0-components of the elements of A (resp. B) by A_0 (resp. B_0). If $A_0 \cap B_0 = \emptyset$, then (A_0, B_0) defines a proper cut in E_0, and then A_0 has a cofinal subset of type ω_α, or B_0 has a coinitial subset of type ω_α^*. In the first case also A has a cofinal subset of type ω_α, in the second also B has a coinitial subset of type ω_α^*.

Now we assume that $A_0 \cap B_0$ is non-empty. Then it contains exactly one element $x_0 \in E_0$. Generally we now consider all sequences $(x_\nu)_{\nu < \mu}$ of elements $x_\nu \in E_\nu$ for ordinals $\mu \leq \tau$, that are initial segments of elements of A and of elements of B. Let σ be the supremum of the ordinals μ which here occur. Then we have a sequence $x = (x_\nu)_{\nu < \sigma}$ of which all the before mentioned sequences are initial segments. We distinguish two cases:

Case 1. x is an initial segment of elements of A and of B.

Then σ is $< \tau$ because of $A \cap B = \emptyset$. And according to the definition of σ there is no sequence of length $\sigma + 1$ (which prolongs x with an additional component) which is still a common initial segment of elements of A and of B. Let A' (resp. B') be the set of the σ - components of those elements of A (resp. B) that have x as initial segment. Then $A' \cap B' = \emptyset$, and thus (A', B') defines a cut in E_σ. Again it follows as before, that A is cofinal with ω_α, or B is coinitial with ω_α^*.

Case 2. x is only an initial segment of elements of A, not of B. (The case with B instead of A is symmetric.)

Then A contains all sequences of L, that have x as initial segment and an arbitrary element of E_σ in the σ - component. Then, of course, A is cofinal with ω_α since E_σ is that also.

So finally we have obtained that there are no non-empty neighboring subsets of L which both have a cardinality $< \aleph_\alpha$. It is trivial that L has a cofinal subset of type ω_α and a coinitial subset of type ω_α^* because this also holds for the set E_0. And therefore L is an η_α - set.

7.3 Remark. By the way, the property to be an η_α - set, is also summative in the following sense: If E is an η_α - set, and if E_i is an η_α - set for every $i \in I$, then the ordered sum $\sum_{i \in I} E_i$ is also an η_α - set. This can easily be verified.

7.4 Lemma. *Let ω_α be a regular initial ordinal and σ an ordinal $< \omega_\alpha$. Then the lexicographically ordered set $L := H_\alpha((\sigma))$ has type h_α.*

Proof. Let $s = (s_\nu)_{\nu < \sigma}$ be an element of L, hence $s_\nu \in H_\alpha$ for $\nu < \sigma$. Each of the sequences s_ν has a last digit 1, say at the position $\pi(\nu)$. The set $\{\pi(\nu) | \nu < \sigma\}$ has a cardinality $\leq |\sigma| < \aleph_\alpha$ and is therefore not cofinal in the regular ordinal ω_α. So there exists an ordinal $\kappa < \omega_\alpha$ such that all $\pi(\nu)$, $\nu < \sigma$, are $< \kappa$, and then we have $s \in (H_\alpha[\kappa])((\sigma))$ (see Definition 4.2), and $H_\alpha((\sigma)) = \cup\{(H_\alpha[\kappa])((\sigma)) | \kappa < \omega_\alpha\}$ is a union of $\aleph_\alpha$ subsets which are free from ω_α and ω_α^*. Indeed, $H_\alpha[\kappa]$ is isomorphic

124

to a subset of $2((\kappa))$ which is free from ω_α and ω_α^* by 2.2. The latter also holds for $(H_\alpha[\kappa])((\sigma))$ because of 2.2. The rest follows with 7.2 and 3.12.

More generally we now obtain:

7.5 Theorem. *For every $\sigma < \omega_\gamma = \mathrm{cf}(\omega_\alpha)$ the set $H_\alpha((\sigma))$ has the order type h_α.*

Proof. If ω_α is regular we are done because of 7.4. So we assume that ω_α is singular. Then we have $H_\alpha = \cup_{\nu<\alpha}H_\alpha[\omega_{\nu+1}]$ and further:

(1) $H_\alpha((\sigma)) = \cup_{\gamma<\nu<\alpha}((H_\alpha[\omega_{\nu+1}])((\sigma)))$.

For let $s = (s_\nu)_{\nu<\sigma}$ be an element of $H_\alpha((\sigma))$. For $\nu < \sigma$ let $\mu(\nu)$ be the greatest index μ, for which $s_\mu = 1$. Then the set $\{\mu(\nu)|\nu < \sigma\}$ has a cardinality $\leq |\sigma| < \aleph_\gamma$, and thus it is not cofinal in ω_α, and there exists an ordinal $\kappa < \omega_\alpha$ such that all $\mu(\nu)$, $\nu < \sigma$, are $< \kappa$. And then all $s_\nu, \nu < \sigma$,are in $H_\alpha[\kappa]$, which is $\subseteq H_\alpha[\omega_{\nu+1}]$ for some ν with $\gamma < \nu < \alpha$, so that $s \in (H_\alpha[\omega_{\nu+1}])((\sigma))$, and (1) is proved.

$H_\alpha[\omega_{\nu+1}]$ is isomorphic to $H_{\nu+1}$, and for $\gamma < \nu < \alpha$ the set $H_{\nu+1}((\sigma))$ has the order type $h_{\nu+1}$ according to the regular case and because of $\sigma < \omega_\gamma < \omega_{\nu+1}$. Finally our assertion follows from 4.3.

In connection with the last theorem the question arises how the order type of $H_\alpha((\omega_\gamma))$ is related to other standard order types. It will turn out that it is not $\leq h_\alpha$, but we can prove:

7.6 Theorem. *For $\omega_\gamma = \mathrm{cf}(\omega_\alpha)$ the set $S := H_\alpha((\omega_\gamma))$ is embeddable in $2((\omega_\alpha))$.*

Proof. Let c be a fixed element of H_α. We consider the sequences $s = (s_\nu|\nu < \omega_\gamma) \in H_\alpha((\omega_\gamma))$ for which there exists a first ordinal β such that $s_\nu = c$ for all ν with $\beta \leq \nu < \omega_\gamma$. We denote their set by $S_c(\beta)$ and put $S_c := \cup\{S_c(\beta)|\beta < \omega_\gamma\}$. Since H_α is dense it is easily seen that also S_c is dense in S.

For every $\beta < \omega_\gamma$ now $S_c(\beta)$ is isomorphic to a subset of of $H_\alpha((\beta))$, and so its type is $\leq \mathrm{tp}H_\alpha((\beta)) = h_\alpha$ because of 7.5. Then S_c is the union of $\aleph_\gamma \leq \aleph_\alpha$ subsets $S_c(\beta)$, each of which has a type $\leq h_\alpha$, and then S_c itself has a type $\leq h_\alpha$ by 4.6.

So we have obtained: S_c is dense in S and embeddable in H_α, say by a mapping f. Now $2((\omega_\alpha))$ has no gaps and contains H_α. Then f can be extended to an isomorphic mapping of S in $2((\omega_\alpha))$. Indeed, every x

$\in S \backslash S_c$ determines a gap $(S_c < x, S_c > x)$ in S_c. Then there is an element $x' \in 2((\omega_\alpha))$ which satisfies $f(a) < x' < f(b)$ for all $a \in (S_c < x)$ and $b \in (S_c > x)$. We ascribe this x' to x and obtain the desired extension.

Also the counterpart of 7.6 can be proved:

7.7 Theorem. *For $\omega_\gamma = \mathrm{cf}(\omega_\alpha)$ the set $2((\omega_\alpha))$ is embeddable in $H_\alpha((\omega_\gamma))$.*

Proof. For the regular ordinals ω_α our assertion is trivial since then $\omega_\gamma = \omega_\alpha$ holds. So we presuppose that ω_α is singular. Then there exists a strictly ascending sequence of initial ordinals ω_{α_ν}, $\nu < \omega_\gamma$, with $\sum \{\omega_{\alpha_\nu} | \nu < \omega_\gamma\} = \omega_\alpha$. This entails
$$2((\omega_\alpha)) \simeq 2((\omega_{\alpha_0})) \times 2((\omega_{\alpha_1})) \times \cdots \times 2((\omega_{\alpha_\nu})) \times \cdots | \nu < \omega_\gamma,$$
where the last product is ordered by the principle of first differences.

Here every factor $2((\omega_{\alpha_\nu}))$ is isomorphic to a subset of H_α, and so $2((\omega_\alpha))$ is isomorphic to a subset of $H_\alpha((\omega_\gamma))$.

7.8 Remark. We have seen that from the order types $h_\alpha^{(\omega_\gamma)}$ and $2^{(\omega_\alpha)}$ each is $\leq$ the other. But these order types are not equal. This is an immediate consequence of the fact that $H_\alpha((\omega_\gamma))$ is dense, whereas $2((\omega_\alpha))$ is not dense: Every sequence $(s_\nu | \nu < \omega_\alpha)$ of it which has a last digit 1, say at the position μ, has an immediate predecessor, namely the element (t_ν), for which $t_\nu = s_\nu$ for $\nu < \mu$, $t_\mu = 0$, $t_\nu = 1$ for $\mu < \nu < \omega_\alpha$.

7.9 Definition. If S is a linearly ordered set, then a subset $I \subseteq S$ is called *isolated*, if for every $x \in I$ there exist two elements a_i, b_i in S with $a_i < x \leq b_i$ such that the intervals $(a_i, b_i]$, $i \in I$, are pairwise disjoint.

In this connection the following holds:

7.10 Theorem (Padmavally [133]). *Let S be a linearly ordered set which has no gaps, but a first element x_0 and a last element. And let I be an isolated subset of S. Then there is no $<$ - preserving mapping of S in I.*

Proof. Suppose indirectly that $f : S \longrightarrow I$ is a $<$ - preserving mapping. The first element x_0 of S is not in the isolated set I and so $x_0 < f(x_0) =: x_1$ holds. From $x_0 < x_1$ we obtain $(x_1 =) f(x_0) < f(x_1) =: x_2$ and so on. Precisely: Let τ be an ordinal such that $x_\nu \in I$ is already defined for $1 \leq \nu < \tau$ such that the x_ν strictly increase with

ν and that $x_\nu < f(x_\nu)$ holds for all $\nu < \tau$. If τ is a successor ordinal $\sigma + 1$, then $x_\sigma < f(x_\sigma) =: x_\tau$ satisfies $x_\tau = f(x_\sigma) < f(x_\tau)$.

If τ is a limit ordinal then $x_\tau := \sup_S\{x_\nu | \nu < \tau\}$ exists because of 2.10. And x_τ is not in I because I is isolated.

Now $f(x_\tau)$ is $\geq f(x_\nu) > x_\nu$ for all $\nu < \tau$, and then also $f(x_\tau) \geq \sup_S\{x_\nu | \nu < \tau\} = x_\tau$ holds. But $f(x_\tau) = x_\tau$ is impossible since $f(x_\tau)$ is in I, but x_τ is not. Thus also $x_\tau < f(x_\tau)$ holds. Applying transfinite induction this process of defining elements x_ν could be performed over the whole class of ordinals ν, which of course yields a contradiction.

For the lexicographic products we obtain as a corollary:

7.11 Theorem. *For every ordinal α the set $2((\omega_\alpha))$ is not embeddable in H_α.*

Proof. The set $H_\alpha \times \{e\}$, where e is an arbitrary element of H_α, is an isolated subset of $H_\alpha \times H_\alpha$. For let e_1, e_2 be elements of H_α with $e_1 < e < e_2$. Then the intervals $[(x, e_1), (x, e_2)]$, $x \in H_\alpha$, of $H_\alpha \times H_\alpha$ contain (x, e) and are pairwise disjoint.

Now $H_\alpha \times H_\alpha$ is by 4.5 isomorphic to H_α and then embeddable in $2((\omega_\alpha))$, say by a mapping f. The set $I := f[\{H_\alpha \times \{e\}]$ is then an isolated subset of $2((\omega_\alpha))$, and it follows from 7.10 that there is no $<$ - preserving mapping of $2((\omega_\alpha))$ in I, and I is isomorphic to H_α.

By the way, using the GCH our assertion would be trivial since then $2^{\aleph_\alpha} = |2((\omega_\alpha))| > \aleph_\alpha = |H_\alpha|$ holds.

7.12 Theorem. *The set $S := \omega_\alpha((\omega_\gamma))$ is not embeddable in H_α (where $\omega_\gamma = \mathrm{cf}(\omega_\alpha)$).*

Proof. Our assertion would follow immediately if the GCH is adopted, for then we would have $|\omega_\alpha((\omega_\gamma))| = \aleph_\alpha^{\aleph_\gamma} > \aleph_\alpha = |H_\alpha|$. But we don't need the GCH.

If ω_α is regular, we have $\omega_\gamma = \omega_\alpha$, and then our assertion is an immediate consequence of 7.11. So we assume that ω_α is singular. And then we define ordinals ω_{α_ν} as in the proof of 7.7. Further for $\nu < \omega_\gamma$ we define H'_{α_ν} as the set of all those sequences of H_α that have their last digit 1 at a position $\mu < \omega_{\alpha_\nu}$.

We assume indirectly that there exists a $<$ - preserving mapping f of S in H_α. If $(\tau_\nu)_{\nu<\lambda}$ is a sequence of length $\lambda \leq \omega_\gamma$ of ordinals $< \omega_\alpha$, we define $S[\tau_\nu | \nu < \lambda]$ to be the set of all elements of S that have $(\tau_\nu)_{\nu<\lambda}$ as initial segment. First we consider the sets $S[\tau]$, where $\tau < \omega_\alpha$. There must exist a $\tau_0 < \omega_\alpha$ such that $f[S[\tau_0]] \cap H'_{\alpha_0} = \emptyset$ because H'_{α_0} has no subset of type ω_α.

Suppose now that we have already defined ordinals $\tau_\nu < \omega_\alpha$ for $\nu < \lambda$, where λ is an ordinal with $0 < \lambda < \omega_\gamma$, such that $f[S[\tau_\nu | \nu < \mu]] \cap H'_{\alpha_\mu} = \emptyset$ holds for all $\mu < \lambda$. Then we conclude as before that there exists an ordinal $\tau_\lambda < \omega_\alpha$, such that $f[S[\tau_\nu | \nu \leq \lambda]] \cap H'_{\alpha_\lambda} = \emptyset$, because H'_{α_λ} has no subset of type ω_α.

By transfinite induction we so obtain a sequence of length ω_γ of ordinals $\tau_\nu < \omega_\alpha$, i. e. an element of S, for which $f[S[\tau_\nu | \nu < \omega_\gamma]] \cap \cup \{H'_{\alpha_\nu} | \nu < \omega_\gamma\} = \emptyset$ which is impossible since the last union is H_α.

7.13 Theorem (Hausdorff [82],1908). *Let S be a linearly ordered set which has no gaps, but a first element a and a last element $z > a$. Let μ be an arbitrary ordinal. Then there is no $< $ - preserving mapping of $S((\mu + 1))$ in $S((\mu))$.*

Proof. $S((\mu))$ is, of course, isomorphic to the set S^* of all $x \in S((\mu + 1))$ that have z as last component. Now S^* is an isolated subset of $S((\mu + 1))$. For let $x = (x_\nu)_{\nu < \mu+1}$ be an element of S^*, and thus $x_\mu = z$. Then $x \in (y, x] := I_x$, where $y = (y_\nu)_{\nu < \mu+1}$ has the components $y_\nu = x_\nu$ for $\nu < \mu$, and $y_\mu = a$. The intervals I_x, $x \in S^*$, are pairwise disjoint, and so our statement follows from 7.10, since $S((\mu + 1))$ has no gaps, but a first and a last element.

As a corollary to 7.13 we mention:

7.14 Theorem. *For every ordinal μ the set $2((\mu + 1))$ is not isomorphic to a subset of $2((\mu))$.*

Now we supplement the statement of 7.6. This theorem has the following immediate consequence:

7.15 Remark. *For $\omega_\gamma = \mathrm{cf}(\omega_\alpha))$ the set $\omega_\alpha((\omega_\gamma))$ is embeddable in $2((\omega_\alpha))$.*

This statement is best possible with regard to the ordinal ω_γ. Namely there holds:

7.16 Theorem. *For $\omega_\gamma = \mathrm{cf}(\omega_\alpha)$ the set $M := \omega_\alpha((\omega_\gamma + 1))$ is not embeddable in $2((\omega_\alpha))$.*

Proof. If in 7.13 we take ω_α instead of μ, we obtain that $2((\omega_\alpha+1))$, and then a fortiori $\omega_\alpha((\omega_\alpha+1))$ is not embeddable in $2((\omega_\alpha))$. Thus we are done, if ω_α is regular.

128

Let now ω_α be singular. Then there exists a (with ν) strictly increasing sequence of inital ordinals $\omega_{\alpha_\nu}, \nu < \omega_\gamma$, for which $\sum\{\omega_{\alpha_\nu}|\nu < \omega_\gamma\} = \omega_\alpha$ holds. We assume indirectly that there exists a $<$ - preserving mapping $f : \omega_\alpha((\omega_\gamma + 1)) \longrightarrow 2((\omega_\alpha))$.

If $t = (\tau_\nu|\nu < \mu)$ is a sequence of ordinals $< \omega_\alpha$ and $\mu < \omega_\gamma$, we define $S(t)$ to be the set of all sequences of $\omega_\alpha((\omega_\gamma + 1))$, that *begin* with t, i. e. which have t as initial segment. And if $d = (\delta_\nu|\nu < \kappa)$ is a dyadic sequence, so that $\delta_\nu \in \{0,1\}$ for $\nu < \kappa$, then we denote the set of all sequences of $2((\omega_\alpha))$, that begin with d, by $D(d)$. First we prove:

(I) There is an ordinal $\tau < \omega_\alpha$ and a dyadic sequence $d = (\delta_\nu|\nu < \omega_{\alpha_0})$ such that $f[S(\tau)] \subseteq D(d)$ holds.

There are three consecutive ordinals $\kappa, \kappa + 1, \kappa + 2 < \omega_\alpha$ such that the f - image sets $f[S(\tau)]$, $f[S(\tau + 1)]$, $f[S(\tau + 2)]$ intersect the same segment $D(d)$ with a $d = (\delta_\nu|\nu < \omega_{\alpha_0})$ of $2((\omega_\alpha))$. For if this would be false, then each $D(\delta_\nu|\nu < \omega_{\alpha_0})$ would intersect at most two of the sets $f[S(t)]$, where t is an ordinal $< \omega_\alpha$. And then the set $\{f[S(t)]|t < \omega_\alpha\}$ would have, relative to the natural order of 1.8.1, an order type $\leq \sum\{2_i|i \in 2((\omega_{\alpha_0}))\}$, where each 2_i is the order type 2. But it has the order type ω_α, which is not embeddable in the last ordered sum. If now we put $\tau := \kappa + 1$ the set $f[S(\tau)]$ fulfills (I).

Suppose now that for an ordinal $\mu < \omega_\gamma$ we have already defined for all $\nu < \mu$ ordinals $\tau_\nu < \omega_\alpha$ and for all $\nu < \omega_{\alpha_\mu}$ ordinals $\delta_\nu \in \{0,1\}$ such that

(II) $f[S(\tau_\nu|\nu < \mu)] \subseteq D(\delta_\nu|\nu \in \sum \omega_{\alpha_\nu}|\nu < \mu)$

holds. Then, in the same way as before, we can find an ordinal τ_μ and a dyadic sequence of length ω_{α_μ} such that $f[S(\tau_\nu|\nu \leq \mu)] \subseteq D(\delta_\nu|\nu \in \sum \omega_{\alpha_\nu}|\nu \leq \mu)$. If λ is a limit ordinal for which (II) holds for all $\mu < \lambda$, then also (II) is valid for λ. So, finally, we have by transfinite induction obtained a sequence $t = (\tau_\nu|\nu < \omega_\gamma)$ of ordinals $\tau_\nu < \omega_\alpha$ such that all sequences of $\omega_\alpha((\omega_\gamma + 1))$ that begin with t, are mapped by f on one element of $2((\omega_\alpha))$. This is a contradiction because there are still $\aleph_\alpha$ sequences in $\omega_\alpha((\omega_\gamma + 1))$ which begin with t.

We remark that in [108] it was indicated that $\omega_\alpha((\omega_\alpha))$ is embeddable in $2((\omega_\alpha))$ for all ω_α. But this is by 7.16 only true for the regular ω_α. The citation in the footnote of [70], p. 245 should be corrected adequately.

In analogy to H_α we define sets G_α as follows:

7.17 Definition. Let α be an ordinal. Then the set of all sequences of $2((\omega_\alpha))$ that have a last digit 0 is denoted by G_α.

The following statement is easily verified:

7.18 Theorem. *Every element $s \in G_\alpha$ has an immediate successor t in $2((\omega_\alpha))$, and every element $t \in H_\alpha$ has an immediate predecessor $s \in G_\alpha$. All elements of $2((\omega_\alpha))$ that have an immediate successor (resp. immediate predecessor) are in G_α (resp. in H_α).*

The mapping which ascribes to every $s \in G_\alpha$ its immediate successor in H_α constitutes an isomorphic mapping of G_α onto H_α. Thus also G_α has the order type h_α.

Proof. Let $s = (s_\nu | \nu < \omega_\alpha) \in G_\alpha$ have its last digit 0 at the position τ. We put $t_\nu = s_\nu$ for $\nu < \tau$, $t_\tau = 1$, $t_\nu = 0$ for $\tau < \nu < \omega_\alpha$. Then the sequence $t = (t_\nu | \nu < \omega_\alpha)$ is evidently the immediate successor of s. The rest follows easily.

Later we wish to prove a theorem of Kurepa/Papic [108], according to which all lexicographic products $(\nu + 1)((\omega_\alpha))$ with $1 \leq \nu \leq \omega_\alpha$ have the same order type. We first establish:

7.19 Lemma. *Let ω_α be an initial ordinal and $S := (\nu + 1)((\omega_\alpha))$ where ν is an ordinal with $0 < \nu < \omega_\alpha$. We define D to be the set of all those sequences $a \in S$ satisfying*

$(*)$ *$a = (a_\mu | \mu < \omega_\alpha)$, and there exists a last component $a_\kappa \neq \nu$. (For $\nu = 1$ the set D is $= G_\alpha$.)*

Then D has the order type h_α.

Proof. First we assume that ω_α is regular. Then there holds:

(1) Every $a \in D$ has the element-character $(\omega_\alpha, \omega_\alpha^*)$.

Let a satisfy $(*)$. For $\kappa < \mu < \omega_\alpha$ let b_μ be that sequence of D which arises from a by exchanging the μ - component of a by 0. Then $\{b_\mu | \kappa < \mu < \omega_\alpha\}$ is cofinal in $(D < a)$.

For $\kappa < \mu < \omega_\alpha$ we define sequences $c_\mu = (c_{\mu\tau} | \tau < \omega_\alpha)$ by

$c_{\mu\tau} = a_\tau$ for $\tau < \kappa$,

$c_{\mu\kappa} = a_\kappa + 1$,

$c_{\mu\tau} = 0$ for $\kappa < \tau \leq \mu$

$c_{\mu\tau} = \nu$ for $\mu < \tau < \omega_\alpha$.

Then $\{c_\mu | \mu < \omega_\alpha\}$ is coinitial in $(D > a)$. And thus (1) is proved.

For $\kappa < \omega_\alpha$ let $(s_{\kappa\mu})$ be the sequence where $s_{\kappa\mu} = 0$ for $\mu \leq \kappa$ and $s_{\kappa\mu} = \nu$ for $\kappa < \mu < \omega_\alpha$. Then $\{s_\kappa | \kappa < \omega_\alpha\}$ is a coinitial subset of D of type ω_α^*.

For $\kappa < \omega_\alpha$ we define $t_\kappa = (t_{\kappa\mu} | \mu < \omega_\alpha)$, where $t_{\kappa\kappa} = 0$ and $t_{\kappa\mu} = \nu$ for all $\mu \neq \kappa$. Then $\{t_\kappa | \kappa < \omega_\alpha\}$ is a cofinal subset of D of type ω_α.

Let now (A, B) be a gap of D. According to 2.10 then there exists the supremum $s = (s_\mu | \mu < \omega_\alpha)$ of A in S. Because of $s \in S \backslash D$ and $s \neq (\nu, \nu, \ldots)$ we have two possible cases:

Case 1. The set of indices μ, for which s_μ is $\neq \nu$, is cofinal in ω_α.

If Case 1 does not hold, then the set of indices with $s_\mu \neq \nu$ is not cofinal in ω_α, and then we have:

Case 2. There exists a least ordinal $\lambda < \omega_\alpha$, such that $s_\mu = \nu$ for all μ with $\lambda \leq \mu < \omega_\alpha$. This λ must be a limit ordinal, for otherwise s would at position $\lambda - 1$ have a last component $\neq \nu$, and then s would belong to D with contradiction. And due to $s \notin D$ the set of ordinals $\kappa < \lambda$, for which $s_\kappa \neq \nu$ holds, is cofinal in λ.

In Case 1 let $\{\mu_\rho | \rho < \omega_\alpha\}$ be the set of all indices μ with $s_\mu \neq \nu$. For $\rho < \omega_\alpha$ we define b_ρ to be that sequence which arises from s by replacing all components s_μ with $\mu > \mu_\rho$ by ν. Then b_ρ has a last component $\neq \nu$ at the position μ_ρ and is thus in D. Now $\{b_\rho | \rho < \omega_\alpha\}$ is coinitial in $(S > s)$, and then also coinitial in B.

In Case 2 we define d_μ for $\mu < \omega_\alpha$ as follows: d_μ arises from s by exchanging the $(\lambda + \mu)$ - component $s_{\lambda+\mu}$ by 0. Then $d_\mu \in D$, and the set $\{d_\mu | \mu < \omega_\alpha\}$ is cofinal in $(S < s)$, and then also cofinal in A.

So we have obtained that every gap (A, B) has character $(\omega_\beta, \omega_\alpha^*)$ in Case 1 (resp. character $(\omega_\alpha, \omega_\beta^*)$ in Case 2) where ω_β is an initial ordinal. Thus D is an η_α - set.

Now D is the union of $\aleph_\alpha$ sets D_μ, $\mu < \omega_\alpha$, where D_μ consists of all those sequences of D that have their last component $\neq \nu$ at a position $< \mu$. Then D_μ is embeddable in $(\nu + 1)((\mu))$ and thus free from ω_α and ω_α^* by 2.2. Due to 3.12 our lemma now is proved for regular ω_α.

Let ω_α be singular. Then we have a representation $\omega_\alpha = \sum_{\mu<\alpha} \omega_{\mu+1}$. We define $D_{\mu+1}$ to be the set of all sequences of D, for which the last component which is $\neq \nu$, is at a position $< \omega_{\mu+1}$. Then it follows from the first part of this proof that $D_{\mu+1}$ has the order type $h_{\mu+1}$. Because of $D = \cup_{\mu<\alpha} D_{\mu+1}$ our assertion now follows from 4.3.

7.20 Lemma. *Under the assumptions of 7.19, D is dense in $S = (\nu + 1)((\omega_\alpha))$.*

Proof. Let $a = (a_\mu|\mu < \omega_\alpha)$ and $b = (b_\mu|\mu < \omega_\alpha)$ be sequences of S with $a < b$, τ the first ordinal where their components a_τ and b_τ are different, so that $a_\tau < b_\tau$ holds. Let then $x = (x_\mu|\mu < \omega_\alpha)$ be defined by $x_\mu = a_\mu$ for $\mu \le \tau$, $x_\mu = \nu$ for $\tau < \mu < \omega_\alpha$. Then we have $x \in D$, for its last component $\neq \nu$ is $x_\tau = a_\tau$, and this is $(< b_\tau$, and a fortiori$) < \nu$. Further we have $a \le x < b$.

D is not strictly dense in S since S has elements which are neighboring.

Now we can prove:

7.21. Theorem (Kurepa/Papic [108]) *Let ω_α be an initial ordinal and $1 \le \nu < \omega_\alpha$. Then the lexicographically ordered set*
$S := (\nu + 1)((\omega_\alpha))$ *is isomorphic to* $2((\omega_\alpha))$.

Proof. With the concepts of 7.19 we have an isomorphic mapping $f : D \to G_\alpha$. Let D' be the set of immediate successors of elements of D. Then we extend f by mapping every $x \in D'$ onto the immediate successor (in $2((\omega_\alpha))$) of $f(x)$, which is in H_α. Further we map the first (resp. last) element of $(\nu + 1)((\omega_\alpha))$ onto the first (resp. last) element of $2((\omega_\alpha))$.

Let now x be an element of $S \backslash (D \cup D')$, which is neither first nor last element of S. Then $((D < x), (D > x))$ determines a gap in D of type (∞, ∞) (see Definition 1.10.15), and $(f[(D < x)], f[(D > x)]$) determines the corresponding gap in $f[D] \subseteq 2((\omega_\alpha))$. Since this latter set has no gaps there exists exactly one element $y \in 2((\omega_\alpha))$ with $f[(D < x)] < y < f[(D > x)]$. We map x onto y. By extending f in this way it can be seen that we so obtain an isomorphic mapping of S onto $2((\omega_\alpha))$.

A fortiori we also have:

7.22 Theorem. *Let ν be an ordinal $< \omega_\alpha$. Then $\nu((\omega_\alpha))$ is embeddable in $2((\omega_\alpha))$.*

For even $(\nu + 1)((\omega_\alpha))$ is embeddable in $2((\omega_\alpha))$.

7.23 Remark. Theorem 7.21 cannot be sharpened so that also $\nu((\omega_\alpha))$ would be isomorphic to $2((\omega_\alpha))$. Indeed, if λ is a limit ordinal $< \omega_\alpha$, the set $\lambda((\omega_\alpha))$ is cofinal with λ and with $\mathrm{cf}(\lambda)$. In particular, it has no last element. Moreover, for different ordinals λ, e.g. for different initial ordinals, the cofinality-types $\mathrm{cf}(\lambda)$ can be different.

132

At the end of this paragraph we compile several results. We can give a complete discussion of which lexicographic products $\mu((\nu))$ are embeddable in H_α resp. in $2((\omega_\alpha))$.

7.24 Theorem. *Let μ, ν be ordinals ≥ 2, $\omega_\gamma = \mathrm{cf}(\omega_\alpha)$. Then there holds:*

1) *If $\mu \geq \omega_{\alpha+1}$, then $\mu((\nu))$ is not embeddable in H_α.*
2) *If $\omega_\alpha \leq \mu < \omega_{\alpha+1}$ and $\nu < \omega_\gamma$, then $\mu((\nu))$ is embeddable in H_α.*
3) *If $\omega_\alpha \leq \mu < \omega_{\alpha+1}$ and $\nu \geq \omega_\gamma$, then $\mu((\nu))$ is not embeddable in H_α.*
4) *If $\mu < \omega_\alpha$ and $\nu < \omega_\alpha$, then $\mu((\nu))$ is embeddable in H_α.*
5) *If $\mu < \omega_\alpha$ and $\nu \geq \omega_\alpha$, then $\mu((\nu))$ is not embeddable in H_α.*

Proof. 1) follows from the fact that H_α has no subset of type $\omega_{\alpha+1}$, due to 2.6, b).

2) μ is embeddable in H_α and then $\mu((\nu))$ in $H_\alpha((\nu))$, which has the type h_α by 7.5.

3) $\omega_\alpha((\omega_\gamma))$ is not embeddable in H_α by 7.12. A fortiori this holds for $\mu((\nu))$.

4) If ω_α is regular, then $\mu((\nu))$ is embeddable in $H_\alpha((\nu))$, and this in H_α by 7.4.

If ω_α is singular, then there exists a regular initial ordinal $\omega_\rho < \omega_\alpha$, for which μ and ν are $< \omega_\rho$, and then $\mu((\nu))$ is embeddable in $(\omega_\rho + 1)((\omega_{\rho+1}))$, the latter in $2((\omega_{\rho+1}))$ by 7.22, and this in H_α.

5) $2((\omega_\alpha))$ is not embeddable in H_α by 7.11. So we are done because of $\mu \geq 2$ and $\nu \geq \omega_\alpha$.

In the cases 2) and 4) of 7.24 the set $\mu((\nu))$ is a fortiori also embeddable in $2((\omega_\alpha))$, so we have only to discuss the cases 1),3),5) in the next theorem:

7.25 Theorem. *Let μ, ν be ordinals ≥ 2, $\omega_\gamma = \mathrm{cf}(\omega_\alpha)$. Then there holds:*

1) *If $\mu \geq \omega_{\alpha+1}$, then $\mu((\nu))$ is not embeddable in $2((\omega_\alpha))$.*
3) a) *If $\omega_\alpha \leq \mu < \omega_{\alpha+1}$ and $\nu = \omega_\gamma$, then $\mu((\nu))$ is embeddable in $2((\omega_\alpha))$.*
3) b) *If $\omega_\alpha \leq \mu < \omega_{\alpha+1}$ and $\nu > \omega_\gamma$, then $\mu((\nu))$ is not embeddable in $2((\omega_\alpha))$.*
5) a) *If $\mu < \omega_\alpha$ and $\nu = \omega_\alpha$, then $\mu((\nu))$ is embeddable in $2((\omega_\alpha))$.*
5) b) *If $\mu < \omega_\alpha$ and $\nu > \omega_\alpha$, then $\mu((\nu))$ is not embeddable in $2((\omega_\alpha))$.*

Proof. 1) $2((\omega_\alpha))$ has no subset of type $\omega_{\alpha+1}$ by 2.6, b).

3) a) $\mu((\nu))$ is embeddable in $H_\alpha((\omega_\gamma))$, and this in $2((\omega_\alpha))$ by 7.6.

3) b) $\omega_\alpha((\omega_\gamma + 1))$ is not embeddable in $2((\omega_\alpha))$ by 7.16. A fortiori this holds for $\mu((\nu))$.

5) a) $\mu((\nu))$ is embeddable in $2((\omega_\alpha))$ by 7.22.

5) b) $2((\omega_\alpha + 1))$ is not embeddable in $2((\omega_\alpha))$ by 7.14. A fortiori this holds for $\mu((\nu))$.

We supplement our investigations by considering which well- (resp. inversely well-ordered) subsets the lexicographically ordered sets $\mu((\nu))$, where μ and ν are ordinals, can have:

7.26 Theorem. *Let μ and ν be ordinals $< \omega_\alpha$, which are not 0. Then the lexicographically ordered set $\mu((\nu))$ has no subset of type ω_α.*

Proof. For $\alpha = 0$ the statement is trivial. So let α be > 0. From $\mu < \omega_\alpha$ we conclude $|\mu| \leq \aleph_\beta$ for a $\beta < \alpha$, and therefore μ is embeddable in H_β, and further $\mu((\nu))$ in $H_\beta((\nu))$. The set H_β is ω_α - free.

If now ω_α is regular we obtain from 2.2, a) that $H_\beta((\nu))$ is ω_α - free, and this a fortiori holds for $\mu((\nu))$.

If ω_α is singular, α is a limit ordinal, and then there exists a regular initial ordinal $\omega_\beta < \omega_\alpha$ which is greater than μ and ν. Now by the first case $\mu((\nu))$ has no subset of type ω_β, and a fortiori no subset of type ω_α.

The Theorem 7.26 is tight:

7.27 Remark. If μ is $\geq \omega_\alpha$ and $\nu = 1$, then $\mu((\nu))$ is isomorphic with ω_α. And if μ is ≥ 2 and $\nu \geq \omega_\alpha$, then already $2((\omega_\alpha)) \supseteq H_\alpha$ contains a subset of type ω_α.

7.28 Theorem. *Let μ be an arbitrary ordinal, $\nu < \omega_\alpha$. Then $\mu((\nu))$ has no subset of type ω_α^*. And this statement is tight.*

Proof. μ has no subset of type ω_0^*, a fortiori no subset of type ω_α^*. If ω_α is regular, then it follows from 2.2, b) that $\mu((\nu))$ is ω_α^* - free.

Let now ω_α be singular, so that α is a limit ordinal. Then there exists a regular initial ordinal $\omega_\beta < \omega_\alpha$ which is $> \nu$. Then by the above case $\mu((\nu))$ has no subset of type ω_β^*, and a fortiori no subset of type ω_α^*.

7.28 is tight: If μ is ≥ 2 and $\nu \geq \omega_\alpha$ then $\mu((\nu))$ contains $2((\omega_\alpha))$ and then also a subset of type ω_α^*.

7.28′ Corollary. *If ω_α is singular and $\omega_\gamma = \mathrm{cf}\,\omega_\alpha$, then $\omega_\alpha((\omega_\gamma))$ has no subset of type $\omega^*_{\gamma+1}$, and a fortiori no subset of type ω^*_α.*

Now we can supplement 7.15 by:

7.29 Theorem. *If ω_α is singular and $\omega_\gamma = \mathrm{cf}\,\omega_\alpha$, then $2((\omega_\alpha))$ is not embeddable in $\omega_\alpha((\omega_\gamma))$.*

Proof. By 7.28′ $\omega_\alpha((\omega_\gamma))$ has no subset of type ω^*_α, but $2((\omega_\alpha))$ does.

In the following we introduce certain subsets of $2((\omega_\alpha))$, that constitute higher-dimensional generalizations of the set $\mathbf{R}$ of real numbers.

The set $2((\omega_\alpha))$ has jumps: There are elements a, b in $2((\omega_\alpha))$, where a is an immediate predecessor of b. Namely we have, recalling a previous result:

7.30 Remark. *Let a and b be elements of $2((\omega_\alpha))$. Then the sequence $a = (a_\nu | \nu < \omega_\alpha)$ is an immediate predecessor of the sequence $b = (b_\nu | \nu < \omega_\alpha)$ iff a has a last digit 0, say $a_\nu = 0$, $b_\mu = a_\mu$ for $\mu < \nu$, $b_\nu = 1$, and $b_\mu = 0$ for $\nu < \mu < \omega_\alpha$.*

In order to obtain a subset of $2((\omega_\alpha))$ which is dense, we eliminate from $2((\omega_\alpha)$ the elements of G_α (of 7.17) and define:

7.31 Definition. For an ordinal α we put $C_\alpha := 2((\omega_\alpha))\backslash(G_\alpha \cup \{\mathfrak{o}, \mathfrak{l}\})$, where $\mathfrak{o}$ is the first and $\mathfrak{l}$ the last element of $2((\omega_\alpha))$.

For these sets we now have:

7.32 Theorem. *Let α be an ordinal. Then C_α is dense, and its subset H_α is strictly dense in C_α. Each open interval of C_α has a segment which is isomorphic to C_α. Further C_α is continuously ordered, that means it has no gaps, and C_α has no first and no last element. The set C_0 is isomorphic to the set $\mathbf{R}$ of real numbers.*

Proof. Let $a = (a_\nu | \nu < \omega_\alpha), b = (b_\nu | \nu < \omega_\alpha) \in C_\alpha$ satisfy $a < b$. At the first index μ, for which $a_\mu \neq b_\mu$ holds, we have $a_\mu = 0$ and $b_\mu = 1$. Since a is not in G_α there is an ordinal $\tau > \mu$ with $a_\tau = 0$, and then the element $z = (z_\nu | \nu < \omega_\alpha)$, which satisfies $z_\nu = a_\nu$ for $\nu < \tau$, $z_\tau = 1$, and $z_\nu = 0$ for $\tau < \nu < \omega_\alpha$, is in H_α and satisfies $a < z < b$.

Let S be the set of all $x = (x_\nu | \nu < \omega_\alpha) \in C_\alpha$ that satisfy $x_\nu = a_\nu$ for $\nu < \tau$ and $x_\tau = 1$. These fulfill $a < x < b$ and form a segment $\subseteq (a, b)$. This is isomorphic to the set F of final segments $(x_\nu | \tau < \nu < \omega_\alpha)$ of the

$x \in S$, and F is isomorphic to one of the sets C_α, $\{o\} \cup C_\alpha$, $C_\alpha \cup \{l\}$, $\{o\} \cup C_\alpha \cup \{l\}$. If we omit the first (resp. last) element of S, if it has one, we obtain a segment $\subseteq S \subseteq (a, b)$, which is isomorphic to C_α.

If A is a non-empty initial segment of C_α, and if the complementary final segment $B := C_\alpha \setminus A$ is also non-empty, then A has a last or B a first element. For A has, due to 2.10, as a subset of $2((\omega_\alpha))$ a supremum s in $2((\omega_\alpha))$. If $s \in C_\alpha$, then s is also the supremum of A in C_α. If $s \notin C_\alpha$, then $s \in G_\alpha$ has the form $(\ldots, 0, 1, 1, 1, \ldots)$ and then the upper neighbor $(\ldots, 1, 0, 0, 0, \ldots)$ of s in $2((\omega_\alpha))$ is the supremum of A in C_α.

B has an infimum i in $2((\omega_\alpha))$. This is not in G_α. For if i would be $= (\ldots 0, 1, 1, 1, \ldots)$ with a last component 0, then its upper neighbor $(\ldots 1, 0, 0, 0, \ldots)$ would be a lower bound of B, and i could not be $= \inf B$ in $2((\omega_\alpha))$. So i is also the infimum of B in C_α.

If now s is the last element of A, we are done. If $s \notin A$, then i must be the first element of B. Indeed, we have $s \leq i$, and $s < i$ is impossible since C_α is dense. Now $s = i \in C_\alpha = A \cup B$, and so $s = i$ belongs to A or B. The rest is clear.

The order type of the set C_α is a subtype of that of $2((\omega_\alpha))$, but the converse also holds:

7.33 Theorem. *For each ordinal α the set $2((\omega_\alpha))$ is embeddable in C_α. And $|C_\alpha| = 2^{\aleph_\alpha}$.*

Proof. If $\delta = (\delta_\nu | \nu < \omega_\alpha)$ is a dyadic sequence $\in 2((\omega_\alpha))$, we ascribe to δ an element $f(\delta) \in C_\alpha$ as follows: We enlarge the sequence δ by putting an additional digit 0 directly behind each digit 1 which occurs in the sequence δ. The sequences which so arise are not in G_α, and thus in $C_\alpha \cup \{o\}$. If then $\delta < \varepsilon$ are sequences $\in 2((\omega_\alpha))$, their image sequences satisfy $f(\delta) < f(\varepsilon)$, so that $2((\omega_\alpha))$ is isomorphic to a subset of $C_\alpha \cup \{o\}$.

Let now g be the mapping which ascribes to every $\delta \in 2((\omega_\alpha))$ that sequence $\in 2((\omega_\alpha))$, which arises from δ by putting the digit 1 before δ. Then the composition $f \circ g : 2((\omega_\alpha)) \to C_\alpha$ is $< \text{-}$ preserving, and this proves our assertion.

Since H_α is dense in C_α we can state that, using GCH, there is a linearly ordered set of cardinality $\aleph_{\alpha+1}$, in which a subset of cardinality $\aleph_\alpha$ is dense. The question arises, what can be said if we don't presuppose the GCH. In this context we can now apply the previous theorems for a short proof of a theorem of Sierpinski [163], with which he gave a positive

136

answer to questions of Knaster (resp. Kuratowski). We formulate it in a more general form, which was already indicated in [163]:

7.34 Theorem [163]. *Let m be an infinite cardinal. Then there is a linearly ordered set S of cardinality $> m$, which has a subset D of cardinality $\leq m$, which is strictly dense in S.*

Proof. For $m = \aleph_0$ the sets $D := \mathbf{Q}$ and $S = \mathbf{R}$ satify the assertion. So we suppose $m > \aleph_0$. There exist cardinals $k \leq m$ with $2^k > m$, (e.g. $k = m$). Let $\aleph_\alpha$ be the least cardinal k with this property, so that we have:

(1) $\aleph_\alpha \leq m$.

(2) $2^{\aleph_\alpha} > m$.

(3) $2^k \leq m$ for all cardinals $k < \aleph_\alpha$.

Now $S := C_\alpha$ (of 7.31) satisfies the assertion; C_α has by 7.33 and (2) the cardinality $2^{\aleph_\alpha} > m$, and $D := H_\alpha$ is strictly dense in C_α. So we are done if we have proved $|H_\alpha| \leq m$. We distinguish two cases:

Case 1. α is a successor number $\beta + 1$. Then $|H_\alpha| = \mathfrak{k}_\alpha = 2^{\aleph_\beta} \leq m$ by (3).

Case 2. α is a limit ordinal. Then $|H_\alpha| = \mathfrak{k}_\alpha = \sum_{\nu < \alpha} 2^{\aleph_\nu} \leq m \cdot |\alpha|$ by (3) because of $\aleph_\nu < \aleph_\alpha$ for $\nu < \alpha$. For $\nu < \alpha$ we have $\nu \leq \omega_\nu$ and thus $|\nu| \leq |\omega_\nu| = \aleph_\nu$ and $|\nu| < 2^{\aleph_\nu} \leq m$ by (3). This yields $\nu < \omega(m)$ for all $\nu < \alpha$. Then $\alpha = \sup\{\nu | \nu < \alpha\}$ is $\leq \omega(m)$. This entails $|\alpha| \leq m$, so that $|H_\alpha| \leq m \cdot m = m$ follows.

As a consequence of 7.34 we also obtain a positive answer to the following question of Kuratowski: Does there exist a chain of subsets of $(\mathbf{R}, \subseteq)$ which has a cardinality $> 2^{\aleph_0}$? First we prove in general:

7.35 Theorem [163]. *Let m be an infinite cardinal, $|M| = m$. Then $(\mathfrak{P}(M), \subseteq)$ has a chain of cardinality $> m$.*

Proof. For the cardinal m we determine sets S and D as in 7.34, so that $|D| \leq |M| = m$ holds. The mapping which assigns to each $x \in S$ the initial segment $(D < x)$ of D, is injective since D is strictly dense in S. And the set of all initial segments $(D < x), x \in S$, ordered by $\subseteq$, is a chain of $(\mathfrak{P}(D), \subseteq)$ of cardinality $|S| > m$. Since $(\mathfrak{P}(D), \subseteq)$ is isomorphic to a subset of $(\mathfrak{P}(M), \subseteq)$ our assertion follows.

If in the last theorem we take $m := 2^{\aleph_0}$ and $M := \mathbf{R}$, the above question is answered.

4.8 Cantor's normal form. Indecomposable ordinals

In the following we need some concepts and theorems concerning the exponentiation of ordinals. We extract here only what we use in the sequel.

8.1 Definition. For the least infinite ordinal ω we define $\omega^0 := 1$ and $\omega^1 := \omega$. If we have already defined ω^ν for an ordinal ν we put $\omega^{\nu+1} := \omega^\nu \cdot \omega \; (= \omega^\nu + \omega^\nu + \cdots | \omega$ many summands$)$. If for a limit ordinal λ the power ω^ν is already introduced for all $\nu < \lambda$ we put $\omega^\lambda := \cup\{\omega^\nu | \nu < \lambda\} \; (= \sup\{\omega^\nu | \nu < \lambda\})$. By transfinite induction we so have defined the powers ω^ν of ω for all exponents ν which are ordinals. (This concept must not be confused with the exponentiation of cardinal numbers. So ω^ω has only cardinality $\aleph_0$, whereas $\aleph_0^{\aleph_0}$ has cardinality $> \aleph_0$.)

An ordinal ζ is said to be *indecomposable,* if it is not the sum of two smaller ordinals:

$$\zeta = \alpha + \beta \Longrightarrow \alpha = \zeta \text{ (and then } \beta = 0) \text{ or } \beta = \zeta \text{ (and then } \alpha < \zeta).$$

8.2 Theorem. *Every power ω^ν is indecomposable.*

Proof. For $\nu = 0$ or 1 this is trivial. Let ν be an ordinal for which the statement is true for all ordinals $\leq \nu$. Suppose indirectly $\omega^{\nu+1} = \alpha + \beta$, where α and β are $< \omega^{\nu+1}$. Then there exist numbers $a, b \in \mathbf{N}_0$ with $\alpha < \omega^\nu \cdot a$ and $\beta < \omega^\nu \cdot b$. For if α would be $> \omega^\nu \cdot n$ for all $n < \omega$, a would also be $\geq \cup\{\omega^\nu \cdot n | n \in \mathbf{N}\} = \omega^{\nu+1}$, and similar for β. Now $\alpha + \beta$ would be $< \omega^\nu \cdot a + \omega^\nu \cdot b = \omega^\nu \cdot (a + b) < \omega^\nu \cdot \omega = \omega^{\nu+1}$, a contradiction.

8.3 Theorem. *Let μ, ν be ordinals with $\mu < \nu$. Then there holds $\omega^\mu < \omega^\nu$. Further we have $\omega^\mu \geq \mu$ for all ordinals μ.*

Proof. $\omega^{\mu+1} > \omega^\mu$ follows immediately from the definition; also for limit ordinals λ we have $\omega^\lambda > \omega^\mu$ for all $\mu < \lambda$. And $\omega^\mu \geq \mu$ follows easily by induction.

Here we remark without proof that it can happen that $\omega^\mu = \mu$ holds.

8.4 Lemma. *Let μ be an ordinal > 0. Then there exists a greatest ordinal α for which ω^α is $\leq \mu$. And then further there exists a greatest number $n \in \mathbf{N}_0$ for which $\omega^\alpha \cdot n \leq \mu$.*

Proof. Let S be the set of all ordinals ν which satisfy $\omega^\nu \leq \mu$. This set contains with an ordinal ν also all ordinals $\tau < \nu$. Then the supremum α of S is an ordinal for which still $\omega^\alpha \leq \mu$ holds, and then α is the greatest exponent with this property.

Now we consider the sequence $\omega^\alpha \cdot 0$, $\omega^\alpha \cdot 1, \ldots$. Here the first member is $\leq \mu$. But not all are $\leq \mu$, because in this case we would have that also $\omega^{\alpha+1} = \cup\{\omega^\alpha \cdot m \mid m \in \mathbf{N}_0\}$ would be $\leq \mu$. So the number n exists.

Now we can construct Cantor's normal form:

8.5 Theorem and **Definition.** *For every ordinal $\mu > 0$ we have a unique representation $\mu = \omega^{\alpha_0} \cdot a_0 + \cdots + \omega^{\alpha_n} \cdot a_n$, where n is a non-negative integer, $\alpha_0 > \cdots > \alpha_n$ is a strictly decreasing sequence of ordinals, and $a_0, \ldots, a_n$ are natural numbers. This is called Cantor's normal form (or representation) of μ.*

Proof. First we prove the existence. Suppose that for all ordinals $< \mu$ that are $\neq 0$ we have already proved that they have a uniquely determined representation of the above kind. Then by 8.4 there exists a greatest ordinal α_0 with $\omega^{\alpha_0} \leq \mu$ and further a greatest number $n_0 \in \mathbf{N}$ with $\omega^{\alpha_0} \cdot a_0 \leq \mu$. Then the ordinal ρ, which is uniquely determined by $\mu = \omega^{\alpha_0} \cdot a_0 + \rho$, is 0 or has, due to the induction hypothesis, a representation $\rho = \omega^{\alpha_1} \cdot a_1 + \cdots + \omega^{\alpha_n} \cdot a_n$ of the kind under consideration, where in particular $\alpha_1 < \alpha_0$ holds. So the existence is proved by induction.

Uniqueness follows so: Let this be proved for all ordinals $< \mu$. In every representation of μ in the above form α_0 is the greatest ordinal for which ω^{α_0} is $\leq \mu$, and so α_0 is uniquely determined. The same holds for a_0, due to 8.4. Since by induction hypothesis also ρ is uniquely determined, the uniqueness of the representation now easily follows by induction on μ.

Using Cantor's normal form a possibility arises to introduce an addition in the class of ordinals which is different from the usual addition, but has nice properties:

8.6 Definition. Let α and β be ordinals > 0 with normal form
(1) $\alpha = \omega^{\alpha_0} \cdot a_0 + \cdots + \omega^{\alpha_n} \cdot a_n$, $\beta = \omega^{\beta_0} \cdot b_0 + \cdots + \omega^{\beta_m} \cdot b_m$.
We can choose here a common system of powers of ω by putting several coefficients a_ν resp. b_ν equal 0. So we can assume w.r.o.g. that (1) holds with $m = n$ and $\alpha_\nu = \beta_\nu$ for $\nu = 0, \ldots, n$, but where now the a_ν and b_ν are only ≥ 0.

And then we define the *Hessenberg natural sum* of α and β to be the ordinal

$$\alpha \oplus \beta := \omega^{\alpha_0} \cdot (a_0 + b_0) + \cdots + \omega^{\alpha_n} \cdot (a_n + b_n).$$

For the sake of completeness we further put $\alpha \oplus \beta := \alpha + \beta$, if α or β is 0.

One can immediately see that this operation $\oplus$ is commutative and associative.

In the following, sets of ordinals are always considered as ordered (in the obvious sense) by magnitude. We prove some properties of indecomposable ordinals:

8.7 Lemma. *Let α be an ordinal > 0, $\lambda < \omega^\alpha$, $L \subseteq \omega^\alpha$ a subset of type λ. Then the set $\omega^\alpha \backslash L$ has the order type ω^α.*

Proof. For $\alpha = 1$ the statement is trivial. We make the assumption that it is satisfied for a fixed $\alpha > 0$. Let then $\lambda < \omega^{\alpha+1}$ and L a subset of $\omega^{\alpha+1}$ of type λ. We have $\omega^{\alpha+1} = \omega^\alpha + \omega^\alpha + \cdots$ (ω many summands), so that $\omega^{\alpha+1}$ is partitioned in segments $S_1, S_2, \ldots$, which each have type ω^α, and where all elements of a segment S_m are less than all elements of a segment S_n if $m < n$ holds. The set F of those n, for which the set $S_n \cap L$ has type ω^α, is finite because otherwise L would have type $\omega^{\alpha+1}$. For the n of the infinite set $\mathbf{N} \backslash F$ then $S_n \backslash L$ has by induction hypothesis type ω^α. These sets together form a subset of $\omega^{\alpha+1}$ of type $\omega^{\alpha+1}$.

As a reformulation of 8.7 and a generalization of 8.2 we have:

8.8 Lemma. *If $\omega^\alpha = A \cup B$, then A or B has the type ω^α.*

8.9 Definition. Let μ be an ordinal and $\mu = A \cup B$, where A, B are disjoint subsets of μ, and where A has the type α, and B the type β. Then μ is said to be a *mixed sum of the ordinals α and β*.

Different ordinals can be a mixed sum of the same two ordinals α and β. E.g. if a and β are both $= \omega$, then $\omega + \omega$ is a mixed sum of α and β, but also ω, for ω can be represented as the union of the set of odd and the set of even numbers, which both have type ω.

8.10 Lemma. *Let a and b be positive integers, α an ordinal. Then every mixed sum S of $\omega^\alpha \cdot a$ and $\omega^\alpha \cdot b$ is $\leq \omega^\alpha \cdot (a + b)$.*

Proof. Let S be a mixed sum $A \cup B$, where A and B are disjoint sets with types $\omega^\alpha \cdot a$ resp. $\omega^\alpha \cdot b$. The initial segment W of type ω^α of

S is $= (W \cap A) \cup (W \cap B)$. By 8.8 one of these summands has type ω^α. W.r.o.g. we assume that $W \cap A$ has type ω^α. Then we construct a new order $\preceq$ in S. We shift the elements of $W \cap B$ behind those of $W \cap A$, so that S with $\preceq$ has an initial segment consisting of the elements of $W \cap A$. And behind this segment follows a segment consisting of the elements of $W \cap B$. The rest is unchanged. Then the type of S with its original order is $\leq$ the type of S with $\preceq$.

Now the type of $S \backslash (W \cap A)$ with $\preceq$ is a mixed sum of $\omega^\alpha \cdot (a - 1)$ and $\omega^\alpha \cdot b$. If we apply induction on the sum $a + b$ we can conclude that the last mixed sum is $\leq \omega^\alpha \cdot (a - 1 + b)$. And then S has a type $\leq \omega^\alpha + \omega^\alpha \cdot (a - 1 + b) = \omega^\alpha \cdot (a + b)$.

Now we obtain the main theorem on mixed sums of ordinals. A proof of this was given by Neumer [125], where he remarked that this theorem was also observed by Carruth [12].

8.11 Theorem. *Every mixed sum of ordinals* α, β *is* $\leq \alpha \oplus \beta$.

Proof. If α or β is 0, the assertion is trivial. So we assume that both are > 0. Using the normal form of ordinals we can represent α and β in the form

$$\alpha = \omega^{\alpha_0} \cdot a_0 + \cdots + \omega^{\alpha_n} \cdot a_n \text{ and } \beta = \omega^{\alpha_0} \cdot b_0 + \cdots + \omega^{\alpha_n} \cdot b_n,$$

where $\alpha_0 > \cdots > \alpha_n$ are ordinals and where $a_0, \ldots, a_n, b_0, \ldots, b_n$ are non-negative integers. The above representations show that α (resp. β) is isomorphic to an ordered sum A (resp. B) of disjoint linearly ordered sets $A_0, \ldots, A_n$ (resp. $B_0, \ldots, B_n$) over the argument $\{0, \ldots, n\}_<$, where A_ν (resp. B_ν) has the order type $\omega^{\alpha_\nu} \cdot a_\nu$ (resp. $\omega^{\alpha_\nu} \cdot b_\nu$) for $\nu = 0, \ldots, n$.

We apply induction on the number n. For $n = 0$ our assertion is proved by 8.10. Suppose now that n is > 1 and that the theorem is proved for $n - 1$. Let then μ be a mixed sum of α and β. Then the ordinal μ is isomorphic to a linearly ordered set $(M, \preceq)$, where M is the union of two disjoint sets A and B, where A has the order type α and B the order type β, both with respect to the restriction of $\preceq$ onto A resp. B.

Let now T be the set of all elements of $M \backslash (A_0 \cup B_0)$ that are $\preceq$ some element of $A_0 \cup B_0$. Then we define an order $\preceq^*$ in M by: The union $A_0 \cup B_0$ is an initial segment of $(M, \preceq^*)$, T is an initial segment of $M \backslash (A_0 \cup B_0)$, and the rest of M forms a final segment of M: Here all three segments shall have the order induced by $\preceq$. Then it is clear that the order type of $(M, \preceq)$ (and then also that of μ) is $\leq$ the type of

$(M, \preceq)$, for the elements of T were "absorbed" before by $A_0 \cup B_0$. By the induction hypothesis it now follows that the set $M \backslash (A_0 \cup B_0)$ with the restriction of $\preceq^*$ has an order type

$\leq \tau := (\omega^{\alpha_1} \cdot a_1 + \cdots + \omega^{\alpha_n} \cdot a_n) \oplus (\omega^{\alpha_1} \cdot b_1 + \cdots + \omega^{\alpha_n} \cdot b_n)$. And so finally M has by 8.10 a type $\leq \omega^{\alpha_0} \cdot (a_0 + b_0) + \tau$. So the theorem is proved for n, and by induction then for all $n \in \mathbf{N}$.

Later we shall need the following simple theorems:

8.12 Theorem. *Let W be a well-ordered set of type $\omega^{\alpha} \cdot n$, where $n \in \mathbf{N}$, and x an element of W. Then $(W > x)$ has an order type $\geq \omega^{\alpha}$.*

Proof. W can be represented as an ordered sum $W = W_1 + \cdots + W_n$, where each W_ν, $\nu = 1, \ldots, n$, is a segment of type ω^{α}. If x is in a W_i with $i < n$, the assertion is trivial. If $x \in W_n$ it follows from 8.7 since $(W_n > x)$ has type ω^{α}.

8.13 Theorem. *Let ρ, μ and α be ordinals and $\mu < \omega^{\alpha}$. Then $\rho \oplus \mu < \rho + \omega^{\alpha}$.*

Proof. We have $\rho = \rho' + \rho''$, where ρ' (resp ρ'') is the sum of those summands ω^{ν} of the representation of ρ in Cantor's normal form for which $\nu \geq \alpha$ (resp. $\nu < \alpha$) holds. Then $\rho \oplus \mu$ is $= (\rho' \oplus \rho'') \oplus \mu = \rho' \oplus \kappa$, where $\kappa = \rho'' \oplus \mu$ is $< \omega^{\alpha}$, as the sum of finitely many ordinals of a form ω^{β} with $\beta < \alpha$. Since $\rho' \oplus \kappa < \rho + \omega^{\alpha}$ holds, we are done.

Chapter 5
Universally ordered sets

In this chapter we treat the question which concerns the existence of $\aleph_\alpha$ - universally ordered sets. These are posets P which have the property: Every poset T of cardinality $\aleph_\alpha$ is isomorphic to a subset of P. In Section *4.3* we considered this question for the linearly ordered case. Here embeddability was reduced to the requirement that the embedding function $\varphi : T \to S$ fulfills: $x < y$ (in T) $\Rightarrow \varphi(x) < \varphi(y)$ (in S).Now, in the general (partially) ordered case also $x \parallel y \Rightarrow \varphi(x) \parallel \varphi(y)$ must hold to make $\varphi[T]$, with the order induced by P, isomorphic to T.

In order to construct $\aleph_\alpha$ - universally ordered sets of cardinality $\aleph_\alpha$ we apply a method of "successively adjoining IF-pairs". These are studied in the next section.

5.1   Adjoining IF-pairs to posets

The following two definitions are essentially the same as certain definitions of Cuesta Dutari [16], [17]:

1.1 Definition. Let X be a poset, A an initial and B a final segment of X, such that $x < y$ holds for all $x \in A$ and $y \in B$. Then we call (A, B) an *IF-pair* of X. The set of all IF-pairs of X is denoted by $P(X)$.

Here A or B or even both can be empty or the whole set X. In a linearly ordered set X every cut (A, B) is also an IF-pair, but of course, not conversely.

In the construction process for universally ordered sets we use an ordering in the set $X \cup P(X)$, whose restriction on X is the given order of X.

1.2 Definition. Let $(X, \leq)$ be a poset, where X is disjoint to the set $P(X)$.Then we extend $\leq$ to an order, which also is denoted by $\leq$, of $X \cup P(X)$ as follows: Let x, y be elements of X, (A, B) and (A', B') IF-pairs of X. Then we define the strict order $<$ of $\leq$ by

(1) $x < y$ in $X \cup P(X) \iff x < y$ in X,

(2) $x < (A, B) \iff x \in A$,

(3) $(A, B) < x \iff x \in B$,

(4) $(A, B) < (A', B') \iff B \cap A' \neq \emptyset$.

Evidently (4) is equivalent to

(4') $(A, B) < (A', B') \iff \exists x \in X$ such that $(A, B) < x < (A', B')$.

From (4) we obtain:

1.3 Theorem. *Suppose $(A, B) < (A', B')$ for two IF-pairs of a poset X, then $A \subseteq A'$ and $B' \subseteq B$ hold.*

Proof. By (4) there exists an element $x \in B \cap A'$. This is $>$ all elements of A, and $A \subseteq A'$ follows from $x \in A'$.

On the other hand B contains x, which is in A' and thus $<$ every element of B'. Then the final segment B also contains B'.

We now have to verify that Definition 1.2 yields indeed an order in $X \cup P(X)$:

1.4 Theorem. *The relation $<$ of 1.2, augmented by the identity relation on $X \cup P(X)$, is an order relation on this set.*

Proof. It follows immediately that $\leq$ is reflexive. It is also antisymmetric: For two IF-pairs (A, B) and (A', B') we cannot have $(A, B) < (A', B')$ and $(A', B') < (A, B)$. For otherwise we would have an element $x \in B \cap A'$ and an element $y \in B' \cap A$. Then $x(\in A')$ $< y(\in B')$ and $y(\in A) < x(\in B)$ would follow with contradiction. To verify the transitivity let x, y, z be elements of X and $C_i = (A_i, B_i)$ for $i = 1, 2, 3$ IF-pairs of X. We have

(5) $x \leq y \leq z$ (in $X \cup P(X)$) $\implies x \leq z$ by (1),

(6) $x \leq y \leq C_1 \implies x \leq y \in A_1 \implies x \in A_1 \implies x < C_1$,

(7) $x \leq C_1 \leq y \Rightarrow x \in A_1$ and $y \in B_1 \Rightarrow x < y$,

(8) $C_1 \leq x \leq y \Rightarrow x \in B_1 \Rightarrow y \in B_1 \Rightarrow C_1 < y$,

(9) $x \leq C_1 \leq C_2 \Rightarrow x \in A_1 \subseteq A_2 \Rightarrow x \in A_2 \Rightarrow x < C_2$,

(10) $C_1 \leq x \leq C_2 \Rightarrow x \in B_1 \cap A_2 \Rightarrow C_1 < C_2$,

(11) $C_1 \leq C_2 \leq x \Rightarrow x \in B_2 \subseteq B_1 \Rightarrow x \in B_1 \Rightarrow C_1 < x$,

(12) $C_1 < C_2 < C_3 \Rightarrow$ By (4') there exist elements a, b in X with $C_1 < a < C_2$ and $C_2 < b < C_3$. Then by (7) we obtain $a < b$, and then by (6) $a < C_3$. This and $C_1 < a$ yield $C_1 < C_3$ by (10).

5.2 Construction of an $\aleph_\alpha$- universally ordered set

2.1 Definition. A poset P is said to be $\aleph_\alpha$ - *universal* or $\aleph_\alpha$ - *universally ordered*, if every poset of cardinality $\aleph_\alpha$ is isomorphic to a subset T of P, where T, of course, is equipped with the restriction of the order of P.

B. Jonsson [92] proved : For every ordinal α there exists an $\aleph_\alpha$ - universally ordered set of cardinality $\mathfrak{k}_\alpha$. His proof is non-constructive. A constructive proof was published by Crawley and Dean [15]. They proved that certain free lattices $F_\alpha(3)$ are $\aleph_\alpha$ - universally ordered sets of cardinality $\mathfrak{k}_\alpha$. In this section we also shall present a constructive proof of the existence of such sets, which was published in [79].

In this connection Kurepa [107] had introduced the following notion of $\eta_{\alpha\alpha}$ - set:

2.2 Definition. A poset E is said to be an $\eta_{\alpha\alpha}$ - *set*, if every subset $X \subseteq E$ with $|X| < \aleph_\alpha$ has in E an extension in every direction. This means precisely: For every $X \subseteq E$ with $|X| < \aleph_\alpha$ there holds: If x_0 is an element which is not in X, and if $X_0 := X \cup \{x_0\}$ is a poset, whose restriction on X coincides with the restriction of the order from E onto X, then there exists an element $x_0^* \in E$, such that the mapping f, which puts $f(x_0) = x_0^*$ and $f(x) = x$ for $x \in X$, is an isomorphism of X_0 onto a subset of E.

From the last definition some elementary properties of $\eta_{\alpha\alpha}$ - sets can immediately be seen: An $\eta_{\alpha\alpha}$ - set is dense, and so there are no neighboring elements. It has no first and no last element.

For these $\eta_{\alpha\alpha}$ - sets Kurepa formulated the following two theorems:

2.3 Theorem [107]. *An $\eta_{\alpha\alpha}$ - set is $\aleph_\alpha$ - universally ordered.*

Proof. Let E be an $\eta_{\alpha\alpha}$ - set and S a poset of cardinality $\aleph_\alpha$. We take a well-ordering of S such that S is representable as $S = \{s_\nu | \nu < \omega_\alpha\}$. Suppose that for an ordinal $\mu < \omega_\alpha$ we have already constructed an isomorphic mapping f_μ of $\{s_\nu | \nu < \mu\}$ (where this set is equipped with the induced order of S) into E. Then there exists an extension $f_{\mu+1} : \{s_\nu | \nu \leq \mu\} \to E$ because E is an $\eta_{\alpha\alpha}$ - set. If for a limit ordinal λ for all $\nu < \lambda$ mappings f_ν are already constructed such that for $\mu < \nu < \lambda$ the function f_ν extends f_μ, we define f_λ to be their limit mapping. By transfinite induction we so construct a tower of mappings $f_\mu : \{s_\nu | \nu \leq \mu\} \to E$, which extend with μ. The mapping f_{ω_α} yields an isomorphism of S onto a subset of E.

2.4 Theorem [107]. *Every maximal chain of an $\eta_{\alpha\alpha}$ - set E is an η_α - set.*

Proof. Let M be a maximal chain of an $\eta_{\alpha\alpha}$ - set E, and A, B subsets of M which are neighboring : Every element of A is less than every element of B, and there is no element x of M with $a < x < b$ for all $a \in A$, $b \in B$. Then A or B must have cardinality $\geq \aleph_\alpha$ because otherwise there would exist an element between A and B in the $\eta_{\alpha\alpha}$ - set E. This would not be in M since A and B are neighboring, and then M would not be maximal. Thus M fulfills the condition for an η_α - set.

In the following we present two further properties of $\eta_{\alpha\alpha}$ - sets. First we define (resp. recall):

2.5 Definition. If S and T are arbitrary sets and $f : S \to T$ a mapping, $x \in S$, we say: *f leaves x fixed* if $f(x) = x$. If $A \subseteq S$ and $f(x) = x$ holds for all $x \in A$ we say: *f leaves A fixed*.

If P is a poset, a subset $D \subseteq P$ is said to be *strictly dense* in P if there holds: For every two elements a,b of P with $a < b$ there exists an element $d \in D$ with $a < d < b$.

If T is a subset of a poset P, and if a and b are elements of $P \backslash T$, we say: *a and b are in the same position to T* iff $(T < a) = (T < b)$ and $(T > a) = (T > b)$. Then, of course, also $(T \parallel a) = (T \parallel b)$ holds.

Now there follows:

2.6 Lemma. *Let T be a subset of a poset P, and let a,b be elements of $P \backslash T$ with $a < b$, which are in the same position to T. Then every element $z \in P$ which satisfies $a < z < b$ is also in the same position to T as a and b.*

Proof. Let $z \in P$ satisfy $a < z < b$, then z is not in T, for otherwise we would have the contradiction $z \in (T < b) \backslash (T < a) = \emptyset$. Now $(T < z) \subseteq (T < b) = (T < a) \subseteq (T < z)$ holds. Then $(T < z) = (T < a)$ follows, and analogously $(T > z) = (T > a)$.

2.7 Theorem [79]. *Let $(E, \leq)$ be an $\eta_{\alpha\alpha}$ - set and D a subset of E, which is dense in E. Then D (with the restriction of $\leq$ onto D) is also an $\eta_{\alpha\alpha}$ - set.*

Proof. First we see that the set D is also strictly dense in E since E is dense. Let $X \subseteq D$ and $|X| < \aleph_\alpha$, $x_0 \notin X$, and let an order $\leq_0$ be given in $X \cup \{x_0\}$ whose restriction onto X coincides with the restriction of $\leq$ onto X. Since E is an $\eta_{\alpha\alpha}$ - set there now exists an $e_0 \in E$ and an isomorphic mapping

$f_0 : X \cup \{x_0\} \to X \cup \{e_0\}$ with respect to the orders $\leq_0$ and $\leq \restriction (X \cup \{e_0\})$,

which leaves X fixed and maps x_0 onto e_0.

Let now e_1 be an element which is not in E. We introduce an order $\leq_1$ in $X \cup \{e_0\} \cup \{e_1\}$ as follows:

$\leq_1 \restriction (X \cup \{e_0\})$ is the same order as $\leq \restriction (X \cup \{e_0\})$. Further we put $e_0 <_1 e_1$, and we define the rest of $\leq_1$ so that e_0 and e_1 are in the same position to X. We have $|X \cup \{e_0\}| < \aleph_\alpha$, and since E is an $\eta_{\alpha\alpha}$ - set there exists an $e_1' \in E$ and an isomorphic mapping

$f_1 : (X \cup \{e_0\}) \cup \{e_1\} \to (X \cup \{e_0\}) \cup \{e_1'\}$ with respect to the orders $\leq_1$ and $\leq \restriction (X \cup \{e_0, e_1'\})$,

which leaves $X \cup \{e_0\}$ fixed and maps e_1 onto e_1'. Then e_1 and e_1' are in the same position to $X \cup \{e_0\}$ and a fortiori in the same position to X.

We had $e_0 <_1 e_1$ and, since f_1 is isomorphic, then also $e_0 = f_1(e_0) < f_1(e_1) = e_1'$. Since D is dense in E there now follows the existence of an element $d \in D$ with $e_0 < d < e_1'$. Now e_0 (and e_1) and e_1' are in the same position to X. By 2.6 then also d is in the same position to X as e_0. Then the mapping $f_2 : X \cup \{e_0\} \to X \cup \{d\}$, which leaves X fixed and maps e_0 onto d, is an isomorphism with respect to the order $\leq$. The mapping $f_2 \circ f_0 : X \cup \{x_0\} \to X \cup \{d\}$ is now an isomorphism with respect to $\leq_0$ and $\leq$, which leaves X fixed and maps x_0 onto d. Thus D is an $\eta_{\alpha\alpha}$ - set.

2.8 Theorem [79]. *Let $(E, \leq)$ be an $\eta_{\alpha\alpha}$ - set, and a,b elements of E with $a < b$. Then the open interval $(a,b) := \{x \in E | a < x < b\}$ is again an $\eta_{\alpha\alpha}$ - set.*

Proof. Let $X \subseteq (a,b)$ and $|X| < \aleph_\alpha$, $x_0 \notin X$, and let $X \cup \{x_0\}$ be a poset with order $\leq_0$, whose restriction onto X coincides with the restriction of $\leq$ onto X. First we assume that x_0 is different from a and b. We consider the set $X' := X \cup \{x_0, a, b\}$ and order it with an order $\leq'$, which is defined by

$\leq' := \leq_0 \cup \{(a,x) | x \in X \cup \{x_0\}\} \cup \{(x,b) | x \in X \cup \{x_0\}\}$.

(Then a is the smallest element in X' and b the greatest.)

Now $|X \cup \{a,b\}|$ is $< \aleph_\alpha$, and since E is an $\eta_{\alpha\alpha}$ - set, there exists an element $e \in E$ and an isomorphism

$f : (X \cup \{a,b\}) \cup \{x_0\} \to (X \cup \{a,b\}) \cup \{e\}$ with respect to $\leq'$ and $\leq$ which leaves $X \cup \{a,b\}$ fixed and maps x_0 onto e. Since f is $<$ - preserving, $a <' x_0 <' b$ implies $a = f(a) < f(x_0) = e < f(b) = b$ so that $e \in (a,b)$. The restriction $f \upharpoonright X \cup \{x_0\}$ is then an isomorphic mapping with respect to $\leq'$ and $\leq$ into the interval (a,b), which leaves X fixed and maps x_0 onto $e \in (a,b)$.

Now we consider the case where x_0 is one of the elements a,b. Suppose $x_0 = a$; the case $x_0 = b$ is analogous. We choose an element x_0^* which does not belong to E and order $X \cup \{x_0^*\}$ with $\leq^*$, where the restrictions of $\leq$ and $\leq^*$onto X coincide, and where x_0^* is with $x_0(= a)$ in the same position to X. Now x_0^* is different from a and b, and so the first part of our proof yields that there exists an element $e \in (a,b)$ and an isomorphic mapping $f^* : X \cup \{x_0^*\} \longrightarrow X \cup \{e\}$ with respect to $\leq^*$ and $\leq$, which leaves X fixed and maps x_0^* onto e. Now we replace here x_0^* by x_0. Since these elements are in the same position to X we so obtain an isomorphic mapping of $X \cup \{x_0\}$ with $\leq_0$onto $X \cup \{e\}$ with $\leq$, which leaves X fixed and maps x_0 onto $e \in (a,b)$.

Kurepa [107] had posed the question, whether there exists an $\eta_{\alpha\alpha}$ - set of cardinality ℓ_α, if $\aleph_\alpha$ is regular. Subsequently such a set is constructed. First we introduce some new concepts:

2.9 Definition. Let A be an initial segment of a poset X, and $A' \subseteq A$ a subset that generates A, i. e. $A = \{x \in X | \exists a \in A'$ with $x \leq a\}$.

Then A' is called a *basis* of the initial segment A. If $|A'| \leq k$ for a cardinal k, we call A' a k - *basis* of A.

Analogously we define: If B is a final segment of X, which is generated by a subset B', i. e.: $B = \{x \in X | \exists b \in B'$ with $x \geq b\}$, B' is said to be a *basis* of B, and a k - *basis*, if in addition to this $|B'| \leq k$.

Finally let (A, B) be an IF-pair of X. Then a pair (A', B') of subsets $A' \subseteq A$, $B' \subseteq B$ is called a *basis* of the IF-pair (A, B), if A' (resp. B') is a basis of A (resp. B).It is said to be a k - *basis* of (A, B), if A' is a k-basis of A and B' a k-basis of B.

Initial segments, final segments, and IF-pairs, for which there exists a basis of cardinality $\leq k$, are called *k-generated*.

Now we construct the $\aleph_\alpha$ - universal set U_α.

2.10 Definition. If τ is an ordinal and $(T_\tau, \leq_\tau)$ a poset, we define $P_\tau(T_\tau)$ to be the set of all $|\tau|$ - generated IF-pairs of T_τ and then further $S(T_\tau) := \{(\tau, P)|\ P \in P_\tau(T_\tau)\}$. The reason why we list the pairs P together with an index τ is that this ensures that the sets $S(T_\tau)$ are pairwise disjoint.

Let T_0 be the empty set $\emptyset$. It is ordered by the empty relation. Let now τ be an ordinal such that we have already defined a poset $(T_\sigma, \leq_\sigma)$ for all $\sigma \leq \tau$. Then we put

(1) $T_{\tau+1} := T_\tau \cup S(T_\tau)$. So e.g. we have $T_1 = \{(0, (\emptyset, \emptyset))\}$, $T_2 = T_1 \cup \{(1, P)|\ P \in P_1(T_1)\}$. We order $T_{\tau+1}$ by $\leq_{\tau+1}$ in the sense of 1.2:

(2) For $x, y \in T_\tau$ we put
$x \leq_{\tau+1} y \Longleftrightarrow x \leq_\tau y$.

(3) For $x \in T_\tau$, $y = (\tau, P)$ with $P = (A, B) \in P_\tau(T_\tau)$ we put
$x \leq_{\tau+1} y \Longleftrightarrow x \in A$.

(4) For $x = (\tau, P)$ with $P = (A, B) \in P_\tau(T_\tau)$ and $y \in T_\tau$ we put
$x \leq_{\tau+1} y \Longleftrightarrow y \in B$.

(5) For $x = (A, B)$ and $y = (C, D) \in P_\tau(T_\tau)$ we put
$(\tau, x) \leq_{\tau+1} (\tau, y) \Longleftrightarrow B \cap C \neq \emptyset$.

If λ is a limit ordinal for which posets $(T_\sigma, \leq_\sigma)$ have been defined for all $\sigma < \lambda$, then we put $T_\lambda := \cup_{\sigma<\lambda} T_\sigma$ and define $\leq_\lambda := \cup_{\sigma<\lambda} \leq_\sigma$. So by transfinite induction a poset T_ν is defined for every ordinal ν.

Then we have $T_1 = (T_0 = \emptyset \cup)\ S(T_0)$, $T_2 = S(T_0) \cup S(T_1), \ldots$. In general there holds for every ordinal τ :

(6) $T_\tau = \cup\{S(T_\sigma)|\sigma < \tau\}$.

Since the sets $S(T_\sigma)$ are pairwise disjoint by construction it follows from (1) and (6) that for every ordinal τ the sets T_τ and $S(T_\tau)$ are disjoint. By transfinite induction we have so obtained a tower of posets $(T_\tau, \leq_\tau)$. And finally we define for every ordinal α :

(7) $U_\alpha := T_{\omega_\alpha}$ and equip it with the order $\leq_{\omega_\alpha}$.

Our definition yields immediately that the U_α form a tower of posets:

2.11 Theorem. *If α and β are ordinals with $\alpha < \beta$, then U_α is a subset of U_β, and its order is the restriction of the order of U_β to U_α. And for limit numbers λ one has $U_\lambda = \cup_{\nu<\lambda} U_\nu$.*

2.12 Theorem. *Let α be an ordinal and $\omega_\gamma = cf(\omega_\alpha)$. Then U_α is an $\eta_{\gamma\gamma}$ - set.*

Proof. Let $X \subseteq U_\alpha$ with $|X| < \aleph_\gamma$ and $x_0 \notin X$. Let further $(X \cup \{x_0\}, \leq_0)$ be a poset, where the restriction $\leq_0 \restriction X$ coincides with the restriction of U_α onto X. We have to show: There exists an isomorphic mapping (with respect to $\leq_0$ and $\leq$) f from $X \cup \{x_0\}$ onto a subset of U_α with $f(x) = x$ for $x \in X$.

Let $\omega(|X|)$ be the initial ordinal for the cardinal $|X|$. Because of $|X| < \aleph_\gamma$ there exists an ordinal $\mu < \omega_\alpha$ such that $X \subseteq T_\mu$. And since the T_ν increase with ν, there exists also a μ which in addition to this satisfies $\omega(|X|) \leq \mu$. The element x_0 determines in X the IF-pair $(X <_0 x_0, X >_0 x_0)$. Then $(X <_0 x_0)$ (resp. $(X >_0 x_0)$) generates an initial segment A (resp. final segment B) in T_μ. The IF-pair (A, B) is thus $|X|$ - generated and thus also $|\mu|$ - generated, hence in $P_\mu(T_\mu)$.

By construction of $T_{\mu+1}$ the element $y_0 := (\mu, (A, B)) \in T_{\mu+1}$ satisfies $A = (T_\mu < y_0)$ and $B = (T_\mu > y_0)$.

Then we map x_0 onto y_0 and every element of X onto itself. This yields an isomorphism of $X \cup \{x_0\}$ (with $\leq_0$) onto $X \cup \{y_0\} \subseteq U_\alpha$. And thus U_α is an $\eta_{\gamma\gamma}$ - set.

As corollary we obtain:

2.13 Theorem [79]. *U_α is an $\eta_{\alpha\alpha}$ - set if $\aleph_\alpha$ is regular.*

Now there follows:

2.14 Theorem. *For every ordinal α there holds: U_α is an $\aleph_\alpha$ - universally ordered set.*

Proof. For regular $\aleph_\alpha$ our assertion is already proved by 2.3 because then U_α is an $\eta_{\alpha\alpha}$ - set and thus $\aleph_\alpha$ - universal.

Let now $(X, \leq_x)$ be a poset of singular cardinality $\aleph_\alpha$. Then there exists a bijective mapping of X onto ω_α and then also a representation $X = \{x_\nu | \nu < \omega_\alpha\}$. We put $X_\mu := \{x_\nu | \nu < \omega_\mu\}$ for $\mu < \alpha$. Then $X = \cup \{X_\mu | \mu < \alpha\}$.

There exists an isomorphism f_0 of X_0 (with $\leq_x$) into U_0 since U_0 is an η_{00} - set.

Suppose that τ is an ordinal $< \omega_\alpha$ and that we have already defined a tower of isomorphic mappings $f_\mu : X_\mu \to f[X_\mu] \subseteq U_\mu$ for all $\mu < \tau$. Then we construct an isomorphic mapping f_τ of X_τ (with $\leq_x$) into U_τ as follows:

If τ is a successor ordinal $\sigma + 1$ let $X_{\sigma+1} \backslash X_\sigma = \{a_\iota | \iota < \kappa\}$. Then, since $U_{\sigma+1}$ is an $\eta_{\sigma+1,\sigma+1}$ - set we can successively construct isomorphic mappings φ_0 of $X_\sigma \cup \{a_0\}$ into $U_{\sigma+1}$, φ_1 of $X_\sigma \cup \{a_0, a_1\}$ into $U_{\sigma+1}, \ldots$

and so on, where every φ_ν extends the mappings φ_μ with $\mu < \nu$.The limit of the mappings φ_ι, $\iota < \kappa$, is then an isomorphic mapping $f_{\tau+1}$ of $X_{\sigma+1}$ into $U_{\sigma+1}$.

If τ is a limit ordinal and f_ν defined for all $\nu < \tau$, then we define f_τ to be the limit mapping of the $f_\nu, \nu < \tau$. By transfinite induction we so obtain a tower of isomorphic mappings $f_\tau : X_\tau \to U_\tau$, $\tau < \alpha$. Their limit mapping $f_\alpha : X(= X_\alpha) \to U_\alpha$ is then an isomorphism of X onto a subset of U_α.

Before we determine the cardinality of U_α we mention two lemmas.

2.15 Lemma. *Let X be a poset with $|X| \leq 2^k$, where k is an infinite cardinal. Then the set P of k-generated IF-pairs of X has cardinality $|P| \leq 2^k$.*

If Y is a poset with $|Y| = \aleph_0$, then the set of all finitely generated IF-pairs of Y is countable.

Proof. P has at most so many elements as there are pairs (A, B) of subsets A, B of X with cardinalities $\leq k$. The set of all subsets $A \subseteq X$ with cardinality $|A| \leq k$ has a cardinality $\leq$ the cardinality of the set of all mappings of $\omega(|k|)$ into X, which means its cardinality is $\leq (2^k)^k = 2^k$. Analogously we have for B also only at most 2^k possibilities. Then $|P| \leq 2^k \cdot 2^k = 2^k$ follows. The rest is an easy consequence of the fact that the set of all finite subsets of a countable set is countable.

2.16 Lemma. *For the sets T_ν of Definition 2.10 there holds: For $\tau \geq \omega_0$ we have*

(1) $|T_\tau| \leq 2^{|\tau|}.$

Proof. (1) is satisfied for $\tau = \omega_0$ because of $|T_{\omega_0}| = \aleph_0 < 2^{\aleph_0}$. This follows immediately from the fact that all T_ν with $\nu \in \mathbf{N}$ are finite. Assume now that (1) holds for some $\tau \geq \omega_0$. From Definition 2.10 and 2.15 we conclude $|T_{\tau+1}| \leq |T_\tau| + |S(T_\tau)| \leq 2^{|\tau|} + 2^{|\tau|} = 2^{|\tau+1|}$, and so (1) also holds for $\tau + 1$ instead of τ.

Let now λ be a limit ordinal and let (1) be proved for all $\tau < \lambda$. Then $|T_\lambda| \leq \sum_{\tau < \lambda} |T_\tau| \leq \sum_{\tau < \lambda} 2^{|\tau|} \leq |\lambda| \cdot 2^{|\lambda|} = 2^{|\lambda|}$. And so the proof follows by induction.

We determine the cardinality of U_α :

2.17 Theorem. $|U_\alpha| = \mathfrak{k}_\alpha$ $(= \sum_{\nu < \alpha} 2^{\aleph_\nu})$.

152

Proof. We consider the representation $U_\alpha = \cup\{T_\nu | \nu < \omega_\alpha\}$. For $\alpha = 0$ then $|U_0| = \aleph_0$ immediately follows. For $\alpha > 0$ (1) yields $|U_\alpha| \leq \sum\{|T_\nu| | \nu < \omega_\alpha\} \leq \sum\{2^{|\nu|} | \nu < \omega_\alpha\} = \mathfrak{k}_\alpha$.

On the other hand, if ω_α is regular, then U_α is an $\eta_{\alpha\alpha}$ - set, and then it contains an η_α - set which by 4.3.7 has a cardinality $\geq \mathfrak{k}_\alpha$. If ω_α is singular, then α is a limit number and we have by 2.11 $U_\alpha = \cup\{U_{\nu+1} | \nu < \alpha\}$. Now $|U_\alpha| \geq |U_{\nu+1}| = 2^{\aleph_\nu}$ holds for all $\nu < \alpha$, and this entails $|U_\alpha| \geq \mathfrak{k}_\alpha$.

2.18 Remark. Subsequently to the construction of the sets U_α we remark that for ordinals of the form $\omega_{\alpha+1}$ one could have performed a somewhat simpler construction, which also would have led to an $\aleph_\alpha$ - universally ordered set $V_{\alpha+1}$. Namely one can alter Definition 2.10 so that the sets $P_\tau(T_\tau)$ are replaced by the sets of the $\aleph_\alpha$ - generated IF-pairs of T_τ. If all other things are unchanged we would arrive at the sets $V_{\alpha+1}$ in quite the same way as before to the sets $U_{\alpha+1}$. Our construction had the advantage that one could immediately see Theorem 2.11.

5.3 Construction of an injective $<$-preserving mapping of U_α into H_α

We supplement Definition 4.1.9 by the following:

3.1 Definition and Theorem. *Let α be an ordinal. We define O_α (resp. E_α) to be the set of all transfinite sequences $s = (s_\nu | \nu < \omega_\alpha) \in 2((\omega_\alpha))$ for which there exists an index μ such that all s_ν with $\mu \leq \nu < \omega_\alpha$ are $= 0$ (resp. $= 1$). We put $F_\alpha := O_\alpha \cup E_\alpha$.*

Then F_α is embeddable in H_α.

Proof. $O_\alpha = \cup\{M_\mu | \mu < \omega_\alpha\}$, where M_μ is the set of those sequences $(s_\nu) \in F_\alpha$, for which $s_\nu = 0$ for $\nu \geq \mu$. Of course, every M_μ is embeddable in H_α, and then also O_α is that by 4.4.6. Analogously also the set E_α is embeddable in H_α, and then also F_α.

3.2 Lemma. *Let α be an ordinal, T a subset of H_α with cardinality $|T| < \aleph_\gamma = cf(\aleph_\alpha)$. Then T has a supremum $s := \sup T \in O_\alpha$ and an infimum $i := \inf T \in E_\alpha$.*

Proof. The existence of s and i follows from 4.2.10. Let $s = (s_\nu | \nu < \omega_\alpha)$. If s would not be in O_α there would exist a strictly increasing sequence $\nu_\kappa, \kappa < \omega_\gamma$, which is cofinal in ω_α and for which $s_{\nu_\kappa} = 1$ for $\kappa < \omega_\gamma$. Because of $s = \sup T$ there exists for every $\kappa < \omega_\gamma$ an element $t_\kappa = (t_{\kappa\nu} | \nu < \omega_\alpha)$ in T with $t_{\kappa\nu} = s_\nu$ for all $\nu \leq \nu_\kappa$, Then the set $C := \{t_\kappa | \kappa < \omega_\gamma\}$ is cofinal in $(H_\alpha < s)$, and $|C| \leq |T| < \aleph_\gamma$ holds.

On the other hand $(H_\alpha < s)$ has a cofinal subset V of order type ω_γ, namely the set of sequences $b_\kappa = (b_{\kappa\nu} | \nu < \omega_\alpha)$,where $b_{\kappa\nu} = s_\nu$ for $\nu \leq \nu_\kappa$ and $b_{\kappa\nu} = 0$ for $\nu_\kappa < \nu < \omega_\alpha$. By 3.1.10 the regular ω_γ is the least ordinal which is cofinal in $(H_\alpha < s)$, and this contradicts $|C| < \aleph_\gamma$.

Analogously to this it follows that, if $i = (i_\nu | \nu < \omega_\alpha)$ would not be in E_α, the set of indices ν with $i_\nu = 0$ would be cofinal in ω_α. Then we could construct a subset of T which is coinitial in $(H_\alpha > i)$ and of cardinality $\leq |T| < \aleph_\gamma$ which is impossible.

The next two lemmas contain most of the proof idea of the main theorem of this section:

3.3 Lemma. *In the following T_τ, $S(T_\tau)$, $\leq_\tau$ are used in the meaning of 2.10. Let τ be an ordinal $< \omega_{\nu+1}$, $f : T_\tau \to H_{\nu+1}$ an injective $< $- preserving mapping (with respect to $\leq_\tau$ and the standard order $\leq$ by first differences of $H_{\nu+1}$).*

Let k be the cardinal $\aleph_\nu$, We define $k(H_{\nu+1}) := H_{\nu+1} \cup P_k(H_{\nu+1})$, where $P_k(H_{\nu+1})$ is the set of all k-generated IF-pairs of $H_{\nu+1}$, and order it according to 1.2. ($H_{\nu+1}$ is disjoint to $P_k(H_{\nu+1})$.) In the following we denote this order of $k(H_{\nu+1})$ by $\leq_k$ to distinguish it from the other involved orders.

Then there exists a $< $- preserving (with respect to $\leq_{\tau+1}$ and $\leq_k$) mapping $f^* : T_{\tau+1} \to k(H_{\nu+1})$, which extends f.

f^* has the additional property: If two different elements of $T_{\tau+1}$ have the same f^*- image, they are in $S(T_\tau)$, and if (τ, x) and (τ, y) are elements with the same f^*- image, then x and y are incomparable IF-pairs of T_τ.

Proof. We put $f^*(x) := f(x)$ for $x \in T_\tau$. Let now (τ, P) be an element of $S(T_\tau)$ where $P = (A, B)$ is a $|\tau|$ - generated IF-pair of T_τ. Then there are subsets $A' \subseteq A$ and $B' \subseteq B$ with $|A'| \leq |\tau| \geq |B'|$, such that A' generates the initial segment A and that B' generates the final segment B. Then we define $f^*(\tau, P)$ to be the IF-pair $(IS(f[A], FS(f[B]))$. We have

(1) $IS(f[A']) = IS(f[A])$ and $FS(f[B']) = FS(f[B])$.

Indeed, $IS(f[A']) \subseteq IS(f[A])$ is trivial, and if $x \in IS(f[A])$ holds,

154

then we have $x \leq f(a)$ for an element $a \in A$. This is $\leq_\tau a'$ for an element $a' \in A'$, and then $f(a) \leq f(a')$ and $x \in IS(f[A'])$ results. The rest follows analogously, and we have obtained

$f^*(\tau, P) = (IS(f[A]), FS(f[B])) \in P_{|\tau|}(H_{\nu+1}) \subseteq P_k(H_{\nu+1})$, for $|\tau| \leq \aleph_\nu = k$ holds.

So f is extended from T_τ to the function f^* on the set $T_{\tau+1}$. We verify that f^* is $<$ - preserving with respect to $\leq_{\tau+1}$ and $\leq_k$. (It is not necessarily injective.)

For $x \in T_\tau$ and $P = (A, B) \in P_\tau(T_\tau)$ we have (compare with 2.10, (3)):

$x <_{\tau+1} (\tau, P) \Rightarrow x \in A \Rightarrow f(x) \in f[A] \Rightarrow f(x) \in IS(f[A]) \Rightarrow f^*(x)(= f(x)) <_k (IS(f[A]), FS(f[B])) = f^*(\tau, P)$.

If $(\tau, P) < x$ holds, it follows that $x \in B \Rightarrow f(x) \in f[B] \Rightarrow f(x) \in FS(f[B]) \Longrightarrow (IS(f[A]), FS(f[B])) < (f(x) =) f^*(x)$, hence $f^*(\tau, P) < f^*(x)$.

Let now $P = (A, B)$ and $P_1 = (A_1, B_1)$ be $|\tau|$ - generated IF-pairs of T_τ with $(\tau, P) <_{\tau+1} (\tau, P_1)$. Then we have $B \cap A_1 \neq \emptyset$, and so there exists an $x \in B \cap A_1$. Let $A_1' \subseteq A_1$ be a subset of cardinality $\leq |\tau|$, that generates the initial segment A_1, and let $B' \subseteq B$ be a subset of cardinality $\leq |\tau|$, that generates the final segment B. Then there exist elements $b \in B'$ and $a \in A_1'$, with $b \leq_{\tau+1} x \leq_{\tau+1} a$. This implies $f(b) \leq f(x) \leq f(a)$. Then $f(x) \in IS(f[A_1']) \cap FS(f[B'])$. Hence this set is non-empty and because of $IS(f[A_1']) = IS(f[A_1])$ and $FS(f[B']) = FS(f[B])$ (by (1)) we also obtain $IS(f[A_1]) \cap FS(f[B]) \neq \emptyset$, so that $f^*(\tau, P) = (IS(f[A]), FS(f[B])) < (IS(f[A_1]), FS(f[B_1])) = f^*(\tau, P_1)$ holds. So we have obtained that $f^* : T_{\tau+1} \to k(H_{\nu+1})$ is a $<$ - preserving (with respect to $\leq_{\tau+1}$ and $\leq_k$) mapping which extends f.

Different elements of $T_{\tau+1}$ can have the same f^*- image. But this can only happen if these elements are in $S(T_\tau)$ because f^* is injective on T_τ and by construction. And there holds:

If (τ, P) and (τ, P_1) are elements of $S(T_\tau)$ with the same f^*- image, then P and P_1 must be incomparable since f^* is $<$ - preserving.

3.4 Lemma. *Let $k := \aleph_\nu$, $k(H_{\nu+1}) = H_{\nu+1} \cup P_k(H_{\nu+1})$, equipped with the order $\leq_k$. Then there exists a linear extension $\leq_l$ of $\leq_k$, such that $(k(H_{\nu+1}), \leq_l)$ is embeddable in $H_{\nu+1}$.*

Then we have further: There exists a mapping $g : k(H_{\nu+1}) \to H_{\nu+1}$ which is injective and $<$ - preserving (with respect to $\leq_l$ and the order of $H_{\nu+1}$).

Proof. We define a linear order $\leq_l$ in $k(H_{\nu+1})$ by :

a) For $x, y \in H_{\nu+1} \subseteq k(H_{\nu+1})$ we put $x \leq_l y \iff x \leq y$ in $H_{\nu+1}$.

b) For $x \in H_{\nu+1}$, $(A, B) \in P_k(H_{\nu+1})$ we put $(A, B) <_l x$, if $x \notin A$, resp. $x <_l (A, B)$, if $x \in A$.

c) For (A, B) and $(C, D) \in P_k(H_{\nu+1})$ we put $(A, B) <_l (C, D) \iff A \subset C$ or $(A = C$ and $D \subset B)$.

Then $\leq_l$ is the union of $<_l$ and the identity relation on $k(H_{\nu+1})$.

We verify that $\leq_l$ is an order relation. If this is proved this order is automatically also linear. Only transitivity is non-trivial. We verify it for all possible cases:

Let x, y, $z \in H_{\nu+1}$ and (A, B), (C, D), $(E, F) \in P_k(H_{\nu+1})$. Then we have

1) $x <_l y <_l z \Rightarrow x <_l z$ trivially,

2) $x <_l y <_l (A, B) \Rightarrow x <_l y \in A \Rightarrow x \in A \Rightarrow x <_l (A, B)$,

3) $x <_l (A, B) <_l y \Rightarrow x \in A$ and $y \notin A \Rightarrow x <_l y$,

4) $(A, B) <_l x <_l y \Rightarrow x \notin A \Rightarrow y \notin A \Rightarrow (A, B) <_l y$,

5) $x <_l (A, B) <_l (C, D) \Rightarrow x \in A \subseteq C \Rightarrow x \in C \Rightarrow x <_l (C, D)$,

6) $(A, B) <_l x <_l (C, D) \Rightarrow x \notin A$ and $x \in C \Rightarrow A \subset C \Rightarrow (A, B) <_l (C, D)$,

7) $(A, B) <_l (C, D) <_l x \Rightarrow x \notin C \supseteq A \Rightarrow x \notin A \Rightarrow (A, B) <_l x$,

8) $(A, B) <_l (C, D) <_l (E, F) \Rightarrow A \subseteq C \subseteq E$. If one of the last two $\subseteq$ can be replaced by $\subset$, then we have $A \subset E$ and further $(A, B) <_l (E, F)$. In the remaining case, where $A = C = E$ holds we also have $D \subset B$ and $F \subset D$. Then $F \subset B$ holds and thus $(A, B) <_l (E, F)$.

Next we verify that the linear order $\leq_l$ is an extension of the order $\leq_k$. Let $x \in H_\alpha$, (A, B) and (C, D) in $P_k(H_\alpha)$. Then we have

α) $x <_k (A, B) \Rightarrow x \in A \Rightarrow x <_l (A, B)$,

β) $(A, B) <_k x \Rightarrow x \in B \Rightarrow x \notin A \Rightarrow (A, B) <_l x$,

γ) $(A, B) <_k (C, D) \Rightarrow$ (by 1.2, $(4')$ $\exists x \in H_{\nu+1}$ with $(A, B) <_k x <_k (C, D)$. This entails also $(A, B) <_l x <_l (C, D)$ by α) and β),and thus $(A, B) <_l (C, D)$.

We still have to prove that $(k(H_{\nu+1}), \leq_l)$ is embeddable in $H_{\nu+1}$. If $(A, B) \in P_k(H_{\nu+1})$ then A (resp. B) is a k - generated initial (resp. final) segment of $H_{\nu+1}$. Then by 3.2 sup $A \in O_{\nu+1}$ and inf $B \in E_{\nu+1}$. The set of all k - generated initial segments of $H_{\nu+1}$ with sup $A \in A$, analogously with sup $A \notin A$, is with respect to the order $\subseteq$ isomorphic to a subset of $O_{\nu+1}$ and thus by 3.1 embeddable in $H_{\nu+1}$. By 4.4.6 then also the set of all k - generated initial segments of $H_{\nu+1}$, equipped with the order $\subseteq$, is embeddable in $H_{\nu+1}$.

156

Analogously, again by 3.1, the set of all k - generated final segments B of $H_{\nu+1}$, equipped with the order $\supseteq$, is embeddable in $H_{\nu+1}$. Then, if A is a fixed k - generated initial segment of $H_{\nu+1}$, the set A^* of all k - generated IF$_k$-pairs $(A, B) \in P_k(H_{\nu+1})$ has an order type $\leq h_{\nu+1}$. Let $\mathfrak{J}_k$ be the set of all these A^*, ordered by $\subseteq$. Now $P_k(H_{\nu+1})$ is the union of the sets A^* of type $\leq h_{\nu+1}$ over the linearly ordered argument $\mathfrak{J}_k$, which also has order type $\leq h_{\nu+1}$. By 4.4.6 then $P_k(H_{\nu+1})$ has an order type $\leq h_{\nu+1}$, and finally this also holds for $k(H_{\nu+1}) = H_{\nu+1} \cup P_k(H_{\nu+1})$. The existence of the function g is now clear.

Since f^* and g are both $<$ - preserving we obtain, combining the results of 3.3 and 3.4:

3.5 Lemma. *The composition* $h := g \circ f^* : T_{\tau+1} \to H_{\nu+1}$ *is* $<$ - *preserving with respect to the orders* $\leq_{\tau+1}$ *and* $\leq$.

h is not yet injective and therefore we shall alter it to obtain a function which is also injective.

3.6 Lemma. *With the concepts of 3.3, 3.4, 3.5 it follows: There exists an injective* $<$ - *preserving (with respect to* $\leq_{\tau+1}$ *and* $\leq$*) mapping* $f_{\tau+1} : T_{\tau+1} \to H_{\nu+1}$, *which satisfies*

(1) For $x, y \in T_{\tau+1}$ *we have* $h(x) < h(y) \implies f_{\tau+1}(x) < f_{\tau+1}(y)$.

Proof. For $y \in h[T_{\tau+1}]$ we define $C(y) := h^{-1}(y)$. Then $\{C(y)| y \in h[T_{\tau+1}]\}$ is a partition of $T_{\tau+1}$ into subsets with the same h - image. Since h is $<$ - preserving, comparable elements of $T_{\tau+1}$ have different h - images, and so every class $C(y)$ is an antichain of $T_{\tau+1}$. By the way : The classes $C(h(x))$ with $x \in T_\tau$ contain only one element, namely x.

Every class $C(y)$ has at most the cardinality $\aleph_{\nu+1}$ because of $\tau < \omega_{\nu+1}$ and 2.15 which entails $|T_{\tau+1}| \leq 2^{\aleph_\nu} = |H_{\nu+1}|$, and so there exists an injective mapping $\varphi_y : C(y) \longrightarrow H_{\nu+1}$, which, of course, is $<$ - preserving since $C(y)$ is an antichain. For $y \in h[T_{\tau+1}]$ we now define $M_y := \{(y, x)| x \in H_{\nu+1}\}$ and with these sets the ordered sum $\Sigma = \sum\{M_y | y \in h[T_{\tau+1}]\}$. Let $\leq_\sigma$ be its order. Roughly speaking: We construct the linearly ordered sum of copies of $H_{\nu+1}$ over the linearly ordered set $h[T_{\tau+1}]$ as index set. Then Σ has an order type $\leq h_{\nu+1}$ by 4.4.6.

We now define a mapping $f'_{\tau+1} : T_{\tau+1} \longrightarrow H_{\nu+1}$ as follows: Every $x \in T_{\tau+1}$ is in the class $C(h(x))$, and then we put $f'_{\tau+1}(x) = (h(x), \varphi_{h(x)}(x)) \in \Sigma$. Then $f'_{\tau+1}$ is an injective $<$ - preserving (with respect to $\leq_{\tau+1}$ and $\leq_\sigma$) mapping of $T_{\tau+1}$ into the linearly ordered set

$(\Sigma, \leq_\sigma)$, which has an order type $\leq h_{\nu+1}$. There further exists an injective $<$ - preserving mapping $\varepsilon : \Sigma \to H_{\nu+1}$. Then finally $f_{\tau+1} := \varepsilon \circ f'_{\tau+1} : T_{\tau+1} \longrightarrow H_{\nu+1}$ is injective and $<$ - preserving with respect to $\leq_{\tau+1}$ and $\leq$.

If elements $x, y \in T_{\tau+1}$ satisfy $h(x) < h(y)$, then $f'_{\tau+1}(x) <_\sigma f'_{\tau+1}(y)$ follows, and then further $f_{\tau+1}(x) < f_{\tau+1}(y)$ (in $H_{\nu+1}$).

We can resume, renaming now the f of 3.3 by f_τ :

3.7 Lemma. *Let τ be an ordinal $< \omega_{\nu+1}$, $f_\tau : T_\tau \to H_{\nu+1}$ an injective and $<$ - preserving (with respect to $\leq_\tau$ and $\leq$) mapping. Then there exists an injective $<$ - preserving (with respect to $\leq_{\tau+1}$ and $\leq$) mapping $f_{\tau+1} : T_{\tau+1} \to H_{\nu+1}$, for which (1) of 3.6 holds.*

In 3.7, $f_{\tau+1}$ is not necessarily an extension of f_τ. But we have the following:

3.8 Lemma. *We introduce a strict linear order $\ll_\tau$ in T_τ by $x \ll_\tau y \iff f_\tau(x) < f_\tau(y)$ in $H_{\nu+1}$ and analogously in $T_{\tau+1} : a \ll_{\tau+1} b \iff f_{\tau+1}(a) < f_{\tau+1}(b)$ in $H_{\nu+1}$. Then $\ll_\tau$ is a linear extension of $\leq_\tau$, and $\ll_{\tau+1}$ a linear extension of $\leq_{\tau+1}$ since f_τ resp. $f_{\tau+1}$ are $<$ - preserving. Then $\ll_\tau$ is the restriction of $\ll_{\tau+1}$ onto T_τ.*

Proof. For $x, y \in T_\tau$ we have $x \ll_\tau y \iff f_\tau(x) < f_\tau(y)$ in $H_{\nu+1}$. This yields, since f^* extends f_τ, $f^*(x) <_k f^*(y)$ in $H_{\nu+1} \cup P_k(H_{\nu+1})$, and since g is $<$ - preserving this entails $h(x) = g(f^*(x)) < g(f^*(y)) = h(y)$. By 3.6 this finally leads to $f_{\tau+1}(x) < f_{\tau+1}(y)$, which means to $x \ll_{\tau+1} y$.

Now we can prove the main result of this section:

3.9 Theorem [79]. *For every ordinal α there exists an injective $<$ - preserving mapping φ_α of U_α into H_α.*

Proof. First we prove the theorem for successor ordinals $\alpha = \nu + 1$. In Definition 2.10 we introduced the sets T_τ for all ordinals τ. For $T_0 = \emptyset$ we trivially have a linear order $\ll_0$ in T_0 whose order type is $\leq h_{\nu+1}$. If for a fixed ordinal $\tau < \omega_{\nu+1}$ we have a linear order $\ll_\tau$ in T_τ of type $\leq h_{\nu+1}$ which extends $\leq_\tau$, we obtain by 3.7 and 3.8 that there also exists a linear order $\ll_{\tau+1}$ in $T_{\tau+1}$ of type $\leq h_{\nu+1}$ which extends $\leq_{\tau+1}$ and $\ll_\tau$.

If λ is a limit ordinal $\leq \omega_{\nu+1}$, and if $\ll_\mu$ is a linear order in T_μ for all $\mu < \lambda$ which extends the order $\leq_\mu$ of T_μ, so that $(T_\mu, \leq_\mu)$ has an order type $\leq h_{\nu+1}$, where the relations $\ll_\mu$ increase with μ, then we define $\ll_\lambda$

on T_λ as the union of the relations $\ll_\mu$, $\mu < \lambda$. It follows directly from the construction that this yields a linear order $\ll_\lambda$ in the set T_λ which extends the order $\leq_\lambda$ and all $\ll_\tau, \tau < \lambda$. And from 4.4.6 we obtain that the order type of $(T_\lambda, \leq_\lambda)$ is $\leq h_{\nu+1}$.

Then we define $f_\lambda : T_\lambda \longrightarrow H_{\nu+1}$ in such a way that $x \ll_\lambda y$ implies $f_\lambda(x) < f_\lambda(y)$. Then f_λ is injective and $<$ - preserving with respect to $\ll_\lambda$ and $\leq$.

So we have defined by transfinite induction linear orders $\ll_\tau$ in the sets T_τ and mappings $f_\tau : T_\tau \longrightarrow H_{\nu+1}$ for $\tau \leq \omega_{\nu+1}$, where $\ll_\tau$ extends $\leq_\tau$ and has order type $\leq h_{\nu+1}$. Now $T_{\omega_{\nu+1}} = U_{\nu+1}$ is equipped with a linear order $\ll_{\omega_{\nu+1}}$ which extends the given order of $U_{\nu+1}$, and with which it has an order type $\leq h_{\nu+1}$. Now $\varphi_{\nu+1} := f_{\omega_{\nu+1}}$ satisfies the assertion for $\alpha = \nu + 1$, and the theorem is proved for successor ordinals $\alpha = \nu + 1$.

Let now α be a limit number. Then by 2.11, $U_\alpha = \cup\{U_{\nu+1} | \nu < \alpha\}$ holds. In every $U_{\nu+1}$ we have the linear order $\ll_{\omega_{\nu+1}}$ of type $\leq h_{\nu+1}$, which extends $\leq_{\omega_{\nu+1}}$ (of $T_{\omega_{\nu+1}}$). Then $\cup\{\ll_{\omega_{\nu+1}} | \nu < \alpha\}$ is a linear order in U_α which extends $\leq_{\omega_\alpha}$. It has a type $\leq h_\alpha$ because of 4.4.6. Here the limit mapping φ_α of the $\varphi_{\nu+1}, \nu < \alpha$, satisfies the assertion.

The content of 3.9 can now also be formulated in the following way: One can make a partial order destruction in H_α such that H_α with this reduced order is $\aleph_\alpha$ - universal, more exactly:

3.9′ Theorem. *For every ordinal α there exists an $\aleph_\alpha$ - universal partial order $\leq'$ in H_α, which has the order by first differences of H_α as a linear extension.*

Proof. The mapping φ_α of 3.9 maps U_α bijectively onto a subset U'_α of H_α. Then we define $\leq'$ so: It contains all pairs (x, x) with $x \in H_\alpha$ and all pairs (x, y) with $x, y \in U'_\alpha$ for which $\varphi_\alpha^{-1}(x) < \varphi_\alpha^{-1}(y)$ (in U_α).

Chapter 6
Applications of the splitting method

The splitting method is one of the most frequently used methods of combinatorial set theory. It was treated in1930 in detail in Ramsey's famous paper "On a problem of formal logic" ([150]). Some years before Urysohn ([174]) had already used a method of that kind. With this he solved a problem of Sierpinski (Fund. Math. 2 (1921), p. 286). The splitting method is now a central item of combinatorial set theory, which was systematically developed in pioneering works of Erdös, Rado, Hajnal, Milner and others (see e.g. [38] and [2]). This method has also several applications in order theory, and so we present here some of its fundamental concepts.

6.1 The general splitting method

First we treat several concepts which concern sets in general, which need not be ordered.

If S is a set which contains at least two elements, one can split S into two or eventually more non-empty disjoint subsets. Those which have more than one element can be split again and so on, until finally we arrive at the one-element subsets (or *atoms*) of S. This "atomization" is possibly reached only after transfinitely many steps.

The simplest way to split a set until one arrives at its atoms, is if in every splitting of a set we obtain exactly two subsets of it. This leads to the so-called dyadic partitions which will be treated in more detail. Of course we now have to give exact definitions for the notions which were mentioned before. To this purpose we introduce some more concepts.

In 1.4.7 we already have defined the notion "partition", and we remarked that a partition of a set is in a bijective correspondence to the notion of equivalence relation on this set.

1.1 Definition. A partition of a set S is said to be *proper*, if it contains at least two subsets of S.

Recall: If $\mathcal{Z}_1$ and $\mathcal{Z}_2$ are partitions of a set S, we call $\mathcal{Z}_1$ *finer* than $\mathcal{Z}_2$, if every element of $\mathcal{Z}_1$ is a subset of an element of $\mathcal{Z}_2$. We denote this by $\mathcal{Z}_1 \prec \mathcal{Z}_2$.

1.2 Theorem. *The set of all partitions of a set S is completely ordered by the relation $\prec$.*

Proof. If $\{\mathfrak{M}_i, i \in I\}$ is a set of partitions of S, then for each $x \in S$ we put $C_x := \cap\{M_i | i \in I\}$, where M_i is that class of $\mathfrak{M}_i$, which contains x. Then the set of all $C_x, x \in S$, forms a partition of S which is the greatest $\prec \mathfrak{M}_i$ for all $i \in I$. And C_x is that class which contains x. So the set of all partitions of S is inf - complete and then also complete.

More general than the concept "partition" is the notion "block-system":

1.3 Definition. A subset $\mathfrak{B}$ of the powerset $\mathfrak{P}(S)$ of a set S is called a *block-system* on S, if for every two sets $A, B \in \mathfrak{B}$ there holds $A \cap B \in \{\emptyset, A, B\}$. In another formulation: Two arbitrary sets of $\mathfrak{B}$ are disjoint or comparable (with respect to the subset relation $\subseteq$).

1.4 Example. Every partition $\mathcal{Z}$ of a set S is also a block-system on S. And here we even have $A \cap B = \emptyset$ for every two different sets A, B of $\mathcal{Z}$.

Another, also very special type of block-system, is described in:

1.5 Example. Let $(S, \leq)$ be a linearly ordered set, $\mathfrak{B}$ the set of its initial segments. Then $\mathfrak{B}$ is a block-system, and here we even have $A \cap B \in \{A, B\}$ for every two sets $A, B \in \mathfrak{B}$.

If $\mathfrak{B}$ is an arbitrary block-system on a set S, then $\mathfrak{B}$, augmented by the set of all atoms of S, is also a block-system on S. Therefore every maximal block-system on S contains among others all atoms of S.

Thus the partition $\mathcal{Z}$ of 1.4 is no maximal block-system, if $\mathcal{Z}$ contains a set with at least two elements.

By the way, one can prove (see [65]) that every maximal block-system on a set has common features with Examples 1.4 and 1.5.

1.6 Definition. We call a poset $(P, \leq)$ a *generalized tree*, if for every $x \in P$ the principal ideal $(P \leq x)$ is linearly ordered.

We call $(P, \leq)$ a *tree*, if it has a least element, and each principal ideal is well-ordered.

A *subtree* of a tree $(T, \leq)$ is a subset $S \subseteq T$ with the induced order $\leq \restriction S$, which has a least element. It is clear that then S is again a tree.

If a is an element of a tree T, the set $(T < a)$ is well-ordered. Its order type is an ordinal, and this is said to be the *height* of a (in T). It is denoted by $h(a)$. We define the height $h(T)$ of a tree T as the least ordinal which is $> h(a)$ for all $a \in T$. The same: $h(T) := \{h(a)|a \in T\}$.

One easily verifies that the concepts $h(a)$ and $h(T)$ are compatible with the definitions of 1.7.1.

Of course, every non-empty well-ordered set is also a tree. More interesting and typical examples are furnished by splittings, which we now define:

1.7 Definition. Let S be a non-empty set. A subset $\mathfrak{A}$ of $\mathfrak{P}(S)\backslash\{\emptyset\}$ is said to be a *splitting* of S, if the following three conditions are satisfied:

1) $S \in \mathfrak{A}$.

2) If $B \in \mathfrak{A}$ and $|B| \geq 2$ holds, then there exists a proper partition $\mathcal{Z}(B)$ of B, whose elements are in $\mathfrak{A}$, such that for every $X \in \mathfrak{A}$, that is a proper subset of B, there holds $X \subseteq C$ for a set $C \in \mathcal{Z}(B)$. (In another formulation: There are no sets of $\mathfrak{A}$ between B and a set $C \in \mathcal{Z}(B)$.)

3) If λ is a limit number and if sets $B_\nu \in \mathfrak{A}$, $\nu < \lambda$, are given which form a descending tower, which means $B_\mu \supset B_\nu$ for $\mu < \nu < \lambda$, then $\cap_{\nu<\lambda}B_\nu$ is also in $\mathfrak{A}$, provided it is non-empty.

The elements $B \in \mathfrak{A}$ are called the *blocks* of the splitting $\mathfrak{A}$.

The condition 2) provides that $\mathfrak{A}$ has not unnecessarily many elements. If it would be dropped, then also $\mathfrak{P}(S)\backslash\{\emptyset\}$ would be a splitting of S. Moreover it has the consequence that $\mathcal{Z}(B)$ is uniquely defined by $\mathfrak{A}$, as can easily be verified:

1.8 Lemma and **Definition.** *Let $\mathfrak{A}$ be a splitting of S, $B \in \mathfrak{A}$ and $|B| \geq 2$. Then the partition $\mathcal{Z}(B)$ of 2) is uniquely determined.*

The mapping $\mathcal{Z}$ which ascribes to every $B \in \mathfrak{A}$ the partition $\mathcal{Z}(B)$ is called the *splitting function* of $\mathfrak{A}$.

Proof. Let $\mathcal{Z}_1$ and $\mathcal{Z}_2$ both satisfy 2) of 1.7. Then every element of $\mathcal{Z}_2(B)$ is a proper subset of B and therefore $\subseteq C$ for a subset $C \in \mathcal{Z}_1(B)$. And so we have $\mathcal{Z}_2(B) \prec \mathcal{Z}_1(B)$, analogously $\mathcal{Z}_1(B) \prec \mathcal{Z}_2(B)$ and then $\mathcal{Z}_1(B) = \mathcal{Z}_2(B)$.

The sets $\mathcal{Z}(B)$ have at least two elements, and then the situation is of special interest where they all have exactly two elements. This leads us to the following concept:

1.9 Definition. Let S be a non-empty set, $\mathfrak{A}$ a splitting of S, where for every $B \in \mathfrak{A}$ with $|B| \geq 2$ the set $\mathcal{Z}(B)$ contains exactly two blocks.

Then $\mathfrak{A}$ is called a *dyadic splitting*.

Later we deal with the question under which assumptions one can obtain the existence of a dyadic splitting. Of course it is trivial that for every set $S \neq \emptyset$ always a splitting exists: The set $\mathfrak{A}$, which contains S and all one-element subsets of S, is a splitting of S.

In order to establish the relation between the concepts "splitting" and "tree" we introduce the concept height also for the blocks of a splitting:

1.10 Definition. Let $\mathfrak{A}$ be a splitting of a non-empty set S. Then we call S the block of *height* 0 of $\mathfrak{A}$. If $|S| > 1$ holds, the elements of $\mathcal{Z}(S)$ are said to be the blocks of *height* 1.

Let now τ be an ordinal > 0 such that for all $\nu < \tau$ the notion block of height ν is already defined. If τ has an immediate predecessor σ, we consider the blocks B of height σ with $|B| \geq 2$ and define the sets of $\cup\{\mathcal{Z}(B)|B$ as a block of height σ and $|B| \geq 2\}$ as the blocks of *height* τ.

If τ is a limit ordinal, we consider all (transfinite) sequences $(B_\nu)_{\nu < \tau}$ where B_ν is a block of height ν and where $\cap_{\nu < \tau} B_\nu$ is $\neq \emptyset$. (This implies that the sets B_ν form a decreasing tower $B_0 \supset B_1 \supset \cdots$.) These intersections are then called blocks of *height* τ.

Recall: If a is a cardinal we denote by a^+ the first cardinal which is $> a$ and by $\omega(a)$ the least ordinal which has cardinality a.

We list some elementary facts:

1.11 Theorem and **Definition.** *Let $\mathfrak{A}$ be a splitting of a set S with $|S| \geq 2$.*

1) For every ordinal β the set of all blocks of height $< \beta$ forms a block-system and, if $\beta > 0$, at the same time a tree with respect to the order $\leq$, which is the reverse inclusion $\supseteq$. The height of a block $B \in \mathfrak{A}$ is the same ordinal as the height of B as an element of the tree $(\mathfrak{A}, \supseteq)$. (See 1.6.)

2) Different blocks of the same height are disjoint. If B_i is a block of height δ_i for $i = 1, 2$ with $\delta_1 < \delta_2$ then $B_1 \cap B_2$ is $\emptyset$ or B_2.

3) There exist ordinals μ for which no blocks of a height $\geq \mu$ are defined. The least ordinal τ with this property shall be called the height of the splitting $\mathfrak{A}$. It is $< \omega(|S|^+) =: \rho$. And the height of the splitting $\mathfrak{A}$ is also the height of the tree $(\mathfrak{A}, \supseteq)$. (See 1.6.)

Proof. The first two statements follow easily from the definition and by transfinite induction on β.

3) Let $S = \{x_\kappa | \kappa < \omega(|S|)\}$ be a representation of S in a well-ordered form. To every x_κ there exists a least ordinal $\mu_\kappa < \rho$ such that x_κ is not contained in blocks B_ν of height ν with $\nu > \mu_\kappa$. For otherwise we would have a sequence of blocks B_ν, $\nu < \rho$, where B_ν has the height ν and contains x_κ. But then the B_ν, $\nu < \rho$, form a decreasing tower and there exist elements $b_\nu \in B_\nu \backslash B_{\nu+1}$ for all $\nu < \rho$. Their set $\{b_\nu | \nu < \rho\}$ would have the cardinality $|\rho| = |S|^+$, which is impossible.

Now ρ is a regular initial ordinal $> \omega(|S|)$, and so $\sigma := \sup\{\mu_\kappa | \kappa < \omega(|S|)\}$ is still $< \rho$. And there don't exist blocks of a height higher than σ. The rest follows from 1).

1.12 Theorem. *Let $\mathfrak{A}$ be a splitting of a non-empty set S. Then for every $x \in S$ the set $\{x\}$ is the intersection of all those blocks of $\mathfrak{A}$ which contain x.*

This entails $|\mathfrak{A}| \geq |S|$.

Proof. Let x be a given element of S and β the least ordinal number such that there is no block of height β which contains x. Then for every $\nu < \beta$ there exists exactly one block B_ν of height ν which contains x. The intersection $D := \cap_{\nu < \beta} B_\nu$ is non-empty and thus in $\mathfrak{A}$. Now D must be a one-element set, for otherwise $\mathcal{Z}(D)$ would contain a subset B_β of D of height β which still contains x, and this contradicts the definition of β.

From 1.12 we obtain nearly immediately the following separation property:

1.13 Theorem. *Let $\mathfrak{A}$ be a splitting of a set S, x and y two different elements of S. Then there exists a least ordinal σ such that there exist disjoint blocks X and Y in $\mathfrak{A}$ of the same height σ with $x \in X$ and $y \in Y$.*

Proof. The blocks that contain x and y form a well-ordered chain $\mathfrak{C}$ of $(\mathfrak{A}, \supseteq)$ (with S as first element). Then $\cap \mathfrak{C}$ is the last ($=$ smallest) block B which contains x and y. Now $\mathcal{Z}(B)$ has a block X, which contains x, but not y, and a block Y, which contains y, but not x.

The question arises how great the cardinality of a splitting of a set S can be in comparison with $|S|$. It is trivial that it is $\geq |S|$ because it contains all one-element subsets of S, and it is $\leq 2^{|S|}$ because it is a subset of the power set $\mathfrak{P}(S)$ of S.

164

In connection with the question concerning the existence of dyadic splittings we introduce the following notion:

1.14 Definition. We say: A set S satisfies the *generalized axiom of choice* (in the sense of Kinna/Wagner [96]), if there exists a function f, which ascribes to every subset $T \subseteq S$ that has at least two elements, a proper non-empty subset of T. Then f is called a *generalized choice function* on S.

A mapping $\mathcal{Z}$ which ascribes to every subset $T \subseteq S$, that has at least two elements, a partition into at least two proper subsets, shall be called a *subdividing function* to S.

These concepts are related by the following trivial facts which we mention for the sake of completeness:

1.15 Theorem. a) *If there exists a choice function to S, then there also exists a generalized choice function to S.*

b) *If there exists a generalized choice function to S, then there also exists a subdividing function to S which partitions every subset of S, which has at least two elements, into two proper subsets.*

The statements of a) and b) cannot be reversed without more assumptions. E.g. it can easily be proved that the set $\mathbf{R}$ of real numbers with their natural order satisfies the generalized axiom of choice: First we take a well-ordering of the set $\mathbf{Q}$ of rational numbers: $\mathbf{Q} := \{x_\nu | \nu \in \mathbf{N}\}$. This is possible without using the axiom of choice. If now T is a set of at least two real numbers there is a first index ν such that there exists at least one number of T which is $< x_\nu$ and one which is $\geq x_\nu$. The function which ascribes to T the set $(T < x_\nu)$ is then a generalized choice function on $\mathbf{R}$.

If we have a subdividing function $\mathcal{Z}$ on S we cannot immediately construct a generalized choice function from it, even if for every $T \subseteq S$ with $|T| \geq 2$ we have $|\mathcal{Z}(T)| = 2$. Indeed, for every at least two-element subset T of S we have at least two proper subsets, but there is no rule in sight to declare one of them as the value for a generalized choice function on S.

A subdividing function on S generates a splitting of S in the following natural way:

1.16 Theorem and **Definition.** *Let $\mathcal{Z}$ be a subdividing function on a non-empty set S. Then there exists a smallest subset $\mathfrak{A}$ of $\mathfrak{P}(S)$ which satisfies the following three conditions:*

1) $S \in \mathfrak{A}$.

2) *If $B \in \mathfrak{A}$ and $|B| \geq 2$ holds, all sets of $\mathcal{Z}(B)$ belong to $\mathfrak{A}$.*

3) *If $B_\nu, \nu < \mu$, are sets of $\mathfrak{A}$ with $\cap_{\nu<\mu} B_\nu \neq \emptyset$, then this intersection also belongs to $\mathfrak{A}$.*

Then $\mathfrak{A}$ is a splitting of S, and we call it the splitting of S which is induced by $\mathcal{Z}$.

The proof is easily verified. And we only have made use of the values of $\mathcal{Z}$ for the blocks of $\mathfrak{A}$. In particular 1.17 contains a sufficient condition for the existence of a dyadic splitting of a set:

1.17 Theorem. *Let $\mathcal{Z}$ be a subdividing function on a non-empty set S, where for every $T \subseteq S$ with $|T| \geq 2$ the set $\mathcal{Z}(T)$ has exactly two elements, then there exists a dyadic splitting of S.*

A fortiori this contains the statement: If there exists a generalized choice function to S, then there also exists a dyadic splitting of S.

If we have a splitting $\mathfrak{A}$ of a set S we have the opportunity to obtain a description of the blocks of $\mathfrak{A}$ and of the elements of S by sequences of ordinals, which reflect the position of the blocks, in particular the one-element blocks, which are practically the elements of S, relative to $\mathfrak{A}$.

1.18 Definition. Let S be a non-empty set and $\mathfrak{A}$ a splitting of S. If $B \in \mathfrak{A}$ has height β, we call the set $\{B_\nu | \nu \leq \beta\}$, where B_ν is the block of height ν which contains B, the *block-degression ending in B*. If B is a one-element block $\{x\}$, it is also called the *block-degression ending in x*.

Let $\mathcal{Z}$ be the splitting function of $\mathfrak{A}$. For every block $B \in \mathfrak{A}$ with $|B| \geq 2$ we choose a bijective mapping of the set $\mathcal{Z}(B)$ onto an initial segment $\zeta(B)$ of the class of ordinal numbers.

Then we ascribe to every block $B \in \mathfrak{A}$ a transfinite sequence $\varphi(B)$ of ordinals as follows: Let $\varphi(S)$ be the empty sequence. Let now μ be an ordinal such that $\varphi(C)$ is already defined for all blocks C with height $\nu < \mu$ and this so that, for the blocks C of height ν, the sequence $\varphi(C)$ has the length ν.

Let now B be a block of height μ. If μ is a successor number $\kappa + 1$, then we have $B \in \mathcal{Z}(C)$ for a block $C \in \mathfrak{A}$ of height κ, and then $\varphi(C)$ is already defined. According to our induction hypothesis there is an ordinal $\beta \in \zeta(C)$ which corresponds to B. Then $\varphi(B)$ shall be the sequence which arises from $\varphi(C)$ by attaching β.

If μ is a limit ordinal, we consider the blocks $B_0 \supset B_1 \supset \cdots \supset B_\mu = B$, where B_ν is the block of height ν which contains B_μ for $\nu < \mu$.

Now we define $\varphi(B_\mu)$ to be that sequence of length μ, of which the $\varphi(B_\nu)$, $\nu < \mu$, are initial segments. By transfinite induction we so ascribe to every block B of height μ a transfinite sequence $\varphi(B)$ of length μ of ordinals.

If we consider the last construction we can observe the following:

1.19 Theorem. *Let $\mathfrak{A}$ be a splitting in a non-empty set S, and let $\mathfrak{F}$ be the set of all transfinite sequences, which correspond to the blocks $B \in \mathfrak{A}$ (according to 1.18). For f_1, $f_2 \in \mathfrak{F}$ we put*
     $f_1 \prec f_2$ iff f_1 is an initial segment of f_2.
     Then $\prec$ is an order on $\mathfrak{F}$, and there holds: If B_1, B_2 are in $\mathfrak{A}$ and if $B_1 \supseteq B_2$ holds, then for the corresponding φ - values we have $\varphi(B_1) \prec \varphi(B_2)$.
     And so the mapping $\varphi : \mathfrak{A} \longrightarrow \mathfrak{F}$ is an isomorphic mapping of $(\mathfrak{A}, \supseteq)$ onto $(\mathfrak{F}, \prec)$. Therefore also $(\mathfrak{F}, \prec)$ is a tree.

6.2 Embedding theorems based on the order types of the well- and inversely well-ordered subsets

In this section we shall prove an embedding theorem which sharpens Hausdorff's embedding theorem 3.3.7 and a theorem of Erdös/Rado [36].

2.1 Definition. Let λ be an ordinal. Then we define $\omega_\alpha((\lambda))_0$ to be the subset of all those sequences $s = (s_\nu)_{\nu < \lambda} \in \omega_\alpha((\lambda))$ which have the property that there exists an ordinal $\mu(s) < \lambda$ such that $s_\nu = 0$ for all ν with $\mu(s) \leq \nu < \lambda$.

The set $\omega_\alpha((\lambda))_0$ has in many cases, if GCH is assumed, a lower cardinality than $\omega_\alpha((\lambda))$.

We begin with the following embedding theorem, which will be generalized a bit later:

2.2 Theorem. *Let S be a linearly ordered set which is free from $\omega_{\alpha+1}$ and from ω_β^*. Then S is embeddable in $\omega_\alpha((\omega_\beta))_0$.*

Proof. In every non-empty segment of S there exists a cofinal subset whose order type is a regular initial ordinal. We choose such a set for each of these segments and call it its cofinality set, which is now uniquely determined. Since S has no subset of type $\omega_{\alpha+1}$ all these cofinality subsets have an order type $\leq \omega_\alpha$.

If M is a segment of S which has at least two elements, we define its *successor-segments* as follows: If M has a last element l, we define $M[0] := (M < l)$ and $M[1] := \{l\}$.

If M has no last element, we consider its cofinality sequence $m_0, m_1, \ldots, m_\nu, \ldots | \nu < \omega_\delta$, where δ is an ordinal $\leq \alpha$. Here we put $M[0] := (M \leq m_0)$ and $M[\nu] := (M \leq m_\nu) \setminus \cup_{\mu < \nu}(M \leq m_\mu)$. The $M[\nu]$ are then called the *successor-segments* of M.

We define $\mathfrak{M}_0 := \{S\}$, and then $\mathfrak{M}_1$ as the set of successor-segments of S. If generally $\mathfrak{M}_\nu$ is already defined as a non-empty set of segments of S, from which at least one has more than one element, we define $\mathfrak{M}_{\nu+1}$ to be the set of all segments which are a successor-segment of one of the segments of $\mathfrak{M}_\nu$.

If λ is a limit ordinal and $\mathfrak{M}_\nu$ already defined for all $\nu < \lambda$, $\mathfrak{M}_\lambda$ shall be the set of all non-empty intersections $\cap_{\nu < \lambda} S_\nu$ where $S_\nu \in \mathfrak{M}_\nu$ for $\nu < \lambda$.

We denote the elements of the $\mathfrak{M}_\nu$ in the following way, using transfinite induction: If such a segment has the notation $M = M[\varphi_\nu | \nu < \lambda]$ and if it has more than one element, we consider its successor-segments. Such a one obtains the notation $M[\varphi_\nu | \nu \leq \lambda]$, if it is the segment $(M[\varphi_\nu | \nu < \lambda])[\varphi_\lambda]$. (It is the φ_λ^{th} segment of the partition of $M[\varphi_\nu | \nu < \lambda]$ which is given by the cofinality set of $M[\varphi_\nu | \nu < \lambda]$.)

If λ is a limit ordinal and $M[\varphi_0, \ldots, \varphi_\nu, \ldots | \nu < \rho] \in \mathfrak{M}_\rho$ is already defined for all $\rho < \lambda$, we denote the intersection $\cap_{\rho < \lambda} M[\varphi_0, \ldots, \varphi_\nu, \ldots | \nu < \rho]$, if it is in $\mathfrak{M}_\lambda$, i. e. if it is non-empty, by $M[\varphi_\nu | \nu < \lambda]$.

If now x is an element of S we ascribe to x a sequence $F(x)$ of ordinals according to the process which is described in 1.18 and 1.19: The ordinals ν for which x is in an element of $\mathfrak{M}_\nu$, form an initial segment $\{\nu | \nu < \sigma\}$ of the class of ordinals because of 1.11, 3). Here σ must be a successor number $\tau + 1$ as can easily be seen. Then we put $F(x) = [\varphi_\nu | \nu < \sigma]$, where the last sequence is determined by $x \in M[\varphi_\nu | \nu < \sigma]$. We call σ the length of $F(x)$ and denote it by $l(F(x))$. Our construction now yields immediately:

(I) $\varphi_\nu < \omega_\alpha$ for every $\nu < \sigma$. Hence: Every $F(x)$, $x \in S$, is a sequence of ordinals $< \omega_\alpha$.

Next we prove:

(II) For every $x \in S$ there holds $l(F(x)) < \omega_\beta$.

To see this we assume indirectly, that there exists an $x \in S$, such that $F(x)$ has a length $\geq \omega_\beta$. Let $F(x)$ be the sequence $[\varphi_\nu | \nu < \sigma]$ with an ordinal $\sigma \geq \omega_\beta$. Then we have $M[\varphi_\xi | \xi < \nu] \in \mathfrak{M}_\nu$ for $\nu < \omega_\beta$. For every $\nu < \omega_\beta$ we obtain the relation:

(III) There exists an element $x_\nu \in M[\varphi_0, \ldots, \varphi_\nu]$, which is greater than all elements of $M[\varphi_0, \ldots, \varphi_\nu, \varphi_{\nu+1}]$.

If $M[\varphi_0, \ldots, \varphi_\nu]$ has a last element l, then $M[\varphi_0, \ldots, \varphi_\nu, \varphi_{\nu+1}]$ is either $= \{l\}$ or $= \{x | x \in M[\varphi_0, \ldots, \varphi_\nu] \text{ and } x < l\}$. In the last case the assertion (III) is obvious. And the first case cannot happen. Indeed, if $M[\varphi_0, \ldots, \varphi_{\nu+1}]$ would have only one element, it has no successor-segments, which yields a contradiction since $\nu + 1$ is $< \omega_\beta$.

If $M[\varphi_0, \ldots, \varphi_\nu]$ has no last element, then the set of its successor-segments has no last element in its natural order, and then also (III) follows.

We consider now the set of all $x_\nu, \nu < \omega_\beta$. From (III) we obtain for every $\nu < \omega_\beta$ the relation $x_\nu > y$ for every $y \in M[\varphi_0, \ldots, \varphi_{\nu+1}]$, and so it follows that the $x_\nu, \nu < \omega_\beta$, form a subset of S of type ω_β^* which contradicts our assumption. So (II) is proved.

Further there holds:

(IV) For $x < y$ in S none of the image sequences $F(x)$, $F(y)$ is an initial segment of the other. Therefore we can order the set $\mathfrak{F} = \{F(x) | x \in S\}$ according to the principle of first differences, and then $x < y$ entails $F(x) < F(y)$.

This is a rather immediate consequence of 1.13. By (I) and (II) $\mathfrak{F}$ is $\subseteq \omega_\alpha((\omega_\beta))_0$, and since S is isomorphic to $\mathfrak{F}$ the theorem is proved.

The Theorem 2.2 can still be sharpened by exchanging the first occurrence of ω_β^* by an inverse ordinal τ^* where $\tau < \omega_{\beta+1}$. Before we present a lemma on indecomposable ordinals:

2.3 Lemma. *Let λ be an indecomposable ordinal and I a linearly ordered set that is embeddable in $\omega_\alpha((\lambda))_0$. Let further $S_i, i \in I$, be disjoint linearly ordered sets, which are embeddable in $\omega_\alpha((\lambda))_0$. Then also the ordered sum $\Sigma := \sum_{i \in I} S_i$ is embeddable in $\omega_\alpha((\lambda))_0$.*

Proof. We take sets $I', S_i' \subseteq \omega_\alpha((\lambda))$ which are isomorphic to I, S_i respectively. If now we have an element of Σ, i. e. an $m_i \in S_i$ with

$i \in I$, we ascribe a sequence $F(m_i)$ to it as follows: We take the image i' of i in I' and the image m_i' of m_i in S_i' and put together the sequences i' and m_i', where i' precedes m_i'. Then $F(m_i)$ is the sequence which so arises. Its length is the sum of the lengths of i' and m_i'. This is $< \lambda$ since λ is indecomposable. Then it is easily seen that F yields an embedding of Σ in $\omega_\alpha((\lambda))_0$.

2.4 Theorem. *Let $(S, \leq)$ be a linearly ordered set with $|S| \geq 2$, where every open interval of S contains a subset of type ω_α. Let τ be an ordinal $< \omega_{\alpha+1}$. Then every open interval of S also contains a subset of type τ.*

Dually: If here ω_α is replaced by its inverse type ω_α^*, then every open interval of S contains a subset of type τ^*.

Proof. Let λ be the supremum of all ordinals τ for which there holds:

Every open interval of S has a subset of type τ.

Then we shall prove $\lambda = \omega_{\alpha+1}$. We assume indirectly that λ is $< \omega_{\alpha+1}$. (By assumption $\lambda \geq \omega_\alpha$ is trivial.) First we see that λ must be a limit ordinal. For let I be an open interval of S, and suppose indirectly that λ is a successor number $\kappa + 1$. Then I has a subset $\{a_0, a_1, a_2\}_<$. Now the intervals (a_0, a_1) and (a_1, a_2) would contain subsets of type κ, and then I would contain a subset of type $\kappa + 1 + \kappa + 1 > \lambda$, which contradicts the supremum- property of λ.

λ is supposed to be $< \omega_{\alpha+1}$ and therefore it has a cofinal subset of regular order type $\omega_\delta \leq \omega_\alpha$, and thus λ is representable in the form $\lambda = \sum\{\zeta_\nu | \nu < \omega_\delta\}$, where the ζ_ν, $\nu < \omega_\delta$, are ordinals $< \lambda$. Every interval $(\zeta_\nu, \zeta_{\nu+1})$ has a subset of type ζ_ν, and then I has a subset of type $\sum\{\zeta_\nu | \nu < \omega_\delta\} = \lambda$.

For elements $a_0 < a_1 < a_2$ of I also (a_0, a_1) and (a_1, a_2) contain subsets of type λ, so that I has a subset of type $\lambda + \lambda > \lambda$. This yields a contradiction to the supremum-property of λ.

Now we can prove the main theorem of this section. First we present it in its negative formulation:

2.5 Main theorem [70]. *Let S be a linearly ordered set which is not embeddable in $\omega_\alpha((\omega_\beta))_0$. Then S has a subset of type $\omega_{\alpha+1}$, or S contains for every ordinal $\tau < \omega_{\beta+1}$ a subset of type τ^*.*

Proof. The proof makes use of a condensation principle which has been used in several papers on order theory. For $a, b \in S$ we put $a \sim b$

iff $a = b$ or if the closed interval with ends a and b is embeddable in $\omega_\alpha((\omega_\beta))_0$. Then $\sim$ is an equivalence relation. We prove the transitivity: Let $a \sim b$ and $b \sim c$ hold. We assume $a < b < c$. (The other cases are analogous.) We take an embedding of $[a, b]$ in $\omega_\alpha((\omega_\beta))_0$ and place a 0 before all the image sequences. This yields again an embedding of $[a, b]$ in $\omega_\alpha((\omega_\beta))_0$ because of $1 + \omega_\beta = \omega_\beta$. Analogously we embed $(b, c]$ in $\omega_\alpha((\omega_\beta))_0$,but now so that the image sequences begin with 1. Together we obtain an embedding of $[a, c]$ in $\omega_\alpha((\omega_\beta))_0$. Now S is partitioned by $\sim$ into equivalence classes, where each of them is a segment of S.

If S has a subset of type $\omega_{\alpha+1}$ we are done. So we assume that there is no such subset, and we have to prove that S contains for every $\tau < \omega_{\beta+1}$ a subset of type τ^*.

Now we can prove:

(I) Every class K of $\sim$ is embeddable in $\omega_\alpha((\omega_\beta))_0$.

We choose a fixed element k in a given equivalence class K. First we show:

(II) $K'' := \{x | x \in K$ and $x \geq k\}$ is embeddable in $\omega_\alpha((\omega_\beta))_0$.

If K has a last element this is trivial. So we assume that there is no last element. Then we can find a sequence $k = k_0 < k_1 < \cdots < k_\nu < \cdots | \nu < \lambda$, where λ is a limit ordinal $\leq \omega_\alpha$, which is cofinal in K. We put $I_\nu := [k_0, k_\nu]$ and $I'_\nu := I_\nu \setminus \cup_{\mu < \nu} I_\mu$. Then we have $K'' = \cup_{\nu < \lambda} I'_\nu$ where the I'_ν are disjoint segments of S. We take an embedding of I'_ν in $\omega_\alpha((\omega_\beta))_0$ and enlarge its image sequences by putting the digit ν before them. Together we so obtain an embedding of K'' in $\omega_\alpha((\omega_\beta))_0$.

In a similar way there follows:

(III) $K' := \{x | x \in K$ and $x < k\}$ is embeddable in $\omega_\alpha((\omega_\beta))_0$.

We assume that K' is non-empty and has no first element, otherwise (III) is trivial. Then there exists a coinitial sequence $k = k_0 > k_1 > \cdots > k_\nu > \cdots | \nu < \lambda$, where λ is a limit ordinal $\leq \omega_\beta$. Here we put $I_\nu := [k_\nu, k_0]$ for $\nu < \lambda$ and define I'_ν as before. Then we take an embedding of I_ν in $\omega_\alpha((\omega_\beta))_0$, but enlarge the image sequences now by putting before their beginning a string of ν zeros, which are followed by a digit 1. Since we have $\nu + \omega_\beta = \omega_\beta$ these embeddings form together an embedding of K' in $\omega_\alpha((\omega_\beta))_0$, and so (III) is proved, and then also (I).

There must be more than one $\sim$ - class because by assumption S is not embeddable in $\omega_\alpha((\omega_\beta))_0$. We choose a subset S' of S, which has exactly one element in common with each of the $\sim$ - classes. Then there holds:

(IV) If a, b are elements of S' with $a < b$, then the interval (of S', not of S) $I = [a, b]$ is not embeddable in $\omega_\alpha((\omega_\beta))_0$.

For otherwise by Lemma 2.3, and since ω_β is indecomposable, also the ordered sum $\sum_{i \in I} K_i$, where K_i is the equivalence class which contains i, would be embeddable in $\omega_\alpha((\omega_\beta))_0$, and this then would hold for the whole interval $[a, b]$ of S, in contradiction to the fact that a is not equivalent to b. In particular (IV) implies that S' is dense.

Let now I be an open interval of S'. As a subset of S, I has no subset of type $\omega_{\alpha+1}$. Then I must have a subset of type ω_β^*, otherwise it would by 2.2 be embeddable in $\omega_\alpha((\omega_\beta))_0$, contradicting (IV). This holds for all open intervals I of S'. Now the dual form of 2.4 yields that I has a subset of type τ^* for every $\tau < \omega_{\beta+1}$.

In a positive formulation the last theorem appears in the form:

2.5' Theorem. *Let S be a linearly ordered set which has no subset of type $\omega_{\alpha+1}$ and no subset of type τ^*, where τ is an ordinal $< \omega_{\beta+1}$. Then S is embeddable in $\omega_\alpha((\omega_\beta))_0$.*

We list some more corollaries of the main theorem:

2.6 Theorem. *The set $\omega_\alpha((\omega_0))_0$ has for every $\tau < \omega_{\alpha+1}$ a subset of type τ.*

Proof. Let τ be an ordinal $< \omega_{\alpha+1}$ and S a linearly ordered set of type τ. Then S is $\omega_{\alpha+1}$ - free and, as a well-ordered set, also ω_0^* - free, and thus by 2.2 embeddable in $\omega_\alpha((\omega_0))_0$.

The set $\omega_\alpha((\omega_\beta))_0$ has the cardinality $\sum\{\aleph_\alpha^{|\nu|} \,|\, \nu < \omega_\beta\}$, and so 2.5 yields the following :

2.7 Corollary. *If S is a linearly ordered set with $|S| > \sum\{\aleph_\alpha^{|\nu|} \,|\, \nu < \omega_\beta\}$, then S has a subset of type $\omega_{\alpha+1}$ or for every $\tau < \omega_{\beta+1}$ a subset of type τ^*.*

In particular there follows:

2.8 Corollary. *If S is a linearly ordered set with $|S| > \aleph_\alpha^{\aleph_\beta}$, then S has a subset of type $\omega_{\alpha+1}$ or for every $\tau < \omega_{\beta+2}$ a subset of type τ^*.*

Proof. If in 2.7 we exchange β by $\beta + 1$ we obtain 2.8 from 2.7 because of $\sum\{\aleph_\alpha^{|\nu|} \,|\, \nu < \omega_{\beta+1}\} = \aleph_\alpha^{\aleph_\beta} \cdot \aleph_{\beta+1} = \aleph_\alpha^{\aleph_\beta}$.

Also the main part of a theorem of Erdös and Rado [36] is now a consequence of 2.8. This states:

2.9 Theorem. *If S is a linearly ordered set with $|S| > 2^{\aleph_\alpha}$, then S has a subset of type $\omega_{\alpha+2}$ or for every $\tau < \omega_{\alpha+2}$ a subset of type τ^*.*

Proof. This follows from 2.8 by application of Bernstein's equality $2^{\aleph_\alpha} = \aleph_{\alpha+1}^{\aleph_\alpha}$ and by replacing α by $\alpha + 1$ and β by α.

A corollary of 2.9 is the theorem of Hausdorff/Urysohn ([174],[173]), which states:

2.9′. Theorem. *If S is a linearly ordered set of cardinality $> 2^{\aleph_\alpha}$, then S has a subset of type $\omega_{\alpha+1}$ or a subset of type $\omega_{\alpha+1}^*$.*

Another application is the following embedding theorem:

2.10 Theorem. *Let ω_α be regular, S a linearly ordered set which is free from $\omega_{\alpha+1}$ and free from ω_α^*. Then S is embeddable in H_α.*

Proof. By 2.2, S is embeddable in $\omega_\alpha((\omega_\alpha))_0$. The latter set is now embeddable in H_α, for it is the union of $\aleph_\alpha$ sets which are isomorphic to lexicographic products $\omega_\alpha((\tau))$ with ordinals $\tau < \omega_\alpha$, and which are embeddable in H_α by 4.7.4. Then our statement follows with 4.3.24.

6.3 The change number of dyadic sequences

In this section we investigate the dyadic splittings of Definition 1.9 in more detail. To this purpose we introduce the concept of change number of a dyadic sequence.

3.1 Definition. Let $d = (d_\nu)_{\nu<\mu}$ be a dyadic sequence, i. e.: μ is an ordinal, and every d_ν, $\nu < \mu$, is one of the numbers 0 or 1. Then we define the *change number* of d as follows: Let $M(d)$ be the set of all maximal segments of the (linearly ordered) index set μ, over which d is constant $(= 0$ or $= 1)$. For the set of these segments we have the following natural order: A segment A is $<$ a segment B if for every $a \in A$ and every $b \in B$ there holds $a < b$. Then the set $M(d)$ is well-ordered and has an ordinal c as order type. This is called the *change number* of d. We denote it by $c(d)$.

$M(d)$ is isomorphic to a uniquely defined initial segment of the class of ordinals, and then we call the segment which corresponds to the ordinal

$\nu < c(d)$ the νth *constancy-segment* of the dyadic sequence d and denote it by $C_\nu(d)$.

E.g. if d is the sequence $0010010010......111100$, it has the change number ω_0+2; over its 4th constancy-segment we have the values 00, over its ω_0^{th} constancy-segment the values 1111. The sequence $00000...11111...$ has the change number 2.

3.2 Definition. If $(S, \leq)$ is a linearly ordered set and $\mathfrak{A}$ a splitting of S, this is called a *splitting into segments* if all blocks of $\mathfrak{A}$ are segments of S.

Supplementing Definition 1.18 we proceed as follows:

3.3 Definition. Let $(S, \leq)$ be a linearly ordered set, $\mathfrak{A}$ a dyadic splitting of S into segments, $\mathcal{Z}$ the splitting function of $\mathfrak{A}$. If B is a block of $\mathfrak{A}$ with $|B| \geq 2$, $\mathcal{Z}(B)$ is $= \{B_l, B_r\}$, where B_l and B_r are disjoint segments of S, whose union is S. Here we choose our notations so that all elements of B_l are less than all elements of B_r. We call B_l the *left* and B_r the *right subsegment* of B.

We ascribe to the block S the empty sequence. Let μ be an ordinal such that for all $\nu < \mu$ we have already defined for all blocks B of height ν a dyadic sequence $\varphi(B)$ of length ν. If now C is a block of height μ, we define $\varphi(C)$ as follows:

If μ is a successor ordinal $\kappa + 1$, then we have $C \in \mathcal{Z}(B)$ for a block $B \in \mathfrak{A}$ of height κ.Then we define $\varphi(C)$ as that sequence which arises from $\varphi(B)$ by attaching 0 (resp. 1) if C is the left (resp. right) subsegment of B.

If μ is a limit ordinal, we define $\varphi(B_\mu)$ in the same way as in Definition 1.18. So, by transfinite induction, we have ascribed to every $B \in \mathfrak{A}$ a dyadic sequence $\varphi(B)$. If B is a one-element block $\{x\}$, we also write $\varphi(x)$ instead of $\varphi(\{x\})$. We put $D(\mathfrak{A}) := \{\varphi(x)|x \in S\}$. It is immediately clear that the set $D(\mathfrak{A})$ is completely determined by $\mathfrak{A}$.

And the construction shows again, which was already mentioned in 1.19, that the set $(S, \leq)$ is isomorphic to $D(\mathfrak{A})$, if this set of dyadic sequences is ordered by first differences.

The system $D(\mathfrak{A})$ which so corresponds to a dyadic splitting $\mathfrak{A}$ of a linearly ordered set S into segments, reflects some of the order-theoretical properties of S.

The next theorem appeared in similar formulations in publications of J. Novák [126], Novotný [129], Padmavally [135], Cuesta Dutari [16], [17].

174

3.4 Theorem. *Let* $(S, \leq)$ *be a linearly ordered set,* $\mathfrak{A}$ *a dyadic splitting of S into segments. Let* $d = (d_\nu | \nu < \mu)$ *be a dyadic sequence* $\varphi(B)$ *for some $B \in \mathfrak{A}$.*

Then there holds:

a) *If the set $D_0 := \{\nu < \mu | d_\nu = 0\}$ has type τ, then S has a subset of type τ^*. Analogously there follows:*

If $D_1 := \{\nu < \mu | d_\nu = 1\}$ has type τ, then S has a subset of type τ .

b) *If μ is an indecomposable ordinal, then S has a subset of type μ or of type μ^*.*

c) *Let c be the change number $c(d)$ of d. Then c is uniquely representable as $c = \lambda + n$, where λ is 0 or a limit ordinal, and n a non-negative integer. And then S has a subset of type $\lambda + \lfloor \frac{n}{2} \rfloor$ and a subset of type $(\lambda + \lceil \frac{n}{2} \rceil)^*$, or a subset of type $\lambda + \lceil \frac{n}{2} \rceil$ and a subset of type $(\lambda + \lfloor \frac{n}{2} \rfloor)^*$.*

Proof. a) Let $D_0 = \{\nu_\iota | \iota < \tau\}_<$. For every $\iota < \tau$ we choose an element x_ι in that block which corresponds to the sequence $(d_\nu | \nu < \nu_\iota)$ at which the digit 1 has been attached. Then x_ι is greater than all elements of the block with dyadic sequence $(d_\nu | \nu \leq \nu_\iota)$ since $d_{\nu_\iota} = 0$, and then by construction also greater than all elements x_σ with $\iota < \sigma < \tau$. The x_ι , form a set of type τ^*.

b) We have $D_0 \cup D_1 = \mu$. In this mixed sum of ordinals D_0 or D_1 must have the type μ by 4.8.8, and b) follows from a).

c) c is the order type of the set M of all ordinals which are the first element of a constancy-segment of d. The set of those elements of M, at which d has the component 0 (resp. 1), has then an order type $\lambda + \lfloor \frac{n}{2} \rfloor$ (resp. $(\lambda + \lceil \frac{n}{2} \rceil)$), or conversely. Indeed, a dyadic sequence δ of length λ, where λ is a limit ordinal, and which has the property that for each of its components $\delta_\nu \in \{0, 1\}$ the immediate successor $\delta_{\nu+1}$ is $1 - \delta_\nu$, has the property that the subsequences of δ of all 0's (resp. all 1's) still have length λ. Then c) is a consequence of a).

We supplement Definition 1.18:

3.5 Definition. Let $\mathfrak{A}$ be a splitting of a set. Then we define a *maximal block-degression* of $\mathfrak{A}$ to be a sequence $(B_\nu)_{\nu < \mu}$, where B_ν is a block of height ν for $\nu < \mu$, and $B_{\nu_1} \supset B_{\nu_2}$ for $\nu_1 < \nu_2 < \mu$, and where $\cap_{\nu < \mu} B_\nu$ is a one-element set or empty. In the first case we call the block-degression *point-containing*, in the second *pointless*.

It follows easily that the above block-degression can only be pointless if μ is a limit ordinal.

If in particular $\mathfrak{A}$ is a dyadic splitting of a linearly ordered set, then we assign to every maximal block-degression $(B_\nu)_{\nu<\lambda}$, where λ is a limit ordinal, that dyadic sequence of length λ, of which the sequences $\varphi(B_\nu)$, $\nu < \lambda$, are initial segments.

At the beginning we had only considered point-containing block-degressions, but also the others can give useful insights into the structure of linearly ordered sets.

3.6 Theorem [67]. *Let S be a linearly ordered set of cardinality $\aleph_{\nu+1}$, which is free from $\omega_{\nu+1}$ and from $\omega^*_{\nu+1}$. Let m be the least cardinal for which $\aleph_\nu^m > \aleph_\nu$ holds. (So, in particular, m is $\leq \aleph_\nu$.)*

Let $\mathfrak{A}$ be a dyadic splitting of S into segments, D the set of those sequences which are ascribed to the maximal block-degressions of $\mathfrak{A}$. Let D' be the set of those sequences of D, which have a change-number $< \omega(m)$. Then D' has cardinality $\leq \aleph_\nu$.

Of course D has at least $\aleph_{\nu+1}$ elements. And thus the set of sequences with change-number $\geq \omega(m)$ has also cardinality $|D|$.

(Roughly speaking: "Nearly all" dyadic sequences of D have change number $\geq \omega(m)$.)

Proof. If $d = (x_\nu)_{\nu<\mu}$ is a sequence of D and σ an ordinal $< c(d)$ we define the $\sigma-$trunc $d|\sigma$ of d to be the initial segment of d, whose range of definition is the union of the first σ constancy-segments of d, hence $d|\sigma = (x_\nu|\nu \in \cup_{\tau<\sigma}C_\tau(d))$. Let further D_σ be the set of all $d|\sigma$ with $d \in D$.

Then D_0 contains only the empty sequence. Further we have $|D_1| \leq \aleph_\nu$. For if d is a sequence of D_1, then all components of d are $= 0$ or all are $= 1$, and so D_1 is the union of two sets $D_1(0)$ and $D_1(1)$ of sequences, where those of $D_1(0)$ consist only of 0's and those of $D_1(1)$ only of 1's. Here the supremum γ_0 of the lengths of the sequences of $D_1(0)$ must satisfy the condition $\gamma_0 < \omega_{\nu+1}$. For otherwise there would exist for every $\beta < \omega_{\nu+1}$ a dyadic sequence in $D_1(0)$ which begins with β digits 0. Then, similarly to the proof of 3.4, there would follow that there exists a block degression $B_0 \supset B_1 \supset \cdots$ over $\omega_{\nu+1}$ steps, where every $B_{\tau+1}$ is the left subsegment of B_τ for all $\tau < \omega_{\nu+1}$. And this would entail the existence of a subset of S of type $\omega^*_{\nu+1}$, contradicting our assumption.

Analogously the supremum γ_1 of the lengths of the sequences of $D_1(1)$ is $< \omega_{\nu+1}$. And then $|D_1| = |D_1(0)| + |D_1(1)|$ is $\leq \aleph_\nu$. So the equation

(1) $|D_\tau| \leq \aleph_\nu$

is satisfied for $\tau = 0$ and $\tau = 1$. We assume now that (1) holds for all $\tau < \rho$, where ρ is an ordinal $< \omega(m)$. If ρ is a successor number, then there holds $|D_\rho| \leq |D_{\rho-1}| \cdot \aleph_\nu \leq \aleph_\nu$. For if a sequence of D_ρ is given, it arises from a sequence of $D_{\rho-1}$ by attaching to it only digits 0 resp. only digits 1. And then the set of all sequences of D_ρ which arise in this way from one of $D_{\rho-1}$ has (similarly to D_1 before) a cardinality $\leq \aleph_\nu$, so that indeed $|D_\rho| \leq |D_{\rho-1}| \cdot \aleph_\nu \leq \aleph_\nu$ follows.

If ρ is a limit ordinal, then the sequence $d|\rho$ is uniquely determined by the complex $(d|\sigma)_{\sigma<\rho}$. And then $|D_\rho| \leq \aleph_\nu^{|\rho|} = \aleph_\nu$ follows because of $|\rho| < m$. Now (1) also holds for ρ instead of τ. By transfinite induction then (1) holds for all $\tau < \omega(m)$. So finally D' has a cardinality $\leq |\cup \{D_\tau | \tau < \omega(m)\}| \leq m \cdot \aleph_\nu = \aleph_\nu$.

If instead of $\aleph_{\nu+1}$ we have a regular limit cardinal, we obtain the following variant of 3.6:

3.7 Theorem [67]. *Let S be a linearly ordered set with a regular cardinality $\aleph_\nu$, where ν is a limit ordinal, and where S is free from ω_ν and ω_ν^*. Let m be the least cardinal, such that a product of m cardinals, which are all $< \aleph_\nu$, is $\geq \aleph_\nu$.*

Let $\mathfrak{A}$ be a dyadic splitting of S and $\alpha < \omega(m)$ a given ordinal. Then the set D' of all dyadic sequences of $D(\mathfrak{A})$, which have a change-number $< \alpha$, has a cardinality $|D'| < \aleph_\nu$.

If the GCH is assumed we have $m = \aleph_\nu$.

Proof. The proof is quite analogous to that of 3.6: Like there, we define sets D_τ. Then by transfinite induction we obtain $|D_\tau| < \aleph_\nu$ for all $\nu < \omega(m)$, and then $|D'| \leq |\cup_{\tau<\alpha} D_\tau| < \aleph_\nu$ follows.

As an application of Theorems 3.6 and 3.7 we obtain the following statement, which is a partial result of a theorem of Erdös/Rado [36]:

3.8 Theorem. *Assuming the GCH there holds: Every infinite linearly ordered set S, which has no well- and no inversely well-ordered subsets of cardinality $|S|$, and for which $|S|^-$ is regular, contains for every cardinal $k < |S|$ a well- and an inversely well-ordered subset of cardinality k.*

Proof. If $|S| = \aleph_{\nu+1}$ holds, where $\aleph_\nu$ is regular, then the m of 3.6 is $= \aleph_\nu$ because of the GCH, and there exists a dyadic sequence in the $D(\mathfrak{A})$ of a dyadic splitting $\mathfrak{A}$ of S, which has a change-number $\geq \omega_\nu$, and this yields our assertion by 3.4, c).

If $|S|$ is a regular limit cardinal $\aleph_\nu$, then the m of 3.7 is $= \aleph_\nu$, and thus 3.7 proves our assertion.

Theorems 3.6 and 3.7 can formally be generalized a bit more after the following theorem, which exhibits a connection between arbitrary representations of a linearly ordered set by dyadic sequences and representations which arise from a dyadic splitting.

3.9 Theorem. *Let S be a set of dyadic sequences, of which no one is a proper initial segment of another one, so that we have a linear order in S according to the principle of first differences. Then there exists a dyadic splitting $\mathfrak{A}$ of S into segments, so that the sequence $s' \in D(\mathfrak{A})$, which corresponds (in the sense of 3.3) to an element $s \in S$, is a subsequence of s, and so that S is isomorphic to the set $\{s'|s \in S\}$ by $s \longrightarrow s'$.*

Proof. Let τ be the least ordinal for which there exist sequences $s \in S$, which have different τ - components $s(\tau)$. Then we define blocks $B[0]$, $B[1]$ (of a splitting of S, which we also define):
$$B[0] := \{s \in S | s(\tau) = 0\}, \text{ and } B[1] := \{s \in S | s(\tau) = 1\}.$$

In the same way as we have split S into $B[0]$ and $B[1]$ we now split $B[0]$ into sets $B[00]$ and $B[01]$, and $B[1]$ into sets $B[10]$ and $B[11]$, always provided that the sets to be split have at least two elements.

In general we proceed as follows: Let $(\nu_\mu)_{\mu<\kappa}$ be a dyadic sequence for which a block $B[\nu_\mu|\mu < \kappa]$ and an ordinal $\tau[\nu_\mu|\mu < \kappa]$ are already defined. Let then $\tau := \tau[\nu_\mu|\mu \leq \kappa]$ be the least ordinal such that there exist sequences $s \in B[\nu_\mu|\mu < \kappa]$ with different $\tau-$ components $s(\tau)$.

Then we put $B[\nu_\mu|\mu < \kappa, 0] := \{s \in B[\nu_\mu|\mu < \kappa] \text{ and } s(\tau) = 0\}$ and $B[\nu_\mu|\mu < \kappa, 1] := \{s \in B[\nu_\mu|\mu < \kappa] \text{ and } s(\tau) = 1\}$.

By the way, in this formalism the ordinal τ from the beginning is $\tau[\ ]$, the τ of the empty sequence.

If λ is a limit ordinal, for which blocks $B[\nu_\mu|\mu < \kappa]$ are defined for all $\kappa < \lambda$, which form a descending sequence with increasing κ, we put
$$B[\nu_\mu|\mu < \lambda] := \cap_{\kappa<\lambda}B[\nu_\mu|\mu < \kappa],$$
if the latter intersection is non-empty. By transfinite induction we so obtain a dyadic splitting $\mathfrak{A}$ of S (into blocks $B[\ldots]$).

Our construction shows that the sequence $s' := (\nu_\mu)_{\mu<\rho}$ which we ascribe to s is a subsequence of s, and that $s \longrightarrow s'$ defines an isomorphic mapping of S onto $S' := \{s'|s \in S\}$.

The change number of a subsequence of a dyadic sequence s is of course always $\leq$ that of s. Then we can establish the following generalization of Theorems 3.6 and 3.7, which has no relation to dyadic splittings:

3.10 Theorem [67]. *Let S be a set of dyadic sequences, of which no one is a proper initial segment of another one, so that S is equipped with the linear order of first differences.*

1) If $|S| > \aleph_{\nu+1}$ and if S has no subsets of type $\omega_{\nu+1}$, $\omega^*_{\nu+1}$, and if m is the least cardinal for which $\aleph_\nu^m > \aleph_\nu$ holds, then there exist at most $\aleph_\nu$ sequences in S with a change-number $< \omega(m)$.

2) If $|S| \geq \aleph_\nu$, where $\aleph_\nu$ is a regular limit cardinal, and if S has no subsets of type ω_ν, ω^*_ν, and if m is defined as in 3.7, then for every $\alpha < \omega(m)$ the set of sequences of S, that have a change-number $< \alpha$, has a cardinality $< \aleph_\nu$.

Supplementing Theorems 3.6 and 3.7 we still have a look at the situation where we have a singular cardinal $\aleph_\alpha$. Before we mention a theorem of Hausdorff [82], see also [36], Lemma 1.

3.11 Theorem. *Let $\aleph_\alpha$ be a singular cardinal, $\omega_\gamma := \mathrm{cf}(\omega_\alpha)$, so that $\omega_\alpha = \sum\{\omega_{\alpha_\nu}|\nu < \omega_\gamma\}$, where the ω_{α_ν}, $\nu < \omega_\gamma$, are initial ordinals $< \omega_\alpha$. For $\nu < \omega_\gamma$ we define S_{α_ν} to be a linearly ordered set of type $\omega^*_{\alpha_\nu}$. Then we define a linearly ordered set S as the ordered sum $S := \sum\{S_{\alpha_\nu}|\nu < \omega_\gamma\}$. (We can assume that the S_{α_ν} are pairwise disjoint.) Then every well-ordered subset of S has an order type $\leq \omega_\gamma$, and every inversely well-ordered subset has an order type τ^* with some $\tau < \omega_\alpha$.*

Proof. Let $W \subseteq S$ be well-ordered. Then W intersects every summand S_{α_ν} in only finitely many elements because S_{α_ν} has no infinite ascending chain. This implies that the type of W is $\leq \omega_\gamma$.

Let $I \subseteq S$ be an inversely well-ordered set. Then I has no infinite ascending chain, and thus it intersects only finitely many of the summands S_{α_ν}. So for its type τ^* there holds $\tau < \omega_\alpha$.

Based on the last theorem we can now prove:

3.12 Theorem. *Let $\aleph_\alpha$ be a singular cardinal. Then there exists a dyadic splitting of the set S of 3.11, in which all dyadic sequences have* *a change-number* ≤ 3.

Proof. For $\nu < \omega_\gamma$ we define a set T_{α_ν} as a set of dyadic sequences as follows: T_{α_ν} contains exactly all sequences which have digits 1 at the first ν positions, and behind this block of 1's follows a block of at least one digit 0, but fewer than ω_{α_ν} digits 0, and behind the zeros is just one digit 1.

Illustration of T_{α_ν} :
$$1111111....01$$
$$1111111....001$$
$$1111111....0001$$

Then the set T_{α_ν} has order type $\omega_{\alpha_\nu}^*$, and the ordered sum $T := \sum \{T_{\alpha_\nu} | \nu < \omega_\gamma\}$ has the order type $\sum \{\omega_{\alpha_\nu}^* | \nu < \omega_\gamma\}$, the same as that of S. By 3.9 there exists a dyadic splitting of T (and then also of S) into segments, in which all dyadic sequences (are subsequences of sequences of the T_{α_ν} and thus) have a change-number ≤ 3.

At the end of this section we have a look at the well-ordered sets in connection with possible change-numbers. There holds:

3.13 Theorem. *Every well-ordered set (and by symmetry also every inversely well-ordered set) has a dyadic splitting into segments, in which every corresponding dyadic sequence has the change-number ≤ 2.*

Proof. Let W be a well-ordered set. W.r.o.g. we can assume that W is an ordinal μ. Let then S be the set of all dyadic sequences s of the following type: s begins with $\nu < \mu$ digits 1, and behind them follows exactly one digit 0, (so that S looks like this: 0, 10, 110, 1110,...... .). Then S has the order type μ. By 3.9 then there exists a dyadic splitting of S, which consists of subsequences of the sequences of S, which then all have a change-number ≤ 2, and this also holds for W.

In 4.3.4 we had seen that every linearly ordered set is isomorphic to a set of dyadic sequences which is ordered according to the principle of first differences. Of course then one is interested to find such representations where the dyadic sequences have length as short as possible. In this connection we introduce the notion of dyadic depth :

3.14 Definition. Let $(S, \leq)$ be a linearly ordered set. If D is a set of dyadic sequences which is isomorphic to S, we denote the supremum of the lengths of the sequences of D by $l(D)$. Then we define the *dyadic depth* $\delta(S)$ of $(S, \leq)$ to be the minimum of all $l(D)$, where D is a set of dyadic sequences which is isomorphic to S.

From the definition it is immediately clear that for every ordinal λ the lexicographically ordered set $2((\lambda))$ of dyadic sequences of length λ has a dyadic depth $\leq \lambda$. But it turns out that it is also $\geq \lambda$. Before we prove this we mention a simple fact:

3.15 Remark. *Let S and T be linearly ordered sets, $f : T \to S$ an isomorphism, $S = S_0 \cup S_1$ and $T = T_0 \cup T_1$ with $T_0 \cap T_1 = \emptyset$, where S_0 resp. T_0 is an initial segment of S resp. T. Then $f \upharpoonright T_0$ is an embedding of T_0 in S_0, or $f \upharpoonright T_1$ an embedding of T_1 in S_1.*

For if all values of $f[T_0]$ are in S_0, then the first assertion holds, in the other case the second.

3.16 Theorem. *Let λ be an ordinal, $T := 2((\lambda))$ ordered according to the principle of first differences. Let S be a set of dyadic sequences, ordered by the principle of first differences, which is isomorphic to T. Then S contains a dyadic sequence of length $\geq \lambda$.*

Proof. By 3.9 there exists a dyadic splitting $\mathfrak{A}$ of S, for which the set $D(\mathfrak{A})$ of corresponding dyadic sequences is isomorphic to S, and where every sequence of $D(\mathfrak{A})$ is a subsequence of a sequence of S. So we are done if we have proved that $D(\mathfrak{A})$ has a sequence of length $\geq \lambda$. So for sake of simplicity we can assume $S = D(\mathfrak{A})$.

If $\mu \leq \lambda$ holds and if $(\alpha_\nu | \nu < \mu)$ is a dyadic sequence, we denote the set of all sequences of S (resp. T) which have α_ν as ν- component for $\nu < \mu$, by $S(\alpha_\nu | \nu < \mu)$ (resp. $T(\alpha_\nu | \nu < \mu)$). Then we have
$S = S(0) \cup S(1)$ and $T = T(0) \cup T(1)$.

Here $S(0)$ and $T(0)$ are initial segments of S resp. T, and $S(1)$ and $T(1)$ the corresponding final segments of S resp. T. By 3.15 $T(0)$ is embeddable in $S(0)$ or $T(1)$ in $S(1)$. In the first case we put $\alpha_0 := 0$, in the second $\alpha_0 := 1$.

Suppose now that we have constructed a dyadic sequence
$\alpha_0, \ldots, \alpha_\nu, \ldots | \nu < \mu$ with the property: For every $\tau < \mu$ the set $T(\alpha_\nu | \nu \leq \tau)$ is embeddable in the block $B_\tau := S(\alpha_\nu | \nu \leq \tau)$ of height $\tau + 1$ of $\mathfrak{A}$. For $\mu = 1$ this induction hypothesis is fulfilled due to the choice of α_0.

If μ is a successor ordinal $\kappa + 1$ then $T(\alpha_\nu | \nu \leq \kappa)$ is embeddable in the block $B_\kappa := S(\alpha_\nu | \nu \leq \kappa)$ of height $\kappa + 1$ of $\mathfrak{A}$. Then by 3.15 one of the two sets $T(\alpha_\nu|\nu \leq \kappa, \alpha_{\kappa+1} = 0)$, $T(\alpha_\nu|\nu \leq \kappa, \alpha_{\kappa+1} = 1)$ is embeddable in a block $B_{\kappa+1}$of height $\kappa+2$ of $\mathfrak{A}$, where this embedding is a restriction of the embedding of $T(\alpha_\nu|\nu \leq \kappa)$. And so one can prolong the sequence α_ν, $\nu < \mu$, by an element $\alpha_\mu \in \{0,1\}$, such that $T(\alpha_\nu|\nu \leq \mu)$ is embeddable in a block of height $\mu + 1$ of $\mathfrak{A}$.

If μ is a limit ordinal, then $T(\alpha_\nu|\nu < \mu) = \cap\{T(\alpha_\nu|\nu \leq \kappa)|\kappa < \mu\}$ is embeddable in $\cap\{B_\kappa|\kappa < \mu\}$, and the last intersection is a block of height μ of $\mathfrak{A}$. So by transfinite induction we construct a descending sequence of blocks $B_\nu, \nu < \lambda$, of height $\nu + 1$ of $\mathfrak{A}$. The set of dyadic sequences which correspond to $\mathfrak{A}$ has then also a sequence of length λ, and this finishes our proof.

In connection with the dyadic depth the question arises, whether a linearly ordered set that has dyadic depth λ also has a subset of type λ or λ^*. In the literature several false statements occur in the surrounding of this question, among others some false "proofs" in connection with the Suslin problem. We present a counter-example to the above question:

3.17 Example. Let I be a subset of $\mathbf{R}$ of cardinality $\aleph_1$, say $I = \{i_\nu|\nu < \omega_1\}$ and equip I with the restriction of the usual order of $\mathbf{R}$. For every $\nu < \omega_1$ we put $M(i_\nu) := 2((\nu))$ and equip this with the order by first differences. Let S be the ordered sum $\sum\{M(i_\nu)|i_\nu \in I\}$. Then, of course, the dyadic depth $\delta(S)$ is $\geq \sup \delta(M(i_\nu)) = \sup\{\nu|\nu < \omega_1\} = \omega_1$ by 3.16.

But S has no subset of type ω_1, and symmetrically no subset of type ω_1^*. Indeed, if T would be a subset of type $\omega_1$of $S = \{(i_\nu, x)|\nu < \omega_1$ and $x \in M(i_\nu)\}$ and hence well-ordered, then the set of first components of the elements of T is well-ordered, but countable, since I, like $\mathbf{R}$, has no uncountable well-ordered subset. Also every $2((\nu))$ with $\nu < \omega_1$ has no well- or inversely well-ordered subset of cardinality $\aleph_1$, so that $M(i_\nu) \cap T$ is also countable for every $\nu < \omega_1$. Then we would obtain that $T = \cup\{M(i_\nu)|\nu < \omega_1\}$ is countable, a contradiction.

6.4 An application in combinatorial set theory

In this section we introduce some fundamental concepts of combinatorial set theory, in particular the arrow-relation of Erdös/Rado [35]:

4.1 Definition. Let a, b, c be cardinals. Then $a \longrightarrow (b, c)^2$ means the following: If A is a set of cardinality a, and if the set $[A]^2$ of two-element subsets of A is divided into two disjoint sets K_1, K_2, then there exists a subset $B \subseteq A$ with $|B| = b$ and $[B]^2 \subseteq K_1$ or a subset $C \subseteq A$ with $|C| = c$ and $[C]^2 \subseteq K_2$. If for $x \neq y$ we have $\{x, y\} \in K_i$, we say: x and y are i-related.

This abstract definition can be described also in the following intuitive manner: If for $x \neq y$ of A the set $\{x, y\}$ belongs to K_i, we say: The edge $\{x, y\}$ has color i. The above statement can then be reformulated as: There exists a monochromatic simplex in color 1 with b vertices, or one in color 2 with c vertices.

The notation $a \not\rightarrow (b, c)^2$ means of course the negation of $a \longrightarrow (b, c)^2$.

In this context the following fundamental theorem of Erdös [40] holds:

4.2 Theorem. *Let a and c be infinite cardinals, where c is the least cardinal for which $a^c > a$ holds. Then we have $a^+ \longrightarrow (a^+, c)^2$.*

Proof. Let A be a set of a^+ elements, $[A]^2 = K_1 \cup K_2$, where $K_1 \cap K_2 = \emptyset$. We assume that A has no subset B with cardinality a^+ for which $[B]^2 \subseteq K_1$ holds. Then we are done if we have proved the existence of a set C with cardinality c and $[C]^2 \subseteq K_2$.

First we define a subdividing function for A. If T is a subset of A with at least two elements, we choose a maximal subset $M(T) \subseteq T$ such that all two-element subsets of $M(T)$ belong to K_1. The existence of such a set $M(T)$ follows easily by applying one of the maximal principles.

Now we choose a fixed well-ordering for every $M(T)$:

$M(T) = \{x_\nu | \nu < \tau\}$, where τ is an ordinal depending on T.

We call elements x, y of A *i-related*, if $\{x, y\} \in K_i$. Then every element of $T \backslash M(T)$ is 2-related to at least one element of $M(T)$, since this set is maximal. For $\nu < \tau$ let T_ν be the set of all elements of $T \backslash M(T)$ which are 2-related to x_ν, but 1-related to all x_μ with $\mu < \nu$. Then $\mathscr{Z}(T) := \{\{x_\nu\} | \nu < \tau\} \cup \{T_\nu | \nu < \tau$ and $T_\nu \neq \emptyset\}$ is a partition of T. So we have defined a subdividing function for A, and this induces now a splitting of the set A. Next we prove:

(*) For every $\sigma < \omega(c)$ there holds:

(1) The set $\mathfrak{B}_\sigma$ of blocks of height σ has a cardinality $\leq a$.

For $\sigma = 0$ the set A is the only block of height 0, and for $\sigma = 1$ there exist $\leq 2 \cdot |M(A)| \leq a$ blocks of height 1 due to our construction (and assumption).

Now we assume that μ is an ordinal $< \omega(c)$ for which (1) is already proved for all $\sigma < \mu$.

Case 1. μ is a successor number $\tau + 1$.

Then for every block B of height τ the partition $\mathcal{Z}(B)$ has at most a blocks, so that on the whole there exist $\leq a \cdot a = a$ blocks of height μ.

Case 2. μ is a limit ordinal.

Then every block of height μ is an intersection $\cap_{\nu < \mu} B_\nu$, where B_ν is a block of height ν. For these intersections we have at most $a^{|\mu|} = a$ possibilities, for $|\mu|$ is $< c$. Thus (1) follows by transfinite induction for all $\sigma < \omega(c)$.

We can now conclude that there exist blocks of height $\omega(c)$. For otherwise (*) would yield $|A| \leq \sum\{|\mathfrak{B}_\sigma| \,|\, \sigma < \omega(c)\} \leq a \cdot c = a$, for c is $\leq a$. And so we have obtained a contradiction.

Let now B be a block of height $\omega(c)$. Then there exists a strictly decreasing (with growing index) sequence of blocks $B_0 \supset B_1 \supset \cdots \supset B_\nu \supset \cdots |\nu < \omega(c)$, where B_ν has height ν. In every B_ν there exists an element e_ν, which is not contained in $B_{\nu+1}$, and then also e_ν is contained in no B_μ with $\nu < \mu < \omega(c)$. Now all elements of $C := \{e_\nu | \nu < \omega(c)\}$ are pairwise 2-related, which proves our statement.

Later we need the following partition theorem of Erdös/Rado [35] which is very similar to 4.2:

4.2′ Theorem. *Let a,b be infinite cardinals. Then*
$$(a^b)^+ \longrightarrow (a^+, b^+)^2.$$

Proof. The proof is a variant of that of 4.2. Let A be a set with $|A| = (a^b)^+$ and $[A]^2 = K_1 \cup K_2$. We assume that A has no subset B with cardinality a^+ which satisfies $[B]^2 \subseteq K_1$. Then we prove the existence of a set $C \subseteq A$ with $|C| = b^+$ and $[C]^2 \subseteq K_2$. For every $T \subseteq A$ which has at least two elements we choose a maximal subset $M(T) \subseteq T$ such that all two-element subsets of $M(T)$ belong to K_1. Then $|M(T)| \leq a$. We choose a well-ordering for every $M(T)$. Consider a fixed $M(T) = \{x_\nu | \nu < \mu\}$. We call elements $x \neq y$ of A *i-related* if $\{x, y\} \in K_i$. Then every element of $T \backslash M(T)$ is 2-related to at least one element of $M(T)$. For $\nu < \tau$ let T_ν be the set of all elements of $T \backslash M(T)$ which are 2-related to x_ν, but 1-related to all x_μ with $\mu < \nu$. Then $\mathcal{Z}(T) := \{\{x_\nu\} | \nu < \mu\} \cup \{T_\nu | \nu < \mu$ and $T_\nu \neq \emptyset\}$ is a partition of T. And $\mathcal{Z}$ defines a subdividing function and and a splitting $\mathfrak{A}$ of S. Let τ be the height of the splitting $\mathfrak{A}$. In complete analogy to 4.2 $\tau \geq \omega(b^+)$ follows. Indeed, in the case $\tau < \omega(b^+)$ we would have $|\tau| \leq b$ and $|A| \leq |\mathfrak{A}| \leq \sum\{a^{|\nu|}||\nu < \tau\} \leq a^b \cdot |\tau| \leq a^b \cdot b = a^b$, a contradiction. So there exists a strictly decreasing sequence $B_0 \supset B_1 \supset \cdots \supset B_\nu \supset \cdots$,where B_ν is a block of $\mathfrak{A}$ of height ν for $\nu < \omega(b^+)$. We choose an element $b_\nu \in B_\nu \backslash B_{\nu+1}$ for every $\nu < \omega(b^+)$.Then the set $C := \{b_\nu | \nu < \omega(b^+)\}$ has cardinality $\geq b^+$, and its elements are pairwise 2-related.

Another important partition result, whose substance is due to Sierpinski [162], is:

4.3 Theorem. $2^{\aleph_\nu} \nrightarrow (\aleph_{\nu+1}, \aleph_{\nu+1})$.

Proof. We consider the set $S := 2((\omega_\nu))$ of dyadic sequences of length ω_ν, taken with the order $\leq_1$by first differences. This linearly ordered set has, due to 4.2.6, no well- and no inversely well-ordered subset of cardinality $\aleph_{\nu+1}$. Now we take also a well-ordering $\leq_2$of S. For $x \neq y$ of S we put $\{x, y\} \in K_1$ if $x \leq_1 y$ and $x \leq_2 y$, or $y \leq_1 x$ and $y \leq_2 x$. And we define $K_2 := [S]^2 \backslash K_1$.

Then there is no subset T of S of cardinality $\aleph_{\nu+1}$, such that all its two-element subsets belong to K_1(resp. K_2). For otherwise in T the orders $\leq_1$and $\leq_2$had to be equal, which is impossible since T has like S no well-ordered and no inversely well-ordered subset of cardinality $\aleph_{\nu+1}$ in the order $\leq_1$.

A similar idea can be applied in the proof of the following theorem of Erdös/Rado [36]:

4.4 Theorem. *Let* ω_α *be a singular initial ordinal,* $\omega_\gamma := \mathrm{cf}(\omega_\alpha)$. *Then* $\aleph_\alpha \nrightarrow (\aleph_{\gamma+1}, \aleph_\alpha)$.

Proof. We consider the set S of 3.11 with the order which it had in 3.11, and which we denote here by $\leq_1$. In addition to this we consider a well-ordering $\leq_2$in S. The definition of K_1 and K_2 is the same as in 4.3. Since by 3.11, S with $\leq_1$has no well-ordered subsets of cardinality $\aleph_{\gamma+1}$ and no inversely well-ordered subsets of cardinality $\aleph_\alpha$, our assertion follows.

In the following we present another important theorem of the partition calculus of set theory which also has applications in the theory of posets. First recall the graph-theoretical notions of 2.5.4. We supplement this by:

4.5 Definition. If $(V, \mathfrak{E})$ is a graph and $a \in V$ we put $C(a) := \{x \in V | \{a, x\} \in \mathfrak{E}\}$ and $C^*(a) := C(a) \cup \{a\}$. The cardinality $|C(a)|$ is called the *degree* of the element a. It is the cardinality of the set of edges ($=$ elements of $\mathfrak{E}$) which contain a.

For $M \subseteq V$ we put $C(M) := \cup\{C(a)|a \in M\}$ and $C^*(M) := M \cup C(M)$.

A subset of V, of which all elements are unlinked, is called *independent*. A subset T of a set S which has the same cardinality as S is also called a *full part* of S.

4.6 Lemma. *Let $(V, \mathfrak{E})$ be a graph, where $|V| = k$ is an infinite cardinal, such that every full part F of V has an element of degree k in the graph $(F, \mathfrak{E} \cap [F]^2)$.*

Then V has an infinite subset whose elements are pairwise connected.

Proof. There exists an element $a_1 \in V =: C_0$ with degree k. Then $C_1 := C(a_1)$ has cardinality k. Analogously there exists an element $a_2 \in C_1$, so that the set C_2 of elements of $C(a_1)$ that are connected with a_2 has cardinality k. If generally we have for an $n \in \mathbf{N}$ already defined sets $C_1 \supseteq \cdots \supseteq C_n$ with $|C_n| = k$ and elements $a_{\nu+1} \in C_\nu$ for $\nu = 1, \ldots, n-1$, we have an element $a_{n+1} \in C_n$ which is connected with all elements of a full part $C_{n+1} \subseteq C_n$. By induction we so construct a set $\{a_1, a_2, \ldots\}$ of $\aleph_0$ elements of V, which are pairwise connected.

4.7 Lemma. *Let $(V, \mathfrak{E})$ be a graph, where $k := |V|$ is a regular infinite cardinal, and where every element of V has a degree $< k$. Then V has an independent full part.*

Proof. We choose $a_0 \in V$, $a_1 \in V \backslash C^*(a_0)$, $a_2 \in V \backslash (C^*(a_0) \cup C^*(a_1)), \ldots$. If in general for an ordinal $\mu < \omega(k)$ already elements a_ν are defined for $\nu < \mu$, we choose $a_\mu \in V \backslash \cup \{C^*(a_\nu)|\nu < \mu\}$. This is possible since every summand $C^*(a_\nu)$ has a cardinality $< k$ and because k is regular. By transfinite induction we so obtain a set $\{a_\nu|\nu < \omega(k)\}$, which is an independent full part of V.

The same statement as that of 4.7 appears in another formulation as:

4.7′ Lemma. *Let $(V, \mathfrak{E})$ be a graph where $k := |V|$ is a regular infinite cardinal, and suppose that every full part of V has two connected elements. Then V has an element of degree k.*

Now we can prove the regular case of the theorem of Dushnik/Miller/Erdös [30]:

4.8 Lemma. *Let k be a regular infinite cardinal. Then $k \longrightarrow (k, \aleph_0)^2$ follows.*

For the special case $k = \aleph_0$ this is Ramsey's theorem (for the infinite case).

Proof. Let $(V, \mathfrak{E})$ be a graph with $|V| = k$. If there exists a full part of V of independent elements, we are done. In the other case every full part F of V contains two connected elements. By 4.7′, applied to F, then F has an element x, which is connected with k elements of F. By 4.6 now V has an infinite subset of pairwise connected elements.

A rather immediate consequence of 4.8 is:

4.8′ Theorem. *Let M be an infinite set and $[M]^2 = K_1 \cup \cdots \cup K_n$, where $n \in \mathbf{N}$. Then there is an infinite subset $T \subseteq M$ and an $i \in \{1, \ldots, n\}$ with $[T]^2 \subseteq K_i$.*

Proof. It suffices to prove this for $|M| = \aleph_0$. For $n = 1$ (resp. $n = 2$) the statement holds by 4.8. Suppose that it is proved for a fixed $n \in \mathbf{N}$ and that $[M]^2 = K_1 \cup \cdots \cup K_n \cup K_{n+1}$. Then 4.8 entails that there is an infinite set $T \subseteq M$ with $[T]^2 \subseteq (K_1 \cup \cdots \cup K_n)$ or $[T]^2 \subseteq K_{n+1}$. And by induction hypothesis then there exists an $i \in \{1, \ldots, n+1\}$ and an infinite $T^* \subseteq T$ with $[T^*]^2 \subseteq K_i$.

We shall extend 4.8 also for non-regular cardinals. The next theorem contains most of the proof of that.

4.9 Theorem [30]. *Let $(V, \mathfrak{E})$ be a graph where $|V| = k$ is infinite, and where every element of V has a degree $< k$.*

Then there exist $\aleph_0$ elements, which are pairwise connected, or k independent elements.

Proof. Suppose that there are no $\aleph_0$ elements which are pairwise connected. Then we are done, if we have proved:

(1) There exists an independent full part I of V.

According to 4.7 we only have to consider the case where k is a singular cardinal. Then there exists a representation

(2) $k = \sum\{k_\nu | \nu < \omega_\gamma\}$, where $\omega_\gamma = \mathrm{cf}(k)$, and where the summands k_ν are regular cardinals $> \aleph_\gamma$, which then also must be $< k$.

First we prove the following, which contains the main idea of the proof:

(3) Let m be a regular cardinal with $\aleph_\gamma < m < k$. Then there exists an independent subset $M \subseteq V$ with $|M| = m$ and $|C(M)| < k$.

Let T be a subset of V of cardinality m. Due to 4.8 then there exists an independent subset $A \subseteq T$ with $|A| = m$, for T has, like V, no infinite subset of connected elements. For $\nu < \omega_\gamma$ we then define:

$A_\nu := \{x \in A|$ the degree of x is $\leq k_\nu\}$.

Since every element of V has a degree $< k$, we have by (2)

$A = \cup\{A_\nu | \nu < \omega_\gamma\}$, and thus $(|A| =)\ m \leq \sum\{|A_\nu| | \nu < \omega_\gamma\}$.

Since m is regular and $> \aleph_\gamma$ this entails that one of the summands $|A_\nu|$ must have cardinality m. Let M be such an A_ν. Then $|C(M)| \leq |M| \cdot k_\nu < k$ holds, and (3) is proved.

Applying (3) we obtain: There exists an independent set $I_0 \subseteq V$ of cardinality k_0 with $|C(I_0)| < k$. Further there exists an independent subset $I_1 \subseteq V \backslash C^*(I_0)$ (the last set has cardinality k) of cardinality k_1 with $|C(I_1)| < k$.

If in general μ is an ordinal $< \omega_\gamma$, and if independent sets I_ν of cardinality k_ν are defined for all $\nu < \mu$, such that $V \backslash \cup \{C^*(I_\nu) | \nu < \mu\}$ still has cardinality k, then by (3) there exists an independent subset $I_\mu \subseteq V \backslash \cup \{C^*(I_\nu) | \nu < \mu\}$ of cardinality k_μ for which $|C(I_\mu)| < k$ holds. By transfinite induction we so construct a sequence of pairwise disjoint independent sets I_ν for $\nu < \omega_\gamma$, where also for $\nu < \mu$ all elements of I_μ are not connected with all elements of I_ν. Finally we then have: $I := \cup\{I_\nu | \nu < \omega_\gamma\}$ has cardinality $\sum\{k_\nu | \nu < \omega_\gamma\} = k$, and I is independent, so that (1) is proved.

There follows:

4.10 Theorem of Dushnik/Miller/Erdös [30]: *For every infinite cardinal k we have $k \to (k, \aleph_0)^2$.*

Proof. Let $(V, \mathfrak{E})$ be a graph with $|V| = k$. We have to show:

There exists an independent set of k elements of V, or there exists an infinite subset of V, whose elements are pairwise connected.

If every full part F of V has an element which is connected with k elements of F, our assertion follows from 4.6, according to which we obtain infinitely many elements of V which are pairwise connected.

So we assume that there is a full part F of V for which every element has degree $< k$. From 4.9 (with F instead of V) we obtain the existence of $\aleph_0$ pairwise connected or of k independent elements of $F \subseteq V$.

The before mentioned partition theorems have interesting consequences in the theory of posets. We list some of them:

4.11 Theorem. *Let $(P, \leq)$ be an infinite poset. We call here two elements $a \neq b$ of P connected, if they are comparable, resp. independent, if they are incomparable. Then there hold:*

a) If P has a cardinality a^+, and if c is the least cardinal for which $a^c > a$ holds, then P has a chain of cardinality a^+ or an antichain of cardinality c.

Also: P has an antichain of cardinality a^+ or a chain of cardinality c.

b) If P has an infinite cardinality k, then P has a chain of cardinality k or an infinite antichain.

Also: P has an antichain of cardinality k or an infinite chain.

Proof. This is a direct consequence of 4.2 and 4.10.

Another application yields a theorem of Lindenbaum [112]:

4.12 Theorem. *Let S be an infinite set with two well-orderings $\leq_1$ and $\leq_2$. Then there exists a full part of S, over which these two orders coincide.*

A direct consequence of this is that if in S there are finitely many well-orderings, then there exists a full part, in which they all coincide.

Proof. If a and b are two different elements of S, we put $a \, R \, b \Longleftrightarrow a <_1 b$ and $a <_2 b$ hold both, or none of them holds. Then there is no infinite set of elements which are unrelated in R. For this would entail the existence of elements $a_1, a_2, \ldots$ of S with $a_1 <_1 a_2 <_1 a_3 <_1 \cdots$ and $a_1 >_2 a_2 >_2 a_3 >_2 \cdots$, which is impossible since a well-ordered set has no infinite strictly decreasing sequence.

We mention without proof a generalization of 4.12 (see [74]):

4.13 Remark *Let S be a set of cardinality $\aleph_{\nu+1}$ and c the least cardinal for which $\aleph_\nu^c > \aleph_\nu$ holds. And let W be a set of well-orderings of S where $|W| < c$ holds. Then there exists a full part of S, over which all well-orderings, which belong to W, coincide.*

This statement is tight.

6.5 Cofinal subsets

After we have proved 2.5, which considered linearly ordered sets, we now address the (only partially) ordered sets and study how their cardinality depends on the cardinality of their well- (resp. inversely well) -ordered chains and antichains. In this context the cofinal (resp. coinitial) subsets of a poset are of importance. First we introduce several concepts (similar to those of [55]):

5.1 Definition. Let $(P, \leq)$ be a non-empty poset. We define $\delta^+(P)$ (resp. $\delta^-(P)$) to be the supremum of all ordinals τ such that S has a subset of type τ (resp. τ^*, the reverse of τ), and put $d^+(P) := |\delta^+(P)|$ and $d^-(P) := |\delta^-(P)|$.

A subset $C \subseteq P$ is said to be *cofinal* in P, if for every $x \in P$ there exists a $c \in C$ with $x \leq c$. Dually C is called *coinitial*, if for every $x \in P$ there exists a $c \in C$ with $c \leq x$. (So cofinal, resp. coinitial, sets of non-empty sets are never empty.)

Then $C \subseteq P$ is cofinal in P iff $P = \cup\{(x]|x \in C\}$.

A set $B \subseteq P$ is called a π - *base* of P, if B is cofinal and coinitial in P. Let $\pi(P)$ denote the smallest cardinality of a π - base of P. We define $\pi_0(P)$ (resp. $\pi_1(P)$) to be the least cardinal k such that P has a coinitial (resp. cofinal) subset of cardinality k.

Then it is trivial that $\pi(P)$ is $\geq \pi_0(P)$ and $\geq \pi_1(P)$, and further $\pi(P) \leq \pi_0(P) + \pi_1(P)$.

So, if one of the cardinals $\pi(P)$, $\pi_0(P)$, $\pi_1(P)$ is infinite, then we obtain $\pi(P) = \max\{\pi_0(P), \pi_1(P)\}$.

Finally we put $\sigma_0(P) := \sup\{\pi_0(T)|T \subseteq P\}$, $\sigma_1(P) := \sup\{\pi_1(T)|T \subseteq P\}$, and $\sigma(P) := \sup\{\pi(T)|T \subseteq P\}$.

If P has a first (resp. last) element a, then $\{a\}$ is a coinitial (resp. cofinal) subset of P, and then $\pi_0(P) = 1$ (resp. $\pi_1(P) = 1$). These cases are not very interesting, and so the importance of the concepts $\pi_0(P)$, $\pi_1(P)$ results mainly in connection with posets P which have no first (resp. last) element.

The substance of the following theorem is due to John Ginsburg [55]:

5.2 Theorem. *Let $(P, \leq)$ be a poset, which is not linearly ordered. Then $|P| \leq \sigma_1(P)^{d^-(P)}$ and dually $|P| \leq \sigma_0(P)^{d^+(P)}$.*

Proof. In every subset $T \subseteq P$ with $|T| \geq 2$ we choose a cofinal subset $C(T)$ of minimal cardinality. Then we have

190

(1) $T = C(T) \cup \bigcup\{(T < c)|c \in C(T)\}$.

Here some summands $(T < c)$ could be empty, and the sets $(T < c)$ need not be pairwise disjoint. But, starting from (1), one can find a partition $\mathcal{Z}(T)$ of T into the singletons of $C(T)$ and into non-empty sets $T'_c \subseteq (T < c)$, $c \in C' \subseteq C(T)$ which are pairwise disjoint. If we have done so for all $T \subseteq P$ with at least two elements, we have constructed a subdividing function $\mathcal{Z}$ to P (see 1.14). By 1.16 $\mathcal{Z}$ induces a splitting $\mathfrak{A}$ of P.

P is the block of height 0 of the splitting, the elements of $\mathcal{Z}(P)$ are the blocks of height 1 and so on. Let τ be the height (see 1.11) of the splitting $\mathfrak{A}$. We prove $\tau \leq \delta^-(P)$. For this purpose suppose that $B_0 \supset B_1 \supset \cdots \supset B_\nu \supset \cdots$, $\nu < \mu$, is a strictly decreasing sequence, where every B_ν, $\nu < \mu$, is a block of $\mathfrak{A}$ of height ν. For $\nu + 1 < \mu$ all elements of $B_{\nu+1}$ are $<$ some x_ν of $C(B_\nu) \subseteq B_\nu$, and we choose for every $\nu + 1 < \mu$ such an element x_ν. And if μ is a successor ordinal $\rho + 1$ we further choose an arbitrary element $x_\rho \in B_\rho$. For every $\nu + 1 < \mu$ all elements of B_κ with $\nu + 1 \leq \kappa < \mu$ are $< x_\nu$. Then also all x_σ with $\nu < \sigma < \mu$ are by construction less than the chosen element x_ν. And so the $x_\nu, \nu < \mu$, form an inversely well-ordered set of type μ^*. So we have $\mu \leq \delta^-(P)$, and therefore we obtain $\tau \leq \delta^-(P)$.

We abbreviate $\sigma_1(P) =: k$. First we consider the case where k is infinite. Now P is the only block of height 0, and the set of blocks of height 1 has cardinality $\leq 2 \cdot k = k$, since k is infinite. For it contains the singletons of $C(P)$ and the at most $|C(P)| \leq k$ many sets, which each contain strict lower bounds to an element of $C(P)$. So for $\nu = 1$ the following induction hypothesis is satisfied:

(2) The set of blocks of height ν has cardinality $\leq k^{|\nu|}$.

If ν is $< \tau$, then every block of height ν is split into $\leq k$ blocks of height $\nu + 1$, and so the total number of blocks of height $\nu + 1$ is $\leq k^{|\nu|} \cdot k = k^{|\nu+1|}$. If (2) holds for all ν with $0 < \nu < \lambda$, where λ is a limit ordinal $< \mu$, then every block of height λ is an intersection $\cap\{B_\nu|\nu < \lambda\}$, where B_ν is a block of height ν. For every $\nu < \lambda$ we have at most $k^{|\nu|} \leq k^{|\lambda|}$ possibilities, and so there exist at most $(k^{|\lambda|})^{|\lambda|} = k^{|\lambda|}$ such intersections, and thus (2) also holds for λ. By transfinite induction then (2) is proved for all ν with $0 < \nu < \tau$. Finally there follows by 1.12:

(3) $|P| \leq |\mathfrak{A}| \leq \sum\{k^{|\nu|}|\nu < \tau\}$.

For $\nu < \tau$ we have $|\nu| \leq |\tau| \leq d^-(P)$. Then the last sum in (3) is $\leq k^{d^-(P)} \cdot |\tau| \leq k^{d^-(P)} \cdot d^-(P) = k^{d^-(P)}$.

Suppose now that k is finite. Every antichain A of P is its unique cofinal subset, so that $\pi_1(A) \leq k$. So every antichain of P has at most k elements, and then the width w of P is $\leq k$. Also P has no subset of type ω_0. Indeed, such a set would have a cofinal subset of cardinality $\aleph_0 \leq k$, a contradiction. By Dilworth's theorem now P is a union of w disjoint chains, that are inversely well-ordered. Let λ^* be the type of the longest of them. Then we have $\lambda \leq \delta^-(P)$, and it follows $|P| \leq w \cdot |\lambda| \leq k \cdot d^-(P)$. This is $\leq k^{d^-(P)}$ because of $k \geq 2$; here we use that P is no chain. So the theorem is proved.

The question arises why we excluded the case where P is linearly ordered. Here one has simple counter-examples. Let e.g. P be a finite chain with at least two elements. Then $|P| > 1 = \sigma_1(P)^{d^-(P)}$ since here $\sigma_1(P) = 1$.

The last theorem has several interesting corollaries. First we recall the concept *well-founded poset* of 1.9.8:

5.3 Definition. A poset satisfies the *descending* (resp. *ascending*) *chain condition* if it has no subset of type ω_0^* (resp. ω_0). So a poset is well-founded iff it satisfies the descending chain condition.

5.4 Theorem [55]. *Let P be a well-founded poset, which is no chain, and $k := \sigma_1(P)$ infinite. Then $|P| = \sigma_1(P)$. And dually:*

If P is a poset which is no chain and satisfies the ascending chain condition, and if $\sigma_0(P)$ is infinite, then $|P| = \sigma_0(P)$.

Proof. In the proof of 5.2 we obtain from (3) with the additional information that P has no subset of type ω_0^*, $|P| \leq |\mathfrak{A}| \leq \sum \{k^{|\nu|} | \nu < \omega_0\} = k$ since $\tau \leq \delta^-(P) \leq \omega_0$. That $|P| \geq k$ holds, is trivial.

5.5 Theorem [55]. *Let P be an infinite poset, but no chain. Then there follows $|P| \leq 2^{\sigma(P)}$.*

Proof. First we mention that $\sigma(P)$ is infinite. For the infinite set P has an infinite antichain or an infinite chain, and in the latter case a subset of type ω_0 or ω_0^*. In all of these cases there exist subsets $T \subseteq P$ where $\pi(T)$, and then also $\sigma(P)$, is infinite. Further it is immediately clear from the definition that $\sigma_0(P)$ and $\sigma_1(P)$ are $\leq \sigma(P)$.

Next we prove:

(1) $d^+(P) \leq \sigma_1(P) =: k$.

Assume the contrary, $k < d^+(P)$. Then P contains a subset T of type $\omega(k^+)$. This T has a cofinal subset of type $\omega(k^+)$, and we obtain

192

the following contradiction: $k^+ \leq \pi_1(T) \leq \sigma_1(P) = k$. The dual of (1)
is

(2) $d^-(P) \leq \sigma_0(P)$.

Now there follows by 5.2 and (2): $|P| \leq \sigma_1(P)^{d^-(P)} \leq 2^{\sigma_1(P) \cdot d^-(P)} \leq$
$2^{\sigma_1(P) \cdot \sigma_0(P)} \leq 2^{\sigma(P)}$.

5.6 Theorem [55]. *Let P be a non-empty poset. Then there exists
a cofinal subset $C \subseteq P$ which is well-founded. And dually: There exists
a coinitial subset $D \subseteq P$, which satisfies the ascending chain condition.*

Proof. If M is a maximal antichain of P, we denote the set of those
elements of P which are greater than some element of M by $U(M)$.

Now we choose a maximal antichain A_0 of S. If $U(A_0)$ is empty we
are done, since then A_0 satisfies the conditions for C. In the other case
we choose a maximal antichain A_1 of $U(A_0)$, and so on. In general we
proceed as follows: Suppose that for an ordinal μ we have already defined
non-empty and pairwise disjoint antichains A_ν for $\nu < \mu$ such that the
following holds: If $\nu + 1 < \mu$ holds, every $A_{\nu+1}$ is a maximal antichain
of $U(A_\nu)$, and for limit ordinals $\lambda \leq \mu$, for that $\cap\{U(A_\nu)|\nu < \lambda\}$ is non-
empty, A_λ is a maximal antichain of this set. By transfinite induction
we so construct a set of pairwise disjoint antichains $A_\nu, \nu < \tau$, of P. This
construction procedure must indeed stop at some ordinal τ because in
every step of the construction the rest of P is diminished.

There cannot exist elements in P which are not surpassed by ele-
ments of $\cup\{A_\nu|\nu < \tau\}$, for then $\cap\{U(A_\nu)|\nu < \tau\}$ would be non-empty,
and the construction had to go on. So $C := \cup\{A_\nu|\nu < \tau\}$ is cofinal in
P.

Also C is well-founded: If $x_0 > x_1 > \cdots > x_n > \cdots$ would be an
infinite decreasing sequence in C, every x_ν is in exactly one antichain
$A_{\kappa(\nu)}$ and there follows:
$$\mu < \nu \ (< \omega_0) \implies x_\mu > x_\nu \implies \kappa_\mu > \kappa_\nu,$$
and then the $\kappa_\nu, \nu < \omega_0$, would form an infinite decreasing subset of τ,
which is impossible.

Now we can prove the following lemma of Ginsburg [55]:

5.7 Theorem. *For every poset P there holds:*
a) $\pi_0(P) \leq w(P)^{d^-(P)}$, *and dually* b) $\pi_1(P) \leq w(P)^{d^+(P)}$.

Proof. It suffices to prove b). By 5.6 P contains a cofinal subset
C which is well-founded. Then every chain in C is well-ordered, and

all chains in C have cardinality $\leq d^+(P) := b$. Further all antichains of C have cardinality $\leq w(P) := a$. We partition the set $[P]^2$ of all two-element subsets of P into the set A (resp. B) of those sets $\{x, y\}$ where x is comparable (resp. incomparable) with y. From 4.2' we obtain $(a^b)^+ \longrightarrow (a^+, b^+)^2$. Then C must by 4.11 have a cardinality $\leq a^b$, and this entails $\pi_1(P) \leq |C| \leq a^b$.

Using the last theorem Ginsburg [55] gave a proof for the following theorem which contains a theorem of Kurepa [106]:

5.8 Theorem. *Let P be a poset with width $w(P) \geq 2$, $d(P) :=$ $\sup\{k | P$ contains a well- or inversely well-ordered subset of cardinality $k\}$. Then*
$$|P| \leq w(P)^{d(P)}.$$

Proof. For subsets $S \subseteq P$ one has of course $w(S) \leq w(P)$, $d^-(S) \leq d^-(P)$, $d^+(S) \leq d^+(P)$. By 5.7 a) there follows $\pi_0(S) \leq w(S)^{d^-(S)} \leq w(P)^{d^-(P)}$. Then we have $\sigma_0(P) = \sup\{\pi_0(S) | S \subseteq P\} \leq w(P)^{d^-(P)}$.

From 5.2 we obtain $|P| \leq \sigma_0(P)^{d^+(P)} \leq (w(P)^{d^-(P)})^{d^+(P)}$. If one of the cardinals $d^-(P)$, $d^+(P)$ is infinite their product is $\leq d(P)$, and we are done.

If both are finite, P has only finite chains, so that all of them have cardinality $\leq d := d(P)$. Let M_1 be the set of minimal elements of P, M_2 the set of minimal elements of $P \backslash M_1$, and so on. Then P is a union of antichains $M_1, \ldots, M_d$, and then its cardinality is $\leq w(P) \cdot d \leq w(P)^{d(P)}$.

5.9 Corollary. $|P| \leq 2^{w(P) \cdot d(P)}$.

5.10 Corollary [55]. *Let P be a poset which is not linearly ordered. If GCH is assumed and $d(P) < \mathrm{cf}(w(P))$, then $|P| = w(P)$ holds.*

If here in addition $|P| = k^+$ holds, then P contains an antichain of cardinality k^+.

Proof. $|P| \leq w(P)^{d(P)} = w(P)$ follows by assumption. Of course, also $w(P) \leq |P|$ holds.

If $|P| = k^+$ and if P would have no antichain of cardinality k^+, 5.8 would entail $k^+ = |P| = w(P) \leq k$, a contradiction.

6.6 Scattered sets

A dense linearly ordered set, which has at least two elements, contains a subset of type h_0. This simple fact follows e.g. by 4.3.13. As a counterpart to this concept we now consider posets which have no subsets of type h_0. In this context Hausdorff ([81], p. 95) had established the following:

6.1 Definition. A poset is said to be *scattered*, if it has no subset of type h_0.

An order type is called *scattered* if it has a scattered realization.

An example for scattered sets are the well- (resp. inversely well)- ordered sets because they already don't contain subsets of type ω_0^* (resp. ω_0).

Hausdorff had investigated only linearly ordered scattered sets. A more general study was performed in the paper [1] of Abraham/Bonnet.

We put the last definition into a more general context, using the concept of forbidden subtype:

6.2 Remark. For an order type τ we defined in 1.9.6 a poset to be τ - free if it has no subset of type τ. The property τ - free is of course *hereditary*. That means: If a poset is τ - free, then every subset of it is also τ - free (with respect to its induced order).

6.3 Theorem. *Let τ be a linear order type, such that for a realization T of τ the following holds: Every open interval of T has a subset of type τ. Then the class of τ -free linearly ordered sets is closed under ordered sums. More precisely:*

Let I and sets M_i, $i \in I$, be τ - free linearly ordered sets, where the M_i are pairwise disjoint, then the ordered sum $S := \sum_{i \in I} M_i$ is also τ - free.

Proof. We assume indirectly that S has a subset T of type τ. Then there holds:

(I) T has at most one point in common with every summand M_i.

For if T has at least two elements in common with an M_i, say a, b with $a < b$, then the whole open interval (a, b) of T must be a subset of M_i, and since (a, b) contains a subset of type τ, also the set M_i would contain a subset of type τ with contradiction.

Now we map every element $x \in T$ onto the uniquely defined $i \in I$ for which $x \in M_i$ holds. Due to (I) this yields an injective and $<$ - preserving mapping of T onto a subset of I. This contradicts the fact that I has no subset of type τ.

The sets H_α of Definition 4.3.3 have by 4.6.3$'$ the property, that every open interval has once again the type h_α. And so 6.3 yields:

6.4 Corollary. *For every ordinal α the class of h_α - free linearly ordered sets is closed under ordered sums.*

In the special case $\alpha = 0$ this is a theorem of Hausdorff ([81], p.95), which states:

6.5 Theorem. *An ordered sum of scattered linearly ordered sets over a scattered linearly ordered argument is again scattered.*

In connection with the last theorems we introduce the *ring* concept:

6.6 Definition. A class R of posets is called a *ring* if the following holds: If I and M_i, $i \in I$, are elements of R, where the M_i, $i \in I$, are pairwise disjoint, then the ordered sum $\sum_{i \in I} M_i$ also belongs to R.

A class R of order types forms a *ring*, if there holds: If I and M_i, $i \in I$, are posets for which the types tp$I$ and tpM_i, $i \in I$, belong to R, and where the M_i are pairwise disjoint, then also the type of the ordered sum $\sum_{i \in I} M_i$ belongs to R.

So we have obtained that the class of h_α- free linearly ordered sets, and in particular the class of scattered linearly ordered sets, forms a ring.

Another interesting theorem of Hausdorff ([81], p. 95) is the following:

6.7 Theorem. *Every linearly ordered set S is an ordered sum of scattered sets over a dense argument I.*

We prove 6.7 together with the next theorem, of which 6.7 is a special case:

6.8 Theorem. *For every ordinal α there holds: Every linearly ordered set S is an ordered sum of h_α - free sets over an argument I, which has the property that for every two elements a, b of I with $a < b$ the open interval (a,b) contains a subset of type h_α.*

Proof. For elements a, b of a linearly ordered set S we put $a \sim b$ if $a = b$ or if the closed interval with ends a and b is h_α - free. Then $\sim$ is an equivalence relation: If $a < b < c$ are elements of S with $a \sim b$ and $b \sim c$, then also $a \sim c$ holds, for if $[a,c]$ would contain a subset M of type h_α, then already $[a,b]$ or $[b,c]$ would contain at least two elements of M and then also a subset of type h_α with contradiction. The other cases are trivial. Further we have:

(I) Every equivalence class is h_α - free.

Let C be an equivalence class, x one of its elements. Then $(C \geq x)$ is h_α - free. This is trivial, if C has a greatest element. If it has none we choose a well-ordered cofinal subset of $(C \geq x)$, which has x as first element, say $\{x_\nu | \nu < \tau\}$, where $x_\nu < x_\mu$ for $\nu < \mu < \tau$. Then $(C \geq x)$ is by 6.4 h_α - free because it is an ordered sum of h_α - free intervals $[x_\nu, x_{\nu+1})$ over the well-ordered (and thus h_α - free) set of all $\nu < \tau$. Analogously it follows that $(C < x)$ is h_α - free, and then also C is h_α - free. For if a set $M \subseteq C$ would have type h_α, this also would by 4.6.3$'$ hold for $M \cap (C < x)$ or $M \cap (C > x)$.

We have obtained that S is partitioned into a set I of h_α - free classes $S_i, i \in I$, which are segments of S. Then I has the natural order given by: For $S', S'' \in I$ we put $S' < S'' \iff a < b$ for all $a \in S', b \in S''$. Now I has the property that everyone of its closed intervals $[a, b]$ contains a subset of type h_α. For if an interval $[a, b]$ of I would not do that, the ordered sum $\sum \{S_i | i \in [a, b]\}$ would be h_α - free due to 6.4. Then the elements of the classes S_a would be equivalent to those of S_b, which means they had to be in the same class, but S_a and S_b are different. This yields a contradiction.

Hausdorff gave another characterization of scattered linear order types. They form the least class of order types, which contains the ordinals and is closed under order-summation and inversions. First we mention:

6.9 Definition. If S is an arbitrary class of order types, we define $\sum(S)$ as the class of all order types $\sum_{i \in I} \tau_i$, where the τ_i, $i \in I$, are order types of S, and where I is a poset, whose type is in S.

We define the *ring-closure $R(S)$* of S as the least class of order types which contains S and is a ring. It always exists because the class of all order types is a ring, and then $R(S)$ is the intersection of all rings that contain S.

Of course the ring-closure of a class of linear order types contains only linear order types.

One can obtain all linearly ordered scattered sets by starting from the ordinals and their inverses by successively adjoining certain ordered sets which are ordered sums of sets which were already defined in earlier steps. Precisely there holds the following structure theorem:

6.10 Theorem. *Let ω_α be a regular initial ordinal. Let C_0 be the class of all linearly ordered sets which have a type τ or τ^* for some regular ordinal $\tau < \omega_\alpha$. If for an ordinal $\mu < \omega_\alpha$ we have already defined a class C_μ of linearly ordered sets, we define $C_{\mu+1}$ to be that class of all linearly ordered sets that are isomorphic to an ordered sum $\sum\{M_i | i \in I\}$, where I and the $M_i, i \in I$, are in C_μ, and where the $M_i, i \in I$, are pairwise disjoint.*

If λ is a limit ordinal $\leq \omega_\alpha$, for which classes $C_\nu, \nu < \lambda$, are already defined, we put $C_\lambda := \cup\{C_\nu | \nu < \lambda\}$. By transfinite induction we so have defined classes C_ν for all $\nu \leq \omega_\alpha$, which form an ascending (with ν) tower of classes.

Now there follows: C_{ω_α} is the class of all scattered linearly ordered sets of cardinality $< \aleph_\alpha$.

Proof. We consider the statement:

(I) All sets of C_ν are linearly ordered, scattered, and have cardinality $< \aleph_\alpha$.

For $\nu = 0$ then (I) is satisfied. Suppose now that (I) holds for a fixed $\nu < \omega_\alpha$ and that M is a set of $C_{\nu+1}$. Then M is isomorphic to an ordered sum of scattered sets of C_ν over an index set, which is in C_ν and thus M is also scattered by 6.5. Since $\aleph_\alpha$ is regular, M has a cardinal $< \aleph_\alpha$. So (I) also holds for $\nu + 1$.

For limit ordinals λ (I) holds, if it is valid for all $\nu < \lambda$, since then every element of C_λ is also in a C_ν with a $\nu < \lambda$. And the construction shows that the classes $C_\nu, \nu < \omega_\alpha$, form an ascending tower. Finally we have obtained by induction:

(II) C_{ω_α} has only scattered linearly ordered sets of cardinality $< \aleph_\alpha$.

We still have to show:

(III) C_{ω_α} contains all scattered linearly ordered sets of cardinality $< \aleph_\alpha$.

Suppose therefore that T is a scattered linearly ordered set of cardinality $< \aleph_\alpha$. For elements $a, b \in T$ we put $a \sim b$ iff $a = b$, or if the interval with ends a, b is embeddable in a set of C_{ω_α}. Now $\sim$ is an equivalence relation: Let a, b, c be elements of T with $a < b < c$, the other cases are analogous or trivial. Then $[a, b]$ is embeddable in a set of C_{ω_α}, and then also in a set C_{κ_1} with $\kappa_1 < \omega_\alpha$, and similarly $[b, c]$ is embeddable in a set of a C_{κ_2} with $\kappa_2 < \omega_\alpha$. Now $[a,c] = [a,b] \cup [b,c]$ is embeddable in a set of $C_\kappa \subseteq C_{\omega_\alpha}$ where $\kappa = \max\{\kappa_1, \kappa_2\} + 1$. Before proving (III) we need the following:

(IV) Every equivalence class K of $\sim$ is a segment of T and isomorphic to a set of some C_ν with $\nu < \omega_\alpha$.

The first is clear. Let now x be an element of K. If K has a last element l, then $x \sim l$ holds, and $(K \geq x) = [x,l]$ is embeddable in a set of some C_μ with $\mu < \omega_\alpha$. Suppose now that K has no last element, and that $\{x = x_0 < x_1 < \cdots < x_\tau < \cdots | \tau < \omega_\delta\}$ is a cofinal subset of K, where ω_δ is a regular initial ordinal. Here $\delta < \alpha$ holds since $(T$ and$)$ K has cardinality $< \aleph_\alpha$. Every interval $[x_\tau, x_{\tau+1})$, $\tau < \omega_\delta$, is embeddable in a set of a $C_{\nu(\tau)}$ with a $\nu(\tau) < \omega_\alpha$. Then all intervals $[x_\tau, x_{\tau+1})$ are embeddable in sets of C_σ where $\sigma := \sup\{\nu(\tau) | \tau < \omega_\delta\}$ which is still $< \omega_\alpha$ because this is regular. Also ω_δ is in $(C_0 \subseteq)$ C_σ, and then the ordered sum $\sum\{[x_\tau, x_{\tau+1}) | \tau < \omega_\delta\} = (K \geq x)$ is in $C_{\sigma+1}$.

Analogously $(K < x)$ is in a class C_ρ with a $\rho < \omega_\alpha$. Now $K = (K < x) \cup (K \geq x)$ is in C_ν where $\nu = \max\{\sigma + 1, \rho\} + 1$, and (IV) holds.

Finally there cannot be two equivalence classes of $\sim$ which are neighboring segments of T for their union would be isomorphic to a set of some C_ζ with $\zeta < \omega_\alpha$, and then both segments had to be subsets of the same equivalence class with contradiction. So we have obtained that the set of equivalence classes, which is a partition of T into segments, is dense (with respect to the natural order inherited by $T)$. But then this set can have only one equivalence class, since otherwise it would contain a subset of type h_0. This would also hold for T in contradiction to the fact that T is scattered. So the unique equivalence class is T. With (IV) the proof is complete.

Another characterization of scattered linearly ordered sets via dyadic sequences was given in [78]. First we mention the following:

6.11 Lemma. *Let $\mathfrak{A}$ be a dyadic splitting of the set $\mathbf{Q}$ of rational numbers (with their natural order) into segments. Let B be a block of $\mathfrak{A}$ which has more than one element. Then there exists a block $C \subseteq B$ such that both blocks of $\mathscr{Z}(C)$ (according to 1.8) have more than one element.*

Proof. 1) If B has no first and no last element, then already both blocks of $\mathscr{Z}(B)$ have more than one element.

2) Suppose that B has a first, but no last element. If the left subsegment (see 3.3) of B has a last element, then the right subsegment

R of B has no first and no last element. By 1) then both blocks of $\mathcal{Z}(R)$ have more than one element, and $C := R$ fulfills our assertion.

3) The case where B has no first, but a last element, is symmetric to 2).

4) Suppose now that B has a first and a last element. Then one of the blocks of $\mathcal{Z}(B)$ has no first or no last element, and so we are done by 2) and 3).

Now there follows:

6.12 Theorem. *Let $\mathfrak{A}$ be a dyadic splitting of the set $\mathbf{Q}$ of rational numbers into segments. Then there exist $2^{\aleph_0}$ block degressions (see 3.5) $B_0 \supset B_1 \supset \cdots \supset B_\nu \supset \cdots$, $\nu < \omega_0$, to $\mathfrak{A}$, whose corresponding dyadic sequences $\alpha_0, \alpha_1, \ldots$ have a change number $\geq \omega_0$.*

Proof. By 6.11 there exists a block $C \in \mathfrak{A}$ with $\mathcal{Z}(C) = \{C(0), C(1)\}$, where $C(0)$ and $C(1)$ both have more than one element, and where $C(0)$ (resp. $C(1)$) is the left (resp. right) subsegment of C. Suppose now, that for a fixed $n \in \mathbf{N}$ we have for every finite dyadic sequence $\alpha_0, \ldots, \alpha_m$ with $m \leq n$, already defined blocks $C(\alpha_0, \ldots, \alpha_m) \in \mathfrak{A}$, which all have more than one element, and which satisfy $C(\alpha_0) \supset \cdots \supset C(\alpha_0, \ldots, \alpha_n)$. If now $\alpha_0, \ldots, \alpha_n$ is a fixed dyadic sequence, then by 6.11 we can choose a block $C^* \subseteq C(\alpha_0, \ldots, \alpha_n)$ for which $\mathcal{Z}(C^*) = \{C_l, C_r\}$, where C_l (resp. C_r) is the left (resp. right) subsegment of C^*, and where C_l and C_r both have more than one element. Then we put $C(\alpha_0, \ldots, \alpha_{n+1}) := C_l$ (resp. C_r) if $\alpha_{n+1} = 0$ (resp. $= 1$). By induction we so have defined for every dyadic sequence $(\alpha_\nu | \nu < \omega_0)$ blocks

$$C(\alpha_0) \supset C(\alpha_0, \alpha_1) \supset \cdots \supset \cdots \supset C(\alpha_0, \ldots, \alpha_n) \supset \cdots .$$

These form a part of a block degression to $\mathfrak{A}$, whose corresponding dyadic sequence s has $(\alpha_\nu | \nu < \omega_0)$ as subsequence. If the latter has change number $\geq \omega_0$, then this also holds for s. Since there exist $2^{\aleph_0}$ dyadic sequences of length ω_0 which have a change number $\geq \omega_0$, the proof is complete.

An easy consequence of 6.12 is:

6.13 Theorem. *Let $(S, \leq)$ be a linearly ordered set which is not scattered. Then for every dyadic splitting of S into segments there holds: We have at least $2^{\aleph_0}$ block degressions whose corresponding change number is infinite.*

Proof. S has a subset T which is isomorphic to the set $\mathbf{Q}$ of rational numbers. If $\mathfrak{A}$ is a dyadic splitting of S into segments, $\{T \cap B | B \in \mathfrak{A}\}$

defines a dyadic splitting $\mathfrak{A}_T$ of T. Every block degression of $\mathfrak{A}_T$ has a corresponding dyadic sequence which is a subsequence of the dyadic sequence which corresponds to some block degression of $\mathfrak{A}$. And so 6.12 entails 6.13.

Now we can establish another characterization of scattered linearly ordered sets:

6.14 Theorem [78]. *L is a scattered linearly ordered set iff there exists a dyadic splitting of L such that all dyadic sequences of the splitting have finite change number.*

Proof. We define the classes C_ν of linearly ordered sets inductively as in the proof of 6.10, but now for all ordinals ν, not only for $\nu < \omega_\alpha$. To this purpose we only have to supplement the former definition by : For every limit ordinal λ, for which C_ν is already defined for the $\nu < \lambda$, we put $C_\lambda := \cup\{C_\nu|\nu < \lambda\}$. We consider the following statement:

$(*)$ For every linearly ordered set of C_ν there exists a dyadic splitting such that all its dyadic sequences have finite change number.

Then by 3.13 $(*)$ holds for $\nu = 0$. Assume now that $(*)$ holds for a fixed ordinal ν. If then T is a linearly ordered set of $C_{\nu+1}$, T has a representation $T = \sum_{i \in I} T_i$ as an ordered sum, where I and the T_i, $i \in I$, are in C_ν. By induction hypothesis, I and the T_i have dyadic splittings in which all dyadic sequences have finite change number. Then we can attach to each sequence $s(i)$, which corresponds to an $i \in I$ in the splitting of I, the sequences of the splitting of T_i.This furnishes a dyadic splitting of T in which all dyadic sequences have finite change number, and thus $(*)$ holds for $\nu + 1$. Finally $(*)$ also holds for limit ordinals λ, if it is proved for all $\nu < \lambda$. And so $(*)$ is proved for all ordinals ν by transfinite induction.

6.15 Remark. One can pose the question whether 6.14 could be sharpened so that every dyadic splitting of a scattered set into segments has only dyadic sequences of finite change-number. But this is false. Already very simple scattered sets furnish counter-examples: We consider the set $S := \{-\frac{1}{n}|n \in \mathbf{N}\} \cup \{0\} \cup \{\frac{1}{n}|n \in \mathbf{N}\}$. Its order type is $\omega_0 + 1 + \omega_0^*$, and so it is scattered.

We can define a dyadic splitting $\mathfrak{A}$ of S into segments, of which the following sets S_i, $i \in \omega$, are blocks of height i. We put $S_0 := S, S_1 := (S_0 < 1), S_2 := (S_1 > -\frac{1}{2}), S_3 = (S_2 < \frac{1}{2}), S_4 = (S_3 > -\frac{1}{3}),$

$S_5 = (S_4 < \frac{1}{3})$, $S_6 = (S_5 > -\frac{1}{4})$, $S_7 = (S_6 < \frac{1}{4})$ and so on. The idea of this construction is that we have:

If ν is even (resp. odd), then $S_{\nu+1}$ is the left (resp. right) block of $\mathcal{Z}(S_\nu)$.

The sequence S_0, S_1, S_2, $S_3,\ldots$ then determines a block-degression of $\mathfrak{A}$ (with $\cap\{S_\nu | \nu < \omega_0\} = \{0\}$), whose corresponding dyadic sequence is $0, 1, 0, 1, 0, 1, \ldots$ which has change number ω_0.

In the class of scattered linearly ordered sets we have a remarkable structure theorem of Laver. Before we formulate it we define in the class of linearly ordered sets a quasi-order $\preceq$, supplementing Definitions 1.9.2 and 1.9.6:

6.16 Definition. For two linearly ordered sets $(L_1, \leq_1)$ and $(L_2, \leq_2)$ we put $(L_1, \leq_1) \preceq (L_2, \leq_2)$ iff there is a $<$ - preserving function from L_1 in L_2. It follows immediately that $\preceq$ is a quasi-order.

In analogy to 1.9.8 we call a quasi-ordered class *well-quasi-ordered*, abbreviated wqo, if it has no subset of type ω^*, (i. e. no infinite strictly descending chain) and no infinite antichain.

Now Laver's theorem states:

6.17 Theorem [111]. *The class of scattered linearly ordered sets is wqo (with respect to the quasi-order of 6.16).*

The proof of this theorem is long and complicated, and so we refer the interested reader to the original paper.

Chapter 7
The dimension of posets

Due to the fact that the linearly ordered sets are much more intuitive than the (only partially) ordered sets one is interested to establish relations between the general ordered sets and linearly ordered sets. One possibility was already treated by Dilworth's theorem, which considered coverings of posets by chains. Another possibility is given by studying linear extensions of posets. This leads to the dimension theory of posets which was founded by Dushnik/Miller in the paper [30] in 1941. We begin with some remarks on the topology of posets.

7.1 The topology of linearly ordered sets and their products

Linearly ordered sets are generalizations of the real line **R,** which besides its order-theoretical aspects also has topological features. So it is not surprising that also general linearly ordered sets give rise to the introduction of corresponding topological concepts. First we compile several notions of general topology and consider some connections between this theory and the theory of posets. We recall the concept of topological space:

1.1 Definition. A topological space is a pair $(X, \mathfrak{O})$, where X is a set and $\mathfrak{O}$ a set of subsets of X, which contains $\emptyset$ and X, and has the following properties:

1) For any two sets O_1, O_2 of $\mathfrak{O}$ the intersection $O_1 \cap O_2$ also belongs to $\mathfrak{O}$.

2) If O_i, $i \in I$, are elements of $\mathfrak{O}$, then their union $\cup_{i \in I} O_i$ also belongs to $\mathfrak{O}$.

E.g. there holds: If $(P, \leq)$ is a poset, $\mathfrak{I}$ (resp. $\mathfrak{F}$) the set of its initial (resp. final) segments, then $(P, \mathfrak{I})$ (resp. $(P, \mathfrak{F})$) are topological spaces.

In the linearly ordered sets one has a very natural topology which is completely analogous to the usual topology of the real line:

1.2 Definition. Let $(S, \leq)$ be a linearly ordered set. Sets which are open intervals (a, b) of S, or of type $(S < x)$ or $(S > x)$ for elements

$x \in S$, or which are $= \emptyset$ or $= S$ are called *open*. Further a subset of S is said to be *open*, if it is a union of segments of the before mentioned kind. (So open intervals are also open sets.) Let $\mathfrak{O}$ be the set of open subsets of S. It is easily verified that $(S, \mathfrak{O})$ is a topological space: If $O_1, O_2 \in \mathfrak{O}$, we have representations $O_1 = \cup\{S_i | i \in I\}$, $O_2 = \cup\{T_j | j \in J\}$ where the S_i and T_j are open segments of S. Then $O_1 \cap O_2 = \cup\{(S_i \cap T_j)| i \in I, j \in J\}$. Here each summand $S_i \cap T_j$ is open as can easily be verified. So condition 1) is satisfied, and 2) is trivial. $\mathfrak{O}$ is called the *order topology* of $(S, \leq)$.

1.3 Definition. A subset T of a topological space $(X, \mathfrak{O})$ is called *compact*, if there holds: If $T \subseteq \cup_{i \in I} O_i$, where the O_i are in $\mathfrak{O}$, then there exists a finite subset $F \subseteq I$, such that $T \subseteq \cup_{i \in F} O_i$. We formulate this also as follows: Every open covering of T has a finite subcovering.

1.4 Theorem. *Let $(S, \leq)$ be a dense linearly ordered set which has no gaps, but a first element a and a last element z, $\mathfrak{O}$ its order-topology. Then $(S, \mathfrak{O})$ is a compact space.*

Proof. The proof runs quite analogously to the proof for the compactness of closed intervals of the real line: If S is covered by the union of open sets $\cup_{i \in I} O_i$ then there exists the supremum s of the values $x \in S$, for which the initial segment $(S \leq x)$ can be covered with finitely many sets of the O_i. Also s is contained in one of the O_i, and then it must be $= z$. And so we have obtained a finite subcovering.

We now define a kind of linearly ordered sets which can be considered as generalizations of the real line $\mathbf{R}$:

1.5 Definition. A dense linearly ordered set which has at least two, but no first and no last element and no gaps, is called a *linearly ordered continuum*. We call a cartesian product $C = C_1 \times \cdots \times C_n$ of $n \in \mathbf{N}$ linearly ordered continua $C_1, \ldots, C_n$ in short an *n-continuum*.

If S_ν is an open segment of C_ν for $\nu = 1, \ldots, n$, the cartesian product $P := S_1 \times \cdots \times S_n$ is called an (*n*- dimensional) *open segment* of C. Every subset of C which is a union of open segments of C, is called an *open set* of C. We denote their set by $\mathfrak{O}$. Again it follows easily that $(C, \mathfrak{O})$ is a topological space.

If $I_\nu = [a_\nu, b_\nu]$, $\nu = 1, \ldots, n$, is a closed interval of C_ν for $\nu = 1, \ldots, n$, the product $I_1 \times \cdots \times I_n$ is said to be an *n-dimensional closed interval* of C.

If M is a subset of C, a point $h \in C$ is an *accumulation point* of M, if every open set which contains h also intersects M. A subset M of C is *closed* if all of its accumulation points belong to M. It follows immediately that M is closed iff its complement $C \backslash M$ is open. We call M *bounded*, if there exists an n- dimensional closed interval $X\{[a_\nu, b_\nu] | \nu = 1, \ldots, n\}$ which contains M.

If T is a subset of an n-continuum C_1, and C_2 an m-continuum, where n and m are natural numbers, a mapping $f : T \longrightarrow C_2$ is said to be *continuous*, if for every point $x \in T$ and every open set O of C_2 which contains $f(x)$ there exists an open set U of C_1, which contains x, with $f[U \cap T] \subseteq O$.

An open set O of an n-continuum C is *connected* if it has no representation $O = O_1 \cup O_2$, where O_1 and O_2 are non-empty disjoint open sets of C. And a set K is a *connectivity component* or *region* of an open set O if it is a maximal connected open subset of O.

Now a lot of classical invariance theorems of the topology of the euclidean n-space $\mathbf{R}^n$ can be generalized to n-continua. We list some of them without proofs (and don't need them in the rest of the book):

First we mention the following generalization of the separation theorem of Jordan/Brouwer/Alexander:

1.6 Theorem [80]. *Let K be a bounded closed subset of an n-continuum C and $f : K \longrightarrow C$ an injective and continuous mapping. Then the sets $C \setminus K$ and $C \setminus f[K]$ have the same cardinal number of connectivity components.*

A special case of 1.6 is the following generalization of the Jordan–Brouwer separation theorem, which for $n = 2$ was previously proved by Löttgen/Wagner [113]:

1.7 Theorem. *Let B be the boundary of a closed n-dimensional interval $X_{\nu=1}^n [a_\nu, b_\nu]$ of an n-continuum $C = C_1 \times \cdots \times C_n$, i.e. the subset of those points $(x_1, \ldots, x_n)$ of it, for which at least one x_ν is $= a_\nu$ or $= b_\nu$. And let $f : B \longrightarrow C$ be an injective and continuous mapping. Then $C \backslash f[B]$ has exactly two connectivity components. In another formulation: $f[B]$ separates C into two regions.*

Also Brouwer's theorem of the invariance of the open set can be generalized to n-continua:

1.8 Theorem [76]. *Let O be an open subset of an n-continuum C,*

$f : O \longrightarrow C$ injective and continuous. Then $f[O]$ is an open subset of C.

The last theorem entails a generalization of Brouwer's theorem on the invariance of dimension:

1.9 Theorem [175]. *If C_m is an m-continuum and C_n an n-continuum, where m,n are natural numbers with $m > n$, then there doesn't exist a bijective and continuous mapping of C_m into C_n.*

Brouwer's fixed point theorem was generalized to n-continua by Dyer [31]. Another proof was given in [76]:

1.10 Theorem. *Let P be an n-dimensional closed interval of an n-continuum C and $f : P \longrightarrow P$ a continuous mapping. Then there exists a point $x \in P$ with $f(x) = x$.*

7.2 The dimension of posets

Now we introduce the concept of dimension for posets, which was established by Dushnik/Miller [30].We mention first:

2.1 Theorem. *Let $(P, \leq)$ be a poset. Then there exists a set of linear orderings $(P, \leq_i)$, $i \in I$, such that $\leq$ is the intersection of the order-relations $\leq_i$, $i \in I$.*

Proof. For every pair (a, b) of incomparable elements of P we choose two linear extensions $\leq_{ab}$ and $\leq_{ba}$ for which there holds $a \leq_{ab} b$ and $b \leq_{ba} a$. Such linear extensions exist due to the theorem of Szpilrajn (2.3.2). Then the intersection of all linear orderings $\leq_{ab}$ and $\leq_{ba}$,where (a, b) runs through the set of all pairs of incomparable elements of $(P, \leq)$ is the relation $\leq$.

Generally the set of linear extensions, which was used in the last proof, has many more elements than necessary. And since every non-empty set of cardinal numbers has a least one, we can now define:

2.2 Definition. Let $(P, \leq)$ be a poset. Then the least cardinal d for which there exists a set of d linear extensions of $\leq$, whose intersection is $\leq$, is called the *dimension* of $(P, \leq)$, in symbols dim $(P, \leq)$ or in short dim P.

It can be expected that there are some common features between the above order-theoretic concept of dimension and the topological one. In this connection there holds:

2.3 Theorem. *Let $\mathbf{R}^n$ be the euclidean n-space, equipped with the product order $\leq$:*

For $x = (x_1 \ldots, x_n)$ and $y = (y_1, \ldots, y_n)$ of $\mathbf{R}^n$ we have $x \leq y \iff x_\nu \leq y_\nu$ for $\nu = 1, \ldots, n$. Then $\dim(\mathbf{R}^n) = n$.

This theorem is contained below in Theorem 2.11.

The following is immediately clear:

2.4 Theorem.a) *A non-empty poset has dimension 1 iff it is linearly ordered.*

b) *If P is a totally unordered set (each two elements of P are incomparable) and $|P| \geq 2$, then $\dim P = 2$.*

Proof. b) Let $\leq$ be an arbitrary linear order on P and $\leq^*$ its reverse order. Then their intersection is the identity relation $=$ on P.

One could guess that the dimension of a poset P is a kind of measure of how much the order of P deviates from a linear one. But 2.4 b) shows that the order of a set can differ very much from a linear one, whereas the dimension nethertheless can be very small.

Concerning the dimension of products we have:

2.5 Theorem. *Let $(P, \leq)$ be the order-product of linearly ordered sets $(S_i, \leq_i)$, $i \in I$. Then there holds $\dim (P, \leq) \leq |I|$.*

Proof. For $i \in I$ we define an order O_i on P as follows: For $x = (x_i)_{i \in I}$ and $y = (y_i)_{i \in I}$ of P we put:

(1) $xO_i y \iff x_i <_i y_i$, or ($x_i = y_i$ and $x \leq y$).

One easily verifies that O_i is an order which contains $\leq$. Let now L_i be a linear order on P with $L_i \supseteq O_i$ for $i \in I$. Then we prove:

(2) $\leq$ is the intersection $\cap_{i \in I} L_i$.

Let $x = (x_i)_{i \in I}$ and $y = (y_i)_{i \in I}$ be elements of P with $x \leq y$. For every $i \in I$ we then have $x_i \leq y_i$; and in both of the cases, where $x_i < y_i$ resp $x_i = y_i$ holds, $xO_i y$, and then also $xL_i y$ is valid, so that $\leq$ is a subset of $\cap_{i \in I} L_i$.

On the other hand, let x, y be elements of P which satisfy $xL_i y$ for all $i \in I$. If there would exist an index $j \in I$ for which $x_j > y_j$ holds, this would by (1) entail yO_jx and then also yL_jx. Together with $xL_j\,y$ we would obtain $x = y$, contradicting $x_j > y_j$. So we must have $x_i \leq y_i$ for all $i \in I$, and then also $\cap_{i \in I} L_i$ is a subset of $\leq$, and both orders are equal.

The last theorem cannot be sharpened so that $\dim\,(P, \leq) = |I|$ results. If e.g. all S_i are one-element sets, P also has only one element and therefore the dimension 1 whatever I is. But if all S_i have at least two elements we obtain equality, as we shall see a bit later.

A simple statement is expressed in:

2.6 Theorem. *Let $(P, \leq)$ be a poset and T a subset of P. Then* $\dim(T, \leq \restriction T) \leq \dim(P, \leq)$ *holds.*

One can pose the question whether every cardinal number can be the dimension of a poset. This is true indeed and will result from the following reflection of Dushnik/Miller [30]:

2.7 Lemma. *Let S be an arbitrary set, A the set of its atoms (= one-element subsets) and B the set of its antiatoms $c_x = S\backslash\{x\}$, where $x \in S$. Then for the set $A \cup B$, equipped with the order by inclusion, there holds $\dim(A \cup B) \geq |S|$.*

Proof. Let $|S|$ be ≥ 2, the other case is trivial. The order $\subseteq$ of $A \cup B$ is the intersection of linear orderings $\leq_i$, $i \in I$, where $|I|$ is the dimension of $A \cup B$. For every $x \in S$ the sets $\{x\}$ and c_x are disjoint and thus incomparable. Therefore there must exist an $i \in I$ with $c_x <_i \{x\}$. Let $y \in S$ be an element $\neq x$. Then there is an index $j \in I$ with $c_y <_j \{y\}$. Now i and j are different. Otherwise, because of $x \in c_y$ and $y \in c_x$, we would have $c_x <_i \{x\} <_i c_y <_i \{y\} <_i c_x$ and thus the contradiction $c_x < c_x$.

So we have obtained that every $\leq_i$, $i \in I$, contains at most one pair $(c_x, \{x\})$ with $x \in S$. And thus there exists an injective mapping f, which ascribes to every $x \in S$ an $i = i(x) \in I$ with $(c_x, \{x\}) \in \leq_i$. Therefore $|S| \leq |I|$ follows.

A bit later the statement of 2.7 will be sharpened. First we determine the dimension of power sets. These are isomorphic to certain ordered products:

2.8 Theorem. *Let S be a non-empty set. For every $i \in S$ we put $S_i := \{0, 1\}$ and equip this set with its natural order $(0 < 1)$. Let then*

$X := \mathsf{X}_{i \in I} S_i$ *be the cartesian product of the* S_i, $i \in I$, *with the product order. Then the power set* $\mathfrak{P}(S)$ *with the inclusion order* $\subseteq$ *is isomorphic to* X. *An order-isomorphism between these sets is given by the mapping* f_T, *which ascribes to every subset* $T \subseteq S$ *its characteristic function* f_T *on* S, *which is given by*

$f_T(x) = 0$ *if* $x \notin T$, $f_T(x) = 1$ *if* $x \in T$.

Proof. f_T is of course bijective. And if we have sets $T_1 \subseteq T_2 \subseteq S$, then $f_{T_1}(x) = 1$ implies $f_{T_2}(x) = 1$, and then $f_{T_1} \leq f_{T_2}$ follows.

An easy consequence of the previous theorems is now the theorem of Komm [100]:

2.9 Theorem. *For every non-empty set* S *there holds*
$\dim(\mathfrak{P}(S), \subseteq) = |S| \; (= \dim(\{0,1\}^S))$.

Proof. By 2.7, 2.6, 2.8 and 2.5 we have $|S| \leq \dim(A \cup B, \subseteq) \leq \dim (\mathfrak{P}(S), \subseteq) = \dim(\{0,1\}^S) \leq |S|$.

The last proof now also yields, sharpening 2.7, the following theorem of Dushnik/Miller [30]:

2.10 Theorem. *Let* S *be an arbitrary set,* A *the set of its atoms,* B *the set of its anti-atoms. Then the set* $A \cup B$, *equipped with the inclusion order, has the dimension* $|S|$.

Also 2.5 can now be sharpened to:

2.11 Theorem. *Let* $(P, \leq)$ *be the ordered product of the linearly ordered sets* $(S_i, \leq_i)$, $i \in I$, *where every* S_i *has at least two elements. Then* $\dim(P, \leq) = |I|$.

Proof. Let T_i be a two-element subset of S_i for $i \in I$. Applying 2.5, 2.6 and 2.9 then there follows $|I| \geq \dim \mathsf{X}_{i \in I} S_i \geq \dim \mathsf{X}_{i \in I} T_i = \dim (\{0,1\}^I) = |I|$.

With 2.11 we obtain a new possibility of geometrical flavour, mentioned by Ore [132], to characterize the dimension of a poset:

2.12 Theorem. *Let* $(P, \leq)$ *be a poset. Then its dimension is the least cardinal* k, *such that* P *is embeddable in an ordered product of* k *linearly ordered sets.*

Proof. Let $\leq_i$, $i \in I$, be linear orderings on P, where $|I|$ is the dimension d of P, such that $\leq$ is the intersection $\cap_{i \in I} \leq_i$. Then $(P, \leq)$

is embeddable in the ordered product $X_{i \in I}(P, \leq_i)$ by mapping $x \in P$ onto $(x) := (x_i)_{i \in I}$, where $x_i = x$ for every $i \in I$.

This is of course injective. If x and y are elements of P with $x \leq y$, then also $x \leq_i y$ holds for all $i \in I$, so that $(x) \leq (y)$ follows. If x and y are incomparable, then there exist two different indices i and j in I with $x <_i y$ and $y <_j x$, so that also (x) and (y) are incomparable.

So we have obtained that $(P, \leq)$ is embeddable in an ordered product of d linearly ordered sets. If it would also be embeddable in a product Q of $k < d$ linearly ordered sets, then this last one would have a dimension $\leq k$ by 2.5, and $k < d = \dim(P, \leq) \leq \dim Q \leq k$ would yield a contradiction.

In the next theorem we treat a relation between the chain-covering-number $\mathrm{CCN}(P)$ of a poset P and the dimension of certain subsets of its power set $\mathfrak{P}(P)$. For this purpose we first define a poset $\mathfrak{D}(P)$, which was introduced by Pouzet in [143], where he proved a generalization of 2.10:

2.13 Definition. Let $(P, \leq)$ be a poset. In a former definition we introduced for $x \in P$ the principal ideal $(x]$ as the set of all elements of P which are $\leq x$, and dually $[x)$. Now we define the *Pouzet-set* $\mathfrak{D}(P)$ of P by

$$\mathfrak{D}(P) := \{(x] | x \in P\} \cup \{P \setminus [x) | x \in P\},$$

and equip it with the order $\subseteq$.

So $\mathfrak{D}(P)$ contains all principal ideals of P and the complements of the principal filters.

Further we denote the set of initial segments of P by $\mathfrak{I}(P)$ and consider it as ordered by the inclusion $\subseteq$.

It can easily be seen that every set $P \setminus [x)$, $x \in S$, is an initial segment of P and hence in $\mathfrak{I}(P)$. Then $\mathfrak{D}(P) \subseteq \mathfrak{I}(P)$ follows.

Now the theorem of Pouzet [143] states:

2.14 Theorem. *Let* $(P, \leq)$ *be a poset. Then* $\dim \mathfrak{D}(P) = \dim \mathfrak{I}(P) = \mathrm{CCN}(P)$.

The proof results from the following four lemmas:

Lemma 1. *Let* $\leq_i$ *be a linear extension of* $\mathfrak{D}(P)$. *Then* $C_i := \{x \in P | (P \setminus [x), (x]) \in \leq_i\}$ *is a chain of* $\mathfrak{D}(P)$.

Proof. Let x and y be elements of C_i, i.e. $(P \setminus [x), (x])$ and $(P \setminus [y), (y])$ are in $\leq_i$. We have to show that x is comparable with y. We assume the contrary $x \parallel y$. Then $y \in (P \parallel x) \subseteq (P\backslash[x))$ and further, since $P\backslash[x)$ is an initial segment of P, $(y] \subseteq P\backslash[x) \leq_i (x]$. Then also $(y] \leq_i (x]$ holds because $\leq_i$ extends $\subseteq$. Analogously $(x] \leq_i (y]$ follows and thus $(x] = (y]$. This implies $x = y$ and yields a contradiction to our indirect assumption $x \parallel y$.

Lemma 2. *Let $\leq_i, i \in I$, be linear extensions of $\mathfrak{D}(P)$ whose intersection $\cap_{i \in I} \leq_i$ is the order of $\mathfrak{D}(P)$. Then for the sets C_i of Lemma 1 we have*

$$P = \cup_{i \in I} C_i.$$

Proof. We have to show: For every $x \in P$ there holds $x \in C_i$ for at least one $i \in I$. The same:

For at least one $i \in I$ there holds $(P\backslash[x), (x]) \in \leq_i$.

If this would be false, we would have $(P\backslash[x), (x]) \notin \leq_i$ for all $i \in I$, and thus $((x], P\backslash[x)) \in \leq_i$ because $\leq_i$ is a linear order. Then also $((x], P\backslash[x)) \in \cap_{i \in I} \leq_i$, and this is the inclusion order of $\mathfrak{D}(P)$. So we would have $(x] \subseteq (P\backslash[x))$, which yields a contradiction, for $x \in (x]$, but $x \notin P\backslash[x)$.

Lemma 3. $\mathrm{CCN}(P) \leq \dim \mathfrak{D}(P)$.

Proof. Let now the order of $\mathfrak{D}(P)$ be the intersection of linear orders $\leq_i$, $i \in I$, where $|I|$ is the dimension of $\mathfrak{D}(P)$. Then by Lemma 2, S is covered by the union of $|I|$ chains C_i, $i \in I$, so that $\mathrm{CCN}(P) \leq (|I| =) \dim \mathfrak{D}(P)$ holds.

Lemma 4. *Let P be covered by chains C_i, $i \in I$, where $|I| = \mathrm{CCN}(P)$. Then the set $\mathfrak{I}(P)$ of all initial segments of P is isomorphic to a subset of the ordered product $\mathsf{X}_{i \in I} \mathfrak{I}(C_i)$, where $\mathfrak{I}(C_i)$ is the set of all initial segments of C_i, which is a chain because of Lemma 1.*

The latter product has by 2.5 a dimension $\leq |I|$, and so we further obtain $\dim \mathfrak{I}(P) \leq |I| = \mathrm{CCN}(P)$.

Proof. We consider an initial segment $B \in \mathfrak{I}(P)$ and map it onto the family $(B \cap C_i)_{i \in I}$. Then each component $B \cap C_i$ of this is an initial segment of C_i, and so we have a mapping $f : \mathfrak{I}(P) \longrightarrow \mathsf{X}_{i \in I} \mathfrak{I}(C_i)$.

f is injective. For if B and B' are different elements of $\mathfrak{I}(P)$, then for at least one index $i \in I$ there holds $B \cap C_i \neq B' \cap C_i$, otherwise we would have $B = \cup_{i \in I}(B \cap C_i) = \cup_{i \in I}(B' \cap C_i) = B'$ with contradiction.

f is also isotone: If $B \subseteq B'$ holds, then for all $i \in I$ we have $B \cap C_i \subseteq B' \cap C_i$.

Finally f also preserves incomparability: Let B and B' be incomparable. Then there exists an element $b \in B \backslash B'$,and a component C_i with $b \in B \cap C_i$, but $b \notin B' \cap C_i$, so that $B \cap C_i \nsubseteq B' \cap C_i$. Thus $f(B)$ is not $\leq f(B')$. Analogously $f(B')$ is not $\leq f(B)$.

Now the rest of the proof of Theorem 2.14 follows: By Lemma 3 we have $\mathrm{CCN}(P) \leq \dim \mathfrak{D}(P)$; this is $\leq \dim \mathfrak{I}(P)$ because of $\mathfrak{D}(P) \subseteq \mathfrak{I}(P)$, and $\dim \mathfrak{I}(P)$ is $\leq \mathrm{CCN}(P)$ by Lemma 4.

2.15 Remark. Pouzet's Theorem 2.14 contains the theorem 2.10 of Dushnik/Miller as a special case:

Let P be a set with $=$ as order relation, so that P is totally unordered in the sense of Definition 1.9.8. Then $\mathrm{CCN}(P) = |P|$. Further we have, with the definitions of 2.13 $\mathfrak{I}(P) = \mathfrak{P}(P)$. And now 2.14 yields $\dim \mathfrak{D}(P) = \dim \mathfrak{I}(P) = \dim \mathfrak{P}(P) = |P|$ by 2.9. Here $\mathfrak{D}(P)$ is the set which contains all one-element subsets of P and their complements in P.

From 2.14 we now can easily derive the following theorem of Hiraguchi [86]:

2.16 Theorem. *Let $(P, \leq)$ be a poset. Then* $\dim P \leq \mathrm{CCN}(P)$ *holds. If in particular P is finite,* $\dim P$ *is* $\leq$ *the width $w(P)$ of P.*

Proof. P is isomorphic to a subset of $\mathfrak{I}(P)$ by 1.9.9. Then $\dim P \leq \dim \mathfrak{I}(P) = \mathrm{CCN}(P)$ follows by 2.14. The rest follows with Dilworth's Theorem 2.5.6.

Using 2.14 Pouzet gave an answer to a question which was posed by Wolk [179]. First we define a special kind of posets, which will be treated later in more detail:

2.17 Definition. A poset is called *partially well-ordered,* in short a *pwo-set,* if all of its antichains and all of its descending chains $a_0 > a_1 > \cdots$ are finite.

Wolk had asked: Is it true that all pwo-sets have a dimension $\leq \aleph_0$? The following theorem of Pouzet [143] gives a negative answer:

2.18 Theorem. *We consider the ordered product $P = \omega_\alpha \times \omega_\alpha$ of the set ω_α, equipped with its natural order, with itself. Then the Pouzet-set $\mathfrak{D}(\omega_\alpha \times \omega_\alpha)$ is a pwo-set and has dimension $\aleph_\alpha$.*

Proof. First we show that P has no infinite antichain. Otherwise P would have a denumerable antichain $\{(a_1, b_1), (a_2, b_2), \ldots\}$ where its elements are enumerated in such a way that the first components increase (in the wide sense), and so we can assume that $a_1 \leq a_2 \leq a_3 \leq \cdots$ holds. Since the pairs (a_i, b_i) are incomparable we then must have $b_1 > b_2 > b_3 > \cdots$, which is impossible, since there is no infinite decreasing chain in the ordinals. Also P has for the same reason evidently no infinite decreasing chain, and thus P is a pwo-set.

Let now A be an antichain of $\mathfrak{D}(P)$. Then A has only finitely many sets $(x]$ with $x \in P$, because also their elements x form an antichain. Also A contains only finitely many sets $P \backslash [x)$ with $x \in P$. For again their corresponding elements x form an antichain. ($x_1 \leq x_2$ implies $P \backslash [x_1) \subseteq P \backslash [x_2)$). Thus A is finite.

In the set of the $(x]$ there is no infinite descending chain in $\mathfrak{D}(P)$, for this would imply that the corresponding x would form an infinite descending chain in P. Also in the set of the $P \backslash [x)$ there does not exist an infinite descending chain $P \backslash [x_1) \supset P \backslash [x_2) \supset \cdots$, for this would imply $[x_1) \subset [x_2) \subset \cdots$ and $x_1 > x_2 > \cdots$, which is impossible. So we obtained that $\mathfrak{D}(P)$ is a pwo-set.

By 2.14 and 2.5.8 there follows dim $\mathfrak{D}(\omega_\alpha \times \omega_\alpha) = \mathrm{CCN}(\omega_\alpha \times \omega_\alpha) = \aleph_\alpha$.

7.3 Relations between the dimension of a poset and certain subsets

In 2.12 we characterized the dimension of a poset P as the least cardinal d such that P is embeddable in an ordered product of d linearly ordered sets. Now every linearly ordered set of cardinality $\aleph_\alpha$ is embeddable in H_α of 4.3.3, and then we can reformulate the content of 2.12 in the form:

3.1 Theorem. *If $(P, \leq)$ is a poset of dimension d with cardinality $|P| = \aleph_\alpha$, then P is embeddable in the ordered product H_α^d (of d factors H_α).*

In the following we need a set which contains H_α and has a compact order-topology. The easiest way to obtain such a set is to fill the gaps of H_α and introduce a first and a last element.

3.2 Definition. For an ordinal α we define K_α as the set C_α, which is augmented by the first and the last element of $2((\omega_\alpha))$. It can easily be verified that K_α is dense and without gaps (compare with 4.7.32), and H_α is dense in K_α. With its order-topology now K_α is a compact space by 1.4.

Of course the posets of finite dimension are of special interest. It turns out that the property to have a dimension $\leq n$, where $n \in \mathbf{N}$, is of finite character. We present two proofs of this. The first one of [68] consists of two embedding theorems and gives some additional insight.

3.3 Theorem [68]. *Let $(P, \leq)$ be a poset, such that every finite subset of P has a dimension $\leq n$, where $n \in \mathbf{N}$. Then also P has a dimension $\leq n$.*

Proof. Let P be infinite, otherwise the assertion is trivial. Let $\aleph_\alpha$ be a regular cardinal $> 2^{|P|}$. We consider the sets H_α^n and K_α^n, the ordered products of n factors H_α resp. K_α. Within this proof we understand under an *embedding* a function $f : T \longrightarrow H_\alpha^n$ which is an isomorphic mapping of a subset $T \subseteq P$ onto a subset of H_α^n, where for each $\nu = 1, \ldots, n$ and different elements $s, t \in T$ the ν-components $(f(s))_\nu$ and $(f(t))_\nu$ of $f(s)$ resp. $f(t)$ are different.

We call here an embedding $f : T \to H_\alpha^n$ *finitely extendable*, if there holds: For every finite subset $E \subseteq P \backslash T$ there exists an embedding $f_E : T \cup E \to H_\alpha^n$, which extends f.

3.3′ Theorem. *Let $f : T \to H_\alpha^n$ be a finitely extendable embedding, e an arbitrary element of $P \backslash T$. Then there exists a finitely extendable embedding $f_e : T \cup \{e\} \to H_\alpha^n$, which extends f.*

Proof. Let $\mathfrak{E}$ be the set of those finite subsets of $P \backslash T$ which contain e. For every $E \in \mathfrak{E}$ there exists an embedding f_E of E which extends f. Then we define:

An element $a \in K_\alpha^n$ is called an *accumulation point* of $\{f_E(e) | E \in \mathfrak{E}\}$ if for every open subset U of K_α^n, which contains a and for every $E \in \mathfrak{E}$ there exists a set $E^* \supseteq E$ in $\mathfrak{E}$ with $f_{E^*}(e) \in U$. The set of the E^*, $E \in \mathfrak{E}$, is then cofinal in $(\mathfrak{E}, \subseteq)$. And now the last definition can, roughly speaking, also be formulated as: For every open U with $a \in U$ there are "cofinally many" sets $E \in \mathfrak{E}$, for which $f_E(e) \in U$.)

First we prove:

(I) The set $\{f_E(e) | E \in \mathfrak{E}\}$ has an accumulation point $a = (a_1, \ldots, a_n)$ in K_α^n.

If (I) would be false, there would exist for every $x \in K_\alpha^n$ an open set U_x, which contains x, and a set $E_x \in \mathfrak{E}$ such that $f_{E^*}(e) \notin U_x$ for all

$E^* \supseteq E_x$ of $\mathfrak{E}$. Since K_α^n is compact, there exist finitely many points $x_1, \ldots, x_k$ in K_α^n such that K_α^n is covered by $U_{x_1}, \ldots, U_{x_k}$. For the union $E^* := E_{x_1} \cup \cdots \cup E_{x_k}$ of the corresponding sets of $\mathfrak{E}$ we then would have $f_{E^*}(e) \notin K_\alpha^n$, which of course is impossible. Thus (I) holds.

Case 1. $a \in H_\alpha^n$. Then we have:

(II) For each $E \in \mathfrak{E}$ there is an $E' \supseteq E$ in $\mathfrak{E}$ with $f_{E'}(e) = a$.

Otherwise we would have:

(III) There is an $E \in \mathfrak{E}$ such that $f_{E'}(e) \neq a$ for all $E' \supseteq E$ of $\mathfrak{E}$,

and then there would exist an open subset $U(a)$, which contains a, but none of the (due to $|\mathfrak{E}| \leq 2^{|P|} < \aleph_\alpha$ less than $\aleph_\alpha$ many) $f_{E'}(e)$, $E' \supseteq E$ of $\mathfrak{E}$. This follows from the fact that each element-character of the components a_ν of a is $(\omega_\alpha, \omega_\alpha^*)$. And this is a contradiction to the property of a to be an accumulation point of the $f_{E'}(e)$. So (II) entails that the function f_e, given by $f_e \restriction T = f$ and $f_e(e) = a$, satisfies **3.3'**.

Case 2. At least one component a_ν of a is in $K_\alpha \backslash H_\alpha$, and then $a \neq f_E(e)$ for all $E \in \mathfrak{E}$. We put

$$X := f[T] \cup \cup \{f_E[E] | E \in \mathfrak{E}\}.$$

Then for each $x = (x_1, \ldots, x_n) \in X$ we have $x_\nu \neq a_\nu$. By construction a_ν is neither least nor greatest element of K_α, and then a_ν has an element-character (see 4.3.8 and 4.3.9) $(\omega_\alpha, \omega_\beta^*)$ or $(\omega_\beta, \omega_\alpha^*)$ with a regular initial ordinal ω_β. W.r.o.g. we assume the first case. Then there exists an $a_\nu' < a_\nu$ in H_α such that the interval $[a_\nu', a_\nu]$ of K_α contains none of the $x_\nu, x \in X$. Otherwise a set of fewer than $\aleph_\alpha$ many x_ν would be cofinal in $(H_\alpha < a_\nu)$ with contradiction.

Now the point a', which results from a by exchanging the component a_ν by a_ν' has the same order position to the elements of X as a. In finitely many steps we so can alter all components of a which are not in H_α to elements of H_α, but in such a way that the point $a'' \in H_\alpha^n$, which is so obtained, is still in the same position to the elements of X as a. Now the function f_e, which is defined by $f_e \restriction T = f$ and $f_e(e) := a''$, satisfies **3.3'**.

3.3″ Theorem. *Let λ be a limit ordinal, $T_\nu \subseteq P$ for $\nu < \lambda$ and $T_\mu \subseteq T_\nu$ for $\mu < \nu < \lambda$. Let further $f_\nu : T_\nu \to P$ be a tower of embeddings which each are finitely extendable. Let $f : T_\lambda := \cup\{T_\nu | \nu < \lambda\} \to P$ be the limit mapping of the $f_\nu, \nu < \lambda$. Then also f is finitely extendable.*

Proof. Let $E = \{e_1, \ldots, e_k\} \subseteq P \backslash T$ be a finite set. For each $\nu < \lambda$ there is an embedding $f_\nu^* : T_\nu \cup E \to H_\alpha^n$ which extends f_ν. First we

216

prove:

(I) There is a cofinal subset C of λ such that for $\mu_1 < \mu_2 < \lambda$, and all $\nu = 1, \ldots, n$ the following holds: The mapping, given by
$$(f_{\mu_1}(e_\kappa))_\nu \to (f_{\mu_2}(e_\kappa))_\nu \text{ for } \kappa = 1, \ldots, k$$
is $<$-preserving and hence an isomorphism.

Roughly speaking: The linear order in the set $S := \{(f_\mu(e_\kappa))_\nu | \kappa = 1, \ldots, k\}$ is the same for all $\mu \in C$.

There are at most $k!$ linear orders in the k-element set S, and so one of them must belong to cofinally many $\mu < \lambda$, and (I) follows.

We call here (other than before) a point $(p_1, \ldots, p_k)$ of $(K_\alpha^n)^k$ an *accumulation point* if for each system of open subsets $O_1, \ldots, O_k$ of K_α with $p_\kappa \in O_\kappa$ for $\kappa = 1, \ldots, k$ there is a cofinal subset $A \subseteq \lambda$ such that $f_\mu^*(e_\kappa) \in O_\kappa$ for $\kappa = 1, \ldots, k$ and all $\mu \in A$. Since K_α^n and $(K_\alpha^n)^k$ is compact (with the product topology) there is an accumulation point $(a_1, \ldots, a_k)$.

Let $a_{\kappa\nu}$ be the ν-component of a_κ. Since a_κ is an accumulation point of $f_\mu^*(e_\kappa)$ its ν-component $a_{\kappa\nu}$ is an accumulation point of the set (of ν-components) $\{f_\mu(e_\kappa))_\nu | \mu \in C\}$. Then we obtain:

(II) If for one (and then also all by (I)) $\mu \in C$ we have $(f_\mu^*(e_{\kappa_1}))_\nu < (f_\mu^*(e_{\kappa_2}))_\nu$ then also $a_{\kappa_1\nu} \leq a_{\kappa_2\nu}$ holds.

This follows immediately from the accumulation point property by indirect assumption.

If all $a_{\kappa\nu}$, $\kappa = 1, \ldots, k$, $\nu = 1, \ldots, n$, would be in H_α we would be done. For an $a_{\kappa\nu} \in H_\alpha$ is $= (f_\mu(e_\kappa))_\nu$ for cofinally many $\mu \in C$. If an $a_{\kappa\nu}$ is in $K_\alpha \backslash H_\alpha$ we shift if to a position $a'_{\kappa\nu} \in H_\alpha$ such that the $a'_\kappa, \kappa = 1, \ldots, k$, preserve the position (coordinatewise) to the elements of $f[T]$, which the a_κ had before and so that $e_{\kappa_1} < e_{\kappa_2} \iff a_{\kappa_1} < a_{\kappa_2}$ holds. This can be done so that in (II) $a'_{\kappa_1\nu} < a'_{\kappa_2\nu}$ is valid. That this is possible follows in an analogous manner as in the proof of 3.3$'$, because each $a_{\kappa\nu}$ has, as an element of K_α, an element character $(\omega_\alpha, \omega_\beta^*)$ or $(\omega_\beta, \omega_\alpha^*)$. Let f^*be that mapping which coincides with f over T and puts $f(e_\kappa) := a_\kappa$ for $\kappa = 1, \ldots, k$. It satisfies the statement of our theorem 3.3$''$.

In some more detail we show:

For $t \in T$ and $\kappa \in \{1, \ldots, k\}$ there holds:

(III) $t < e_\kappa \Rightarrow f(t) < a_\kappa$.

For then for $\mu \in C$ we have $(f_\mu(t) =) f(t) \leq \inf\{f_\mu(e_\kappa) | \mu < \lambda\} \leq a_\kappa$. For $a_\kappa < \inf\{f_\mu(e_\kappa) | \mu \in C\}$ is not possible by construction of a_κ. So we already have $f(t) \leq a_\kappa$. If $a_\kappa \notin H^n_\alpha$ also $f(t) < a_\kappa$ follows. If $a_\kappa \in H^n_\alpha$ holds, a_κ has to be $= f_\nu(e_\kappa)$ for cofinally many ν, otherwise a_κ could not have the above mentioned accumulation point property. And then $f(t) = f_\nu(t) < f_\nu(e_\kappa) = a_\kappa$ follows for cofinally many ν, so that (1) holds.

(IV) $e_i < e_j \Rightarrow a_i \leq a_j$.

For $e_i < e_j$ implies $f_\mu(e_i) < f_\mu(e_j)$ for all $\mu \in C$, and for each $\nu = 1, \ldots, n$ the $\nu-$ component of a_i cannot be greater than that of a_j.

If finally we choose a representation $P = \{p_\iota | \iota < \omega_\beta\}$, where β is $\leq \alpha$. The Theorems 3.3′ and 3.3″ together yield by transfinite induction a tower of embeddings $f_\mu : \{p_\iota | \iota \leq \mu\} \rightarrow H^n_\alpha$, $\mu < \omega_\beta$. Their limit mapping f_{ω_β} is now an embedding of P into H^n_α.

We present now a proof of 3.3 which was suggested by H.A.Jung:

Second Proof of **Theorem 3.3**. If S is a subset of P we call an n-tuple $(\leq_1, \ldots, \leq_n)$ of linear orders $<_\nu$ on P an *n-order* on S, iff their intersection is $\leq \upharpoonright S$.

Let $\mathfrak{I}$ be the set of all finite subsets of P. For the finite subsets $\mathfrak{L}$ of $\mathfrak{I}$ we define $f_\mathfrak{L}$ as follows:

We consider the union $\cup \mathfrak{L}$ (of finitely many finite sets $L \in \mathfrak{L}$) and equip it with n linear orders $\leq_1, \ldots, \leq_n$, whose intersection is $\leq \upharpoonright \cup \mathfrak{L}$. For $I \in \mathfrak{L}$ let then $f_\mathfrak{L}(I)$ be that n-order on I which results from $\leq_1, \ldots, \leq_n$ by restricting these linear orders on I.

According to Rado's Selection Lemma (see 8.1.6) there is a mapping $f^* : \mathfrak{I} \rightarrow A = \cup\{A_I | I \in \mathfrak{I}\}$, where A_I is the set of all n-orders on $I\}$, such that there holds:

For each finite subset $\mathfrak{L}$ of $\mathfrak{I}$ there is a finite set $\mathfrak{M}$ with $\mathfrak{L} \subseteq \mathfrak{M} \subseteq \mathfrak{I}$ for which

(*) $\quad f^*(I) = f_\mathfrak{M}(I)$ for $I \in \mathfrak{L}$.

For every finite $I \subseteq P$ (i.e. $I \in \mathfrak{I}$) an n-order on I is uniquely defined by f^*. Now also f^* defines an n-order for the whole set P, since the definitions for the finite subsets of P are compatible:

Let I_1 and I_2 be finite subsets of P. Then there is a finite set $\mathfrak{M}$ satisfying $\mathfrak{I} \supseteq \mathfrak{M} \supseteq \{I_1, I_2\} =: \mathfrak{L}$ for which (*) holds. In particular we have $f^*(I_1) = f_\mathfrak{M}(I_1)$ and $f^*(I_2) = f_\mathfrak{M}(I_2)$. Herewith $f_\mathfrak{M}(I_1)$ and $f_\mathfrak{M}(I_2)$ are those n-orders on I_1 resp. I_2 which result as restrictions of the n-order of $\cup\mathfrak{M}$, which is defined by $f_\mathfrak{M}$ and also by f^*. So for $a, b \in I_1 \cap I_2$ we have $a \leq_\nu b$ in $I_1 \iff a \leq_\nu b$ in I_2 since both $\leq_\nu$ are

218

the restrictions of the $\leq_\nu$ of $\cup\mathfrak{M} \supseteq I_1 \cup I_2$. And so f^* defines an n-order on P.

Another proof of 3.3 was given in the book [45]; this makes use of the ultrafilter theorem.

The Theorem 3.3 yields the following corollary:

3.4 Theorem. *For every natural number n there exists a countable set B_n of order types of dimension n, such that every poset of dimension n contains a subset of a type $\tau \in B_n$.*

Proof. It is obvious that the set B_n of all order types that belong to finite posets of dimension $n \in \mathbf{N}$ is countable, for even the set of all order types that belong to finite posets is countable. Then by 3.3 every poset of dimension n contains a finite poset of dimension n, whose type is in B_n.

Now the question arises whether the last theorem can be sharpened so that B_n can be replaced by a finite set of order types. For $n = 2$ this is trivial. For every poset of dimension 2 contains two incomparable elements, and so the order type of a two-element antichain forms such a finite set B_2 of order types. But already for $n = 3$ the answer to the above question is negative, as we shall prove a bit later. First we consider the "descending staircase"-set:

3.5 Theorem. *Let P be the set of all points $(x, -x)$ and $(x, -x+1)$ of $\mathbf{R}^2$ where x is an integer. We order P with the restriction of the product order of $\mathbf{Z} \times \mathbf{Z} : (a, b) \leq (c, d) \iff a \leq c$ and $b \leq d$. Then $(P, \leq)$ has the dimension 2. See the illustrations in Figure 8 and 9.*

Proof. $\mathbf{Z} \times \mathbf{Z}$ has the dimension 2, and then its subset P has a dimension ≤ 2. It is also ≥ 2, because P contains incomparable elements.

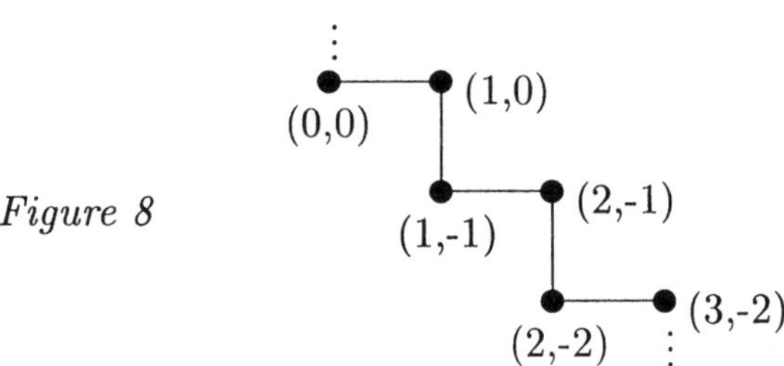

Figure 8

The diagram of $(P, \leq)$ (in the sense of 1.7.6) is

Figure 9

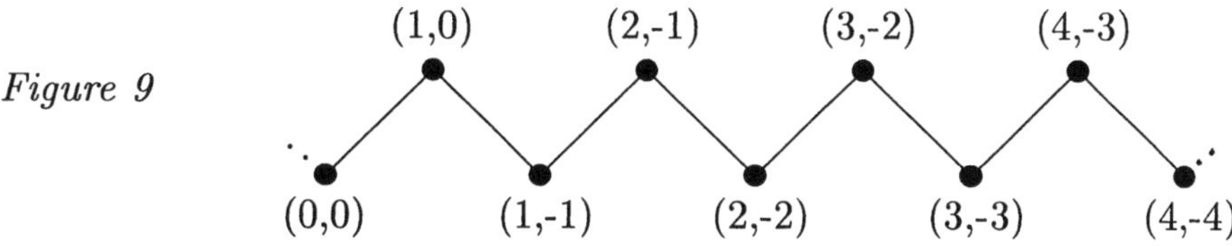

Now the following theorem (Baker/Fishburn/Roberts [7], see also [68], -in this paper n should be replaced by $2n$ on page 188 line 12-) gives an answer to the question which was raised before:

3.6 Theorem. *Let n be a natural number ≥ 3 and let C be the set $\{a_1, \ldots, a_n, b_1, \ldots, b_n\}$, which is equipped with the following order-relation $\leq$, which contains all (a_ν, b_ν), $\nu = 1, \ldots, n$, and all $(a_\nu, b_{\nu-1})$ with $\nu = 2, \ldots, n$, and (a_1, b_n) and of course all (p, p) with $p \in C$.*

So in the sequence $a_1, b_1, a_2, b_2, \ldots, a_n, b_n, a_{n+1} := a_1$ every element is comparable exactly with its predecessor and successor. (The comparability graph, which is defined later, is then a circle with $2n$ vertices and $2n$ edges.)

Then $(C, \leq)$ has the dimension 3.

The following Figure 10 (for $n = 5$) illustrates the structure of the order of C:

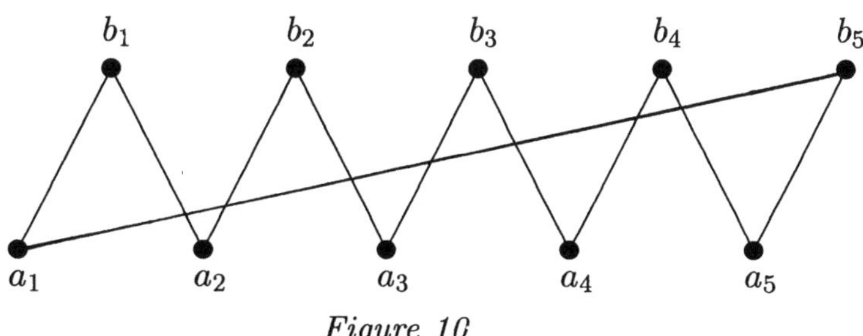

Figure 10

Proof. Since C has incomparable elements, its dimension is at least 2. We now assume that it is 2 and construct a contradiction. Then we can further assume that C is a subset of the euclidean plain $\mathbf{R}^2$ with the order induced from the product order of $\mathbf{R}^2$, where all $2n$ points $p \in C$ have pairwise different first and pairwise different second components.

Let $w_1, w_2, \ldots, w_n$ be an enumeration of the set $\{a_1, .., a_n\}$ of the $a's$ such that the first components of the elements $w_1, \ldots, w_n$ strictly increase. This implies that the set of corresponding second components strictly decreases. Then there holds:

(1) For $1 \leq i < j - 1 \leq n$ the elements w_i and w_j have no common upper neighbor.

For if z would be an upper neighbor of w_i and w_j, it would also be an upper neighbor of w_{i+1}. (In the following Figure 11, z would be situated in the hatched area.) And no element of C has three lower neighbors.

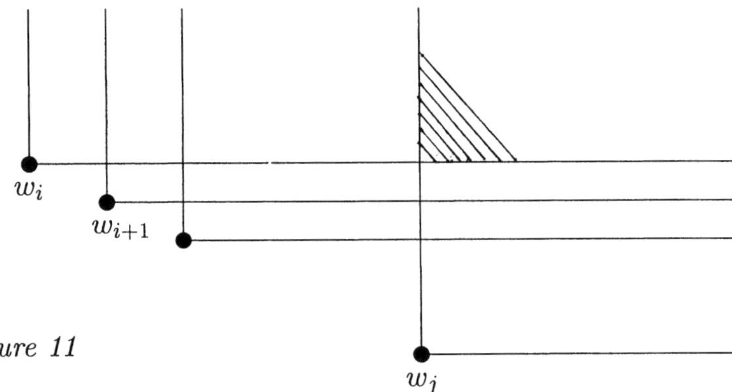

Figure 11

Let v_1 be one of the two upper neighbors of w_1. According to (1) the lower neighbor of v_1 which is different from w_1 must be w_2. Now w_2 has exactly one upper neighbor which is different from v_1, say v_2. This has besides w_2 exactly one lower neighbor, and this must be w_3. For it cannot be w_1 because this would entail $C = \{w_1, v_1, w_2, v_2\}$. Continuing in this way we finally obtain that w_n has besides v_{n-1} exactly one upper neighbor v_n. This has exactly one lower neighbor different from w_n in the set $\{w_1, \ldots, w_{n-1}\}$, and this can only be w_1 since to the elements $w_2, \ldots, w_{n-1}$ we have already ascribed two upper neighbors. But now w_1 and w_n would have v_n as a common upper neighbor, and this contradicts (1).

We have proved that the dimension of C is ≥ 3. It remains to prove that it is also ≤ 3. This (can easily be verified but it) is a consequence of the following theorem of Hiraguchi, according to which the dimension of a poset P increases by at most 1 if a new element is added to P. Indeed, if one of the points of C is omitted, the remaining poset has dimension 2 according to 3.5.

3.7 Theorem [86] (Hiraguchi 1956) *Let $(P^*, \leq^*)$ be a poset, $a \in P^*$, $P := P^* \backslash \{a\}$ ordered by $\leq$, which is the restriction $\leq^* \upharpoonright P$. Then*

$\dim (P^*, \leq *) \leq \dim (P, \leq) + 1.$

Proof. Let d be the dimension of $(P, \leq)$ and $P \neq \emptyset$. Then $\leq$ is the intersection of linear orderings L_i, $i \in A$, on P, where A is an initial segment of the class of ordinals and $|A| = d$ is the dimension of P. In particular $0 \in A$.

We consider the initial (resp. final) open segment of a in P^* :

$I := \{x \in P \mid x <^* a\}$ and $F := \{x \in P \mid a <^* x\},$

and then further the initial (resp. final) segment of $C_0 := (P, L_0)$ which is generated by I (resp. F):

$I^* := \{x \in P \mid x \; L_0 \; y$ for some $y \in I\}$ and $F^* := \{x \in P \mid y \; L_0 \; x$ for some $y \in F\}$.

Let further $B := P \setminus (I^* \cup F^*)$ be the set of all elements which in C_0 are between I^* and F^*. The following Figure 12 illustrates this: C_0 with order L_0 :

Figure 12

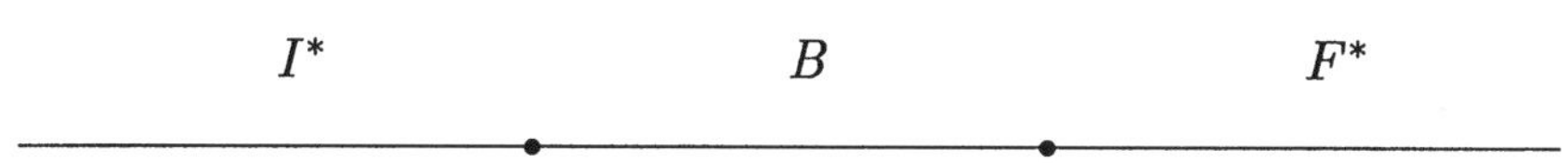

We now define two linear orders L^*, L^{**} on P^* : L^* and L^{**} each consist of five consecutive segments, each of which is linearly ordered by (the restriction of) L_0 :

(P^*, L^*) has the consecutive segments I, $\{a\}$, $I^* \backslash I$, B, F^*, and

(P^*, L^{**}) has the consecutive segments I^*, B, $F^* \backslash F$, $\{a\}$, F.

For $0 \neq i \in A$ we further choose orders L_i^* on P^*, which are linear extensions of $\leq^*$, and for which $L_i^* \upharpoonright P$ is the order L_i. (There are usually more than one possibility to find a suitable position for a in L_i^*.)

For $0 \neq i \in A$ now (P^*, L_i^*) has the three consecutive segments

$\{x \in P \mid x \; L_i^* \; a\}, \{a\}, \{x \in P \mid a \; L_i^* \; x\},$

222

where in the first and third of these segments L_i^* coincides with L_i.

The linear extensions L^*, L^{**}, and the L_i^*, $0 \neq i \in A$, contain $\leq^*$, and so $\leq^*$ is a subrelation of $\leq_\cap := L^* \cap L^{**} \cap \cap \{L_i^* | 0 \neq i \in A\}$.

But also $\leq_\cap$ is a subrelation of (and then $=$) $\leq^*$:

Let x, y be different elements of P with $x \leq_\cap y$. Then xL^*y and $xL^{**}y$ hold and this entails $x \, L_0 \, y$, as can easily be verified in all cases ($x \in I^*$ resp. $x \in B$ resp. $x \in F^*$, $y \in I^*$ resp. $y \in B$ resp. $y \in F^*$).

$x \leq_\cap y$ also entails xL_i^*y and then further $x \, L_i \, y$ for all i with $0 \neq i \in A$. Then $x \leq y$ and $x \leq^* y$ hold.

If $x \leq_\cap a$ holds for an $x \in P$, we have xL^*a and thus $x \in I$, so that $x <^* a$ holds. If $a \leq_\cap x$ holds for an $x \in P$ we have $aL^{**}x$. This yields $x \in F$ and $a <^* x$.

So we have obtained $\leq_\cap$ ($\subseteq$ and then also) $= \leq^*$, and this order is now the intersection of the $d + 1$ linear orders L^*, L^{**}, $L_i^*, i \in A \setminus \{0\}$.

In the dimension theory of posets one mainly investigates relations between the cardinality of the ground set, the dimension of the poset itself, and the dimension resp. cardinality of certain subsets. We present some of the most important ones. First we introduce a concept:

3.8 Definition. Let $(P, \leq)$ be a poset, $a \neq b$ two elements of P. We say: a and b are *position-equivalent*, iff there holds: For $P' := P \setminus \{a, b\}$ we have $(P' < a) = (P' < b)$ and $(P' > a) = (P' > b)$ (and then of course also $(P' \| a) = (P' \| b)$). In this context we have the following reduction lemma of Trotter [170]:

3.9 Lemma. *Let $(P, \leq)$ be a finite poset, $a \neq b$ elements of P, which are position-equivalent.*

a) If $P \setminus \{a\}$ is not linearly ordered, then $\dim(P \setminus \{a\}) = \dim(P)$.

b) If $P \setminus \{a\}$ is linearly ordered, and a and b are comparable, then P is also linearly ordered, and then both sets have dimension 1.

c) If $P \setminus \{a\}$ is linearly ordered, and a and b are incomparable, then $1 = \dim(P \setminus \{a\}) < \dim(P) = 2$.

Proof. a) Suppose $\dim(P \setminus \{a\}) = d$, and the order $\leq \restriction P \setminus \{a\}$ is the intersection of the d linear orders $L_1, \ldots, L_d$ on $P \setminus \{a\}$. We have $d \geq 2$, and then we define linear orders $L_1^*, \ldots, L_d^*$ on P as follows: For $\nu = 1, \ldots, d - 1$ we take the order of L_ν and insert a directly above b to obtain L_ν^*. In L_d^* we take the order of L_d and insert a directly below b. Then the order $\leq$ of P is $= \cap_{\nu=1}^d L_\nu^*$, so that $\dim P$ is $\leq d$. It is $\geq d$ by 2.6.

b) is trivial, and c) follows with 3.7.

The last lemma enables us to delete certain elements in posets without lowering the dimension. By such reductions often situations become more transparent. Another lemma of that kind is:

3.10 Lemma. *Let $(P, \leq)$ be a poset with $|P| \geq 2$ and x an isolated element of P, which means one that is incomparable with all elements of $P \backslash \{x\} := T$. Suppose that T with its induced order is not linearly ordered. Then $d := \dim T = \dim P$.*

Proof. Let $\{L_i | i \in I\}$ be a set of linear orders on T with $|I| = d$, for which $\leq \restriction T$ is $= \cap \{L_i | i \in I\}$. We have $|I| \geq 2$. We choose a fixed $j \in I$ and define a linear order $L_j^* := L_j \cup \{(t, x) | t \in T\}$ on P. (Roughly speaking: One puts x as last element behind (T, L_i).) For $i \neq j$ we put $L_i^* := L_i \cup \{(x, t) | t \in T\}$. Then $\leq$ is the intersection $\cap \{L_i^* | i \in I\}$, and so $\dim P$ is $\leq d$. Of course $\dim P$ is also $\geq (\dim T =) d$.

These lemmas are now applied in the next theorem of Trotter:

3.11 Theorem [170]. *Let X be a finite poset which is no chain and no antichain, A an antichain of X with $|X \backslash A| = 2$. Then $\dim X = 2$.*

Proof. $\dim X \geq 2$ is trivial. Let $X \backslash A = \{a, b\}$. If one of these two elements, say a, is incomparable with all elements of A, but b not, let $x_1, \ldots, x_i$ be all elements of the antichain $A \cup \{a\}$ which are comparable with b and $x_{i+1}, \ldots, x_n$ the rest of the elements of X. Our notation assumes that this rest is not empty, but in this case things become easier. Then $x_1, \ldots, x_i$ are all $< b$ or all $> b$. For if there would exist elements y, z in $A \cup \{a\}$ with $y < b < z$ then we would have $y < z$, and this contradicts the fact that $A \cup \{a\}$ is an antichain. W.r.o.g we assume that b is greater than $x_1, \ldots, x_i$. Then we take the following two linear orders L_1, L_2 of X:

We define L_1 by the sequence $x_1, \ldots, x_i, b, x_{i+1}, \ldots, x_n$. And L_2 is given by the sequence $x_n, \ldots, x_1, b$. Then $L_1 \cap L_2$ is the order of X, which now has dimension 2.

So we still have to discuss the situation where each of a and b is comparable with at least one element of A. Now a (and analogously b) cannot be greater than an element of A, and at the same time smaller than an element of A, and so we have to treat, due to symmetry, only two cases:

1) a is greater than an element of A, and b is greater than an element of A.

2) a is greater than an element of A, and b is less than an element of A.

If A has elements which are isolated in X, we can omit them as long as at least two elements of A remain, without changing the dimension. Further we apply the reduction process of 3.9: If a is greater than three elements x_1, x_2, x_3 of A, then, due to the pigeonhole principle, there are two of them which are $< b$, or two which are $> b$, or two which are incomparable with b, for b cannot be less than one of the three and greater than another of them. In each of these three subcases there are two elements of $\{x_1, x_2, x_3\}$ which are posisition-equivalent, and then one of the three can be omitted by 3.9. Iterating this reduction process we can assume that a is greater than exactly one or exactly two elements of A, and by symmetry (between a and b) we can reduce the Cases 1) and 2) as follows:

Case 1. a is greater than exactly one or exactly two elements of A, and the same for b.

Case 2. a is greater than exactly one or exactly two elements of A, and b is smaller than eactly one or exactly two elements of A.

Now we see that after these reductions A has at most four elements. For each of its elements is comparable with a or b, and these two elements are each comparable with at most two elements of A. If A would have exactly four elements we could make another reduction, for then two of its elements had to be comparable with a, the other two with b, and then the two elements, which are comparable with a, are position-equivalent, so that one of them could be omitted. So finally we can assume w.r.o.g. that A has two or three elements. If $|A| = 2$ holds we have $|X| = 4$, and then $\dim X \leq 2$ easily follows. Let now $A := \{x_1, x_2, x_3\}$. If a is greater than two elements, say x_1, x_2 of A, and if b is greater than these both, or smaller than both, or incomparable with both, then again x_1 and x_2 are position-equivalent, and we can omit one of them without increasing the dimension. After all reductions we then have obtained that in case1 (resp. case 2) X has the same dimension as a set which is isomorphic to a subset with the following diagram of Figure 13 (resp. Figure 14):

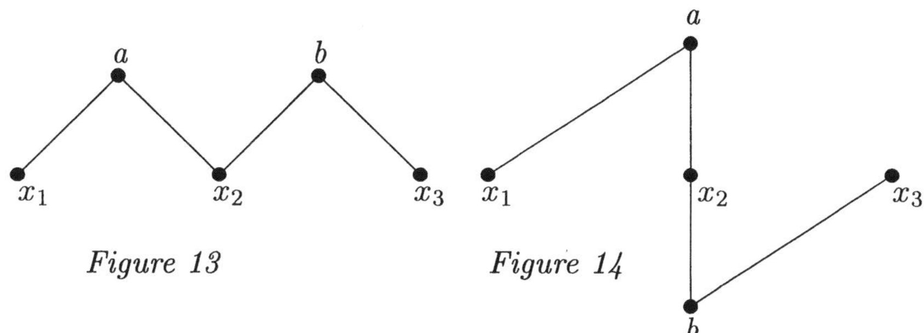

Figure 13   Figure 14

These sets have dimension 2, the first one by 3.5. The order of the second set is the intersection of two linear orders, namely x_1, b, x_2, a, x_3 and b, x_3, x_2, x_1, a.

We mention an easy property of posets:

3.12 Lemma. *Let $(X, \leq)$ be a poset, A an antichain of X with $|A| \geq 2$ for which $|X \backslash A| = 1$ holds. Then $\dim X = 2$.*

Proof. Let x be the element of $X \backslash A$. If x is incomparable with all elements of A, then X is an antichain and we have $\dim X = 2$. So we assume that x is comparable with an element of A, w.r.o.g. we can assume that x is greater than an element of A. Then all elements of A are smaller or incomparable with x. We put $A_1 :=$ (resp. A_2) the set of those elements of A, which are $< x$ (resp. $\parallel x$). Then we define a linear order L_1 on X as follows: L_1 has A_1 as initial segment, then follows x, and A_2 is a final segment. We define a linear order L_2 on X by: L_2 has A_2 as initial segment, then follows A_1 as a segment, and x is the last element of (X, L_2). But $L_2 \restriction A_1$ resp. $L_2 \restriction A_2$ are the reverse orders of $L_1 \restriction A_1$ resp. $L_1 \restriction A_2$. See the Figure 15:

Figure 15

Now $L_1 \cap L_2$ is the order $\leq$.

3.13 Theorem (Trotter [170]). *Let X be a finite poset, A an antichain of X with $|X \backslash A| \geq 2$. Then* $\dim X \leq |X \backslash A|$.

Proof. If X is a chain or antichain the assertion is trivial. In the other case we put $X \backslash A = \{x_1, \ldots, x_n\}$. Then by 3.11 the poset $\{x_1, x_2\} \cup A$ has dimension 2. Now we add successively to this set the remaining elements $x_3, \ldots, x_n$ if $n \geq 3$. Then the set $\{x_1, x_2, x_3\} \cup A$ has a dimension $\leq \dim(\{x_1, x_2\} \cup A) + 1$ by 3.7. After $n - 2$ steps, always applying 3.7, we obtain: $X = \{x_1, \ldots, x_n\} \cup A$ has a dimension $\leq \dim(\{x_1, x_2\} \cup A) + n - 2 = 2 + n - 2 = n = |X \backslash A|$.

Now a theorem of Hiraguchi [86] follows in an elegant way (see [170]):

3.14 Theorem (Hiraguchi 1955). *Let X be a finite poset with $n \geq 4$ elements. Then its dimension is* $\leq \lfloor \frac{n}{2} \rfloor$.

Proof. Let A be an antichain of X with w elements, where w is the width of X. For $X = A$ the statement is trivial.

If $|X \backslash A| = 1$ we have $|A| \geq 3$, and then by 3.12 $\dim X = 2 \leq \lfloor \frac{n}{2} \rfloor$.

If finally $|X \backslash A| \geq 2$ holds we have by 2.16 $\dim X \leq w = |A|$ and by 3.13 $\dim X \leq |X \backslash A|$. Summing up the left and the right sides of these inequalities we obtain $2 \cdot \dim X \leq |A| + |X \backslash A| = |X| = n$.

Before stating the next theorem we need a new concept, which was introduced in a paper of Bogart [10]:

3.15 Definition. Let $(X, \leq)$ be a finite poset, C a chain of X. Then a linear extension L of $\leq$ is said to be a *lower* (resp. *upper*) linear extension of $\leq$ with respect to C, if there holds: If $x \in C$, and if y is incomparable with x, then $y \, L \, x$ (resp. $x \, L \, y$) holds.

Concerning the existence of lower linear extensions we have:

3.16 Theorem. *Let $(X, \leq)$ be a finite poset, C a chain of X. Then there exists a lower (and analogously an upper) linear extension of $\leq$ with respect to C.*

Proof. Let $x_1, \ldots, x_n$ be the elements of $X \backslash C$. Then the set $I := (C < x_1)$ is an initial segment, $F := (C > x_1)$ a final segment, and $B := (C \parallel x_1)$ a segment of C, all elements of which are between those of I and those of F. We adjoin x_1 to C and put it between I and B. Then $C_1 := C \cup \{x_1\}$ is linearly ordered and a lower linear extension of $\leq \restriction C$ with respect to C. If for an $i < n$ already a lower linear extension $C_i := C \cup \{x_1, \ldots, x_i\}$ of C is constructed, we put x_{i+1} between

$(C_i < x_{i+1})$ and the rest of C_i. Then $C_{i+1} := C \cup \{x_1, \ldots, x_{i+1}\}$ is a lower linear extension of $\leq$ (on C_i) with respect to C. After n steps $X = C \cup \{x_1, \ldots, x_n\}$ is a lower linear extension of $\leq$ with respect to C.

Trotter [170] proved:

3.17 Theorem. *Let X be a finite poset, A an antichain of X, and $X \backslash A \neq \emptyset$. Then* $\dim X \leq 2 \cdot \text{width}(X \backslash A) + 1$.

Proof. *Let* $w := \text{width}(X \backslash A)$. Because of Dilworth's theorem there exists a representation of $X \backslash A = C_1 \cup \cdots \cup C_w$ as union of w disjoint chains. For every $i = 1, \ldots, w$ there exists a lower linear extension L_{2i-1} and an upper linear extension L_{2i} of $\leq$ with respect to C_i. Further we define a linear extension L_{2w+1} of $\leq$ to X, in which A is a segment (see 2.3.3) and where $L_{2w+1} \restriction A$ is the reverse order of $L_1 \restriction A$, the rest is arbitrary. Then $\leq$ is the intersection $\cap \{L_i | i = 1, \ldots, 2w + 1\} := I$.

Indeed, if $a \leq b$, then also $a \, I \, b$ holds trivially. Let now a and b be incomparable in $\leq$. We distinguish three cases:

1) If a and b are both in $X \backslash A$ we have $a \in C_i$ for some $i \in \{1, \ldots, w\}$. Then $b \, L_{2i-1} \, a$ and $a \, L_{2i} \, b$ hold, so that a, b are also incomparable in I.

2) If a and b are both in A, then in L_1 and in L_{2w+1} they appear in reverse order and are thus incomparable in I.

3) If one of a, b is in $X \backslash A$, the other in A, say w.r.o.g. $a \in A$, $b \in X \backslash A$, then b is in some C_i. It follows $a \, L_{2i-1} b$ and $b \, L_{2i} \, a$, and again a and b are incomparable in I.

So I is the order $\leq$ of X, whose dimension is now $\leq 2w + 1$.

3.18 Theorem (Trotter [170]). *Let $(X, \leq)$ be a finite poset, M the set of its maximal elements, and $M \subset X$. Then*
$\dim X \leq \text{width}(X \backslash M) + 1$.

Proof. Let $t := \text{width}(X \backslash M)$. Then by 2.16 the dimension of $X \backslash M$ is $\leq t$, and so its order is the intersection of t linear orders $L_1, \ldots, L_t$ on $X \backslash M$. We extend every L_i to a linear order L_i^* of X. Further we define a linear order L_{t+1} on X as follows: We take some linear extension of $\leq \restriction X \backslash M$ and place all elements of M on its top in the order which is reverse to that one in which they appear in L_t. It follows immediately that $\leq$ is the intersection $\cap \{L_i | i = 1, \ldots, t + 1\}$.

Finally we mention a theorem of V. Novák [127], see also [75]:

3.19 Theorem. *Let $(P, \leq)$ be a poset, $DM(P)$ its Dedekind–MacNeille completion. Then $DM(P)$ has the same dimension as $(P, \leq)$.*

Proof. Let d be the dimension of $(P, \leq)$ and $\{\leq_i \mid i \in I\}$ a set of linear orders on P such that $|I| = d$ and $\leq \, = \cap_{i \in I} \leq_i$. By 1.10.9' every linearly ordered set $(P, \leq_i)$ can be enlarged to a complete linearly ordered set $(K_i, \leq_i)$, we use the same sign $\leq$ for the enlarged relation. Then $X := X\{K_i | i \in I\}$ is a completely ordered set by 4.1.18. And it has a dimension $\leq d$. The mapping f, which for $x \in P$ satisfies $f(x) = (x_i)_{i \in I}$, where $x_i = x$ holds for all $i \in I$, is an isomorphism of P onto a subset P^* of X. By 1.10.11 $DM(P^*)$, and then also $DM(P)$,is embeddable in X, and so its dimension is $\leq d$. Since P is embeddable in $DM(P)$ the latter also has a dimension $\geq d$.

3.20 Remark. In [127] Novák proves a more general theorem, which we mention here without proof, namely: If P is a poset and D a subset of P which is σ-dense and δ-dense in P (see 1.10.8), then $\dim P = \dim D$.

7.4 Interval orders

Starting from a linearly ordered set one can investigate the set of its closed intervals with the inclusion order $\subseteq$. Using this order Dushnik/Miller [30] obtained a new characterization for posets of dimension ≤ 2.

4.1 Definition. Let $(P, \leq)$ be a poset, $a < b$ and $c < d$ elements of P. Then we define $[a, b] \leq [c, d] \iff [a, b] \subseteq [c, d]$. (The same: $\iff c \leq a < b \leq d$). It is clear that this defines an order in the set of closed intervals of P.

4.2 Theorem ([30])**.** *Let $(P, \leq)$ be a poset. Then the following two conditions are equivalent:*

   1) P has a dimension ≤ 2.

   2) P is isomorphic to a set of closed intervals of a linearly ordered set, which is ordered by inclusion.

Proof. Suppose 1) holds. Then the order $\leq$ is $= L_1 \cap L_2$, where L_1 and L_2 are linear orders on P. Let $f : P \to P'$ be a bijective mapping of P onto a set P' which is disjoint to P. Instead of $f(x)$ we also write x'.

We order P' by L' where for $x, y \in P$ we put $x'L'y' \iff yL_2x$. Then we define a linearly ordered set C as an ordered sum $C := (P', L') + (P, L_1)$. Let $\preceq$ denote the linear order of C.

Now we map every $x \in P$ onto the interval $I(x) := [x', x]$ of C. This yields an isomorphism of $(P, \leq)$ onto a set of intervals of C. We verify this:

Suppose $x \leq y$ (in P). Then $x L_1 y$ and $x L_2 y$ hold, and the last relation implies $y' L' x'$ (in P'). Now there follows $y' \preceq x' \preceq x \preceq y$ and then $[x', x] \subseteq [y', y]$. So the mapping $x \to I(x)$ is $\leq$ - preserving.

Let now x and y be incomparable elements of P. W.r.o.g. we can assume that $x L_1 y$ and $y L_2 x$ hold, where the last relation implies $x' L' y'$. Then we have $x' \preceq y'$. Here x' is strictly less than y' with respect to $\preceq$ because of $x' \neq y'$, and so $[x', x]$ is no subset of $[y', y]$. Also x is strictly less than y in C, and thus $[y', y]$ is no subset of $[x', x]$. Therefore the image-intervals of x and y are incomparable. And $1) \to 2)$ is verified.

Now we prove $2) \to 1)$: Let $\mathfrak{J}$ be a set of closed intervals of a linearly ordered set $(T, \leq_T)$. Let then A (resp.B) be the set of first (resp. last) elements of the intervals of $\mathfrak{J}$. We equip A with the order $\leq_A$, which is the inverse order of the restriction of $\leq_T \restriction A$, and B with the order $\leq_B := \leq_T \restriction B$. Now we map every interval $[a, b]$ of $\mathfrak{J}$ onto the pair $\varphi([a, b]) := (a, b) \in A \times B$, where $A \times B$ is equipped with the product order $\leq_\pi$. Then we prove that φ is an isomorphism of $\mathfrak{J}$ onto a subset of $A \times B$, so that $\mathfrak{J}$ and P have a dimension ≤ 2.

Case 1. Let $[a, b] \subseteq [c, d]$ be intervals of $\mathfrak{J}$. Then $c \leq_T a <_T b \leq_T d$ holds. Now $c \leq_T a$ implies $a \leq_A c$ and then also $(a, b) \leq_\pi (c, d)$.

Case 2. Let $[a, b]$ and $[c, d]$ be incomparable intervals of $\mathfrak{J}$. W.r.o.g. we can assume $a \leq_T c$, and then also $b <_T d$. Here $a = c$ is impossible because then $[a, b] \subseteq [c, d]$ would hold. Thus we have $a <_T c$ and further $c <_A a$ and $b <_B d$. Then (a, b) and (c, d) are incomparable in $\leq_\pi$.

Among the sets of dimension ≤ 2 is an important class of posets, namely the generalized trees, which have been defined in 6.1.6.

4.3 Theorem. *If $(T, \leq)$ is a generalized tree, it is either a chain or it has dimension 2. In particular every tree has dimension ≤ 2.*

Proof. First we prove our statement for finite trees and apply induction on the cardinality of T. For trees $(T, \leq)$ with $|T| \leq 2$ the assertion is trivial. Let now n be a natural number ≥ 3, such that the theorem is already proved for all trees with $\leq n$ elements, and let T be a tree with $n + 1$ elements. Let a be the least element of T and $T^- := T \backslash \{a\}$, and let $a_1, \ldots, a_k$ be the immediate successors of a. For each $\kappa = 1, \ldots, k$ the set T_κ of all elements of T^-, that are $\geq a_\kappa$ is again a tree, and since T_κ has $\leq n$ elements, its dimension is ≤ 2, so that its order $\leq_\kappa$ is the intersection of two linear orders $O_{\kappa 1}$ and $O_{\kappa 2}$. Then we consider the following two linear orders L_1, L_2 on T: *(Our notation assumes $k \geq 2$, but the case $k = 1$ follows similarly.)*

(T, L_1) is the linearly ordered sum $\{a\} + (T_1, O_{11}) + \cdots + (T_\kappa, O_{k1})$,

(T, L_2) is the linearly ordered sum $\{a\} + (T_\kappa, O_{k2}) + \cdots + (T_1, O_{12})$.

In L_2 the segments $T_1, \ldots, T_k$ appear in the reverse order to that, in which they appear in L_1, so that the elements of T_i are in the order $L_1 \cap L_2$ incomparable with all elements of T_j if i, j are different elements of $\{1, \ldots, k\}$. Further for every $\kappa = 1, \ldots, k$ the intersection $(L_1 \cap L_2) \restriction T_\kappa$ is equal to $\leq_\kappa$. So the intersection $L_1 \cap L_2$ is exactly the order of T.

Let now $(T, \leq)$ be an infinite tree. Then all finite subtrees of it have a dimension ≤ 2, and then also T has dimension ≤ 2 by 3.3.

If finally $(T, \leq)$ is a generalized tree, it suffices for the same reason to prove the assertion for the case where T is finite. Then each principal ideal $(T \leq x)$ with $x \in T$ is (not only linearly ordered, but also) well-ordered. We add a new element e to T, which is less than all $t \in T$, so that e becomes the minimum of the enlarged set. Now $T \cup \{e\}$ is a finite tree; so it has a dimension ≤ 2. And this holds a fortiori for T.

Chapter 8
Well-founded posets, pwo-sets and trees

In this chapter we consider a class of posets which fulfill certain finiteness conditions. For this reason they have a lot of features in common with finite posets. A forerunner of this class is the algebraic concept of Hilbert and others of the descending chain condition in algebra, which is satisfied e.g. in the set of ideals of a ring, if every descending sequence of ideals is finite.

8.1 Well-founded posets

1.1 Definition. A quasi-ordered set $(W, \leq)$ is called *well-founded*, if W has no infinite descending chain $\{a_0, a_1, .., a_n, ..|n < \omega\}$, i. e. with $a_0 > a_1 > a_2 > \cdots$.

If W is well-founded, and if in addition every antichain of W is finite, W is called *well-quasi-ordered*, abbreviated *wqo*, or a *wqo-set*.

If here W is not only quasi-ordered but also ordered, W is said to be a *partially well-ordered set*, abbreviated *pwo-set*.

Another possibility to define the notion "well-founded" for posets is given by the following characterizations, which are immediately clear:

1.2 Remark. *A poset is well-founded iff each of its chains is well-ordered.*

A poset is well-founded iff each of its non-empty subsets has a minimal element.

Every finite quasi-ordered set is well-founded, also every well-ordered set. The property to be well-founded is of course hereditary. And it turns out that some important notions for finite posets, which have been defined in 1.7.1 (see also 1.7.3), can be generalized to well-founded sets:

1.3 Definition. Let W be a non-empty well-founded poset, L_0 the set of its minimal elements. Since W is non-empty also L_0 is non-empty. If for an ordinal ν sets L_μ, $\mu < \nu$, have already been constructed, we define L_ν to be the set of all minimal elements of $W \setminus \cup\{L_\mu|\mu < \nu\}$ and

232

call it the ν - *level* of W. This is also denoted by $W(\nu)$. There exists a least ordinal α for which L_α is empty (and then also $L_\beta = \emptyset$ for all $\beta > \alpha$). This α is denoted by $h(W)$ or $\mathrm{Ht}(W)$ and called the *height* of W. And so we have $h(W) = \{\nu | L_\nu \neq \emptyset\}$.

The elements x of L_ν are said to have *height* $h(x) = \nu$. So we have $h(W) = \{h(x) | x \in W\}$. We also write $h(x, W)$ for $h(x)$, if there are more sets in consideration, of which x is an element, and in which x has possibly different heights..

A maximal chain of W is called a *branch* of W, and an initial segment of a branch is said to be a *path* of W. Then every branch and every path of a well-founded poset is well-ordered. The height of a branch or path is also called its *length*.

If β is an ordinal $\leq h(W)$ we define the β - *stump* of W to be the union $\cup\{L_\nu | \nu < \beta\}$.

We compile several simple facts about well-founded sets, which are partially in analogy to 1.7.2:

1.4 Theorem. *Every level L_ν of a well-founded poset W is an antichain of W. The levels $L_\nu, \nu < h(W)$, are pairwise disjoint, and so they form a partition of W into antichains.*

Let μ and ν be ordinals with $\mu < \nu < h(W)$, and $x \in L_\nu$. Then there holds:

a) *There exists an $x_\mu \in L_\mu$ with $x_\mu < x$.*

b) *For every $y \in L_\mu$ we have $y < x$ or $y \parallel x$.*

c) *For elements x, y of W there follows: $x < y \Longrightarrow h(x) < h(y)$.*

d) *For every $a \in W$ there holds $h(a) = \cup\{(h(x) + 1) | x < a\} =: R$.*

e) *For $a \in W$ we have $h(a) = h(W < a)$.*

f) *If two different elements a, b of W have the same set of predecessors $(W < a) = (W < b)$, then $h(a) = h(b)$, and a and b are incomparable.*

Proof. a) L_μ is the set of minimal elements of $W \setminus \cup\{L_\kappa | \kappa < \mu\} =:$ S, and we have $x \in S$, which implies a).

b) An element $y \in L_\mu$ is minimal in S, and so $x \in S$ cannot be less than y.

c) By definition y is a minimal element of $W \setminus \cup\{L_\nu | \nu < h(y)\}$. Then an element $x < y$ cannot be in this set, and so $x \in L_\nu$ must hold for some $\nu < h(y)$. This ν is $= h(x)$, and c) is proved.

d) For $x < a$ in W we have $h(x) < h(a)$ and thus $h(x) + 1 \leq h(a)$, so that $h(a) \geq R$ holds. On the other hand, let β be an ordinal $< h(a)$.

Then by a) we have an element $x < a$ in W with height $h(x) = \beta$. Then $h(x) + 1 > \beta$, and thus R is $> \beta$ for all ordinals $\beta < h(a)$, which means $R \geq h(a)$.

e) $a \in W$ holds by definition: $h(W < a)$ is the least ordinal which is $> h(x)$ for all $x < a$, and so $h(W < a)$ is the least ordinal which is $\geq h(x) + 1$ for all $x < a$. This implies $h(W < a) = \cup\{(h(x) + 1)|x < a\} = h(a)$ by d).

f) This follows immediately by e).

Here we recall 1.7.5, which shows that it can happen that elements a,b of a well-founded poset can satisfy $a \lhd b$ but $h(b) > h(a) + 1$.

Every time a new class of ordered sets is introduced the question arises of what can be said about the structure of the chains and antichains of the sets of the class. In 1.2 we already had mentioned a fundamental property of the chains of well-founded posets. Now we investigate them in greater detail to learn more about the order types of the well-ordered chains.

A first question which suggests itself, is: If W is a well-founded poset, does there exist a chain passing through all levels of W ? It can easily be seen that this is not true. If W is a union of disjoint chains C_n, $n \in \mathbf{N}$, where each C_n has n elements, and where for $m \neq n$ all elements of C_m are incomparable with all elements of C_n, then W is well-founded, and has height ω, but no chain of type ω.

A weakened form of the above question has a positive answer, namely there holds:

1.5 Theorem. *Let W be a well-founded poset, $x \in W$ an element of height ξ. Let $h_1 < \cdots < h_n = \xi$ be finitely many ordinals $< h(W)$. Then there exists a chain $\{x_1, \ldots, x_n = x\}$ of W, such that x_ν has height h_ν for $\nu = 1, \ldots, n$.*

In particular there holds: Let $x_{\tau+n} \in W$ be an element with height $\tau + n$, where τ is an ordinal and $n \in \mathbf{N}$. Then there exists a chain $C \subseteq W$, which contains $x_{\tau+n}$ as last element and for every non-negative integer $m < n$ an element $x_{\tau+m}$ of height $h(x_{\tau+m}) = \tau + m$.

So C passes through all finitely many levels L_α of W with $\tau \leq \alpha \leq \tau + n$.

Proof. The assertion is an immediate consequence of 1.4, a).

A class of well-founded sets W, in which a chain exists which passes through all levels of W, is that of those well-founded sets in which each

234

level is finite. For the proof of this we use Rado's Selection Lemma (see [147] and [148]), which has several useful applications in set theory. In the following proof of it we use the notation $X \ll Y$ for: X is a finite subset of Y.

1.6 Theorem. *Let A and I be non-empty sets and $A_i \ll A$ for $i \in I$. Suppose that for each $L \ll I$, we are given a choice function $f_L : L \to A$ such that $f_L(i) \in A_i$ for $i \in L$.*

Then there exists a choice function $f^* : I \longrightarrow A$ such that, given any $L \ll I$, there is an M with $L \subseteq M \ll I$ and $f^*(i) = f_M(i)$ for $i \in L$.

Proof. If I is finite then $f^* = f_I$ satisfies our assertion. Now let I be infinite. Let Ω be the set of all families $(X_i | i \in I)$ such that $X_i \subseteq A_i$ for $i \in I$ and, given any $L \ll I$, there is an M with $L \subseteq M \ll I$ and $f_M(i) \in X_i$ for $i \in L$. (This entails that the X_i are non-empty.) By hypothesis, $(A_i | i \in I) \in \Omega$ (put $M = L$).

We consider the following (natural) order in Ω :
$$(Y_i | i \in I) \leq (Z_i | i \in I) \Longleftrightarrow Y_i \subseteq Z_i \text{ for } i \in I.$$

We prove that $(\Omega, \leq)$ has a minimal element by using Zorn's lemma. Let $\mathfrak{K} = \{\mathfrak{C}_k | k \in K\}$ be a chain of Ω, i.e. K a linearly ordered index set where $i < j$ (in K) $\Longrightarrow \mathfrak{C}_i < \mathfrak{C}_j$ (in Ω). Let $\mathfrak{C}_k = (X_{ki} | i \in I)$ for $k \in K$. Then there holds for every $i \in I$: The family $(X_{ki} | k \in K)$ is monotonically decreasing with k, and since the sets X_{ki} are finite this entails that for every $i \in I$ there exists an index $k^*(i)$ such that all X_{ki} with $k \leq k^*(i)$ are equal, say $= X_i^*$, and this is non-empty.

The family $(X_i^* | i \in I)$ belongs to Ω : Let $E \ll I$. For $k \leq \min\{k^*(i) | i \in E\}$ we have $X_{ki} = X_i^*$ and so there exists a set F_k with $E \subseteq F_k \ll I$ and $f_{F_k}(i) \in (X_{ki} =) X_i^*$ for $i \in E$. Of course, $(X_i^* | i \in I)$ is a lower bound of $\mathfrak{K}$, and thus by Zorn's lemma, Ω has a minimal element $P = (P_i | i \in I)$.

Next we prove that all sets P_i, $i \in I$, contain exactly one element. Let j be a fixed element of I. By construction P_j is non-empty. Let now a and b be elements of P_j. We wish to show $a = b$.

We consider the following statement (1), which will turn out to be false:

(1) For every $L \ll I$ there exists an M with $L \subseteq M \ll I$, such that $f_M(i) \in P_i$ for all $i \in I$ and $f_M(j) \neq a$.

If (1) would be true, then the family $(P_i' | i \in I)$, for which $P_i' = P_i$ for all $i \neq j$ of I and $P_j' = P_j \setminus \{a\}$ holds, would also be in Ω, and it would be less than a minimal element of Ω with contradiction. So (1) is false, and thus we have:

(2) There exists an $(L =)\ I_a \ll I$ such that for every M with $I_a \subseteq M \ll I$ and $f_M(i) \in P_i$ for $i \in I$ there holds $f_M(j) = a$.

Symmetrically there holds:

(2′) There exists an $I_b \ll I$ such that for every M with $I_b \subseteq M \ll I$ and $f_M(i) \in P_i$ for $i \in I$ there holds $f_M(j) = b$.

But now $I_a \cup I_b$ is a finite set M with $I_a \cup I_b \subseteq M \ll I$, and so by (2) and (2′) we have $f_M(j) = a = b$. So all $P_i, i \in I$, have exactly one element p_i. Finally we put $f^*(i) := p_i$ for $i \in I$. Due to $(P_i|\ i \in I) \in \Omega$ now f^* satisfies our assertion.

If we make use of topological concepts, in particular of Tychonoff's product theorem, we can present another elegant proof of 1.6, which was found by Gottschalk [59]:

Second proof of 1.6: Let $\mathcal{L}$ be the set of all finite subsets of I. For $L \in \mathcal{L}$ let E_L be the set of all points $p = (p_i|i \in I) \in X :=\text{X}\{A_i|i \in I\}$ for which there exists a $B \in \mathcal{L}$ with $B \supseteq L$ where for each $i \in L$ we have $p_i = f_B(i)$. (For $i \in I \setminus L$ the element p_i is arbitrary.) By assumption every E_L is non-empty.

Provide each A_i, $i \in I$, with the discrete topology. (Every subset of A_i is open and also closed.) With this A_i is a compact space, since A_i is finite. The set $X :=\text{X}\{A_i|i \in I\}$, equipped with the product topology, is then also compact by Tychonoff's theorem.

E_L is a closed subset of X. For every cartesian product $P :=\text{X}\{Y_i|i \in I\}$, where for $i \notin L$ we have $Y_i = X_i$, and where Y_i is a one-element subset $\{p_i\}$ of A_i for $i \in L$, is closed; and E_L is a union of finitely many such sets P.

(I) The set $\{E_L|L \in \mathcal{L}\}$ has the finite intersection property, which means the intersection of finitely many of its elements is non-empty.

Indeed, let $L_1, \ldots, L_n$ be finitely many finite subsets of I. Then
$$E' := E_{L_1} \cap \cdots \cap E_{L_n} = \{p = (p_i)_{i \in I} \in X|\ \text{For every } \nu = 1, \ldots, n$$
and each $i \in L_\nu$ we have $p_i = f_{B_\nu}(i)$ for a $B_\nu \supseteq L_\nu$ of $\mathcal{L}\}$.

Now $L := L_1 \cup \cdots \cup L_n$ is finite and a superset of the $L_1, .., L_n$. We put $p_i := f_L(i)$ for $i \in L$, and p_i arbitrary $\in A_i$ for $i \in I \setminus L$. Then $(p_i)_{i \in I}$ belongs to E' which is hence non-empty, and (I) is proved.

A compact space has the property that if an intersection of a set $\mathfrak{C}$ of closed sets is empty, then already a finite subset of $\mathfrak{C}$ has an empty intersection. Therefore (I) entails $\cap\{E_L|L \in \mathfrak{L}\} \neq \emptyset$.

Let $q = (q_i)_{i \in I}$ be a point of this intersection. We define $f^*(i) := q_i$ for $i \in I$. Now f^* satisfies our assertion: If $L \ll I$ is given and $i \in L$, then, since $q \in E_L$, $q_i = f_M(i)$ for some $M \in \mathcal{L}$ with $M \supseteq L$.

We can now prove the theorem of Wolk which was mentioned before:

1.7 Theorem [179]. *Let W be a well-founded poset of height h with level-decomposition $\{L_\alpha|\alpha < h\}$, in which every level L_α, $\alpha < h$, is finite. Then there is a chain containing precisely one element from each level L_α of P.*

Proof. Let $H = \{h_1, \ldots, h_n\}_<$ be a non-empty finite subset of the ordinal h. Then by 1.5 there exists a chain $\{x_1, \ldots, x_n\}$ of W with $h(x_\nu) = h_\nu$ for $\nu = 1, \ldots, n$. In this manner we construct a mapping

$f_H : H \to W$, where for every $\alpha \in H$ we have $f_H(\alpha)$ $(= x_\alpha) \in L_\alpha$, and where the $f_H(\alpha), \alpha \in H$, form a chain.

Due to 1.6 there exists a function $f^* : h \to P$, such that for every $H \ll h$ there is an M with

(*) $H \subseteq M \ll h$ and $f^*(\alpha) = f_M(\alpha)$ for $\alpha \in H$.

Now $f^*(\alpha) \in L_\alpha$ for $\alpha \in h$, and these elements form a chain: If $a, b \in h$ we can take $H = \{a, b\}$ and an M satisfying (*). Then $f^*(x) = f_M(x)$ for $x \in H$, and then $f^*(a)$ and $f^*(b)$ are comparable since $f_M[\{a, b\}]$ is a chain.

The last theorem can be considered as a generalization of König's Infinity Lemma, which corresponds to the case $h = \omega$.

In the last theorem the assumption that all levels of W are finite cannot be omitted, as the example before 1.5 shows. Pouzet [141] exhibited the following counterexample in which additionally all maximal chains have the same height as W:

1.8 Example. Let W be the set of pairs (n, m) of non-negative integers n, m with $m < n$. This set is ordered by $(n, m) \leq (n', m') \Longleftrightarrow n = n'$ and $m \leq m'$, or $n < n'$ and $m + 2 \leq m'$. Then $(W, \leq)$ is well-founded and of height ω. But if C is a chain of W, it cannot intersect each level. Indeed, let (n, m) be the least element of C, and (n, k) the greatest element of C which has n as first component. Then k is $< n$, and every immediate successor (n', m') of (n, k) in C must have $m' \geq k + 2$. Then

C has no element with second component $k+1$ and thus no element of height $k+1$.

In the following we investigate in more detail the types of chains and antichains of well-founded posets. In this context we introduce a useful notation which was defined by Milner and Sauer. In the pioneering paper [120] they give a very detailed discussion on this subject.

1.9 Definition [120]. Let α, β, γ be ordinal numbers. Then $\alpha \to [\beta, \gamma]$ means that the following statement is true: Whenever $(W, \leq)$ is a well-founded poset of height α, then either there is a chain $C \subseteq W$ of type β, or there is an antichain $A \subseteq W$ such that the order type of the set of ordinals $\{\nu \in \alpha | A \cap W(\nu) \neq \emptyset\}$ is γ, where $W(\nu)$ is the set of those elements of W which have height ν.

The negation of $\alpha \to [\beta, \gamma]$ is, of course, denoted by $\alpha \nrightarrow [\beta, \gamma]$.

We remark already here that the main theorem of the rest of this section will be a theorem of Milner/Sauer and Pouzet, stating that $\alpha \to [\alpha, \omega]$ holds for all ordinals α. For the case of denumerable ordinals α this was first proved by Diana Schmidt [155].

The symbol $\alpha \to [\beta, \gamma]$ is suggested by the symbol $\alpha \to (\beta, \gamma)^2$ of the partition calculus [38], (with which it should not be confused). This is defined as follows, supplementing Definition 6.4.1 which concerned cardinals:

1.10 Definition. For ordinals α, β, γ the symbol $\alpha \to (\beta, \gamma)^2$ means that the following is true: Whenever $(S, \leq)$ is a well-ordered set of type α, and the set $[S]^2$ of two-element subsets of S is partitioned into two sets K_0, K_1, then either there is a set $B \subseteq S$ of order type β such that $[B]^2 \subseteq K_0$, or there is a set $C \subseteq S$ of order type γ such that $[C]^2 \subseteq K_1$.

The symbol $\alpha \to (\beta, \gamma)^2$ is stronger than $\alpha \to [\beta, \gamma]$:

1.11 Theorem. *Let α, β, γ be ordinals for which $\alpha \to (\beta, \gamma)^2$ holds. Then also $\alpha \to [\beta, \gamma]$ is valid.*

Proof. Let W be a well-founded poset of height α. Then for every $\nu < \alpha$ we choose an element x_ν of height ν. For $\mu < \nu < \alpha$ we put $\{\mu, \nu\} \in K_0$ iff x_μ, x_ν are comparable, and this means $x_\mu < x_\nu$ since x_μ is on a lower level than x_ν. And we put $\{\mu, \nu\} \in K_1$ iff x_μ and x_ν are incomparable. Due to our assumption and 1.10 with α for S we obtain:

There is a set $B \subseteq \alpha$ of order type β such that all x_ν with $\nu \in B$ are comparable, and then they form a chain of type β, or there exists a

238

set $C \subseteq \alpha$ of type γ such that all x_ν with $\nu \in C$ are incomparable. And then the indices of the levels of W which contain elements of $\{x_\nu | \nu \in C\}$ form a set of order type γ.

By the way, the partition relation $k \rightarrow (k, \aleph_0)^2$ of 6.4.10 entails $\alpha \rightarrow (\alpha, \omega)^2$ and thus by 1.11 $\alpha \rightarrow [\alpha, \omega]$ for the case where α is an initial ordinal. This will later be generalized in 1.16' on all ordinals α.

Now we mention several elementary facts on the height function, which are rather obvious, but should nevertheless be checked carefully:

1.12 Lemma. *Let W be a well-founded poset, and $V \subseteq W$ a subset which is equipped with the restriction of the order of W. Let $x \in V$. Then for the heights of x in W resp. V we have*

(1) $h(x, W) \geq h(x, V)$.

Proof. If $h(x, V) = 0$ holds, (1) is trivial. Let now (1) be proved for all values $h(x, V) < \mu$, and let $y \in V$ have height $h(y, V) = \mu$. Then by 1.4 a) for all $\nu < \mu$ there exists a y_ν in the ν - level $L_\nu(V)$ of V with $y_\nu < y$. By induction hypothesis then $y_\nu \in L_{\mu(\nu)}(W)$ for some $\mu(\nu) \geq \nu$. Since $y > y_\nu$ holds for all $\nu < \mu$ we obtain by 1.4 c) $h(y, W) > h(y_\nu, W) = \mu(\nu) \geq \nu$, and thus $h(y, W) \geq \mu$, so that (1) also holds for $h(x, V) = \mu$. And our statement is proved.

1.13 Lemma. *Under the assumptions of 1.12 we have $h(V) \leq h(W)$.*

Proof. $h(V)$ is the least ordinal which is $> h(x, V)$ for all $x \in V$, and $h(W)$ is the least ordinal which is $> h(x, W)$ for all $x \in W$. So 1.12 implies 1.13.

If we consider a well-founded set W and a subset V of it with the induced order, so that V is also well-founded, usually the heights of an element $x \in V$ in W and in V are different. In the following lemma we have a situation where they are equal:

1.14 Lemma. *Let W be a well-founded poset, $x \in W$ an element with height h. We denote the ν - level of W by L_ν, that of the subset $(W < x)$ by l_ν, for $\nu < h$. Then there holds $l_\nu = (L_\nu < x)$ for all $\nu < h$.*

Proof. For $\nu = 0$ the statement is trivial. Let now μ be an ordinal $< h$ such that the assertion holds for all $\nu < \mu$. Let $m \in l_\mu$. Then $m < x$ holds, further $m \notin \cup\{L_\nu | \nu < \mu\}$, because otherwise by induction hypothesis m would be in an L_ν with $\nu < \mu$, and then also in l_ν with contradiction. Now m is a minimal element of $T := W \setminus \cup \{L_\nu | \nu < \mu\}$ and thus in L_μ. For if an element y of T would be $< m \in l_\mu$, then also $y < x$ would hold, and then by 1.4 c) y would already be in some l_ν with $\nu < \mu$ and then, due to the induction hypothesis, also in the corresponding L_ν, contradicting $y \in T$. Hence $m \in (L_\mu < x)$.

On the other hand, let $m \in (L_\mu < x)$. Then $m \notin l_\nu$ for $\nu < \mu$, since otherwise by the induction hypothesis also $m \in L_\nu$ would hold for a $\nu < \mu$, which is impossible. Therefore $m \in (W < x) \setminus \cup \{l_\nu | \nu < \mu\} =: U$, and m is minimal in U and thus in l_μ. For if we would have $u < m$ for an element $u \in U$, then by 1.4 c) $u \in L_\nu$ would follow for a $\nu < \mu$. This would by induction entail $u \in l_\nu$ for some $\nu < \mu$, contradicting $u \in U$.

1.15 Lemma. *Let δ be an indecomposable ordinal ω^λ, where λ is a limit ordinal. Let $(\nu_i | i < cf(\lambda))$ be a strictly ascending sequence of successor ordinals which has λ as supremum, so that also $\sup\{\omega^{\nu_i} | i < cf(\lambda)\} = \omega^\lambda$. Then $S := \sum \{\omega^{\nu_i} | i < cf(\lambda)\} = \omega^\lambda$.*

Proof. $\omega^\lambda \supseteq \cup \{\omega^{\nu_i} \setminus \omega^{\nu_i - 1} | i < cf(\lambda)\}$. Here each $\omega^{\nu_i} \setminus \omega^{\nu_i - 1}$ has order type ω^{ν_i}. Further the summands in the set union are pairwise disjoint, and so its order type is $\geq S$. Of course, it is also $\leq S$ and thus $= S$.

The main theorem of this section, which we address now, was established independently by Milner/Sauer [120] and Pouzet [141] with entirely different proofs. Here we present the proof of Pouzet.

1.16 Theorem. *Let W be a well-founded poset of height δ. Then there exists an infinite antichain of W all of whose elements have different heights, or there exists a chain of W which has the same height δ as W.*

Proof. If $\delta = 0$ holds, W must be empty, and then the statement is trivial. Also the case $\delta = 1$ is trivial. Suppose now $\delta > 1$, and that the assertion is proved for all $\delta' < \delta$. We make the assumption that there is no infinite antichain of W such that all elements of it have different heights. We distinguish two cases:

Case 1. δ is not indecomposable.

Then we have a representation $\delta = \delta' + \delta''$, where δ' and δ'' are both $< \delta$. Let W'' be the set of all $x'' \in W$ for which the height $h(x'', W)$ of x'' in W is $\geq \delta'$. We have $h(W'') = \delta''$, and then by induction hypothesis W'' has a chain C'' of type δ''. Let m be its least element. Then we have $m \geq x$ for an element $x \in W(\delta')$. According to 1.4 e) the well-founded set $(W < x)$ has height δ', and so by induction hypothesis $(W < x)$ contains a chain C' of type δ'. Now $C' \cup C''$ is a chain of W of type $\delta' + \delta'' = \delta$.

Case 2. δ is indecomposable and > 1.

Then by 1.15 we can represent δ, if it is $> \omega$, as a sum $\delta = \sum\{\delta_i | i < cf(\delta)\}$, where $(\delta_i), i < cf(\delta)$, is a strictly ascending sequence of ordinals $\delta_i = \omega^{\nu_i}$ with $\sup\{\delta_i | i < cf(\delta)\} = \delta$. (If $\delta = \omega^{\nu+1}$ we can take $\delta_i = \omega^\nu \cdot i$ for $i < \omega$. If $\delta = \omega$ we put $\delta_i := i$ for $i < \omega$.) Let
$$W_i := \{x \in W \mid \textstyle\sum_{j<i} \delta_j \leq h(x, W) < \sum_{j \leq i} \delta_j\}.$$
Then W_i is well-founded and has height $\leq \delta_i < \delta$. By induction hypothesis there is a chain $C_i \subseteq W_i$ of type δ_i. Now $A := \cup\{C_i | i < cf(\delta)\}$ has height δ, for it is $\geq \delta_i$ for all $i < cf(\delta)$. All elements of A have different height in W.

A has no infinite antichain since otherwise A would intersect infinitely many chains C_i, and then also infinitely many levels of W, which is excluded by our assumption. A fortiori this entails that all levels of A, these are antichains, are finite, and so from 1.7 there follows the existence of a chain C of (A and) W which has exactly one element of each level $A(\nu)$ with $\nu < \delta$. And so also C has height δ.

In the notation of 1.9 the last theorem states:

1.16′ Theorem. $\delta \to [\delta, \omega]$ *for all ordinals* δ.

In connection with 1.16 Pouzet established a theorem, which also implies a proof of 1.16, but gives more insight into the structure of well-founded posets, so that it deserves some interest in itself. The main part of its proof is treated in the following technical Lemma 1.18. First we introduce a new concept.

1.17 Definition. Let δ be an ordinal > 0. Then by 4.8.5, δ has a unique representation as $\delta = \omega^{\alpha_0} + \cdots + \omega^{\alpha_k}$ as a sum of finitely many indecomposable ordinals, where $\alpha_0 \geq \cdots \geq \alpha_k$ are ordinals. Then we define $l(\delta)$ to be the last summand ω^{α_k} in this normal form representation of δ. If δ is a successor number, then $\alpha_k = 0$ and $\omega^{\alpha_k} = 1$.

If W is a well-founded poset and δ an ordinal $< h(W)$, then $W(\delta)$ denotes the δ - level of W.

1.18 Lemma [141]. *Let W be a well-founded poset with height function h, $\delta < h(W)$, $X \subseteq W(\delta)$ and $|X| \leq |l(\delta)|$. Then W has a subset $A \subseteq \cup\{W(\zeta) | \zeta < \delta\}$, which satisfies*
  1) $|A \cap W(\zeta)| \leq 1$ *for all* $\zeta < \delta$,

2) $h(A < x) = \delta$ for all $x \in X$.

Proof. We apply induction on δ. For $\delta = 0$ the set $A = \emptyset$ satisfies 1) and 2), here we have $h(\emptyset) := 0$. Also the case $\delta = 1$ is trivial. Let now δ be > 1 and the lemma be proved for all $\delta' < \delta$. We distinguish three cases:

Case 1. δ is not indecomposable.

Then we can write $\delta = \delta' + \delta''$, where $\delta'' = l(\delta)$. Let W'' be the set of all $x \in W$ which have a height $h(x) \geq \delta'$. Then we have

(1) $X \subseteq W''(\delta'')$.

Indeed, for $x \in X$ we have $h(x, W) = \delta = \delta' + \delta''$ and on the other hand $h(x, W) = \delta' + h(x, W'')$, so that $\delta'' = h(x, W'')$, and thus (1) follows. Further we have $|X| \leq |l(\delta'')|$ $(= |l(\delta)|$ because of $l(\delta'') = l(\delta))$. Now our induction hypothesis yields:

(I) There is a set $A'' \subseteq W''$ which satisfies 1) and 2) (with W, δ replaced by W'', δ''). So by 1) we have $|A''| \leq |\delta''|$.

Now we choose for each $x'' \in A''$ an element $x' \in W(\delta')$ with $x' \leq x''$. Let X' be the set of all x' so chosen. Then $X' \subseteq W(\delta')$ and $|X'| \leq |A''|$.

Further we have $|\delta''| \leq |l(\delta')|$, since in the normal form representation of $\delta = \delta' + \delta''$ the term $l(\delta')$ is a summand which precedes $l(\delta) = \delta''$, so that $\delta'' \leq l(\delta')$ holds. So we have $|X'|(\leq |A''| \leq |\delta''|) \leq |l(\delta')|$. Now, by induction hypothesis, applied on W, δ' and X', we obtain:

(II) There exists a set $A' \subseteq \cup\{W(\zeta)|\zeta < \delta'\}$ which satisfies 1) and 2), with δ, X replaced by δ', X'. In particular we have

(2) $h(A' < x') = \delta'$ for all $x' \in X'$.

Finally $A := A' \cup A''$ satisfies our assertion: 1) is trivially fullfilled, due to (I) and (II). Let now $x \in X$. For every $x'' \leq x$ in A'' and $x' \leq x''$ in $X' \subseteq W(\delta')$ we have by 1.4 e) and (2)

$h(A < x) \geq h(A' < x') + h(x'', A'') = \delta' + h(x'', A'')$.

Then $h(A < x)$ is also $\geq$ the supremum of the values $\delta' + h(x'', A'')$, which is $\delta' + \delta'' = \delta$, and so 2) holds.

Case 2. $\delta = \omega$.

Here we have $|X| \leq |l(\delta)| = |\omega| = \aleph_0$. For every $x \in X$ we choose an infinite subset $Y_x \subseteq \omega$ such that these sets are pairwise disjoint. Since $\omega \times \omega$ is equipotent to ω this is easily done. Next we choose a subset $A_x \subseteq (W < x) \cap \{y|h(y, W) \in Y_x\}$ which contains an element of each level $W(n)$ for $n \in Y_x$ and has height ω. This can be achieved as follows: We represent each Y_x as an ordered sum $Y_x = \sum\{I_m|m \in \mathbf{N}\}$, where

242

I_m is a segment of Y_x with m elements. Then by 1.5 there exists an m-element chain C_m of W which has exactly one element in common with every level $W(n), n \in I_m$. Now the set $A_x := \cup\{C_m | m \in \mathbf{N}\}$ has height $\geq m$ for all $m \in \mathbf{N}$, and so its height is ω. Finally $A := \cup\{A_x | x \in X\}$ satisfies 1) and 2).

Case 3. δ is an indecomposable ordinal $> \omega$, which means $\delta = \omega^\nu$ for an ordinal $\nu > 1$. Here $l(\delta) = \delta$ holds.

Then we choose a strictly increasing sequence $(\delta_i | i < cf(\delta))$ of ordinals $< \delta$ for which there holds $\sup\{\delta_i | i < cf(\delta)\} = \delta$:

If ν is a successor ordinal $\mu + 1$, and hence $\delta = \omega^{\mu+1}$,we take $\delta_i := \omega^\mu \cdot (i + 1)$ for $i < \omega \ (= cf(\delta))$. Here $l(\delta_i) = \omega^\mu$ for all $i < \omega$.

If ν is a limit ordinal, we take a strictly increasing sequence of successor ordinals $\nu_i, i < cf(\nu)(= cf(\delta))$, whose supremum is ν, and $\delta_i := \omega^{\nu_i}$. Here $l(\delta_i) = \delta_i$. In any case we have by 1.15,

$\sum\{\delta_i | i < cf(\delta)\} = \delta$.

Next we choose a system of subsets $X_i, i < cf(\delta)$, of X, which satisfies $X_i \subseteq X_j$ for $i < j < cf(\delta)$, and

a) $\cup\{X_i | i < cf(\delta)\} = X$, and b) $|X_i| \leq |l(\delta_i)|$.

Now we define for each $i < cf(\delta)$ the set $W_i := \{x \in W | \sum_{j<i} \delta_j \leq h(x, W) < \sum_{j \leq i} \delta_j\}$, and put $\widetilde{W}_i := W_i \cup X$. Then we have $X_i \subseteq \widetilde{W}_i(\delta_i)$. Due to b) we can apply our induction hypothesis on $\widetilde{W}_i, \delta_i$ and X_i. So we obtain a subset $A_i \subseteq \widetilde{W}_i$ which satisfies 1) and 2) for δ_i and X_i instead of δ, X. Finally we put $A := \cup\{A_i | i < cf(\delta)\}$, and this now fulfills our assertion: 1) is clear, and since an element $x \in X$ is in some X_i with $i < cf(\delta)$, it is also in all X_j with $i < j < cf(\delta)$, and so $h(A < x)$ is $\geq \delta_i$ for all $i < cf(\delta)$, which means $= \delta$.

In 1.18 the assumption $|X| \leq |l(\delta)|$ cannot be omitted. The following simple example confirms this:

1.19 Example. Let W be the union of $\aleph_1$ disjoint chains of type $\omega + 1$, where the elements of different chains are incomparable. Then W is well-founded and has height $\omega + 1$. Let $X := W(\omega)$. Suppose now indirectly that A is a set $\subseteq \cup\{W(\zeta) | \zeta < \omega\}$, which satisfies 1) and 2) of 1.18. Then, due to 2), for $x \in X$ holds: $(A < x)$ must have height ω, and therefore it contains an element $x' < x$. The set $X' := \{x' | x \in X\}$ has cardinality $\aleph_1 (= |X|)$ because $x \to x'$ is injective. Since the $\aleph_1$ elements of $X' \subseteq A$ are distributed over the countably many levels of W there exist two elements (also $\aleph_1$ many) in X' which have the same height, and this contradicts 1).

The technical lemma 1.18 now implies as a special case:

1.20 Theorem [141]. *Let $W \neq \emptyset$ be a well-founded poset of height $h(W) =: \delta$. Then W has a subset A which (with the induced order) has the same height δ as W, and which for every $\nu < \delta$ has at most one element y of height (in W) $h(y, W) = \nu$.*

Proof. We adjoin to W an element x which is $>$ all $z \in W$ and put $W' := W \cup \{x\}$. We apply 1.18 on W'. We have $\delta < h(W') = h(W) + 1 = \delta + 1$ and $h(x, W') = \delta$, so that $X := \{x\} \subseteq W'(\delta)$. Of course, also $|X| = 1 \leq |l(\delta)|$ holds. Thus by 1.18 there exists a set $A \subseteq W'$ with $h(A < x) = \delta$, which has at most one element of height ν (in W' and in W) for each $\nu < \delta$.

In the last theorem we have simpler assumptions as in Lemma 1.18, but the induction process in the proof of the lemma needed more detailed assumptions.

At the end of this section we list some theorems without proof, which are proved or mentioned in the paper [120] of Milner/Sauer. Here κ denotes infinite initial ordinals and κ^+ the initial ordinal of the successor cardinal of $|\kappa|$. Other greek letters denote ordinal numbers, $n \in \mathbf{N}$.

$\kappa n \to [\kappa n, \beta]$ for $n < \omega, \beta < \omega_1$, cf $\kappa > \omega$.

$\alpha \nrightarrow [\omega, \kappa + 1]$ for $\alpha < \kappa^+$. (For $\alpha < \omega_1$ this was proved by Diana Schmidt [155].)

$\alpha \nrightarrow [\kappa, \text{cf } \kappa + 1]$ for $\alpha < \kappa^+$.

$\alpha \nrightarrow [\kappa, \omega_1]$ for $\kappa \leq 2^{\aleph_0}, \alpha < \kappa^+$.

$\alpha \nrightarrow [\omega_{\gamma+1}, \kappa]$ for $\kappa \leq 2^{\aleph_\gamma}, \alpha < \kappa^+$.

$\alpha \nrightarrow [\kappa\omega, \omega + 1]$ for $\alpha < \kappa^+$. (Galvin [50])

$\alpha < \omega_1 \leq \text{cf } \kappa$ implies $\kappa \to [\alpha, \kappa]$.

1.21 Remark. A simple fact, which is contained in the second of the above theorems, is that there are well-founded posets having arbitrary height but which do not contain any infinite chains.

This follows by induction: For the ordinals $n \in \mathbf{N}$ the sets $\{0, \ldots, n-1\}$ (with their natural order) have height n and no infinite chain. Let now α be an ordinal such that for all $\beta < \alpha$ there is a well-founded poset $(P_\beta, \leq_\beta)$ of height β, which has no infinite chain. W.r.o.g. we can assume that the $P_\beta, \beta < \alpha$, are pairwise disjoint. Then the union $U :=$

$\cup\{P_\beta|\beta < \alpha\}$, equipped with the order $\cup\{\leq_\beta \,|\beta < \alpha\}$, is well-founded and has height $\geq \beta$ for all $\beta < \alpha$. If α is a limit ordinal, then U has height α. If α is a successor ordinal $\beta + 1$ we add a new element to U which is greater than all elements of U. The enlarged set has height α, but no infinite chain.

8.2 The notions well-quasi-ordered and partially well-ordered set

Two specializations of the concept "well-founded poset", which was defined in 1.1, are of great importance, namely *partially well-ordered set* and *tree*. The last notion was already defined in 1.9.8 (see also 6.1.6), but trees will be investigated in detail later. By definition the partially well-ordered sets arise from the well-founded posets by implying a second finiteness condition. The well-founded quasi-ordered sets and the wqo-sets can be characterized by using forbidden subsets: A quasi-ordered set is well-founded, if it has no subset of type ω^*, and it is wqo, if it has no subset of type ω^* and no subset of the type of an antichain with $\aleph_0$ elements. A property which is defined by forbidden subsets is hereditary, and so we can state:

2.1 Remark. If W is a well-founded quasi-ordered set, then every subset of W with the induced quasi-order is also well-founded. And if W is wqo, each of its subsets is also wqo.

The concepts of 2.1 were introduced in the early 1950 s at nearly the same time and independently by several authors (Higman [85], [84], Erdös/Rado [37], Kruskal [102], [101], Michael [119], Rado [149]). We mention here also the paper [124] of Nash-Williams and the two very informative survey articles of Milner [121] and Pouzet [142].

The preoccupation with trees, which are also well-founded, was e.g. promoted in a lot of papers by D. Kurepa, beginning with [104], and in connection with Suslin's problem, which we treat later.

The pwo-sets have many applications in different parts of mathematics. We mention two important ones in graph theory. We refer to the definition of graph which was introduced in 2.5.4. For general informations on graphs see e.g. the book [23] of Diestel.

2.2 Definition. A subgraph of a graph (V, E) is a graph (V', E'), where $V' \subseteq V$ and $E \subseteq E'$ holds. We say: (V', E') arises from (V, E) by *contraction* of an edge $\{a, b\}$ of E, if $V' = V \backslash \{a, b\} \cup \{c\}$, where c is a vertex which is different from all vertices of V, and where E' contains all edges which link elements of $V \backslash \{a, b\}$, and further all two-element subsets $\{c, x\}$, for which $\{a, x\}$ or $\{b, x\}$ are in E.

If a graph G' arises from a finite graph G by successively contracting edges, it is called a *minor* of G. For this we use the sign $G' \prec G$. Also G itself is considered as a minor of G. Then the relation $\prec$ is a quasi-order in every set of finite graphs. For a finite graph H we further put $H \prec G$, if H is isomorphic to a minor of G.

In this context Robertson and Seymour proved between 1986 and 1996 in a series of more than twenty papers the famous minor theorem: The class of finite graphs is well-quasi-ordered by the minor-relation $\prec$. Of course, we cannot go into the details here. It comprises a theorem of Kruskal [102] for graph-theoretical trees: In graph theory a tree $\mathfrak{T}$ is defined as a connected graph which has no circle. Here a graph G is said to be *connected*, if for each two different vertices a, b of G there is a finite sequence $a = a_1, \ldots, a_n = b$, such that $\{a_i, a_{i+1}\}$ is an edge of G for $i = 1, \ldots, n - 1$. And a *circle* is a sequence of vertices $a_1, \ldots, a_n$ with $n \geq 3$, which are pairwise different by exception of $a_1 = a_n$, such that $\{a_i, a_{i+1}\}$ is an edge of G for $i = 1, \ldots, n$, where we have put $a_{n+1} := a_1$.

Then Kruskal's theorem states: The class of finite trees is well-quasi-ordered by the minor-relation. This theorem was generalized by Nash-Williams [122] on infinite trees. Recently Daniela Kühn [103] published a proof of his theorem which is considerably shorter than the original proof.

There are some more possibilities to characterize wqo-sets, and we present the most important ones here:

2.3 Theorem and **Definition.** *Let Q be a quasi-ordered set. Then the following conditions are equivalent:*

1) Q is wqo.

2) For every Q-sequence $(a_n)_{n \in \omega}$, which means for every sequence $(a_n)_{n < \omega}$ of elements of Q, there exists an infinite subsequence $(a_{n_i})_{i < \omega}$ such that $a_{n_i} \leq a_{n_j}$ for $i < j < \omega$. Such a sequence is called a perfect subsequence.

3) For every Q-sequence $(a_n)_{n < \omega}$ there exist two indices $i < j$ such that $a_i \leq a_j$ holds.

A Q-sequence $(a_n)_{n\in\omega}$, for which there exist two indices $i < j$ with $a_i \leq a_j$, is called *good*. If not it is called *bad*. So 3) can be stated as :

Every Q-sequence $(a_n)_{n\in\omega}$ is good.

4) *There is no infinite strictly increasing sequence of final segments of Q (with respect to the order $\subseteq$ by inclusion).*

5) *There is no infinite strictly descending sequence of initial segments of Q.*

6) *Any subset A of Q has a finite coinitial subset.*

Proof. We make a ring-proof. 1) $\Longrightarrow$ 2): Suppose Q is wqo, and $(a_n)_{n<\omega}$ is a Q-sequence. Then there exists an index n_0 such that a_{n_0} is minimal in the set $\{a_n \mid n \in \omega\}$ because Q has no infinite descending chain. For the same reason the set $\{a_n \mid n > n_0\}$ has a minimal element a_{n_1} and so on. Then in the sequence $a_{n_0}, a_{n_1}, \ldots$ we have a_{n_i} ($\leq$ or $\|$) a_j for $i < j < \omega$.

Thus by putting $b_i := a_{n_i}$ for $i < \omega$ we have a subsequence $(b_i)_{i<\omega}$ of $(a_n)_{n<\omega}$ where $b_i(\leq$ or $\|$) b_j holds for $i < j < \omega$.

For $i, j \in \omega$ we put iRj iff $i < j \to b_i \leq b_j$. By Ramsey's theorem (see 6.4.8) there exists an infinite subset $\mathbf{N}' \subseteq \mathbf{N}$ such that iRj for all $i \neq j$ of $\mathbf{N}'$ or $i(non\text{-}R)$ j for all $i \neq j$ of $\mathbf{N}'$. In the last case we would also have $b_i \| b_j$ for all $i \neq j$ of $\mathbf{N}'$ and thus an infinite antichain, which is impossible. Therefore the first case must hold and we are done.

2) $\Longrightarrow$ 3) is trivial.

3) $\Longrightarrow$ 4). Suppose indirectly that $F_0 \subset F_1 \subset \cdots \subset F_n \subset \cdots$ is a strictly increasing sequence of final segments of Q. For $i \in \omega$ we choose an element $x_i \in F_{i+1} \backslash F_i$. Due to 3) there exist $i < j$ such that $x_i \leq x_j$. This implies $x_j \in F_i$, and this contradicts $x_j \in F_{j+1} \backslash F_i$.

4) $\Longrightarrow$ 5). If $(I_\nu)_{\nu<\omega}$ would be a strictly descending sequence of initial segments, $(Q \setminus I_\nu)_{\nu<\omega}$ would be a strictly increasing sequence of final segments in contradiction to 4).

5) $\Longrightarrow$ 6). Let A be a non-empty subset of Q. Then the set M of those elements which are minimal in A (with its induced order) is coinitial in A and satisfies $M \subseteq A \subseteq F(M)$, where $F(M)$ is the final segment generated by M. Also M is finite. For if M has an infinite subset $\{m_\nu \mid \nu < \omega\}$, then the initial segments $\mathrm{IS}(\{m_\nu \mid n \leq \nu < \omega\})$, $n \in \omega$, form an infinite sequence of initial segments of Q which is strictly descending because of $m_n \notin \mathrm{IS}(m_\nu \mid n+1 \leq \nu < \omega)$ contradicting 5).

6) $\Longrightarrow$ 1). Let A be an antichain of Q. Then A has a finite coinitial subset $B \subseteq A$. So for every $a \in A$ there exists an element $b \in B$ with

$b \leq a$. This b must be $= a$ because $b < a$ cannot hold in an antichain, and therefore A is $\subseteq B$ and finite.

If Q would have an infinite descending chain D of elements $a_0 > a_1 > \cdots$, then D had to contain a finite coinitial subset of D, which of course is impossible.

The property, which was mentioned in 2.3.2, can be generalized to more general transfinite sequences:

2.4 Theorem. *Let $(Q, \leq)$ be a well-quasi-ordered set, and let $(u_\nu | \nu < \omega_\alpha)$ be a transfinite sequence of elements of Q, where ω_α is an initial ordinal. Then there exists a subsequence $(x_\nu | \nu < \omega_\alpha)$ of $(u_\nu | \nu < \omega_\alpha)$ such that $\mu < \nu < \omega_\alpha \implies x_\mu \leq x_\nu$.*

Proof. Let $\mathfrak{A}$ be the set of all 2 - element subsets of ω_α, and $\mathfrak{B}$ (resp. $\mathfrak{C}$) that subset of $\mathfrak{A}$, which contains all those $\{\mu, \nu\} \in \mathfrak{A}$ for which
$$\mu < \nu \implies u_\mu \leq u_\nu \text{ (resp. } u_\mu > u_\nu).$$
Due to $\aleph_\alpha \to (\aleph_\alpha, \aleph_0)^2$ (see 6.4.10) there exists a) a subset $B \subseteq \omega_\alpha$ of cardinality $\aleph_\alpha$ for which $[B]^2 \subseteq \mathfrak{B}$ holds, so that for $\mu, \nu \in B$ we have $\mu < \nu \implies u_\mu \leq u_\nu$, or b) there exists a subset $C \subseteq \omega_\alpha$ of cardinality $\aleph_0$ such that $[C]^2 \subseteq \mathfrak{C}$ holds.

The case b) cannot happen. For in this case there would exist an infinite strictly descending sequence in Q, which is impossible.

The importance of the concept pwo stems among others from the fact that every linear extension of a pwo-set P yields a well-ordering of P. We prove this with the following lemma:

2.5 Lemma. *Let $(P, \leq)$ be a poset which has an infinite antichain A. Then there exists a linear extension of $\leq$ which is not a well-ordering.*

Proof. We choose a linear order $\leq_A$ of A, which is no well-ordering, e.g. one in which A has a subset of type h_0. Then by 2.3.3 there exists a linear order on P which extends $\leq$ and whose restriction on A is $\leq_A$. And so it cannot be a well-ordering.

Now we prove the characterization theorem of Wolk [179] for pwo-sets, which justifies the choice of the naming:

2.6 Theorem. *A poset $(P, \leq)$ is pwo iff every linear extension of $\leq$ is a well-ordering of P.*

Proof. Let $(P, \leq)$ be pwo and $\leq_L$ a linear extension of $\leq$. Suppose indirectly that $x_1 >_L x_2 >_L \cdots >_L x_n >_L \cdots$, $n \in \mathbf{N}$, is a strictly decreasing sequence in $\leq_L$. Then we consider the set $X := \{x_\nu | \nu < \omega\}$ with the relation R (of strict comparability with respect to $\leq$) : For $x \neq y$ of X we put

$$x \, R \, y \iff x < y \text{ or } y < x.$$

Then from Ramsey's theorem there follows the existence of an infinite subset $I \subseteq X$ such that every two elements of I are in relation R to one another, or every two elements of I are not in relation R. The first case cannot happen because this would yield the existence of an infinite chain of $(P, \leq)$ with elements in $\{x_\nu | \nu < \omega\}$. This would be strictly descending, which is impossible. Also the second case is impossible because then we would have an infinite antichain in $(P, \leq)$.

Suppose now that every linear extension of $\leq$ is a well-ordering of P. Then, of course, also $(P, \leq)$ has no infinite strictly descending chain. And P has no infinite antichain because of 2.5.

A similar characterization theorem can be formulated for well-founded posets:

2.7 Theorem. *A poset $(P, \leq)$ is well-founded iff there exists a linear extension of $\leq$ which is a well-ordering of P.*

Proof. If we have a linear extension of $\leq$ which is a well-ordering, then a fortiori $(P, \leq)$ cannot have an infinite strictly decreasing sequence, and is thus well-founded.

On the other hand let $(P, \leq)$ be well-founded. And let L_ν, $\nu < h(P)$, be the levels of P. Each of them is an antichain of P, and we choose a well-ordering $\leq_\nu$ in L_ν for each $\nu < h(P)$. Then we form the ordered sum $\sum \{(L_\nu, \leq_\nu) | \nu < h(P)\}$. This furnishes a well-ordering on P which extends the order $\leq$ of P. This is an immediate consequence of 1.4 c).

Now we consider the question whether the properties well-founded (resp. pwo) are also productive, which means: If all factors of a cartesian product X have such a property, does X have this property as well. Here we only have that the properties well-founded, wqo and pwo are "finitely productive".

2.8 Theorem. *Let $W_1, \ldots, W_n$ be well-founded quasi-ordered sets. Then the ordered product $X := W_1 \times \cdots \times W_n$ is also well-founded.*

Proof. Let $(x_{i1}, \ldots, x_{in})$, $i \in \mathbf{N}$, be a weakly descending sequence of X, which means for $i < j$ in $\mathbf{N}$ we have $(x_{i1}, \ldots, x_{in}) \geq (x_{j1}, \ldots, x_{jn})$.

Then for every $\nu \in \{1, \ldots, n\}$ the sequence $(x_{i\nu})_{i\in\mathbf{N}}$ is weakly descending, and thus there exists an index $i(\nu) \in \mathbf{N}$ such that all $x_{i\nu}$ with $i \geq i(\nu)$ are equal, say $= x_\nu^*$. Then for all $i \geq \max\{i(\nu)|\nu \in \{1, \ldots, n\}\}$ all $(x_{i1}, \ldots, x_{in})$ are $= (x_1^*, \ldots, x_n^*)$, and thus X is well-founded.

2.9 Remark. If in 2.8 we have more than finitely many well-founded sets, their ordered product is not necessarily also well-founded. If e.g. W_n is the linearly ordered set $\{0, 1\}$ (with $0 < 1$) for $n \in \mathbf{N}$, then the ordered product $W_1 \times W_2 \times \cdots \times W_n \times \cdots$ is not well-founded. The infinite set of sequences $(111\ldots)$, $(01111\ldots)$, $(001111\ldots)$, $(0001111\ldots)$, $\ldots$ is strictly descending.

2.10 Theorem ([119], [123]). *Let* $W_1, \ldots, W_n$ *be finitely many wqo-sets. Then their ordered product* $X := W_1 \times \cdots \times W_n$ *is wqo.*
Then also pwo is finitely productive.

Proof. For $i \in \mathbf{N}$ let $(a_{i1}, \ldots, a_{in})$ be an element of X. By 2.3, 3) we are done if we can prove that there exist two indices $i < j$ such that $(a_{i1}, \ldots, a_{in}) \leq (a_{j1}, \ldots, a_{jn})$ holds. The set $\{a_{i1}|i \in \mathbf{N}\}$ of first components is wqo, and so there exists by 2.3, 2) an infinite subset $N_1 \subseteq \mathbf{N}$, such that the sequence $\{a_{i1}|i \in N_1\}$ is increasing with i.

Analogously there exists an infinite subset $N_2 \subseteq N_1$ such that $\{a_{i2}|i \in N_2\}$ is increasing with i, and so on. After finitely many steps we obtain a subset $N_n \subseteq \cdots \subseteq N_1$ such that for every $\nu = 1, \ldots, n$ the sequence $\{a_{i\nu}|i \in N_n\}$ is increasing with i. This entails that the elements $(a_{i1}, \ldots, a_{in})$, $i \in N_n$, form a sequence in X which increases. By 2.3, 2) then X is wqo.

In connection with 2.8 and 2.10 the question arises whether the property to have only finite antichains is productive. An easy counterexample shows that this is not the case, even if we have only two factors:

2.11 Example. Let $\leq$ be the usual order in the set $\mathbf{N}$ of the natural numbers and $\leq^*$its reverse order. Then the order-product $(\mathbf{N}, \leq) \times (\mathbf{N}, \leq^*)$ has an infinite antichain, e.g. the set of all pairs (x, x) with $x \in \mathbf{N}$. Indeed, if $x < y$ in $\mathbf{N}$, then $(x, x) \parallel (y, y)$ holds because of $y <^* x$.

The following theorem was observed by Wolk [179]:

2.12 Theorem. *Let* $(P, \leq)$ *be a poset of finite dimension* $d > 0$. *Then there holds:*

P is pwo $\iff$ P is embeddable in an ordered product of finitely many well-ordered sets.

Proof. Let P be pwo. The order $\leq$ is $= \cap_{i=1}^{d} \leq_i$ where the $\leq_i, i = 1, \ldots, d$, are linear orders, and then also well-orderings by 2.6, of P. Let X be the ordered product $(P, \leq_1) \times \cdots \times (P, \leq_d)$. Then the mapping $x(\in P) \longrightarrow (x_1, \ldots, x_n)$, where $x_i = x$ for $i = 1, \ldots, n$, is an isomorphism of P onto a subset of X.

$\Longleftarrow$: The product of finitely many well-ordered sets is pwo by 2.10, and this also holds for all of its subsets.

Considering wqo-sets instead of pwo- sets does not yield a real gain in generality, as the following trivial theorem shows:

2.13 Theorem. *Let $(X, \prec)$ be a quasi-ordered set and $(S, \leq)$ its corresponding ordered set (of equivalence classes according to 1.4.9). Then there holds:*

X is wqo $\iff$ S is pwo.

Proof. Let X be wqo and let $C_1, C_2, \ldots$ be a sequence of S. We choose an element $c_i \in C_i$ for every $i \in \mathbf{N}$. Then there exist indices $i < j$ with $c_i \prec c_j$ and this entails $C_i \leq C_j$. And thus S is pwo by 2.3, 3).

Suppose S is pwo, and $c_1, c_2, \ldots$ a sequence in X, C_i the c_i containing class of S. Then there exist indices $i < j$ with $C_i \leq C_j$ and this implies $c_i \prec c_j$.

In well-founded posets $(W, \leq)$ we have the trivial relation $|h(W)| \leq |W|$. In the much more special situation of pwo-sets we even have:

2.14 Theorem. *For the height h of an infinite pwo-set P there holds $|h| = |P|$.*

Proof. P is the union of its $|h|$ levels, which are antichains of P and thus finite. Therefore $|h|$ must be infinite. Each non-empty level contains at least one element of P; so there is an injective mapping of the set of all levels into P, and $|h| \leq |P|$ follows. On the other hand we have $|P| \leq |h| \cdot \aleph_0 = |h|$, and so we obtain $|h| = |P|$.

8.3 Partial ordinals

The class of order types is quasi-ordered by the embedding relation, see 1.9.6. This is reflexive and transitive, but not antisymmetric. However the class of the types of well-ordered sets, which is nothing else than the class of ordinals, is well-ordered. This fact can be embedded into a more general context. Following Wolk [179] we define:

3.1 Definition. An order type of a pwo-set is called a *partial ordinal*. Of course, every ordinal is also a partial ordinal.

If σ and τ are partial ordinals, we put $\sigma \leq_w \tau$ if the following holds: If S (resp. T) is a poset of type σ (resp. τ), then there exists an isomorphism of S onto an initial segment of T.

The relation $\leq_w$ is a subrelation of the relation $\leq$ of Definition 1.9.6, which was not anti-symmetric. Now it follows immediately that $\leq_w$ is reflexive and transitive. We shall prove that it is also antisymmetric. First we show:

3.2 Theorem (Wolk [179]). *Let P be a pwo-set, f an isomorphism of P onto an inital segment $f(P)$ of P. Then $f[P] = P$.*

Proof. Let f^0 be the identity function on P, and if for an $n \in \omega$ we have already defined f^n we put $f^{n+1} := f(f^n)$. We suppose indirectly that $f[P] \subset P$ holds. Then

(1) $f^n[P] \subset f^{n-1}[P]$

holds for $n = 1$. Suppose that (1) holds for a fixed $n \in \omega$. Since f is injective this entails $f[f^n[P]] \subset f[f^{n-1}[P]]$, so that (1) also is valid for $n + 1$ and then by induction for all natural numbers $n \geq 1$. Further there holds for every $n \in \omega$:

(2) $f^n[P]$, $n \in \omega$, is an initial segment of P.

For $n = 1$ this is true by hypothesis. Suppose now that (2) holds for a fixed $n \geq 1$ of ω, and let $x \in f^{n+1}[P]$, $y \in P$ and $y < x$. We have to show that $y \in f^{n+1}[P]$. We have $x \in f[f^n[P]]$ and thus $x \in f[P]$. From $y < x$ we obtain $y \in f[P]$ and then $f^{-1}(y) < f^{-1}(x) \in f^n[P]$. Since $f^n[P]$ is an initial segment of P we also have $f^{-1}(y) \in f^n[P]$. Finally $y = f(f^{-1}(y)) \in f[f^n[P]] = f^{n+1}[P]$.

Now (1) and (2) yield that $(f^n[P]|n \in \omega)$ is a strictly decreasing sequence of initial segments of P, and this contradicts 2.3, 5).

The next two theorems are listed as corollaries in [179]:

3.3 Theorem. *If P and Q are pwo-sets, and each is isomorphic to an initial segment of the other, then P and Q are isomorphic. And therefore the relation $\leq$ of Definition 3.1 is also antisymmetric.*

Proof. We have an isomorphism $f : P \to I_q$, where I_q is an initial segment of Q, and an isomorphism $g : Q \to I_p$, where I_p is an initial segment of P. Then $g[f[P]]$ is a subset of I_p and moreover an initial segment of this set, and then also of P. Since it is isomorphic to P it is $= P$ by 3.2. So we obtain $g[f[P]] = P$. This entails $f[P] = Q$, for otherwise $g[f[P]]$ would be a proper subset of P.

3.4 Theorem. *If A and B are distinct initial segments of a pwo-set P and A is isomorphic to B, then A and B are incomparable sets.*

Proof. If A would be a subset of B, A would also be an initial segment of B. Then B would be isomorphic to its initial segment A and hence $= A$ by 3.2, contradicting $A \neq B$.

Instead of the relation $\leq_w$ which was introduced in 3.1 one could also have a look at the following relation $\preceq$: If σ and τ are partial ordinals we put $\sigma \preceq \tau \iff$ If S and T are posets with types σ resp. τ, then S is isomorphic to a subset of T. Then $\preceq$ is reflexive and transitive and thus a quasi-order. But $\preceq$ is not antisymmetric, as the following example shows:

3.5 Example. We consider the following two pwo-sets A and B, whose diagrams are sketched in Figure 16, where an element x of A resp. B is $\leq$ an element y of the same set, if x is situated strictly left of y:

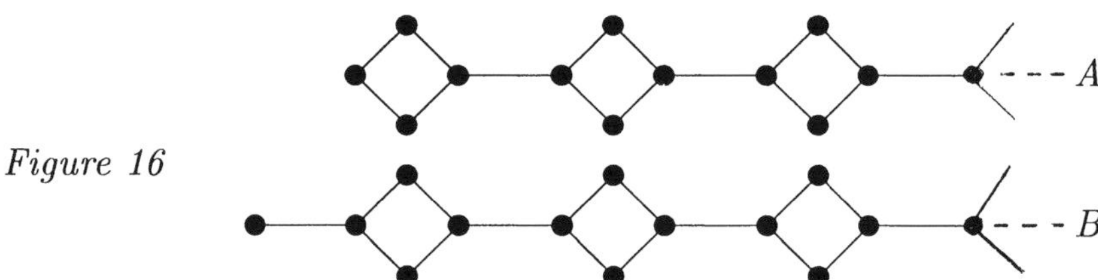

Figure 16

Each of these two sets is embeddable in the other, but they are not isomorphic.

A counterexample with finite sets is, of course, not possible because an isomorphism of a finite set into another finite set and an isomorphism of this into the first set is only possible if both sets have the same cardinality.

The counter-example exhibits a difference between the ordinals and the partial ordinals because for ordinals there holds: If σ and τ are ordinals, and if σ is isomorphic to a subset of τ, and if τ is isomorphic to a subset of σ, then $\sigma = \tau$.

The ordered sum of well-ordered sets over a well-ordered index set is again a well-ordered set. This fact can be generalized to:

3.6 Theorem. *Let $(I, \preceq)$ be a pwo-set and $(P_i, \leq_i)$, $i \in I$, pairwise disjoint pwo-sets. Then the ordered sum (see 4.1.1) $(\Sigma, \leq) := \sum_{i \in I}(P_i, \leq_i)$ is pwo.*

Proof. Let $(a_n | n \in \omega)$ be an ω - sequence of elements of Σ. If the set $A := \{a_n | n \in \omega\}$ intersects infinitely many of the sets P_i, then there exists an infinite subset $B := \{b_n | n \in \omega\}$ of A, such that every two of the $b's$ are in different sets P_i. If then $b_m \in P_i$, $b_n \in P_j$, then $b_m \leq b_n$ holds iff $i \preceq j$, and so the order of Σ, restricted to B, is pwo like that of I, and so there exist two indices $k < l$ with $b_k \leq b_l$.

If A intersects only finitely many of the sets P_i. then one of them, say P_j, contains a_n for infinitely many $n \in \omega$. And since P_j is pwo, there exist again indices $k < l$ with $a_k \leq_j$ (and then also $\leq$) a_l.

8.4   The theorem of De Jongh and Parikh

In 2.6 we saw that every linear extension of a pwo-set is well-ordered, so that its order type is an ordinal. The question arises whether the supremum of the set of the ordinals which occur as such types is also an order type of a linear extension of the pwo-set. The theorem of de Jongh and Parikh [21] gives an affirmative answer to this question. We mention first a theorem which also has interest in itself:

4.1 Theorem [21]. *Let $(Y, \leq_0)$ be a poset, $Z \subseteq Y$ equipped with the restriction $\leq_0' := \leq_0 \restriction Z$. Let further $\leq_1$ be an order on Z which satisfies $\leq_1 \supseteq \leq_0'$. Then the transitive hull T of $\leq_0 \cup \leq_1$ is again an order on Y, and of course the least one, which contains $\leq_0$ and $\leq_1$.*

Proof. It is trivial that T is reflexive and transitive. We prove its antisymmetry: Suppose indirectly that a,b are different elements of Y for which aTb and bTa holds. From aTb follows the existence of a sequence $a = a_1 T a_2 T \cdots T a_n = b$, where n is an integer ≥ 2 and where every sign T is to be replaced by one of the signs $<_0, <_1$. An analogous remark concerns bTa. And so we have a sequence

(*) $a = a_1 T a_2 T \cdots T a_n = b T a_{n+1} T \cdots T a_m = a$, where every T is to be replaced by one of the signs $<_0, <_1$.

If from the signs $<_0, <_1$ which replace the T's in (*) we have one, which occurs twice in succession, we can omit one of them since $<_0$ and $<_1$ are transitive, and so shorten the sequence (*). In this way we can obtain a sequence of type (*) of minimal length. If this would be $a <_i a$ with an $i \in \{0, 1\}$, we would already have a contradiction. In the other case the minimal sequence would have a form $a = a_1 <_i a_2 <_{1-i} a_3 <_i a_4 <_{1-i} \cdots <_j a_n = a$ with $n \geq 3$ and $i, j \in \{0, 1\}$, where the signs $<_0$ and $<_1$ are used alternately, and where the $a_1, \ldots, a_n$ are pairwise different by exception of $a_1 = a_n$. Now for every $i \in \{2, \ldots, n-1\}$ we have $a_i \in Z$ since $a_{i-1} <_1 a_i$ or $a_i <_1 a_{i+1}$ must hold. For in the first case a_{i-1} and a_i must be in Z, in the second a_i and a_{i+1}.

So the elements a_i, $i = 2, \ldots, n-1$, form a chain $\subseteq Z$ with respect to $\leq_1$ since $\leq_0 \restriction Z$ is contained in $\leq_1$. If now a is also in Z, then $a = a_1 <_1 a_n = a$ is the desired contradiction. If $a \notin Z$ we have $a <_0 a_2$ and $a_{n-1} <_0 a$, and so $a_{n-1} <_0 a_2$ due to the transitivity of $<_0$. Now also $a_{n-1} <_1 a_2$ holds in contradiction to $a_2 \leq_1 a_{n-1}$.

4.2 Corollary. *Let $(Y, \leq)$ be a poset, $Z \subseteq Y$. Let L_Z be a linear extension of $\leq \restriction Z$. Then there exists a linear extension of $\leq$ which contains L_Z.*

Proof. By 2.3.2 the transitive hull of $\leq \cup L_Z$ is an order and can be extended to a linear order of Y by 2.3.2.

We now define several concepts which are essential for the proof of the theorem of de Jongh and Parikh:

4.3 Definition. Let $(X, \leq)$ be a pwo-set. We define $o(X, \leq)$, in short $o(X)$, as the supremum of all ordinals τ, such that $(X, \leq)$ has a linear extension (which is well-ordered by 2.6) of type τ.

For $x \in X$ we put $L(x) := \{y \in X \mid y < x \text{ or } y \parallel x\}$ $(=: (X \not\geq x))$ and $l(x) := o(L(x)) = \sup\{\tau \mid L(x)$ (with the order induced from $\leq$) has a linear extension of type $\tau\}$.

An element $x \in X$ is said to be *superfluous*, if $o(L(x)) = o(X)$. Later it turns out that superfluous elements don't exist in pwo's, but for the moment we have to take into account the possibility that they do exist.

In this context we have the following:

4.4 Lemma [21]. *Let $(X, \leq)$ be a pwo-set. Then there exists a subset $Y \subseteq X$ with $o(Y) = o(X)$, which has no superfluous elements.*

Proof. We define inductively sets $X_0, X_1, \ldots$ and elements $x_0, x_1, \ldots$ $\in X$ as follows: $X_0 := X$. If X has a superfluous element x_0, so that $o(X_0) = o(X_0 \not\geq x_0)$ holds, then we put $X_1 := (X_0 \not\geq x_0)$ and have $o(X_1) = o(X_0)$.

If in general we have already defined $X_n \subseteq X$ for an $n \in \omega$ such that $o(X_n) = o(X)$, and if X_n still has superfluous elements, we choose one of them, say x_n, and put $X_{n+1} := (X_n \not\geq x_n)$. Then $o(X_{n+1}) = o(X_n \not\geq x_n) = o(X_n) = o(X)$. This construction stops after finitely many steps. For we have for the sequence $x_0, x_1, \ldots$ the relation: $i < j \implies x_i \not\leq x_j$, because all elements of X_{i+1}, and a fortiori all of X_j are $\not\geq x_i$. According to 2.3, 3) then the construction has to break off, say at $X_m =: Y$, which now has no superfluous elements, but satisfies $o(Y) = o(X_m) = o(X_{m-1}) = \cdots = o(X_0) = o(X)$.

4.5 Lemma. *Let $(P, \leq)$ be a poset, τ an ordinal, and $I_\nu, \nu < \tau$, initial segments of P with $\cup\{I_\nu | \nu < \tau\} = P$, where for $\mu < \nu < \tau$ we have $I_\mu \subseteq I_\nu$. For $\nu < \tau$ we put $B_\nu := I_\nu \setminus \cup \{I_\mu | \mu < \nu\}$ and call it the ν - block. Then we define an extension $\leq^* of \leq$ as follows:*

For $x, y \in P$ we put $x \leq^* y \iff x \leq y$, or $x \not\leq y$ and for the blocks B_ξ and B_η which contain ξ resp. η we have $\xi < \eta$. Then $\leq^*$ is an order relation on P.

Proof. By construction we have

(1) If $x \in P$ is in B_ξ, then ξ is the least ordinal with $x \in I_\xi$.

(2) If $x \in B_\xi$, $y \in B_\eta$ and $x \leq y$, then $\xi \leq \eta$.

Indeed, the initial segment I_η contains y and then also x , and so (2) follows from (1). Also there holds:

(3) If $x \in B_\xi$, $y \in B_\eta$ and $x \leq^* y$, then $\xi \leq \eta$.

For if also $x \leq y$ holds we are done due to (2). And if $x \not\leq y$, then by definition $\xi \leq \eta$ follows. In the rest of the proof we always suppose $x \in B_\xi$ and $y \in B_\eta$. By definition $\leq^*$is reflexive.

$\leq^*$is antisymmetric: Let $x \leq^* y$ and $y \leq^* x$. By (3) we have $\xi \leq \eta \leq \xi$ and thus $\xi = \eta$. Then x and y are in the same block, and we have $x \leq y \leq x$ and $x = y$.

$\leq^*$ is transitive: Let $x \leq^* y$ and $y \leq^* z$. Then by (3) $\xi \leq \eta \leq \zeta$ holds, where $z \in B_\zeta$. If also $\xi = \zeta$ is valid, then x, y, z are in the same block, and $x \leq y \leq z$ and $x \leq z$ holds. If we have $\xi < \zeta$ then by definition $x \leq^* z$ follows.

Now we can prove the main theorem of this section:

4.6 Theorem of de Jongh and Parikh [21]. *Let $(X, \leq)$ be a pwo-set. Then there exists a well-ordering $\leq'$of X extending $\leq$ such that $\mathrm{tp}(X, \leq') = o(X, \leq)$.*

Proof. We use the definitions of 4.3. First we can assume that X has no superfluous elements. For otherwise we can find by 4.4 a subset $Y \subseteq X$ with $o(Y) = o(X)$, which has no superfluous elements. If then our theorem is proved for Y, we have a well-ordering $\leq_Y$of Y extending $\leq \restriction Y$ with type $o(Y) = o(X)$, and this can by 4.2 be extended to a linear order $\leq_X$of X, which by 2.6 is automatically a well-ordering. This has a type $\geq o(X)$. This type is trivially also $\leq o(X)$ and then $= o(X)$. And this shows that it suffices to prove:

There exists a subset $S \subseteq X$ with a well-ordering $\leq_S$which extends $\leq \restriction S$ and has type $o(X)$.

If $o(X)$ is a successor ordinal the assertion is trivial, since here supremum is the same as maximum. So let $\lambda := o(X)$ be a limit ordinal and $\omega_\gamma := \mathrm{cf}(\lambda)$.

From the representation of λ in Cantor's normal form we obtain $\lambda = \rho + \omega^\alpha$, where ρ is a sum of finitely many powers ω^ν which all are $\geq \omega^\alpha$, and where α is > 0 since λ is a limit ordinal.

First we show that we can restrict ourselves to the case $\rho = 0$: To this purpose we choose an element z of X with $l(z) > \rho$. This is possible since evidently $o(X) = \sup\{l(x)|x \in X\}$. Since z is not superfluous we have $l(z) < \lambda = \rho + \omega^\alpha$, and so we have $l(z) < \rho + \zeta$ for a $\zeta < \omega^\alpha$. For z we obtain:

(1) $o(X \geq z) \geq \omega^\alpha$.

Suppose the contrary. Then $o(X \geq z) < \eta$ holds for an $\eta < \omega^\alpha$. Every subset W of X with a well-ordering which extends $\leq \restriction W$ is a union of a subset of $L(z)$ and a subset of $(X \geq z)$, and so its order type is a mixed sum of an ordinal $\leq l(z) < \rho + \zeta$ and an ordinal $\leq o(X \geq z) < \eta$. And so its order type is by 4.8.11 and 4.8.13 $\leq (\rho + \zeta) \oplus \eta \leq \rho \oplus \zeta \oplus \eta = \rho \oplus (\zeta \oplus \eta) < \rho + \omega^\alpha$, since $\zeta \oplus \eta < \omega^\alpha$ holds, and we have a contradiction to $o(X) = \rho + \omega^\alpha$.

So it remains to prove that $(X \geq z)$ has a subset T which can be well-ordered in extension of $\leq \restriction T$ such that it has the type ω^α, since a well-ordered subset of $L(z)$ of type ρ,which extends $\leq$, can then be prolonged by T to furnish a well-ordered subset of X, extending $\leq$, of type $\rho + \omega^\alpha = o(X)$.

Due to (1) we can from now on assume $\lambda = \omega^\alpha$.

Let $(\tau_\nu | \nu < \omega_\gamma)$ be a strictly ascending sequence of ordinals whose supremum is $o(X) = \lambda = \omega^\alpha$. For $\nu < \omega_\gamma$ let u_ν be an element of X with $l(u_\nu) \geq \tau_\nu$. Then by 2.4 the sequence $(u_\nu | \nu < \omega_\gamma)$ has a subsequence $(t_\nu | \nu < \omega_\gamma)$ such that $\mu < \nu < \omega_\gamma \implies t_\mu \leq t_\nu$ holds. In particular then $\{t_\nu | \nu < \omega_\gamma\}$ is a chain of X, and $\sup\{l(t_\nu) | \nu < \omega_\gamma\} = \lambda = \omega^\alpha$. Since the t_ν are not superfluous all $l(t_\nu)$, $\nu < \omega_\gamma$, are $< \lambda = \omega^\alpha$, and their set is cofinal in ω^α.

We finally choose a subsequence $(x_\nu | \nu < \omega_\gamma)$ of $(t_\nu | \nu < \omega_\gamma)$, for which we have:

(2) $\mu < \nu < \omega_\gamma \implies x_\mu < x_\nu$ and $l(x_\mu) < l(x_\nu)$, and $\sup\{l(x_\nu) | \nu < \omega_\gamma\} = \omega^\alpha = \lambda$.

To this purpose let E be the set of all those t_ν (with $\nu < \omega_\gamma$ and) for which $l(t_\nu) > l(t_\mu)$ holds for all $\mu < \nu$. Their set E, ordered by magnitude, yields a sequence $(x_\nu | \nu < \omega_\gamma)$ which satisfies (2).

We distinguish two cases:

Case a) α is a successor number $\beta + 1$, so that $\omega^\alpha = \omega^\beta \cdot \omega$. Then cf $\lambda = \omega$, because the set $\{\omega^\beta \cdot n | n \in \omega\}$ has $\lambda = \omega^\alpha$ as supremum. Using (2) we choose for every $i \in \omega$ an element $y_i \in \{x_\nu | \nu < \omega_\gamma\} := \mathfrak{x}$ and a number $n_i \in \omega$ such that for all $i \in \omega$ there holds:

(3) $\omega^\beta \cdot n_i < l(y_i) < \omega^\beta \cdot n_{i+1}$.

First we choose $n_0 = 0$, then a suitable y_0 in $\mathfrak{x}$, then n_1 so that (3) is fulfilled for $i = 0$, which is possible since X has no superfluous elements. Then $y_1 \in \mathfrak{x}, n_2, y_2 \in \mathfrak{x}, n_3, \ldots$ are determined successively. From (3) we now obtain for every $i \in \omega$ the existence of a subset $W_i \subseteq L(y_i)$ with a well-ordering extending $\leq \upharpoonright W_i$ of type $\omega^\beta \cdot n_i$. We put $Y_{i+1} := L(y_{i+1}) \backslash L(y_i)$ for $i \in \omega$. Then Y_{i+1} contains all elements of X, which are $\geq y_i$ and $\not\geq y_{i+1}$. And for $i < j < \omega$ we have $y_i < y_j$ since the $y's$ are in the chain $\mathfrak{x}$ and due to (3).

We have $W_{i+1} \cap Y_{i+1} \neq \emptyset$. Otherwise $W_{i+1} \subset L(y_i)$ would hold because of $W_{i+1} \subseteq L(y_{i+1})$. But W_{i+1} has type $\omega^\beta \cdot n_{i+1} > l(y_i)$ by (3), a contradiction. So for $i \in \omega$ we can choose an element $p_{i+1} \in W_{i+1} \cap Y_{i+1}$, for which now also $p_{i+1} \geq y_i$ holds. Finally for $i \in \omega$ we put $F_{i+1} := (W_{i+1} > p_{i+1})$ This final segment of W_{i+1} is $\subseteq Y_{i+1}$ and has an order type $\geq \omega^\beta$ by 4.8.12. Now we consider the ordered sum $S := \sum\{F_{i+1} | i \in \omega\}$. The summands are pairwise disjoint, the $L(y_i)$ are initial segments of X, and so S is by 4.5 a well-ordered set of type

$\geq \omega^{\beta} \cdot \omega = \omega^{\alpha}$, whose order is an extension of $\leq\restriction S$. Thus case a) is proved.

Case b) α is a limit ordinal. Then we choose successively for every $i \in \omega_{\gamma}$ an element $y_i \in X$ and an ordinal ω^{ν_i} such that for all $i \in \omega_{\gamma}$ there holds:

(4) $\omega^{\nu_i} < l(y_i) < \omega^{\nu_i+1} \geq \tau_i$.

Similarly to case a) we can determine successively by transfinite induction elements $y_i \in \{x_\nu | \nu < \omega_\gamma\}$ and ordinals ν_i, beginning with $\nu_0 = 0, y_0, \nu_1, y_1, \dots$, which satisfy (4).

Due to (4), there exists for every $i < \omega_\gamma$ a subset $W_i \subseteq L(y_i)$ with a well-ordering on W_i of type ω^{ν_i} which extends $\leq\restriction W_i$. And we have:

(5) For $i < \omega_\gamma$ there holds
$F_{i+1} := W_{i+1}\backslash L(y_i) = W_{i+1}\backslash(W_{i+1} \cap L(y_i))$ has type ω^{ν_i+1}. And

(6) $F_{i+1} \subseteq L(y_{i+1})\backslash L(y_i)$.

Indeed, we have $\mathrm{tp}W_{i+1} = \omega^{\nu_i+1}$ and $\mathrm{tp}(W_{i+1} \cap L(y_i)) \leq l(y_i) < \omega^{\nu_i+1}$, and then (5) follows from 4.8.7. Now, using 4.5 we see that the ordered sum $\sum\{F_{i+1} | i < \omega_\gamma\}$ is a well-ordered extension of $\leq$ on this set. By (6) its summands are pairwise disjoint and thus the sum has the type $\cup\{\omega^{\nu_i+1} | \nu < \omega_\gamma\} = \omega^{\alpha} = o(X)$.

4.7 Remark. We mentioned before that superfluous elements don't exist. Now, where we can use the last theorem, this easily follows: If x would be a superfluous element of X, then there exists a set $W \subseteq L(x)$ with a well-ordering $\leq_w$ which extends $\leq\restriction W$ of type $o(X)$. Then we can introduce an order $\leq_x$ in the set $W \cup \{x\}$ which extends $\leq$ on this set and $\leq_w$, and in which x is the last element. This set would have the type $o(X) + 1$, which is impossible.

8.5 On the structure of $\mathfrak{I}(P)$, where P is well-founded or pwo

In many cases it gives useful insight into the structure of an ordered set if we consider the set of its initial segments with its natural order by inclusion. For this we use the following notation:

5.1 Definition. Let $(P, \leq)$ be a poset. Then the set of all initial segments of P is denoted by $\mathfrak{I}(P)$. And this set is always considered as equipped with the order by inclusion.

Of course there are intensive connections between the order of P and the order of $\mathfrak{I}(P)$, and in general every time, in introducing a new concept, the question arises whether the properties of P are shared by $\mathfrak{I}(P)$. We investigate this now for the notions well-founded and pwo. First we mention a version of König's Infinity Lemma:

5.2 Theorem. *Let S_i, $i \in \omega$, be pairwise disjoint non-empty finite sets and R a relation on $U := \cup\{S_i | i \in \omega\}$, for which the following holds: For every $i \in \omega$ every element of S_{i+1} is in relation R to at least one element of S_i. Then there exist elements $a_i \in S_i$ for $i \in \omega$ such that $a_{i+1}R\,a_i$ holds for all $i \in \omega$.*

Proof. For every $n \in \omega$ and $x_n \in S_n$ there exists a set $\{x_0, \ldots, x_n\}$ with $x_i \in S_i$ for $i = 0, \ldots, n$ such that $x_{i+1}Rx_i$ holds for $i < n$. We call this set a *path* from x_0 to x_n. Since S_0 is finite it follows from the pigeonhole principle that the set of all paths from elements of S_0 to elements which are not in S_0 has an infinite subset of paths which all share the same element, say a_0,of S_0. Then we consider all paths from a_0 to elements which are not in S_0. Infinitely many of them share the same element, say a_1,of S_1. Continuing in this way we can successively find elements $a_i \in S_i$ for $i \in \omega$ such that $a_{i+1}R\,a_i$ holds for all $i \in \omega$.

An initial segment I of a poset P is called *finitely generated*, if there exists a finite subset $F \subseteq I$ which is cofinal in I. Now there follows a theorem of Birkhoff [9]:

5.3 Theorem. *Let $(W, \leq)$ be a well-founded poset. Then the set $\mathfrak{F}$ of its finitely generated initial segments, equipped with the order $\subseteq$, is well-founded.*

Proof. Suppose indirectly that $(I_n | n \in \omega)$ is a strictly descending sequence of elements of $\mathfrak{F}$. For $n \in \omega$ let M_n be a subset of I_n of minimal size which generates I_n. Then M_n is finite, and every element of M_{n+1} is $\leq$ an element of M_n. Next we intend to reduce the system of the M_n to a sequence of pairwise disjoint sets. We call a sequence $S_0, S_1, \ldots$ of subsets of W *special*, if the following holds:

(*) All S_n, $n \in \omega$, are finite and non-empty, and every $x_{n+1} \in S_{n+1}$ is $\leq$ an element $x_n \in S_n$.

So $(M_n | n \in \omega)$ is special. If an element $e \in M_0$ is contained in infinitely many of the M_n, we consider the subsequence $(A_0 = M_0), A_1, A_2, \ldots$ of all those sets M_n which contain e. Then we delete e from all sets A_n and obtain a sequence $B_0 (= M_0 \backslash \{e\})$, $B_1, B_2, \ldots$, which is still special. Indeed, for every $b_{n+1} \in B_{n+1}$ there exists a $b_n \in B_n$ with $b_{n+1} \leq b_n$. Indeed, b_{n+1} is $\leq$ an element $x \in A_n$, and x is $\neq e$, because otherwise $b_{n+1} < e \in A_{n+1}$ holds, and then A_{n+1} would not be minimal, for b_{n+1} could be omitted since all elements which are $< b_{n+1}$ are also $< e$.

Continuing in this manner we can eliminate in finitely many steps all those elements of M_0 which occur in infinitely many of the M_n with $n > 0$. And in every step the new sequence remains special. So we obtain a special sequence $C_0, C_1, \ldots$, which begins with a subset $C_0 \subseteq M_0$, such that every element of C_0 occurs only in finitely many of the C_n. Then $(C_n | n \in \omega)$ has a subsequence $(D_n | n \in \omega)$ with $D_0 = C_0$, which is special, and where D_0 is disjoint to all D_n with $n > 0$.

Iterating this process we finally arrive at a sequence $(E_n | n \in \omega)$ which is special, and where all sets E_n are pairwise disjoint. But now by König's Infinity Lemma 5.2 there exist elements $a_n \in E_n$ with $a_{n+1} < a_n$ for $n \in \omega$. They form an infinite decreasing chain in W, a contradiction.

5.4 Remark. In in 5.3 we replace the set $\mathfrak{F}$ by the set $\mathfrak{I}(W)$ of all initial segments, the statement no longer remains true. Let e.g. $\mathbf{Q}$ be the set of rational numbers with the identity relation $\{(x, x) | x \in \mathbf{Q}\}$ as order, then $(\mathbf{Q}, =)$ is well-founded, but the set of its initial segments contains all subsets of $\mathbf{Q}$, and in this we have infinite decreasing sequences.

With a stronger assumption one can prove that $\mathfrak{I}(W)$ is well-founded. In this context Higman [84] proved:

5.5 Theorem. *A poset P is pwo iff $\mathfrak{I}(P)$ is well-founded.*

Proof. Suppose that P is pwo. We assume indirectly that $\mathfrak{I}(P)$ is not well-founded. Then there exists a sequence of initial segments $(I_n | n \in \omega)$ of P which is strictly descending. Now we choose for $n \in \omega$ an element $x_n \in I_n \backslash I_{n+1}$. Then x_n is also not in I_m for $m > n$, and this entails $x_n \nleq x_m$ because otherwise x_n had to be in I_m. But now P could not be pwo because of 2.3, 3).

Suppose that P is not pwo. Then there exists a strictly descending infinite sequence of elements of P, or P has an infinite antichain. If $(x_n | n \in \omega)$ is a strictly descending sequence of P, then the sets $(P < x_n)$, $n \in \omega$, form a strictly descending sequence of elements of $\mathfrak{I}(P)$, which is thus not well-founded.

And if $\{x_n | n \in \omega\}$ is an infinite antichain, we define I_n to be the set of all elements of P, that are $\leq$ some element of $\{x_n, x_{n+1}, x_{n+2}, \ldots\}$. Then the initial segments I_n, $n \in \omega$, form a strictly descending sequence of $\mathfrak{I}(P)$. For $m < n < \omega$ entails, that x_m is not $\leq$ an element of $\{x_n, x_{n+1}, \ldots\}$ so that $I_m \supset I_n$ holds. And again $\mathfrak{I}(P)$ cannot be well-founded.

The property of an ordered set P to be pwo does not entail that also $\mathfrak{I}(P)$ is pwo. The standard counter-example was given by R.Rado [149] and independently by Kruskal (unpublished):

5.6 Example. Let P be the set $\omega \times \omega$. We introduce in P the following relation $\leq_P$: For every $x \in \omega$ and $y \leq z$ of ω we put $(x,y) \leq_P (x,z)$, and for every $(x,y) \in P$ we put $(u,v) \leq_P (x,y)$ if $u + v \leq x$. In Figure 17 all lattice points (= points with integer components) (u,v) of the hatched area are $\leq_P$ all lattice points of the vertical ray r which starts in that point of the x - axis whose first component is x.

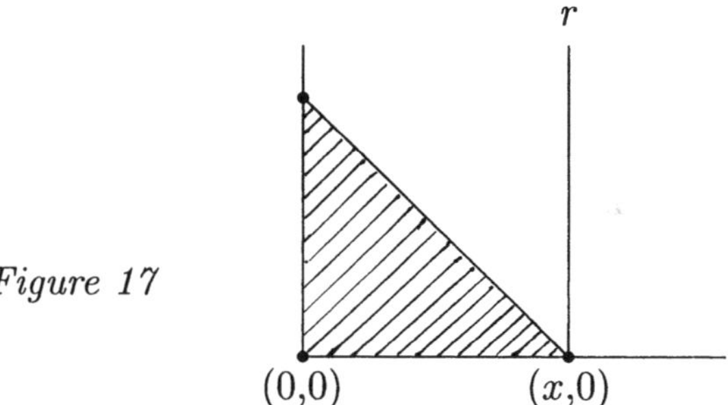

Figure 17

$\leq_P$ is trivially reflexive. And if two points $\mathfrak{p}$ and $\mathfrak{q}$ would satisfy $\mathfrak{p} <_P \mathfrak{q}$ and $\mathfrak{q} <_P \mathfrak{p}$, then they could not lie on the same vertical line, and so the first component of $\mathfrak{p}$ had to be $\leq$ the first component of $\mathfrak{q}$, and conversely, which is impossible. Thus $\leq$ is antisymmetric.

If (a_1, b_1), (a_2, b_2) and (a_3, b_3) are points with $(a_1, b_1) <_P (a_2, b_2)$ and $(a_2, b_2) <_P (a_3, b_3)$, then we have $a_1 \leq a_2 \leq a_3$. If here the signs $\leq$ can be replaced by $=$ we are done. If $a_1 < a_2$ we have $a_1 + b_1 \leq a_2 \leq a_3$. If $a_1 = a_2 < a_3$ holds, it follows $b_1 \leq b_2$ and $a_1 + b_1 \leq a_2 + b_2 \leq a_3$. In both cases $(a_1, b_1) <_P (a_3, b_3)$ follows, and $\leq_P$ is transitive.

Every descending chain of P is finite: This follows a fortiori from the fact that every point (x,y) of P has only finitely many predecessors, namely the lattice points (x,z) with $0 \leq z < y$ and the points of P with component sum $\leq x$.

Finally every antichain of P is finite. If the different points $(a_1, b_1), (a_2, b_2), \ldots$ would form an infinite antichain, where we can assume w.r.o.g. that $a_1 < a_2 < \cdots$ holds, then we would have $a_1 + b_1 > a_n$ for every integer $n \geq 2$. But then the infinitely many a_n could not all be different.

The next theorem presents a situation where pwo is carried over from a set X to $\mathfrak{I}(X)$:

5.7 Theorem (Nash-Williams [123]). *Let k be a positive integer and W_κ, $\kappa = 1, \ldots, k$, well-ordered sets, $X := W_1 \times \cdots \times W_k$ equipped with the product order. Then its set $\mathfrak{I}(X)$ (of initial segments) is pwo.*

Proof. Every initial segment of X has evidently the form $I_1 \times \cdots \times I_k$, where I_κ is an initial segment of W_κ for $\kappa = 1, \ldots, k$. Let $(I_{n1}, \ldots, I_{nk})$, $n \in \omega$, be a sequence of elements of $\mathfrak{I}(X)$. Since $\mathfrak{I}(W_1)$ is pwo there exists by 2.3, 2) an infinite subset $N_1 \subseteq \omega$ such that the sets I_{n1}, $n \in N_1$, increase with n. Analogously there exists an infinite subset $N_2 \subseteq N_1$ such that the sets I_{n2}, $n \in N_2$, increase with n. In k steps we obtain an infinite subset $N_k \subseteq \omega$ such that the k-tuples $(I_{n1}, \ldots, I_{nk})$, $n \in N_k$, increase with n. By 2.3, 2) our statement follows.

5.8 Remark. In [179] Wolk remarked that the last theorem shows that the set P of the counter-example of Kruskal/Rado of 5.6 has infinite dimension. Otherwise P would by 2.12 be embeddable in a product of finitely many well-ordered sets, and then $\mathfrak{I}(P)$ would be pwo, with contradiction.

8.6   Sequences in wqo-sets

If we have two ω - sequences consisting of elements of a poset P, there exist several possibilities to introduce a quasi-order between these sequences which stand in relation to the order of P. In section 4.1 we already studied the ordered product and the lexicographic product of orders. Here we investigate another relation, which, contrary to the before mentioned notions, is only a quasi-order.

6.1 Definition. Let $(Q, \leq)$ be a quasi-ordered set. If α is an ordinal and a_ν, $\nu < \alpha$, are elements of Q (not necessarily pairwise different), we call $A := (a_\nu | \nu < \alpha)$ an α - *sequence* over Q. Its *length* is α. If α

is a finite ordinal, this sequence, of course, is called *finite*. Here $\alpha = 0$ is admitted and describes the *empty sequence*. If $\beta < \alpha$ holds, the sequence $(a_\nu | \nu < \beta)$ is called the β - *segment* of $(a_\nu | \nu < \alpha)$ and a_ν the ν - *component* of it.

A *subsequence* of A is a family $(a_\mu | \mu \in B)$, where $B \subseteq \alpha$ holds.

If now $A = (a_\nu | \nu < \alpha)$ and $B = (b_\nu | \nu < \beta)$ are sequences over P (with possibly different lengths α and β), we put $A \preceq B \iff$ there exists a $<$ - preserving mapping g of α into β such that $a_\nu \leq b_{g(\nu)}$ holds for all $\nu < \alpha$.

Intuitively speaking: $A \preceq B$ holds, iff B has a subsequence S of the same length as A, such that A is $\leq S$ "componentwise". To a certain extent $\preceq$ generalizes the notion of product order.

It is immediately clear that $\preceq$ is reflexive and transitive. But, contrary to the product order, it is not anti-symmetric. If e.g. A and B are ω - sequences of $2((\omega_0))$ which both have infinitely many 0's and 1's, then we have $A \preceq B$ and $B \preceq A$.

If W is a finite sequence over a quasi-ordered set Q, it is in short called a *word* over Q.

6.2 Theorem. *Let P be a well-founded quasi-ordered set. Then the set $\mathfrak{W}(P)$ of all words over P is well-founded.*

Proof. Let S be a non-empty set of non-empty words of $\mathfrak{W}(P)$, S' the subset of those words of S which have minimal length l. The set of the 0 - components of the words of S' is $\subseteq P$ and thus well-founded, and so it has a minimal element m_0. Let M_0 be the set of all words of S' which have m_0 as 0 - component. Then the set of 1 - components (if $l \geq 2$) of the words of M_0 has a minimal element m_1. Let M_1 be the subset of those elements of M_0 which have m_1 as 1 - component, that means which begin with m_0, m_1. After l steps we have obtained a word $(m_0, \ldots, m_{l-1})$, which is minimal in S, and so $\mathfrak{W}(P)$ is well-founded. by 1.2.

An important theorem of Higman [84] is now:

6.3 Theorem. *Let Q be a wqo-set. Then the set $\mathfrak{W}(Q)$ of all words over Q is wqo.*

Proof (by Nash-Williams [123]) Suppose that $\mathfrak{W}(Q)$ is not wqo, so that the set B of bad sequences in $\mathfrak{W}(Q)$ is non-empty. Since $\mathfrak{W}(Q)$ is well-founded by 6.2 the set of 0 - components of the sequences of B has a minimal word W_0. Let now $B(W_0)$ be the set of all sequences $(W_0, \ldots)$

264

of B which begin with W_0. Then the set of all 1 - components of the elements of $B(W_0)$ has a minimal element W_1. Let then $B(W_0, W_1)$ be the set of all sequences of $B(W_0)$ that have W_1 in the 1 - component. By induction we so obtain an ω - sequence $W_0, W_1, \ldots$ of words, for which by construction there follows:

(1) $W_0, W_1, \ldots$ is a bad sequence.

Indeed, for $i < j < \omega$ there exists a sequence S in B whose i-component is W_i and whose j-component is W_j, so that $W_i \preceq W_j$ cannot hold since S is bad.

All words $W_n, n \in \omega$, are non-empty, otherwise their sequence would be good. So for all $n \in \omega$ the 0 - component w_n of the word W_n exists. And then the ω - sequence $w_0, w_1, \ldots$ (of elements of Q) has an increasing subsequence by 1.3, 2), and so we have:

(2) There exists a strictly increasing mapping $f : \omega \to \omega$ such that $(w_{f(n)} | n \in \omega)$ increases with n.

We define for $n \in \omega$ a word $X_{f(n)}$ which arises from $W_{f(n)}$ by omitting from $W_{f(n)}$ its 0 - component $w_{f(n)}$.

Finally there follows:

(3) The sequence $W_0, W_1, \ldots, W_{f(0)-1}, X_{f(0)}, X_{f(1)}, X_{f(2)}, \ldots$ is good.

For $X_{f(0)}$ is strictly shorter, and then also strictly less than $W_{f(0)}$. But this was a minimal word among all words which prolong $W_0, \ldots, W_{f(0)-1}$ to a bad sequence.

Now (3) leads to a contradiction. For (3) implies the existence of two words A, B in the sequence of (3), where A is left from B, but $A \preceq B$. Then A and B cannot be both in the segment $W_0, \ldots, W_{f(0)-1}$ because of (1). Also they cannot be both in the segment $X_{f(0)}, X_{f(1)}, \ldots$. For then also $W_{f(0)}, W_{f(1)}, \ldots$ would be good, contradicting (1). If finally A would be in the segment $W_0, \ldots, W_{f(0)-1}$ and B in the segment $X_{f(0)}, X_{f(1)}, \ldots$, say $B = X_{f(m)}$ for an $m \in \omega$, then a fortiori $A \preceq W_{f(m)}$ would hold, contradicting (1).

By the way, in [84] a problem of Erdös [39] is solved.

6.4 Definition. A word $(x_1, \ldots, x_n)$ over $\mathbf{N}$ is (of course) said to be *of sum s*, if $x_1 + \cdots + x_n = s$.

An application of Higman's theorem yields immediately with 2.3, 2) the following theorem, which will be used later:

6.5 Theorem. *We consider the set* **N** *of natural numbers with its natural order. For $n \in$ **N** let $w_n = (x_{n1}, \ldots, x_{n,l(n)})$ be a word over **N** with sum n, this implies $l(n) \leq n$. We equip the set $\{w_n | n \in$ **N**$\}$ with the order for words according to 6.1. Then there exists an infinite subset* **N*** *of* **N** *such that the set $\{w_n | n \in$ **N**$^*\}$ is strictly increasing with n.*

The question arises whether there is a version of 6.5 for the finite case, where instead of **N** proper initial segments of **N** are considered. Indeed, one can obtain a finitistic version of 6.5 by applying a compactness argument. There holds:

6.6 Theorem. *For every $m \in$ **N** there exists a natural number $k(m)$ such that the following holds: If for every natural number $n \leq k(m)$ a word $w_n = (x_{n1}, \ldots, x_{n,l(n)})$ over **N** of sum n is given, then there exists a strictly increasing set of size m of words $w_{i_1} \preceq \cdots \preceq w_{i_m}$.*

Proof. Suppose indirectly that there exists a natural number m, for which no $k(m)$ exists which satisfies the assertion.Then for every $n \in$ **N** there exists a mapping f_n which ascribes to every natural number $x \leq n$ a word w_x over **N** of sum x, such that there is no m-element subset of $\{1, \ldots, n\}$, over which f_n is strictly increasing. Of course $f_n(1)$ is the word $w_1 = 1$ for all $n \in N_1 :=$ **N**. From the pigeonhole principle it follows that there is an infinite subset $N_2 \subseteq N_1$, such that all words $f_n(2)$, $n \in N_2$, are equal, say $= w_2$. Analogously there exists an infinite subset $N_3 \subseteq N_2$ such that for $n \in N_3$ all words $f_n(3)$ are equal $= w_3$ and so on. Then we define f by $f(n) := w_n$ for $n \in$ **N**. Now $w_1, w_2, \ldots$ has no increasing subset of length m, since this would also be an increasing subset with respect to a function f_n, which by construction is impossible.

The last proof is a pure existence proof. It does not contain any information about the growth of the function $k(m)$. In the paper [62] a constructive proof for the existence of $k(m)$ is given.

Finally we mention a theorem of Laver [111]. Its proof is long and complicated, and so we omit it here:

6.7 Theorem. *There is no infinite set of pairwise incomparable order types of scattered linearly ordered sets. And there is no infinite strictly descending sequence $\tau_0 > \tau_1 > \cdots > \tau_n > \cdots$, $n \in$ **N**, of order types of scattered linearly ordered sets.*

In short: Every set of scattered linear order types is well-quasi-ordered.

8.7 Trees

Now we wish to study trees in more detail. In particular we investigate the relationship between set splittings and trees. Since every principal ideal of a tree is well-ordered, a tree is well-founded. But, contrary to the situation of a pwo-set, a tree can have infinite antichains. E.g. a poset P that has a first element x, which has infinitely many upper neighbors, which are pairwise incomparable, and which has no other elements, is a tree. On the other hand, a pwo-set need not be a tree; every finite poset is pwo.

In 1.7.1 we defined the concept "height" for elements of finite posets P, and in 1.3 we extended this notion to elements of well-founded posets. So also in a tree every element has a well-defined height, and the tree itself has a height. Also the concepts branch (resp. path, resp. $\beta-$stump) of a tree are now special cases of the general definition in 1.3.

A branch of a tree is a maximal chain and thus it contains with an element x also all predecessors of x. In this context we have a simple fact:

7.1 Theorem and **Definition.** *Let $B = \{b_\nu | \nu < \tau\}$ be a branch of a tree T, where for $\mu < \nu < \tau$ we have $b_\mu < b_\nu$. Then b_ν has the height ν and belongs to the νth level L_ν of T.*

For $x \in T$ the set of elements $y \in T$ with $y \leq x$ is a path, which we call the path ending in x. It has the form $\{x_\nu | \nu \leq h(x)\}$ where $x_\nu \in L_\nu$. We call x_ν the ν - predecessor of x. (So x is its $h(x)$ - predecessor.)

Proof. For $\nu = 0$ the assertion holds since b_0 must be the least element of T. Let μ be an ordinal $< \tau$, such that the assertion holds for all $\nu < \mu$. Then b_μ is minimal in the set of those elements of T which are $> b_\nu$ for all $\nu < \mu$ because B is a maximal chain of T. Therefore $b_\mu \in L_\mu$ holds, and our theorem is proved by induction.

If a set $B = \{b_\nu | \nu < \tau\}_<$ is a branch of a set W which is only well-founded, we have no such conclusion: If W is the pentagon (see Figure 18) then $\{b_0, b_1, b_2\}$ is a branch of W, but b_2 has height 3.

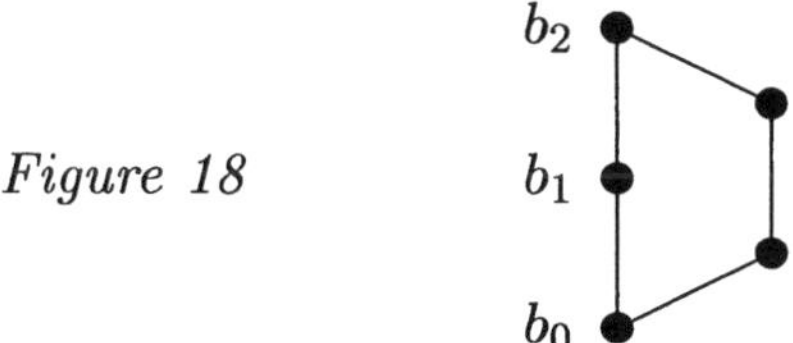

Figure 18

This also shows that in a well-founded set an element x can have more than one path ending in x, which in trees is not possible.

The concept "tree" is closely related with the concept "set-splitting", which was defined in 6.1.7, and this is one of the main reasons for the importance of trees. Now we consider this relationship in detail. To this purpose we first define:

7.2 Definition. A *branching point* of a tree is an element which has at least two immediate successors. We call a tree T a *splitting tree* if the following two conditions are satisfied:

1) Every element of T which is not maximal is a branching point.

2) If λ is a limit ordinal and $\{b_\nu | \nu < \lambda\}_<$ a chain of T, where b_ν has height $h(b_\nu) = \nu$ for $\nu < \lambda$, and if the set A of elements of T that are $> b_\nu$ for all $\nu < \lambda$ is non-empty, then A has a least element.

A tree T which satisfies 2) is called *semi-chain-complete*. So, a tree is semi-chain-complete, iff every non-empty bounded chain of T has a supremum. (It is not required that every non-empty chain of T has a supremum.)

If T is a tree, which is not necessarily a splitting tree, then the set A which is defined in 2) can have many elements which are minimal in A.

Now the equivalence between set-splittings and splitting trees can be described:

7.3 Remark. Let $\mathfrak{A}$ be a splitting (see 6.1.7) of a non-empty set S. In $\mathfrak{A}$ we have the order by inclusion. But in order to establish the analogy between splitting trees and set-splittings we now put for blocks B_1, B_2 of $\mathfrak{A}$:

$$B_1 \le B_2 \iff B_1 \supseteq B_2.$$

Then $(\mathfrak{A}, \leq)$ is a tree, and S is its least element (in the order $\leq$). Indeed, let $B_\nu \in \mathfrak{A}$ be a block of height ν. (For the concept "height" of a block see Definition 6.1.10.). Then for every $\mu < \nu$ there exists exactly one block B_μ of height μ with $B_\mu \supset B_\nu$. And so the set $\{B_\mu | \mu < \nu\}$ of all predecessors (with respect to $\leq$) of B_ν is well-ordered. Thus $(\mathfrak{A}, \leq)$ is a tree. Since $\mathfrak{A}$ satisfies 2) of 6.1.7 also 1) of 7.2 is fulfilled. And since $\mathfrak{A}$ satisfies 3) of 6.1.7 also 2) of 7.2 holds. Thus $(\mathfrak{A}, \leq)$ is a splitting tree.

On the other hand, let T be a splitting tree. Then we define $\mathfrak{A}$ to be the set of all principal final segments $[a)$, $a \in T$, of T. If $B \in \mathfrak{A}$, say $B = [a)$ for an $a \in T$, we define the set $\mathcal{Z}(B)$ (corresponding to 6.1.7, 2)) as the set of all segments $[b)$, where the elements b are immediate successors of a. And then the properties 1) and 2) of Definition 6.1.7 are fulfilled. By 2) of 7.2 $\mathfrak{A}$ contains also all non-empty intersections of final segments $[a)$, so that 3) of 6.1.7 is satisfied.

In 6.1.10 we defined the concept "height" for the blocks of a set-splitting $\mathfrak{A}$. If this is considered as a tree with order $\leq$, then the elements of this tree also have a height as defined in 1.3. Now these heights coincide, so that the corresponding definitions are compatible.

Next we wish to study linear extensions of trees. We begin with a general concept, which is a weakening of the notion position-equivalent of Definition 7.3.8:

7.4 Definition. Let P be a poset. For elements a,b of P we put $a \sim b \iff (P < a) = (P < b)$, (in words: a and b have the same set of predecessors).

It follows immediately that $\sim$ is an equivalence relation. And the set of minimal elements of P is an equivalence class — for all x of it we have $(P < x) = \emptyset$. Equivalent elements of P are incomparable, and so every equivalence class is an antichain of P.

By 1.4. f) equivalent elements of a well-founded poset, in particular of a tree, have the same height. If we choose a linear order in every equivalence class of a tree we have a natural method to extend the order of the tree to a linear one:

7.5 Definition and **Theorem.** Let $(T, \leq)$ be a tree, and suppose that in every equivalence class K of T a linear order $\leq_K$ is given. Then we define a *standard linear extension* $\leq^*$ of $\leq$ in the following way: Let x and y be in T.

If already $x \leq y$ holds we put also $x \leq^* y$.

If x and y are not comparable in $\leq$ we consider the paths $\{x_0,\ldots,x_\nu,\ldots,x_\xi = x\}_<$ and $\{y_0,\ldots,y_\nu,\ldots,y_\eta = y\}_<$ which end in x resp. y, so that x_ν is in the level L_ν for $\nu \leq \xi$, and analogously for y_ν. Due to the fact that x and y are not comparable in $\leq$, there exists a least index ν for which $x_\nu \neq y_\nu$ holds. Indeed, if such a ν would not exist, one of the paths, ending in x (resp. y) would be an initial segment of the other, and this would entail $x \leq y$ or $y \leq x$. So, in the case where x and y are not comparable in $\leq$, we consider the first index ν where the above paths differ. Then x_ν and y_ν are in the same equivalence class K, and we put

$$x <^* y \iff x_\nu <_K y_\nu.$$

If x and y are incomparable elements of T, we define $d(x,y)$ to be the least ordinal ν, for which $x_\nu \neq y_\nu$ holds.

It follows that $\leq^*$ is indeed a linear order. Reflexivity and antisymmetry are trivial. Let now x, y, $z \in T$ satisfy $x <^* y <^* z$. We have to show $x <^* z$. The case $x < y < z$ is trivial.

If $x < y <^* z$ with $y \not< z$ holds, let $\nu := d(y,z)$ so that $y_\nu <_K z_\nu$ holds for the class K which contains y_ν and z_ν. If $x < y_\nu$, then also $x < z_\nu \leq z$ and $x < z$. If $y_\nu \leq x$, then $x_\nu = y_\nu$,and ν is the least index with $x_\nu \neq z_\nu$. Now $x_\nu(= y_\nu) <_K z_\nu$ and thus $x <^* z$.

If $x <^* y < z$ with $x \not< y$ holds, let $\nu := d(x,y)$. Then $d(x,z) = \nu$ and $x_\nu <_K y_\nu = z_\nu$, where K is the class containing x_ν and z_ν. And so $x <^* z$ follows.

Finally we consider the case where neither $x < y$ nor $y < z$ holds. Let $\mu := d(x,y)$ and $\nu := d(y,z)$. If $\mu > \nu$ we have $x_\nu = y_\nu <_K z_\nu$ for the class K which contains y_ν and z_ν. Due to $d(x,z) = d(y,z) = \nu$ this entails $x <^* z$.

If $\mu = \nu$ we have $x_\nu <_K y_\nu <_K z_\nu$ for the class K which contains x_ν, y_ν, and z_ν, and $x <^* z$ follows since $\leq_K$ is transitive.

If $\mu < \nu$ we have $x_\mu <_K y_\mu = z_\mu$ for the class K which contains x_μ and y_μ. Due to $d(x,z) = \mu$ this entails $x <^* z$. So $\leq^*$ is also transitive and a linear extension of $\leq$.

7.6 Remark. It is by no means so that every linear extension of the order of a tree is a standard linear extension of the above kind. To see this we can e.g. use 2.3.3, in which the following was proved: If A is an antichain of a poset P, and if we introduce an arbitrary linear order $\leq_A$ in A, then there exists a linear extension of the order of P which contains $\leq_A$.

The linear order of 7.5 preserves some features of the tree T. First we mention:

7.7 Lemma. *Let $(T, \leq)$ be a tree, $\leq^*$ a standard linear extension of $\leq$ in the sense of Definition 7.5. For $s \in T$ let $[s)$ be the final segment of $(T, \leq)$ containing all $t \in T$ with $s \leq t$. Let a and b be elements of T which are in the same equivalence class K of T and which satisfy $a <_K b$.*

Then for every $x \in [a)$ and every $y \in [b)$ there holds $x <^* y$.

Proof. We consider the paths of x and y. The first ordinal ν for which $x_\nu \neq y_\nu$ holds, is the height of a and b, and so $x_\nu = a$ and $y_\nu = b$. By definition we have $x <^* y \iff a <_K b$, and this implies our assertion.

Now there follows an important relation between the order of a tree and the linear order of a standard linear extension of it:

7.8 Theorem. *Let a be an element of a tree $(T, \leq)$ and $\leq^*$ a standard linear extension of $\leq$. Then the final segment $[a)$ of a in $(T, \leq)$ is a segment in $(T, \leq^*)$ which has a as first element.*

Proof. All elements b of T which are $> a$, are also $>^* a$ by definition, and so $[a) \subseteq (T \geq^* a)$ holds.

Let now x and y be elements of $[a)$ with $x <^* y$. We have to show that every element z which satisfies $x <^* z <^* y$, also fulfills $z \geq a$. Suppose indirectly that $z \geq a$ is false. Then there exists a least index κ such that for the κ - predecessors of a and z we have $a_\kappa \neq z_\kappa$. The elements x and y are $(\geq a) \geq a_\kappa$ and therefore in $[a_\kappa)$. And $z \geq z_\kappa$ is of course in $[z_\kappa)$.

If $a_\kappa <_K z_\kappa$ holds where K is the class containing a_κ and z_κ, then by 7.7 every element of $[a_\kappa)$ is $<^*$ every element of $[z_\kappa)$, and we would obtain that x and y are both $<^* z$, a contradiction.

If $z_\kappa <_K a_\kappa$ holds we conclude analogously that z is $<^* x$ and $<^* y$, which again is a contradiction.

Finally we consider an obvious method to embed a tree into a semi-chain-complete tree:

7.9 Theorem. *Let T be a tree with levels $L_\nu, \nu < h(T)$. If λ is a limit ordinal and $B := \{b_\nu | \nu < \lambda\}_<$ a chain of T with $b_\nu \in L_\nu$ for $\nu < \lambda$, which has upper bounds, but no minimal upper bound in T, we introduce a new element e and put $e >^* b_\nu$ for $\nu < \lambda$, and $e < {}^* x$ for all upper bounds x of B, and e is incomparable with all other elements of T. If a and b are new elements we put $a <^* b$, if there exists a $z \in T$*

with $a <^* z <^* b$. *For* $a, b \in T$ *we of course put* $a \leq^* b \iff a \leq b$. *Then the set* $T^* := T \cup E$, *where* E *is the set of new elements, is a semi-chain-complete tree with respect to the order* $\leq^*$.

The height of a new element in the tree T^**is a limit ordinal.*

Proof. First we have to show that the relation $\leq^*$is an order relation. If a, b, c are in T^* and $a <^* b <^* c$, we have to prove $a <^* c$. For this purpose one has to consider eight cases, depending on the situation, whether a, b, c are in T or E. Each of these is trivially verified.

If a, b are in T^*, we cannot have $a <^* b$ and $b <^* a$. If from a and b at least one is in T, this is trivial. So let a and b be both in E, and suppose that $a <^* b$ and $b <^* a$ hold. Then, by definition, we must have elements y and z in T satisfying $a <^* y <^* b$ and $b <^* z <^* a$. This entails $y < z$ and $z < y$ with contradiction. So $\leq^*$is antisymmetric.

We still have to prove that T^* is a tree. If a new element e is introduced in the tree T, we first see that $T_e := T \cup \{e\}$ is also a tree, in which $(T_e < e) = (T < e)$ holds by construction, so that it is well-ordered. And e has in T_e as height a limit ordinal, namely the height of $(T < e)$. The successors of e in T^*, which have a height $h(T < e) + n$ with $n \in \omega$ in T, obtain in T_e the height $h(T < e) + n + 1$. All other heights remain unchanged.

Let $\{\lambda_\nu, \nu < \kappa\}$, be the set of limit ordinals $< h(T)$ with $\lambda_\mu < \lambda_\nu$ for $\mu < \nu < h(T)$. Then also the set T, augmented by the new elements b, for which $(T < b)$ has type $\omega = \lambda_0$, is a tree $T[\lambda_0]$. Successively we add the new elements b, for which $T[\lambda_0] < b$ has type λ_1, and so on. With transfinite induction we so obtain an ascending sequence of trees, whose union is again a tree, and $= T^*$.

By construction it is clear that T^* is semi-chain-complete, and that the new elements have as height in T^* limit ordinals.

Comparing the size of chains and antichains of T resp. T^* we have the following statement:

7.10 Theorem. *Let* $(T, \leq)$ *be a tree and* $(T^*, \leq^*)$ *as defined in 7.9. Then there holds:*

a) *Every chain of* $(T^*, \leq^*)$ *has a length which is equal to the length of a chain of* $(T, \leq)$, *or such a length* $+1$.

b) *Every antichain of* $(T^*, \leq^*)$ *has a cardinality which is also the cardinality of an antichain of* $(T, \leq)$.

Proof. a) is an easy consequence of the fact that, in a maximal chain $\{a_\nu | \nu < \tau\}_<$ of T^*, only elements a_ν, where ν is a limit ordinal, can belong to $T^* \backslash T$.

b) Let B be an antichain of $T^* \backslash T$, so B contains only new elements. For each $b \in B$ we choose an immediate successor (with respect to $\leq^*$) $b' \in T$ of b. Then we have:

(I) $B' := \{b' | b \in B\}$ is an antichain of cardinality $|B|$.

For if $b_1, b_2 \in B$ we cannot have $b_1' < b_2'$ because this would entail $b_1 <^* b_1' < b_2'$ and thus $b_1 <^* b_2'$. Now b_2 is an immediate predecessor of b_2', and $T^* < b_2'$ is linearly ordered. So $b_1 \leq b_2$ would follow with contradiction to the fact that B is an antichain.

Now every antichain of T^* has the form $A \cup B$, where $A \subseteq T$ and $B \subseteq T^* \backslash T$. Each element a of A is incomparable in $\leq$ with every element b' of B'. Indeed, $a < b'$ would entail ($a <^* b'$ and) $a \leq^* b$, and $a > b'$ would imply $a > b$. Each of these is impossible in the antichain $A \cup B$. So $A \cup B'$ is an antichain of T with the same cardinality as $A \cup B$.

8.8 Aronszajn trees and Specker chains

A description of properties of a tree suggests considering the cardinality of the chains, the antichains and the levels of the tree. A typical result in this area is König's Infinity Lemma, which now can be reformulated in the form:

8.1 Theorem. *Let T be an infinite tree, all of whose levels are finite. Then there exists an infinite branch.*

Subsequently to this the question arises if there exist generalizations to higher cardinalities. The first interesting case is then: Given a tree with $\aleph_1$ elements, all of whose levels are countable, does there exist a branch of type ω_1? Aronszajn found a counterexample, which we present a bit later. (It was communicated by him in a letter to Kurepa.) We mention before some remarks on the set of well-ordered subsets of $\mathbf{R}$.

8.2 Example. Let $\mathfrak{W}$ be the set of all well-ordered subsets of $\mathbf{R}$, each of them is ordered by the restriction of the natural order of $\mathbf{R}$. The set $\mathfrak{W}$ has cardinality $2^{\aleph_0} =: c$.

$|\mathfrak{W}| \geq c$ is trivial since $\mathfrak{W}$ contains all one-element subsets of $\mathbf{R}$. On the other hand, if $W \in \mathfrak{W}$, then W has the form $\{x_\nu | \nu < \alpha\}_<$, where α

is a countable ordinal. And then $\mathfrak{W}$ is the union $\cup\{\mathfrak{W}_\alpha | \alpha < \omega_1\}$, where $\mathfrak{W}_\alpha$ is the set of all well-ordered subsets W of $\mathbf{R}$ with $\mathrm{tp}(W) = \alpha$. Now $\mathfrak{W}$ is a union of $\aleph_1$ sets $\mathfrak{W}_\alpha$, $\alpha < \omega_1$, where each $\mathfrak{W}_\alpha$ has cardinality $\leq |\mathbf{R}|^{|\alpha|} = |\mathbf{R}|^{\aleph_0} = c$, so that also $|\mathfrak{W}| \leq c$ holds.

8.3 Example. A fortiori it follows from 8.2 that the set of all well-ordered subsets of the set $\mathbf{Q}$ of rational numbers has cardinality $\leq c$. Its cardinality is also $\geq c$, for it contains the set of all well-ordered subsets of $\mathbf{N}$, which is identical with the set of all subsets of $\mathbf{N}$, and so has cardinality c.

Contrary to the situation in $\mathbf{N}$, every ordinal $< \omega_1$ is the type of a well-ordered subset of $\mathbf{Q}$. (Later this fact is important in the proof for the existence of an Aronszajn tree.) In this example we define T as the set of all well-ordered subsets of $\mathbf{Q}$. If A and B are elements of T we put $A \preceq B$ if A is an initial segment of B. It can easily be verified that T with this relation $\preceq$ is a tree, since the set of initial segments of a well-ordered set is also well-ordered. It follows easily that all levels L_ν with $\nu \geq \omega$ have cardinality $\geq c$. But in spite of the fact that all levels L_ν, $\nu < \omega_1$, of T are non-empty, there cannot exist a branch of length ω_1. For if $\mathfrak{B} = \{B_\nu | \nu < \omega_1\}_<$ would be such a branch, we could choose an element $b_{\nu+1} \in B_{\nu+1} \backslash B_\nu$ for every $\nu < \omega_1$, and the set $\{b_{\nu+1} | \nu < \omega_1\}$ would be a well-ordered subset of $\mathbf{Q}$ of type ω_1, which is impossible. The Aronszajn tree will be a strict subset of the tree T just defined.

8.4 Definition. If κ is an infinite initial ordinal, an *Aronszajn κ-tree* is a tree of height κ which has no branch of length κ, and is such that each of its levels has cardinality $< |\kappa|$. Instead of Aronszajn ω_1-tree one also says *Aronszajn $\aleph_1$- tree* or in short *Aronszajn tree*.

In order to construct an Aronszajn tree we have to modify the construction of the tree of 8.3 so that it satisfies the additional requirement, that each level is countable, but the height remains ω_1. We recall:

8.5 Definition. If κ is an ordinal and $X_\kappa = \{x_\mu | \mu < \kappa\}_<$ a well-ordered set, and if $\nu \leq \kappa$ we call the well-ordered set $X_\nu = \{x_\mu | \mu < \nu\}_<$ the *ν - segment* of X_κ.

8.6 Theorem. *There is an Aronszajn ω_1- tree T.*

Proof. We construct sets $L_\nu, \nu < \omega_1$, successively by induction. These will become the levels of the tree T which we construct. We put $L_0 := \{\emptyset\}$, and $L_1 := \{\{r\} | r \in \mathbf{Q}\}$. Generally for $\nu < \omega_1$ the set L_ν will

274

be a set of bounded subsets of the set $\mathfrak{W}_\nu$ of all well-ordered subsets of $\mathbf{Q}$ which have type ν.

For ordinals $\nu < \omega_1$ we consider the following property (P_ν), which will be proved for all $\nu < \omega_1$ by induction:

(P_ν) *The sets L_μ are already defined for all $\mu \leq \nu$ as countable bounded (in $\mathbf{Q}$) subsets of $\mathfrak{W}_\mu$ (of 8.2). And further there holds:*

If $\mu < \nu$ and $X_\mu \in L_\mu$, $X_\nu \in L_\nu$ is called a prolongation of X_μ, if for all κ with $\mu \leq \kappa \leq \nu$ the $\kappa-$ segment X_κ of X_ν belongs to L_κ.

If now $\mu < \nu$, $X_\mu \in L_\mu$, and if r is a rational $> \sup X_\mu$, then there exists a prolongation X_ν of X_μ with $\sup X_\nu = r$.

It is clear that (P_ν) holds for $\nu = 0$ and $\nu = 1$ (recall $tp(\emptyset) = 0$). Let now α be an ordinal $< \omega_1$ such that (P_ν) holds for all $\nu < \alpha$. Then we define L_α and prove (P_α). We distinguish two cases:

(I) If α is a successor number $\beta + 1$, we put $L_\alpha := \{X \cup \{r\} | X \in L_\beta,$ and r is a rational $> \sup X\}$. Intuitively speaking: All well-ordered sets $\in L_\beta$, which all have type β, are prolonged by a rational, and this leads to well-ordered subsets of $\mathbf{Q}$ of type α.

Then also (P_α) holds: For let $0 < \mu \leq \beta$, $X_\mu \in L_\mu$, $r > \sup X_\mu$ a rational. If $\mu = \beta$, then $X_\mu (= X_\beta) \in L_\beta$ has by construction a prolongation $X_\beta \cup \{r\} \in L_\alpha$.

Now suppose $\mu < \beta$. Then, due to (P_β), there exists a prolongation $X_\beta \in L_\beta$ of X_μ with $\sup X_\beta = r'$ for some r' with $\sup X_\mu < r' < r$ of $\mathbf{Q}$. Then $X_\alpha := X_\beta \cup \{r\} \in L_\alpha$ is a prolongation of X_μ with $\sup X_\alpha = r$. And thus (P_α) holds.

(II) Let now $\alpha < \omega_1$ be a limit number, $\mu < \alpha$, $X_\mu \in L_\mu$, $r > \sup X_\mu$ a rational. Then we choose a strictly increasing sequence $(\alpha_\nu | \nu < \omega)_<$ of ordinals whose upper limit is α, and with $\mu < \alpha_0$. Further we choose a strictly increasing sequence $(r_\nu | \nu < \omega)_<$ of rationals whose upper limit is r, and where $r_0 > \sup X_\mu$ holds. By induction hypothesis we conclude that there exist sets $X_{\alpha_\nu} \in L_{\alpha_\nu}$ where X_{α_0} is a prolongation of X_μ with $\sup X_{\alpha_0} = r_0$, and where for every $\nu \in \omega$ the set $X_{\alpha_{\nu+1}}$ is a prolongation of X_{α_ν} with $\sup X_{\alpha_{\nu+1}} = r_{\nu+1}$.

Let $X_\alpha := \cup \{X_{\alpha_\nu} | \nu < \omega\}$, and let L_α be the set of all these X_α so obtained. Then L_α is countable since $\cup \{L_\nu | \nu < \alpha\}$ and $\mathbf{Q}$ are countable sets. And by construction (P_α) holds.

By induction we have obtained that each set L_α with $\alpha < \omega_1$ is non-empty and countable. The set $T := \cup \{L_\nu | \nu < \omega_1\}$, equipped with the order "is an initial segment of ", is a tree. For it has $\emptyset$ as least element, and each of its elements is a well-ordered set, whose set of predecessors, which are initial segments of it, is well-ordered. For this tree T we have:

(III) For each $\nu < \omega_1$ every element of L_ν has height ν. So L_ν is indeed the ν - level of T. And thus T has height ω_1.

For $\nu = 0$ this is trivial. If this statement holds for a fixed $\nu < \omega_1$, it holds by construction also for $\nu + 1$. For every element of $L_{\nu+1}$ arises from an element of L_ν, which by induction hypothesis has height ν, by attaching exactly one element to it. And if λ is a limit number, such that the statement holds for all $\nu < \lambda$, then the construction of L_λ ran in such a way that only such elements were included in L_λ, which had height λ. Thus (III) is proved.

Finally it follows (as in 8.3) that T has no branch of type ω_1. If $\mathfrak{B} = \{B_\nu | \nu < \omega_1\}_<$ would be such a branch, we could choose an element $b_{\nu+1} \in B_{\nu+1}\backslash B_\nu$ for every $\nu < \omega_1$, and the set $\{b_{\nu+1} | \nu < \omega_1\}$ would be a well-ordered subset of $\mathbf{Q}$ of type ω_1, which is impossible.

Next we shall construct a Specker chain. With this set Specker was able to solve a problem of Erdös/Rado [36]. This chain has interesting applications in the partition calculus of set theory (see e.g. a result of Galvin/Shelah [48]). We define:

8.7 Definition. A linearly ordered set $(S, \leq)$ is called a *Specker chain*, if $|S| \geq \aleph_1$, and if S has no subset of type ω_1, no subset of type ω_1^*, and no subset of cardinality $\aleph_1$ which is embeddable in $\mathbf{R}$.

Concerning the possible cardinalities of a Specker chain we have:

8.8 Remark. A Specker chain has cardinality $\leq 2^{\aleph_0}$ $(= |\mathbf{R}|)$. This is already a consequence of the fact that a Specker chain has no subsets of type ω_1, ω_1^* and of the embedding Theorem 3.3.7 of Hausdorff according to which a Specker chain is embeddable in H_1, whose cardinality is $2^{\aleph_0}$.

Also it is clear that every subset of a Specker chain that has $\aleph_1$ elements is again a Specker chain.

8.9 Lemma. *Let $(S, \leq)$ be a linearly ordered set which has a countable subset D which is dense in S. Then S is embeddable in $\mathbf{R}$.*

Proof. The set X of elements $x \in S$, which have an immediate successor y in S, is countable. For since D is dense in S, x or y belongs to D. The set X^+ of these $x \in X$ which are in D is countable, and also the set X^- of those $x \in X$ which are not in D is countable, since their corresponding elements y are in D, and form a set equipotent to

X^-. Analogously the set Y of those elements $y \in S$ which have an immediate predecessor in S is countable. Let now D^* be that subset of S which contains $D \cup X \cup Y$ and the least and greatest element of S, if such an element exists. Then D^* is again countable and dense in S, and there exists a $<$ - preserving mapping $f : D^* \to \mathbf{Q}$. If now $x \in S \backslash D^*$, then $(S < x)$ has no greatest, and $(S > x)$ no least element, and the corresponding statement holds for their f - images $f[S < x]$ and $f[S > x]$. Since $\mathbf{R}$ has no gaps there is an element z between these sets. We put $f(x) := z$, and obtain in this way an embedding of S in $\mathbf{R}$.

An essential property of Specker chains is expressed in the following statement, which is an immediate consequence of 8.9:

8.10 Theorem. *A Specker chain S has no countable subset which is dense in S.*

Starting from an Aronszajn ω_1- tree T we can now construct a Specker chain by applying a standard linear extension of 7.5:

8.11 Theorem. *There exists a Specker chain.*

Proof. Let T be an Aronszajn ω_1- tree. (It is not necessarily the same as in 8.6.) We choose a linear order $\leq_K$ in every equivalence class K (see 7.4) of T and define a corresponding standard linear extension $\leq^*$of the order $\leq$ of T according to 7.5. We prove that $(T, \leq^*)$ is a Specker chain. First we show:

(I) $(T, \leq^*)$ has no subset of type ω_1, and symmetrically no subset of type ω_1^*.

Suppose indirectly that a set $W \subseteq T$ has type ω_1 with respect to the linear order $\leq^*$. Let ν be an ordinal $< \omega_1$. Then the set $\{x \in W \mid h(x) < \nu\}$ is countable (as a subset of the union of countably many levels L_μ, $\mu < \nu$, of T) and thus not cofinal in $(W, \leq^*)$. Therefore W has a non-empty final segment F_ν of cardinality $\aleph_1$such that all elements of F_ν have height $> \nu$.

There exists an element a_ν of the level L_ν such that $\aleph_1$ elements of F_ν are in $[a_\nu)$ $(= \{x \in T | x \geq a_\nu\})$. This follows from the pigeonhole principle since each of the $\aleph_1$ elements of F_ν is $\geq$ one of the countably many elements of L_ν. By 7.8 $[a_\nu)$ is also a segment of $(W, \leq^*)$, and moreover a final segment, for otherwise it could not contain $\aleph_1$ elements of F_ν since this has type ω_1. So we have obtained:

(II) For every $\nu < \omega_1$ there exists an $a_\nu \in L_\nu$ such that $[a_\nu)$ contains a non-empty final segment S_ν of $(W, \leq^*)$.

This entails now that for $\mu < \nu < \omega_1$ we have $a_\mu < a_\nu$. For $S_\mu \cap S_\nu$ is $= S_\mu$ or $= S_\nu$, and hence $\neq \emptyset$, and for an element $z \in S_\mu \cap S_\nu$ we have $z \geq a_\mu$ and $z \geq a_\nu$, so that a_μ and a_ν are comparable and then $a_\mu < a_\nu$, so that $\{a_\mu | \mu < \omega_1\}$ would be a chain of type ω_1 of $(T, \leq)$, a contradiction.

Next we show:

(III) Let U be an uncountable subset of T. Then U (with $\leq^* \restriction U$) is not embeddable in $\mathbf{R}$.

Suppose the contrary. Then there also exists an uncountable set $U' \subseteq U$ which is dense and not embeddable in $\mathbf{R}$. To see this we consider (in analogy to the construction in 3.1.15) the relation ρ which is defined by: For $x, y \in U$ with $x < y$ we put

$x \rho y \iff \{z \in U | x < z < y\}$ is countable.

Then ρ is an equivalence relation in U, and every equivalence class is countable. This is an easy consequence of the fact that $\mathrm{cf}(U) < \omega_1$ and $\mathrm{coin}(U) = \gamma^*$ for a $\gamma < \omega_1$. Therefore there must be $\aleph_1$ equivalence classes. We choose an element $u \in U$ in each of these classes. The set U' of these u is uncountable and dense. So for the rest of the proof we can assume that U is $(= U')$ dense and uncountable.

(IV) There exists a countable $D \subseteq U$ such that D is strictly dense in U.

By our indirect assumption U is isomorphic to a subset U^* of $\mathbf{R}$, say by a mapping φ. In $\mathbf{R}$ we have countably many open intervals (r_1, r_2) where $r_1 < r_2$ are rationals. If such an interval contains elements of U^* we choose one of them and collect them in a set D^*, which then is countable.

D^* is strictly dense in U^*. For let $u_1 < u_2$ be elements of U^*. Since U^* is dense, like U, there exists an element $u \in U^*$ with $u_1 < u < u_2$ and further two rationals r_1 and r_2 with $u_1 < r_1 < u < r_2 < u_2$. By construction of D^* there also exists an element $d \in D^*$ with $(u_1 <)$ $r_1 < d < r_2 (< u_2)$, and so D^* is strictly dense in U^*. Then also U has a countable subset $D := \varphi^{-1}(D^*)$ which is strictly dense in U, and (IV) is proved.

The set of heights $\{h(x) | x \in D\}$ is a countable set of ordinals $< \omega_1$, and so its supremum is still $< \omega_1$. This entails that there exists an $\alpha < \omega_1$ such that all $x \in D$ are in the α - stump $T|\alpha$ of the tree T. The set $U^- := U \setminus T|\alpha$ has cardinality $\aleph_1$ since $T|\alpha$ is countable as the union of countably many levels of T.

For $u \in U^-$ let u_α be the α - predecessor of u in T. Then $\{u_\alpha | u \in U^-\} \subseteq L_\alpha$ is countable, and so, by the pigeonhole principle, there exist two elements u, t in U^- with equal α - predecessors $u_\alpha = t_\alpha$, so that u and $t \in [u_\alpha)$. Now there is no $x \in D$ with $u <^* x <^* t$. For this would by 7.8 entail that also $x \in [u_\alpha)$ and thus $h(x) \geq \alpha$ holds. But all $x \in D$ have a height $< \alpha$.

Summarizing we have obtained: There are two elements u, $t \in U$, for which there is no element of D which is strictly between them relative to $\leq^*$. This is a contradiction to (IV), which arose from the indirect assumption that (III) would be false. So (III) and the theorem is proved.

8.9 Suslin chains and Suslin trees

In the first volume of Fundamenta Mathematicae M. Suslin posed in 1920 the following problem: Let C be a linearly ordered set without gaps and steps such that every set of non-overlapping intervals which each have more than one element is countable. Is C embeddable in **R** ? [In the original text: . . . est-il nécessairement un continue linéaire (ordinaire) ?]

This problem had great impact on set theory, and many papers dealt with it and with related questions. This is not astonishing since here properties of the real line are in discussion. The set **R** has a countable subset which is dense in **R,** e.g. the set **Q**. And as a consequence of this **R** also has the property that every set of disjoint segments, which each have at least two elements, is countable. So it makes sense to compare these two properties also in the general case and to determine if one involves the other. One direction is nearly trivial:

9.1 Theorem. *Let $(P, \leq)$ be a linearly ordered set which has a countable subset D which is dense in P. Then every set of disjoint segments, each of which contains at least two elements, is countable.*

Proof. Let $\mathfrak{S}$ be a set of disjoint segments of P that all have at least two elements. In every $S \in \mathfrak{S}$ we choose an element $d(S)$ of D.Then by mapping each $S \in \mathfrak{S}$ onto $d(S)$ we obtain an injective mapping of $\mathfrak{S}$ into the countable set D, and then also $\mathfrak{S}$ is countable.

In connection with the question whether 9.1 admits an inversion we define:

9.2 Definition. A *Suslin chain* (or *Suslin line*) S is a linearly ordered set for which there holds: Every set of disjoint segments of S, that each have at least two elements, is countable, but S has no countable subset which is dense in S.

So one version of the Suslin problem can be formulated as: Does there exist a Suslin chain ? It is known that the existence of a Suslin chain is consistent with ZFC, and even with GCH. And this was also proved for the negation of this statement. In this context we refer to papers of Jech [89], Tennenbaum [169], Solovay and Tennenbaum [165], and Ronald Jensen [91], [90]. Jech and Tennenbaum discovered models of set theory in which a Suslin line exists, and Solovay and Tennenbaum proved that existence of a Suslin line is not provable in ZFC. Jensen proved that a Suslin line exists in the constructible universe. We also refer to the book [22] of Devlin and Johnsbraten.

Concerning the cardinalities of Suslin chains we have:

9.3 Theorem. *A Suslin chain has no subset of type ω_1 and no subset of type ω_1^*. The cardinality of a Suslin chain is $\geq \aleph_1$ and $\leq 2^{\aleph_0}$.*

Proof. Suppose indirectly that $\{a_\nu | \nu < \omega_1\}_<$ is a subset of a Suslin chain S, where for $\mu < \nu$ we have $a_\mu < a_\nu$. Then the intervals $[a_\nu, a_{\nu+1}]$, $\nu < \omega_1$, where ν has an odd number as last summand in the Cantor normal form of ν, are disjoint, and their set is uncountable. This yields a contradiction. The case of ω_1^* is symmetric. Now $|S| \leq 2^{\aleph_0}$ follows by the Theorem 6.2.9' of Hausdorff/Urysohn.

If S would be a countable linearly ordered set, then S itself is a subset of S which is dense in S, and then S could not be a Suslin chain. So $|S| \geq \aleph_1$.

In the preoccupation with the Suslin problem a related concept turned out to be of importance, namely:

9.4 Definition. A tree T is called a *Suslin tree* if T is uncountable but all of its chains and all of its antichains are countable.

A Suslin tree with $\aleph_1$ elements is also an Aronszajn $\aleph_1$- tree. On the other hand we cannot prove that an Aronszajn tree is also a Suslin tree, for it is not excluded that an Aronszajn $\aleph_1$- tree has uncountable antichains.

Another version of the Suslin problem is: Does there exist a Suslin tree. For in the following we shall see that the existence of a Suslin chain implies the existence of a Suslin tree and conversely. The relationship between Suslin chains and Suslin trees was established by Kurepa [104] in 1935.

An elementary property of Suslin trees is:

9.5 Theorem. *All levels L_ν, $\nu < \omega_1$, of a Suslin tree are non-empty.*

Proof. Each level of a Suslin tree T is an antichain and thus countable. If there would be only countably many levels of T, then T would contain only countably many elements.

9.6 Theorem. *Let S be a Suslin chain. Then there exists a subset $\mathfrak{I}$ of the power set $\mathfrak{P}(S)$ of S, where the elements of $\mathfrak{I}$ are closed intervals of S or $= S$, so that $\mathfrak{I}$ is a Suslin tree with respect to the order $\supseteq$ of reverse inclusion.*

Proof. The idea of the proof is to construct a block-system $\mathfrak{I}$ (see 6.1.3) of $\aleph_1$ closed intervals of S. Let $\tau < \omega_1$ and suppose that we have already defined a set of different intervals $I_\nu := [a_\nu, b_\nu]$ of S for $\nu < \tau$ such that each two of these intervals are disjoint or comparable. Then the set $C := \{a_\nu | \nu < \tau\} \cup \{b_\nu | \nu < \tau\}$ is a countable subset of S. It is not dense in S. And so there exists an interval $I_\tau = [a_\tau, b_\tau]$ which contains no element of C. So it satisfies:

(1) For every $\nu < \tau$ holds: I_τ is disjoint to I_ν or a proper subset of it.

By transfinite induction we so obtain a block-system $\mathfrak{I}$ of $\aleph_1$ closed intervals I_ν, $\nu < \omega_1$, of S. Next we prove:

(2) Every chain of $(\mathfrak{I}, \supseteq)$ is well-ordered.

A chain of $(\mathfrak{I}, \supseteq)$ has the form $\{I_\alpha | \alpha \in A\}$, where A is a subset of ω_1, where the I_α, $\alpha \in A$, are pairwise comparable. If $\alpha_1 < \alpha_2$ are in A, then $I_{\alpha_1} \supset I_{\alpha_2}$ must hold by (1) since $I_{\alpha_1} \cap I_{\alpha_2} = \emptyset$ is impossible, for it is $= I_{\alpha_1}$ or $= I_{\alpha_2}$. This implies that $\{I_\alpha | \alpha \in A\}$, equipped with the order $\supseteq$, is isomorphic to A, and thus well-ordered.

Now we can prove:

(3) Every chain of $(\mathfrak{I}, \supseteq)$ is countable.

Suppose the contrary. Then by (2) there exists a well-ordered chain of $\mathfrak{I}$, which can be described in the form $\{I_\nu | \nu < \omega_1\}$, where for $\mu < \nu < \omega_1$ there holds $I_\mu \supset I_\nu$. For every $\nu < \omega_1$ we choose an element x_ν

in $I_\nu \backslash I_{\nu+1}$, and then x_ν is also in none of the sets I_μ with $\mu > \nu$. Now for every $\nu < \omega_1$ one of the following two possibilities holds: a) x_ν is $<$ all elements x_μ with $\mu > \nu$, or b) x_ν is $>$ all elements x_μ with $\mu > \nu$.

Then there are $\aleph_1$ indices ν, for which a) holds, or there are $\aleph_1$ indices ν, for which b) holds. In the first case S would have a chain of type ω_1, in the second a chain of type ω_1^*. By 9.3 each of these is impossible, and (3) is proved.

(4) Every antichain of $\mathfrak{I}$ is countable.

For every antichain A of $\mathfrak{I}$ consists of disjoint segments of S, which each have at least two elements, and so A is countable.

If $\mathfrak{I}$ has a first element, then $\mathfrak{I}$ is already a Suslin tree because $\mathfrak{I}$ is uncountable. If it has not yet a first element, we add to $\mathfrak{I}$ the set S which is $\supseteq$ (and hence $\leq$) all $I \in \mathfrak{I}$. Then the extended set is a Suslin tree.

Before we prove the opposite direction of 9.6 we mention first some lemmas.

9.7 Lemma. *Let T be a Suslin tree. Then there exists a subset T_1 of T, which is still a Suslin tree and satisfies:*

(1) *Every element of T has uncountably many successors.*

Proof. We consider the set S of those elements of T that have only countably many successors. Let M be the set of minimal elements of S. Then M is an antichain and thus countable. Further S is $= \cup\{(S \geq m)|m \in M\}$ and then also countable. So finally $T_1 := T \setminus S$ is still uncountable and, of course, a Suslin tree.

In 7.2 we defined the concept semi-chain-complete tree. We can restrict our consideration to these trees, for there holds:

9.8 Lemma. *Let T be a Suslin tree satisfying (1). Then there exists a Suslin tree $T^* \supseteq T$ which is semi-chain-complete and still satisfies (1).*

Proof. We consider the tree T^* which is defined according to 7.9 by adding to T the set E of new elements. Then T^* is semi-chain-complete, and (1) is satisfied for T^*, since every element of $E = T^* \backslash T$ has an immediate successor in T, which by assumption has uncountably many successors.

After these modifications we can prove the counterpart of 9.6:

9.9 Theorem. *Let T be a Suslin tree. Then there exists a Suslin chain.*

Proof. Due to 9.7 and 9.8 we can assume that T satisfies (1) and is semi-chain-complete. In the following we construct a subtree L of T, which has other levels than T, and so we define:

For $\nu < \omega_1$ let L_ν denote the ν - level of L.

First we show:

(I) Every $x \in T$ has an infinite set of successors which form an antichain.

Let $x \in T$. If the set $S(x)$ of successors of x would have only finite antichains, then $S(x)$ would have a chain of cardinality $\aleph_1$, which is impossible. This follows from the partition relation $\aleph_1 \to (\aleph_1, \aleph_0)^2$ of 6.4.10, applied on the comparability relation in $S(x)$. So we can choose a denumerable antichain in every successor set $S(x)$, and can extend it to a maximal antichain $\subseteq S(x)$ by applying Zorn's lemma, so that we obtain:

(II) For every $x \in T$ there exists a maximal infinite antichain $A(x) \subseteq S(x)$, which, of course, is denumerable.

We put $L_0 := \{m\}$, where m is the least element of the tree T, then $L_1 := A(m)$, $L_2 := \cup\{A(x)|x \in L_1\}$, and so on. In general we proceed as follows: Suppose that λ is an ordinal $< \omega_1$, such that the sets $L_\nu, \nu < \lambda$, have already been defined. If λ is a successor number $\kappa + 1$, we put $L_\lambda := \cup\{A(x)|x \in L_\kappa\}$.

Let now λ be a limit ordinal. We consider all sequences $(x_\nu|\nu < \lambda)_<$, where $x_\nu \in L_\nu$ and where for $\mu < \nu$ we have $x_\mu < x_\nu$. (These are paths in the tree L which is coming into being.) Since T is semi-chain-complete and since (1) holds, there exists $\sup\{x_\nu|\nu < \lambda\}$ (in T). And now we define L_λ as the set which contains all these suprema. With transfinite induction all sets $L_\nu, \nu < \omega_1$, are thus defined, and we put $L := \cup\{L_\nu|\nu < \omega_1\}$.

Next we verify:

(III) For every $\nu < \omega_1$ there holds:
(*) L_ν is an antichain. And $\cup\{L_\mu|\mu \leq \nu\}$ is a subtree of T.

For $\nu = 0$ and $\nu = 1$ (*) is trivial. Let now $\lambda < \omega_1$ be an ordinal such that (*) holds for all $\nu < \lambda$.

Case 1. λ is a successor ordinal $\nu + 1$.

Let a and b be different elements of L_λ. They have immediate predecessors a' (resp. b') in L_ν. If these are equal, $a \parallel b$ follows immediately. If $a' \neq b'$ we have by induction hypothesis that L_ν is an antichain and thus $a' \parallel b'$. And then also their successors a (resp. b) are incomparable. Thus L_λ is an antichain. It is easily seen that now also (*) holds for $\nu = \lambda$.

Case 2. λ is a limit ordinal.

For two different elements a, b of L_λ let $a = \sup A$, $b = \sup B$, where $A = \{a_\nu | \nu < \lambda\}_<$, $B = \{b_\nu | \nu < \lambda\}_<$ with $a_\nu, b_\nu \in L_\nu$ for $\nu < \lambda$. These sequences represent the well-ordered sets of predecessors of a (resp. b) in L. (A, resp. B, is the path of L, which ends in a, resp. b.) From the sequences A and B no one can be an initial segment of the other because they have the same length. And since they are different, there exists a first index $\tau < \lambda$ with $a_\tau \neq b_\tau$. These elements are in the antichain L_τ and so we have $a_\tau \parallel b_\tau$. This, together with $a \geq a_\tau$ and $b \geq b_\tau$ implies $a \parallel b$, and thus L_λ is an antichain. By construction now (*) holds for $\nu = \lambda$, and by induction (III) is proved.

(IV) Each $L_\nu, \nu < \omega_1$, is non-empty.

Let this be proved for all L_ν with $\nu < \mu$, where μ is an ordinal $< \omega_1$. Since the L_ν are antichains and thus countable, their union $U := \cup\{L_\nu | \nu < \mu\}$ is also countable, and for every $x \in U$ the set of its predecessors $(T < x)$ (in T) is countable. This finally also holds for $I := \cup\{(T < x) | x \in U\}$, which is an initial segment of T. This entails that the heights of the elements of I form a countable subset of ω_1, which thus is bounded by an ordinal $\rho < \omega_1$. Therefore there exists an elements t in T with height (in T) $> \rho$.

Now t is $> m =: t(0)$. Also t is $>$ an element of $L_1 = A(m)$ because $A(m)$ is a maximal antichain, and then t is comparable with an element $t(1) \in L_1$ and then also $> t(1)$ because its height exceeds the heights of the elements of L_1. This $t(1)$ is uniquely defined since all predecessors of an element of a tree are comparable. Suppose that we have already defined elements $t(\iota) \in L_\iota$ for $\iota < \nu$ such that $t(\iota) < t$ for $\iota < \nu$ holds. If ν is a successor ordinal $\kappa + 1$, then we have $t > t(\kappa)$, and then t is comparable with an element $t(\kappa + 1) = t(\nu)$ of the maximal antichain $A(t(\kappa))$ of elements $> t(\kappa)$. Since the height of t in T is $> \rho$, it exceeds the heights of the elements of L_κ, so that $t(\nu)$ is also $>$ an element of $A(t(\kappa))$. And $t(\nu)$ is uniquely defined since in a tree an element cannot have two incomparable predecessors.

If ν is a limit ordinal, then we have $t \geq \sup\{t(\iota)|\iota < \nu\} =: t(\nu) \in L_\nu$, which exists since T is semi-chain-complete. In any case L_ν is non-empty, and (IV) is proved by induction.

In the tree L every element x has by construction denumerably many immediate successors. We order their set N_x by a linear order $\leq_x$, so that $(N_x, \leq_x)$ has the order type η of $\mathbf{Q}$. Now we can define a Suslin chain C. Its elements are the maximal chains (= branches) of $(L, \leq)$. These are well-ordered sets, and so we can order their set by the principle of first differences, since no one of these maximal chains can be a proper initial segment of another one. We denote the corresponding order by $\preceq$: If c and c' are different elements of C, there exists a least ordinal κ, for which the κ - predecessors (in L) c_κ and c'_κ of c resp. c' differ. Since T is semi-chain-complete, κ is a successor ordinal $\alpha + 1$. For otherwise c_κ and c'_κ had both to be the supremum of the set of their (common) predecessors. Now we put $c \preceq c' \iff c_\kappa \leq_x c'_\kappa$ for that x of the level L_α of which c_κ and c'_κ are immediate successors. It can easily be verified that $\preceq$ is indeed a linear order on C, this corresponds largely to the ordering by first differences.

(V) $(C, \preceq)$ is dense.

Let $a \neq b$ be elements of C with $a \preceq b$, (a, b) their open interval in C. Then there exists an element $t \in L$ for which the final segment $[t)$ of t in L is $\subseteq (a, b)$. For let κ be the least ordinal for which the κ - predecessors a_κ, b_κ of a (resp. b) in L satisfy $a_\kappa <_x b_\kappa$, where x is their common immediate predecessor in L. Then there exists a t_κ of the same level L_κ with $a_\kappa < t_\kappa < b_\kappa$, for $\mathrm{tp}(N_x) = \eta$. Every $t \in C$ that begins with the path (in L) ending in t_κ then satisfies $[t) \subseteq (a, b)$. Thus $(C, \preceq)$ is dense.

If now (a_i, b_i), $i \in I$, is a set of pairwise disjoint intervals of C, we choose elements t_i, $i \in I$, with $[t_i) \subseteq (a_i, b_i)$. These final segments (in L) $[t_i)$, $i \in I$, are pairwise disjoint, and this implies that the t_i, $i \in I$, form an antichain of L, and thus I is countable.

(VI) C has no countable subset which is dense in C.

Let B be a countable subset of C. Every branch $b \in B$ has as height an ordinal $< \omega_1$, and then there exists an ordinal $\mu \leq \omega_1$ such that all $b \in B$ have a height $< \mu$ in T. Now for every element $x \in L$ with height $\geq \mu$ the final segment $[x)$ of x in L contains by (II) infinitely many elements of L, but no element of B, so that B cannot be dense in C. This proves (VI). And so $(C, \preceq)$ is a Suslin chain.

Chapter 9
On the order structure of power sets

In the class of posets the power sets deserve great interest. They occur in many mathematical considerations, and every poset is isomorphic to a subset of a power set (see 1.9.9). With regard to this, order theory is nothing else than the theory of power sets and their subsets.

When a poset is investigated, a first overview can be obtained by studying the structure of its chains and antichains, in particular the maximal ones. Their order type, resp. cardinality, determines a great deal of the properties of the set under consideration.

9.1 Antichains in power sets

If S is a finite set with $n \in \mathbf{N}$ elements, we can immediately determine the maximal chains of the power set $\mathfrak{P}(S)$ of S:

1.1 Theorem. *Let S be a set with $n \in \mathbf{N}$ elements. Then every maximal chain of $\mathfrak{P}(S)$ has the form $\{\emptyset, \{a_1\}, \{a_1, a_2\}, \ldots, \{a_1, \ldots, a_n\}\}$, so that every linear order $\{a_1, \ldots, a_n\}_<$ of S yields a maximal chain of $\mathfrak{P}(S)$.*

Since an n-element set admits exactly $n!$ permutations, the number of maximal chains of $\mathfrak{P}(S)$ is $n!$.

The question concerning the maximal antichains of $\mathfrak{P}_n := \mathfrak{P}(\{1, \ldots, n\})$ is more complicated. In particular one is interested to know those antichains which are not only maximal, but also of maximal size. In other words: Which antichains have as cardinality the width of $\mathfrak{P}_n$? A maximal antichain of $\mathfrak{P}_n$ is of course every subset which contains exactly all k - element subsets of $\{1, \ldots, n\}$ for a fixed $k \in \{0, \ldots, n\}$. We call this set the *k-level* of $\mathfrak{P}_n$. In a famous paper [166] Sperner determined in 1928 the width of $\mathfrak{P}_n$ and its maximal-sized antichains. His paper is one of the pioneering publications in the field "Combinatorics of finite sets". This discipline has many correlations with order theory. We refer here to the textbooks [5] of Anderson and [32] of Engel/Gronau.We introduce a notation which is analogous to that of $\binom{n}{k}$ for numbers n, k.

1.2 Definition. *If S is a set and k a cardinal, then $\binom{S}{k}$ denotes the set of all k-element subsets of S.*

Before we proceed we recall some elementary facts about the binomial coefficients $\binom{n}{k}$.

1.3 Remark. Let n and $k \leq n$ be natural numbers. Then $\binom{n}{k}$ is defined as $\frac{n \cdot (n-1) \cdots (n-(k-1))}{1 \cdot 2 \cdots k}$. We further put $\binom{n}{0} = 1$. The number $\binom{n}{k}$ is the cardinality of the set of all k - element subsets of an n - element set.

Then $\binom{n}{k} = \binom{n}{n-k}$. This follows e.g. by considering the function f, which maps every k - element subset K of an n - element set S onto its complement $S \setminus K$. This function is bijective, and so $\binom{S}{k}$ and $\binom{S}{n-k}$ have the same number of elements.

For $k > 0$ there holds the formula $\binom{n}{k} = \binom{n}{k-1} \cdot \frac{n-(k-1)}{k}$. It follows immediately from the definition.

The factor $\frac{n-(k-1)}{k}$ is > 1 exactly for $k \leq \frac{n}{2}$. And so we obtain from the last formula, that the values $\binom{n}{0}, \binom{n}{1}, \ldots, \binom{n}{k}$ strictly increase until $k := \lfloor \frac{n}{2} \rfloor$. If n is even, then the numbers of the sequence $\binom{n}{k}, \binom{n}{k+1}, \ldots, \binom{n}{n}$ strictly decrease. If n is odd $= 2k+1$, then $\binom{n}{k} = \binom{n}{k+1}$, and from $k + 1$ on the binomial expressions strictly decrease.

Before we deal with Sperner's theorem we prove the very useful LYM-inequality. First we remark:

1.4 Lemma. *Let S be a set with $n \in \mathbf{N}$ elements, $X \subseteq S$ a subset with $|X| = s \leq n$. Then the number of maximal chains of $\mathfrak{P}(S)$ that contain X is $= s! \cdot (n - s)!$.*

Proof. A maximal chain of $\mathfrak{P}(S)$ which contains X is obtained by a permutation $a_1, \ldots, a_n$ of the elements of S, for which $\{a_1, \ldots, a_s\} = X$. There are $s!$ possibilities to find such an arrangement of the elements of X, and for each of them there are $(n - s)!$ possibilities to prolong it with sequences whose elements are in $S \setminus X$. On the whole we so obtain $s! \cdot (n - s)!$ maximal chains which contain X.

Now we can prove the LYM-inequality. This acronym refers to three mathematicians who had come across related things, Lubell [115], Yamamoto [181] and Meschalkin [117].

1.5 Theorem. *Let S be a set with $n \in \mathbf{N}$ elements and $\mathfrak{A}$ an antichain of $\mathfrak{P}(S)$. Then there holds*

(1) $\sum \{ \binom{n}{|X|}^{-1} | X \in \mathfrak{A} \} \leq 1$ *(LYM-inequality).*

Proof. We have

(2) $\sum\{|X|! \cdot (n - |X|)! \, | \, X \in \mathfrak{A}\} \leq n!$.

For the summand $|X|! \cdot (n - |X|)!$ is the number of maximal chains of $\mathfrak{P}(S)$, which contain X. In the sum (2) no chain is counted twice, since every maximal chain of $\mathfrak{P}(S)$ contains at most one X of the antichain $\mathfrak{A}$. The left side of (2) counts a set of maximal chains of $\mathfrak{P}(S)$, and is hence $\leq n!$ by 1.1. Division by $n!$ yields $\sum\{\frac{|X|! \cdot (n - |X|)!}{n!} | X \in \mathfrak{A}\} \leq 1$, and this is the same as (1).

One can refine the statement of 1.5 in the following way:

1.5′ Theorem. *Let the assumptions of 1.5 be satisfied, and let p_i be the number of elements of $\mathfrak{A}$ which have i elements. Then $\sum\{p_i/\binom{n}{i} | i = 0, \ldots, n\} \leq 1$. Mind: $p_i/\binom{n}{i}$ is the proportion of elements of $\mathfrak{A}$ in the i-level of $\mathfrak{P}(S)$.*

Proof. From (2) there follows $\sum\{i! \cdot (n - i)! \cdot p_i | i = 0, \ldots, n\} \leq n!$, and division by $n!$ yields the assertion.

We can now prove Sperner's theorem [166]:

1.6 Theorem. *Let S be a set with $n \in \mathbf{N}$ elements, $\mathfrak{A}$ an antichain of $\mathfrak{P}(S)$, $k := \lfloor \frac{n}{2} \rfloor$. Then $|\mathfrak{A}| \leq \binom{n}{k}$ holds. In particular:*
  a) If n is even, and then $= 2k$ for some $k \in \mathbf{N}$, then $\mathfrak{A} = \binom{S}{k}$.
  b) If n is odd, say $= 2k + 1$, then $\mathfrak{A} = \binom{S}{k}$ or $\mathfrak{A} = \binom{S}{k+1}$.

Proof. For every $i \in \{0, \ldots, n\}$ we have $\binom{n}{i} \leq \binom{n}{k}$, and then the left side of (1) is $\geq |\mathfrak{A}| \cdot \binom{n}{k}^{-1}$, and this is ≤ 1 by (1). Thus $|\mathfrak{A}| \leq \binom{n}{k}$.

Let now $\mathfrak{A}$ be an antichain of $\mathfrak{P}(S)$ with $\binom{n}{k}$ elements. Then (1) yields for this $\mathfrak{A}$:

$1 \geq \sum\{\binom{n}{|X|}^{-1} | X \in \mathfrak{A}\} \geq |\mathfrak{A}| \cdot \binom{n}{k}^{-1} = 1$. So the sum is $= 1$. It is the sum of $|\mathfrak{A}| = \binom{n}{k}$ summands $\binom{n}{|X|}^{-1}$, each of which is $\geq \binom{n}{k}^{-1}$. But this is only possible if every summand is $= \binom{n}{k}^{-1}$. So every $X \in \mathfrak{A}$ has size k, if n is even, and then $\mathfrak{A} = \binom{S}{k}$ holds, so that a) is proved.

Let now n be odd $= 2k + 1$. Then every $X \in \mathfrak{A}$ must satisfy $\binom{n}{|X|} = \binom{n}{k}$, and thus X has size k or $k+1$. And we have $\mathfrak{A} \subseteq \binom{S}{k} \cup \binom{S}{k+1}$, which means:

(3) $\mathfrak{A} = \mathfrak{A}_k \cup \mathfrak{A}_{k+1}$ with $\mathfrak{A}_k \subseteq \binom{S}{k}$ and $\mathfrak{A}_{k+1} \subseteq \binom{S}{k+1}$.

If $\mathfrak{A}_{k+1}$ is empty, we are done. So suppose that it is $\neq \emptyset$. For the set $\mathfrak{B} := \binom{S}{k} \backslash \mathfrak{A}_k$ we obtain:

(4) All k - element subsets of a set $A \in \mathfrak{A}_{k+1}$ belong to $\mathfrak{B}$.

For such a subset cannot be in $\mathfrak{A}_k$ since $\mathfrak{A}$ is an antichain.

For $B \in \mathfrak{B}$ let r_B be the number of elements $A \in \mathfrak{A}_{k+1}$ with $B \subset A$. Then there holds $r_B \leq k + 1$ for all $B \in \mathfrak{B}$. For such a set A can only have the form $B \cup \{x\}$, where $x \in S \backslash B$, and we have $|S \backslash B| = k + 1$.

Now we count the number of pairs (B, A) with $B \in \mathfrak{B}$, $A \in \mathfrak{A}_{k+1}$ and $B \subseteq A$ in two ways and obtain by (4):

(5) $\sum \{r_B | B \in \mathfrak{B}\} = |\mathfrak{A}_{k+1}| \cdot (k + 1)$.

By (3) we have $|\mathfrak{B}| = \binom{n}{k} - |\mathfrak{A}_k| = |\mathfrak{A}_{k+1}|$, and so the left side of (5) is a sum of $|\mathfrak{B}| = |\mathfrak{A}_{k+1}|$ numbers $r_B \leq k + 1$. But then all r_B, $B \in \mathfrak{B}$, must be $= k + 1$, so that we have:

(6) Each $B \in \mathfrak{B}$ has $k + 1$ supersets in $\mathfrak{A}_{k+1}$.

Now it turns out that $\mathfrak{A}_{k+1}$ contains with some set (and it was supposed to be non-empty) also all sets of $\binom{S}{k+1}$. For let $X \in \mathfrak{A}_{k+1}$. Then every set which arises from X by exchanging an element of X by an element of $S \backslash X$ is again in $\mathfrak{A}_{k+1}$. Indeed, if $X' = (X \backslash \{x\}) \cup \{y\}$ with $x \in X$, $y \in S \backslash X$, then $X \backslash \{x\} \in \mathfrak{B}$ by (4), and then $X' \in \mathfrak{A}_{k+1}$ by (6). By finitely many exchanges one can transform X into every other set of $\binom{S}{k+1}$, so that this set is $\subseteq \mathfrak{A}_{k+1}$. Now $\binom{S}{k+1}$ contains already $\binom{n}{k+1}$ sets, and so it is $= \mathfrak{A}_{k+1} = \mathfrak{A}$.

Sperner's theorem has a lot of generalizations, and there also exist several other proofs of this theorem or parts of it. We mention here a proof of Freese which under the assumptions of 1.6 proves that $|\mathfrak{A}| \leq \binom{n}{k}$ and a) of 1.6 holds.

1.7 Proof. (by Freese [46],1974). We consider the set $\mathcal{A}$ of all maximal-sized antichains of $\mathfrak{P}(S)$. Their size is the width w of $\mathfrak{P}(S)$, and so, by Dilworth's theorem, $\mathfrak{P}(n)$ can be covered with w disjoint chains $\mathfrak{C}_1, \ldots, \mathfrak{C}_w$. Let $\mathfrak{B}$ and $\mathfrak{D}$ be elements of $\mathcal{A}$. Then $\mathfrak{B}$ (resp. $\mathfrak{D}$) has exactly one element B_ν (resp. D_ν) in common with each $\mathfrak{C}_\nu$, $\nu = 1, \ldots, w$. And then $\{\max\{B_\nu, D_\nu\} = B_\nu \cup D_\nu | \nu = 1, \ldots, w\}$ is a w - element antichain of $\mathfrak{P}(S)$ (see 2.6.6 and 2.6.13) which is the supremum in the lattice $\mathcal{A}$ of maximal-sized antichains of $\mathfrak{P}(S)$ of $\mathfrak{B}$ and $\mathfrak{D}$. This lattice is complete (like every finite lattice), and so it has a greatest element $\mathfrak{M}$.

Let π be a permutation of S, and let $\mathfrak{X} := \{X_1, \ldots, X_w\} \in \mathcal{A}$. Then also $\mathfrak{X}_\pi := \{\pi[X_1], \ldots, \pi[X_w]\}$ is a w - element antichain and thus $\in \mathcal{A}$. The mapping $\Pi : \mathcal{A} \to \mathcal{A}$, which puts $\Pi(\mathfrak{X}) := \mathfrak{X}_\pi$ is an isomorphism of $\mathcal{A}$, and so it maps the greatest element $\mathfrak{M}$ of $\mathcal{A}$ onto itself.

If now $M \in \mathfrak{M}$, then $\pi[M]$ must again be in $\mathfrak{M}$, and it has the same cardinality as M. Among the sets $\pi[M]$, where π runs through the set of all permutations of S, are all subsets of S, that have the same cardinality as M, and therefore $\mathfrak{M}$ contains with an element M also all elements of that level of $\mathfrak{P}(S)$, to which M belongs, which means: $\mathfrak{M}$ is a union of levels of $\mathfrak{P}(S)$. But it cannot be the union of at least two levels, since every set of the lower level of these would be a strict subset of a set of the higher level. So $\mathfrak{M}$ is a level of $\mathfrak{P}(S)$, and then of the maximal cardinality $\binom{n}{k} = w$. Since $\mathfrak{M}$ is maximal each w-element antichain contains in case a) no set with more than k elements and symmetrically no with fewer than k elements, and a) of 1.6 follows.

Another proof of 1.6 was given by Lovasz [114].

In Sperner's original proof a concept was used to which the next definition refers:

1.8 Definition. Let $\mathfrak{K}$ be a set of k - element subsets of an n - element set S, where $k \leq n$ are non-negative integers. Then the *upper shadow* $\nabla \mathfrak{K}$ of $\mathfrak{K}$ is the set $\{X \subseteq S \| |X| = k + 1 \text{ and } X \supset K \text{ for a set } K \in \mathfrak{K}\}$.

The *lower shadow* $\Delta \mathfrak{K}$ of $\mathfrak{K}$ is the set $\{X \subseteq S \| |X| = k-1 \text{ and } X \subset K$ for a $K \in \mathfrak{K}$. (Instead of upper (resp. lower) shadow also the name *shade* (resp. *shadow*) is in use.)

So the upper (resp. lower) shadow of $\mathfrak{K}$ consists of all those subsets of S which have exactly one element more (resp. fewer) than a set of $\mathfrak{K}$.

For the upper and lower shadow of a set system we have two elementary inequalities:

1.9 Theorem. [166] *Under the assumptions of 1.8 there holds:*
a) $|\nabla \mathfrak{K}| \geq \frac{n-k}{k+1} \cdot |\mathfrak{K}|$ for $k < n$.
b) $|\Delta \mathfrak{K}| \geq \frac{k}{n-k+1} \cdot |\mathfrak{K}|$ for $k > 0$.

Proof. **a)** We count the number of pairs (K, X) with $K \in \mathfrak{K}$, $K \subset X \subseteq S$ and $|X \backslash K| = 1$. For each $K \in \mathfrak{K}$ there exist $n - k$ elements in $S \backslash K$, such that each of them can be added to K in order to obtain a $(k + 1)$ - element superset $\subseteq S$ of K. The number of the above pairs is thus $= (n - k) \cdot |\mathfrak{K}|$.

On the other hand we have for each X of such a pair $k + 1$ many k - element subsets, which are not necessarily in $\mathfrak{K}$. So the number of the above pairs (K, X) is $\leq |\nabla \mathfrak{K}| \cdot (k + 1)$. And this yields $(n - k) \cdot |\mathfrak{K}| \leq |\nabla \mathfrak{K}| \cdot (k + 1)$, which proves a).

b) In this case we count the number of pairs (X, K) with $K \in \mathfrak{K}$, $X \subset K$ and $|X| = k - 1$. Each $k \in \mathfrak{K}$ has k subsets of size $k - 1$, and so the number of pairs is $= |\mathfrak{K}| \cdot k$.

On the other hand we have for each X of such a pair (X, K) — hence with $X \in \Delta\mathfrak{K}$ —, $n - (k - 1)$ elements in $S \backslash X$, and thus $n - k + 1$ many k - element supersets of X within S which are not necessarily in $\mathfrak{K}$. This entails $|\Delta\mathfrak{K}| \cdot (n - k + 1) \geq |\mathfrak{K}| \cdot k$ and b).

In the following we prove some generalizations of Sperner's theorem. One of these is a lemma of Littlewood and Offord. First we mention a notion of Kleitman [98] and one of its theorems:

1.10 Definition. Let P be a finite poset with a rank function r. Then a set $\mathfrak{C}$ of maximal chains of P is said to form a *regular covering* of P by chains, if there holds: For each rank number k every element of rank k occurs in the same number of chains of $\mathfrak{C}$.

Then there holds the following theorem, which was proved by Kleitman [98] in a more general context:

1.11 Theorem. *Let P be a finite poset with a rank function r, and let $\mathfrak{C}$ be a regular covering of P by chains. For $x \in P$ let $N_{r(x)}$ be the number of elements of P which have rank $r(x)$. Then there holds for every subset $T \subseteq P$:*

$$|T| \leq \max_{C \in \mathfrak{C}} \sum \{N_{r(x)} | x \in C \cap T\}.$$

Proof. $S := \sum_{C \in \mathfrak{C}} \sum \{N_{r(x)} | x \in C \cap T\} = \sum_{x \in T} N_{r(x)} \cdot |\mathfrak{C}'|$, where $\mathfrak{C}'$ is the set of those $C \in \mathfrak{C}$, which contain x. Since $\mathfrak{C}$ is a regular covering of P by chains, the number $|\mathfrak{C}'|$ is the same for all elements of rank $r(x)$, which means it is $= \frac{|\mathfrak{C}|}{N_{r(x)}}$. Now the sum S is $= \sum_{x \in T} |\mathfrak{C}| = |T| \cdot |\mathfrak{C}|$. This implies that at least one of the $|\mathfrak{C}|$ summands of S is $\geq |T|$, and this proves the assertion.

With the last theorem one can now prove a theorem of Erdös [33]:

1.12 Theorem. *Let $\mathfrak{B}$ be a subset of $\mathfrak{P}(S)$, where S is a set with $n \in \mathbf{N}$ elements, such that $\mathfrak{B}$ has no chain with $k+1$ elements. Then $|\mathfrak{B}|$ is $\leq$ the sum of the k greatest binomial coefficients $\binom{n}{i}, (i \in \{0, \ldots, n\})$.*

Proof. In order to apply 1.11 we first construct a regular covering of $\mathfrak{P}(S)$ by chains of maximal length. Let $a_1, \ldots, a_n$ be an arrangement of the elements of S. To this there corresponds the maximal chain $C := \{\emptyset, \{a_1\}, \{a_1, a_2\}, \ldots, \{a_1, \ldots, a_n\}\}$ of the initial segments. We so obtain

$n!$ chains, which form the set $\mathfrak{C}$. Every k - element subset $T \subseteq S$ is in exactly $k! \cdot (n-k)!$ chains of $\mathfrak{C}$ (by 1.4), and therefore $\mathfrak{C}$ is a regular covering of P by chains.

According to 1.11 we have $|\mathfrak{B}| \leq \max_{C \in \mathfrak{C}} \sum \{N_{r(x)} | x \in C \cap \mathfrak{B}\}$. Now, by assumption, every $C \in \mathfrak{C}$ has at most k elements in common with $\mathfrak{B}$, so that in the last sum at most k elements x contribute to $\sum N_{r(x)}$. Therefore $|\mathfrak{B}|$ is $\leq$ the sum of the k greatest $N_{r(x)}$.

1.13 Remark. If we take $k = 1$ in 1.12, then this means that $\mathfrak{B}$ is an antichain. And the conclusion yields $|\mathfrak{B}| \leq \binom{n}{k}$ with $k := \lfloor \frac{n}{2} \rfloor$, which is again a part of the statement of 1.6.

In the paper [98] Kleitman established some conditions which are equivalent to the property that a finite poset has a regular covering by chains. From this we later need the following:

1.14 Theorem. *Let P be a non-empty finite poset with a rank function, and for $i \in \omega$ let n_i be the number of elements of P which have rank i. Suppose that there is a regular covering $\mathfrak{C}$ of P by chains.*

Then P has the so-called LYM-property. That means: If A is an antichain of P with p_i members of rank i, then $\sum_i \frac{p_i}{n_i} \leq 1$.

Proof. Let A be an antichain of P and A_i the set of its members of rank i, so that $|A_i| = p_i$. Since $\mathfrak{C}$ is a regular covering of P every element of A_i is in the same number of chains of $\mathfrak{C}$, which means in $\frac{|\mathfrak{C}|}{n_i}$ chains. Let $\mathfrak{C}_i$ be the set of those chains of $\mathfrak{C}$ which contain an element of A_i. Then $|\mathfrak{C}_i| = p_i \cdot \frac{|\mathfrak{C}|}{n_i}$, since no chain of $\mathfrak{C}$ has more than one element of the antichain A_i. For different indices i the sets $\mathfrak{C}_i$ are disjoint since A is an antichain. Therefore we obtain $\sum_i |\mathfrak{C}_i| = \sum_i (p_i \cdot \frac{|\mathfrak{C}|}{n_i}) \leq |\mathfrak{C}|$, which entails our assertion.

In the following we prove a theorem which has some resemblance with Sperner's theorem, but which is no direct consequence of it.

1.15 Theorem. *Let S be a set with $n \in \mathbf{N}$ elements, j an integer with $j \leq \lfloor \frac{n}{2} \rfloor$. Then the set P of all subsets of S with $\leq j$ elements has width $\binom{n}{j}$, and the only antichain of P which has $\binom{n}{j}$ elements is $\binom{S}{j}$.*

Proof. We consider the set $\mathfrak{C}$ of all chains $\{\emptyset, \{a_1\}, \{a_1, a_2\}, \ldots, \{a_1, \ldots, a_j\}\}$, where the system $(a_1, \ldots, a_j)$ runs through all combinations of j different elements of S. Then $\mathfrak{C}$ is a regular covering of P. Indeed, every $i(\leq j)$ - element subset of S is contained in the same number of chains of $\mathfrak{C}$, namely in $i! \cdot \binom{n-i}{j-i}$ of them. The number of elements of P with rank i is $n_i := \binom{n}{i} \leq w$, where w is the width of P. Now we can apply 1.14:

Let $\mathcal{A}$ be an antichain of P with w elements. And let p_i denote the number of elements of $\mathcal{A}$ which have i elements, and thus rank i. By 1.14 now there follows:

(1) $\sum\{\frac{p_i}{n_i}|i = 0,\ldots,j\} \leq 1$

and a fortiori $\sum\{p_i/\binom{n}{j}|i = 0,\ldots,j\} \leq 1$, and $w = \sum\{p_i|i = 0,\ldots,j\} \leq \binom{n}{j}$.

Of course also $w \geq$ (and then also $=$) $\binom{n}{j}$ holds.

For each i with $0 \leq i < j$ the inequality (1) entails $\frac{p_i}{\binom{n}{i}} + \frac{w-p_i}{\binom{n}{j}} \leq 1$, and so $\frac{p_i}{\binom{n}{i}} + 1 - \frac{p_i}{\binom{n}{j}} \leq 1$ and $\frac{p_i}{\binom{n}{i}} \leq \frac{p_i}{\binom{n}{j}}$. Then $p_i = 0$ must hold since otherwise we can divide by p_i and obtain the contradiction $\binom{n}{j} \leq \binom{n}{i}$. So all p_i, $i = 0,\ldots,j-1$ are $= 0$, and thus $\mathcal{A}$ must be a subset of $\binom{S}{j}$, and of course it is then equal to this set.

From 1.15 we extract the following specialization:

1.16 Corollary. *Let S be a set with $n \in \mathbf{N}$ elements, i an integer $< \lfloor \frac{n}{2} \rfloor$. Then the set $\binom{S}{i} \cup \binom{S}{i+1}$, equipped with $\subseteq$, has width $w = \binom{n}{i+1}$, and $\binom{S}{i+1}$ is the only antichain with w elements.*

With 1.16 we can prove a relation between "neighboring" ranks of a finite power set and a partition theorem:

1.17 Theorem. *Let S be a set with size $n \in \mathbf{N}$, $k := \lfloor \frac{n}{2} \rfloor$, i an integer $< k$. Then there exists an injective mapping g_i of $\binom{S}{i}$ into $\binom{S}{i+1}$.*

Further there exists a partition of $\cup\{\binom{S}{i}|0 \leq i \leq \lfloor \frac{n}{2} \rfloor\}$ into $\binom{n}{k}$ saturated chains.

Proof. According to 1.16 the set $\binom{S}{i} \cup \binom{S}{i+1}$ has width $\binom{n}{i+1}$. By Dilworth's theorem then this set can be decomposed into $\binom{n}{i+1}$ disjoint chains. Each of these must contain exactly one element of $\binom{S}{i+1}$ and no or exactly one element of $\binom{S}{i}$. If an element $x \in \binom{S}{i}$ is in one of the chains, we map it to the greatest element of that chain. So we obtain an injective mapping g_i satisfying the assertion.

With the mappings $g_0,\ldots,g_{k-1}$ we now construct a partition of $\mathfrak{T} := \{T \subseteq S||T| \leq k\}$ into saturated chains. This contains (as longest chain)
$$\{\emptyset, g_0(\emptyset), g_1(g_0(\emptyset)),\ldots,g_{k-1}(g_{k-2}(\ldots,g_0(\emptyset)\ldots))\}.$$

If in general A_{i+1} is an element of $\binom{S}{i+1}$, which is not the g_i - image of an element of $\binom{S}{i}$, we take its g_{i+1}- image A_{i+2}, from this the g_{i+2} -

image, from this the g_{i+3} - image and so on until we have reached an element of $\binom{S}{k}$. In this manner we obtain a decomposition $\mathfrak{D}$ of $\mathfrak{T}$ in disjoint saturated chains.

The statement of 1.17 can be strengthened to a partition of the whole set $\mathfrak{P}(S)$ into saturated chains, which also satisfy a symmetry condition:

1.18 Definition. Let P be a finite ranked poset with rank function r. Then the elements $x_1, \ldots, x_h$ (where h is an integer ≥ 1) of P form a *symmetric chain* C of P, if the following two conditions are satisfied:
1) For $i = 1, \ldots, h-1$ the element x_{i+1} is an immediate successor of x_i. The same: C is a saturated chain.
2) $r(x_1) + r(x_h) = r(P)$ (= the highest rank).

The second property states, roughly speaking, that C is symmetric to the middle rank. For it includes that C contains with an element of rank i also an element of rank $r(P) - i$.

In connection with a prize problem which was posed by the Dutch Wiskundig Genootschap in 1949, de Bruijn, Tengbergen and Kruyswijk [19] proved the following:

1.19 Theorem. *Let S be a set with $n \in \mathbf{N}$ elements. Then the power set $\mathfrak{P}(S)$,equipped with its natural rank function r, can be partitioned in symmetric chains.*

Proof. If S has only one element x, then the set which contains only the chain $\{\emptyset, \{x\}\}$ is a partition of $\mathfrak{P}(S)$ in symmetric chains. Suppose now that the theorem is proved for a fixed number $n \in \mathbf{N}$, and that S is a set with $n+1$ elements. W.r.o.g. we can assume $S = \{1, \ldots, n+1\}$. Let $\mathfrak{C}$ be a partition of $\mathfrak{P}(\{1, \ldots, n\})$ in symmetric chains. Let $\mathfrak{C}_2$ (resp.$\mathfrak{C}_1$) be the set of those chains $C = \{C_1, \ldots, C_t\}_< \in \mathfrak{C}$ for which $t \geq 2$ (resp. $t = 1$) holds. For $C \in \mathfrak{C}_2$ we define two symmetric chains of $\mathfrak{P}(\{1, \ldots, n+1\}$:
$C' := \{C_1, \ldots, C_t, C_t \cup \{n+1\}\}$ and $C'' := \{C_1 \cup \{n+1\}, \ldots, C_{t-1} \cup \{n+1\}\}$.
Indeed $r(C_1) + r(C_t \cup \{n+1\}) = r(C_1) + r(C_t) + 1 = n+1$, and $r((C_1 \cup \{n+1\}) + r(C_{t-1} \cup \{n+1\}) = r(C_1) + 1 + r(C_{t-1}) + 1 = n+1$. The set $\mathfrak{C}_1$ can only be non-empty, if n is even, because of the symmetry of its chains. For $C \in \mathfrak{C}_1$ we put $C' := \{C, C \cup \{n+1\}\}$. Now it can easily be seen that $\cup\{\{C' \cup C''\}|C \in \mathfrak{C}_2\} \cup \{C'|C \in \mathfrak{C}_1\}$ is a partition of $\mathfrak{P}(\{1, \ldots, n+1\})$ in symmetric chains.

1.20 Remark. In [19] the authors proved a more general theorem, namely: Let D be the set of positive divisors of a natural number n. Then the relation $a|b$ (a divides b) defines an order in D, and D is ranked by: For a divisor d of n we put $r(d) :=$ the number of prime factors of d, counted with their multiplicity. So e.g. $r(36) = r(2 \cdot 2 \cdot 3 \cdot 3) = 4$. Then D has a partition in symmetric chains.

With 1.19 we obtain again a proof for a part of Sperner's theorem. Namely the following holds:

1.21 Theorem. *Let $S := \{1, \ldots, n\}$ and let $\mathfrak{D}$ be a partition of $\mathfrak{P}(S)$ in symmetric chains, w the width of $\mathfrak{P}(S)$. Then $|\mathfrak{D}| = \binom{n}{k} = w$ follows for $k := \lfloor \frac{n}{2} \rfloor$.*

Every antichain of $\mathfrak{P}(S)$ of size w has exactly one element in common with every chain of $\mathfrak{D}$.

Proof. Every chain of $\mathfrak{D}$ contains exactly one element of $\binom{S}{k}$, and every element of $\binom{S}{k}$ is in exactly one chain of $\mathfrak{D}$. Thus we have $|\mathfrak{D}| = \binom{n}{k}$.

Let $\mathfrak{A}$ be an antichain of $\mathfrak{P}(S)$ of size w. Each $X \in \mathfrak{A}$ is in exactly one of the chains of $\mathfrak{D}$, so that $\binom{n}{k} \leq w = |\mathfrak{A}| \leq |\mathfrak{D}| = \binom{n}{k}$, and thus $|\mathfrak{A}| = |\mathfrak{D}|$ holds. This proves the assertion.

One of the earlier generalizations of Sperner's theorem was given by Erdös/Ko/Rado [41] in 1961, where the cardinality of antichains is investigated which satisfy additional conditions. In 1972 Katona [94] gave another proof of their theorem which used a new method, namely the application of cyclic permutations:

1.22 Definition and **Lemma.** Let S be a finite set with at least two elements. Then we define a *cyclic permutation* of S as a relation ρ in S, for which there exists an indexation $a_1, \ldots, a_n$ of the elements of S, such that $a_i \rho a_{i+1}$ holds for $i = 1, \ldots, n-1$ and $a_n \rho a_1$.

Intuitively speaking: It is possible to arrange the elements of S in a circle so that, starting from a_1 and moving forwards in the positive sense, one reaches $a_2, \ldots, a_n$ in this succession and comes back to a_1. So a_1 is not distinguished in comparison with the other elements of S.

The number of cyclic permutations of an n - element set S is $(n-1)!$ For there are $n!$ permutations of S, and each cyclic permutation is determined by n permutations:
$(a_1, \ldots, a_n), (a_2, \ldots, a_n, a_1), (a_3, \ldots, a_n, a_1, a_2), \ldots, (a_n, a_1, \ldots, a_{n-1})$ *determine the same cyclic permutation.*

1.23 Theorem of Erdös/Ko/Rado [41]. *Let $\mathfrak{A} := \{A_1, \ldots, A_m\}$ be an antichain of the power set of a set S which has $n \in \mathbf{N}$ elements, where every two sets of the antichain have a non-empty intersection. Further it is assumed that there is a natural number $k \leq \frac{n}{2}$, so that $|A_i| \leq k$ holds for all $i = 1, \ldots, m$. Then we have $m \leq \binom{n-1}{k-1}$, and this is tight.*

Proof (by Katona [94]). First we show that m can be $= \binom{n-1}{k-1}$. To this purpose let $A_1, \ldots, A_m$ be all k - element subsets of S that contain a fixed element x of S. There are as many as there are possibilities to choose $k - 1$ elements in $S \backslash \{x\}$, namely $\binom{n-1}{k-1}$ many.

Before we start with the main part of the proof we first mention that it suffices to assume that $|A_i| = k$ holds for all $i = 1, \ldots, m$. Indeed, if some $|A_i|$ are $< k$, we can replace them by subsets of S which have size k. To this purpose we take a partition of $\mathfrak{P}(S)$ in symmetric chains and replace each A_i with size $< k$ by the uniquely defined subset of size k which contains A_i and which is an element of that chain of the partition which contains A_i. (By the way: Here also 1.17 could be applied.) The chosen supersets of these A_i are distinct since the chains of the partition are disjoint. So it is sufficient to prove 1.23 under the specialized form:

Let $A_1, \ldots, A_m$ be m pairwise distinct k - element subsets of $\{1, \ldots, n\}$, where $k \leq \frac{n}{2}$, and where the A_i have pairwise a non-empty intersection. Then $m \leq \binom{n-1}{k-1}$.

The proof applies the method of counting in two ways. We determine the size p of the set P of all pairs (α, A), where α is a cyclic permutation of $\{1, \ldots, n\}$, and where A is an A_i of $\mathfrak{A}$ which is encompassed by α. That means: A is a segment of α in the circle representation of α with positive orientation.

First we determine the size p^* of the set P^* of all pairs (α, K), where α is a cyclic permutation of $\{1, \ldots, n\}$ and K a k - element subset of $\{1, \ldots, n\}$, which is encompassed by α. We obtain $p^* = (n-1)! \cdot n = n!$. For by 1.22 there are $(n - 1)!$ possibilities for α, and for each α there are n sets K, which are encompassed by α.

By reasons of symmetry here every k - element subset $K \subseteq \{1, \ldots, n\}$ appears in the same number of pairs of P^*, which means in $p^*/\binom{n}{k}$ of them. And so for every $i \in \{1, \ldots, m\}$ the above number p of pairs (α, A_i) is $= n!/\binom{n}{k}$. The number p is the m - fold of this, since for i we have m possibilities. So we have obtained:

$$(1) \quad p = m \cdot n!/\binom{n}{k}.$$

Now we apply a second form of counting. First we prove:

(2) Every cyclic permutation α of $\{1, \ldots, n\}$ encompasses at most k of the sets A_i.

Indeed, let α be a cyclic permutation of $\{1, \ldots, n\}$ and A_h a set of $\{A_1, \ldots, A_m\}$, which is encompassed by α. On the circle representation A_h determines a certain sector. The rest of the sets of $\{A_1, \ldots, A_m\}$, which are encompassed by α, must have sectors, which intersect the sector of A_h. Among these let A_l (resp. A_r) be that set, which in the positive orientation is farmost to the left (resp. right) of A_h. Then the sectors of A_l and A_r must still have a non-empty intersection because of $A_l \cap A_r \neq \emptyset$.

Now the following is clear: All A_i which are encompassed by α, have sectors whose beginning b is between (in the wide sense) the beginning of A_l and the beginning of A_r. Therefore b can only be one of the k numbers of A_l, and this proves (2).

Now (2) yields the following estimation for $|P|$:

(3) $|P| \leq (n-1)! \cdot k$.

From (1) and (3) we conclude $m \cdot n! / \binom{n}{k} \leq (n-1)! \cdot k$, and further $m \leq \frac{k}{n} \cdot \binom{n}{k} = \binom{n-1}{k-1}$.

At the end of this section we also consider antichains in infinite power sets. Before we introduce an obvious concept:

1.24 Definition. Let $\mu > 0$ be an ordinal. It has by 4.8.5 a representation in Cantor's normal form as $\mu = \omega^{\alpha_0} \cdot a_0 + \cdots + \omega^{\alpha_n} \cdot a_n$, where n is a non-negative integer, $\alpha_0 > \cdots > \alpha_n$ is a strictly decreasing sequence of ordinals, and $a_0, \ldots, a_n$ are natural numbers. If $\alpha_n = 0$ holds we have $\omega^{\alpha_n} = 1$ by definition, and then the last summand in the normal form is a_n. Then μ is called *even* (resp. *odd*) if a_n is even (resp. odd). If $\alpha_n > 0$ or $\mu = 0$ we also call μ even. It is clear that even (resp. odd) non-negative integers are also even (resp. odd) ordinals.

1.25 Theorem. *Let S be an infinite set of cardinality $\aleph_\alpha$. Then the power set $\mathfrak{P}(S)$ has antichains of cardinality $2^{\aleph_\alpha} := c$. Moreover $\mathfrak{P}(S)$ has 2^c antichains of cardinality c.*

Proof. Evidently it suffices to consider the case where $S = \omega_\alpha$. We partition the set ω_α in the two-element segments $S_\nu := \{\nu, \nu+1\}$, where the ν are even ordinals $< \omega_\alpha$. Let $\mathfrak{M}$ be the set of all subsets $T \subseteq \omega_\alpha$, which have the property: T contains exactly one element from each S_ν. Evidently we have $|T| = \aleph_\alpha$.

If now A and B are different sets $\in \mathfrak{M}$, there is at least one ν, such that A contains an element $e_\nu \in S_\nu$, which is not in B. Then B contains that element d_ν of S_ν which is $\neq e_\nu$, and d_ν is not in A. Thus we have that $A\backslash B$ and $B\backslash A$ are non-empty. So A and B are incomparable, and $\mathfrak{M}$ is an antichain of $\mathfrak{P}(S)$.

The cardinality of $\mathfrak{M}$ is $2^{\aleph_\alpha}$, for we have $\aleph_\alpha$ segments S_ν, and for each of them we have two possible choices to obtain an element of $\mathfrak{M}$. So $\mathfrak{M}$ is equipotent to the set of all mappings of $\{S_\nu | \nu < \omega_\alpha\}$ in $\{0,1\}$.

There also exist 2^c antichains of $\mathfrak{P}(\omega_\alpha)$ which have cardinality c. For we can represent $\mathfrak{M}$ as $\mathfrak{M} = \{M_\nu | \nu < \omega(c)\}$. Now every subset of $\mathfrak{M}$, which contains (among others) the sets M_ν with even ν, has cardinality c. For there are c sets M_ν with odd ν, and each of their 2^c subsets can be added to $\{M_\nu | \nu < \omega(c)$ and ν even$\}$ to obtain an antichain of $\mathfrak{P}(\omega_\alpha)$ with c elements.

9.2 Contractive mappings in power sets

A *choice function* f in a set S ascribes to every non-empty subset $T \subseteq S$ an element $f(T) \in T$. In 6.1.14 we mentioned a generalization of this concept, which was introduced by Kinna/Wagner [96], namely generalized choice functions on a set S, which ascribe to every $T \subseteq S$, that has at least two elements, a proper subset of T. In this section we consider some combinatorial properties of power sets, in which we have a generalized choice function. First we define for general ordered sets:

2.1 Definition. Let P be a poset. A mapping $f : P \to P$ is said to be *contractive*, if for each $x \in P$ we have $f(x) \leq x$.

We call f *strictly contractive* or *regressing*, if $f(x) < x$ holds for all $x \in P$, which are not minimal elements, and for these we have $f(x) = x$.

We are chiefly interested in the case where P is a set of sets which contains with a set X also all subsets of X, and where the order of P is the set inclusion $\subseteq$.

There are many examples in mathematics, where from given sets certain special kinds of subsets are of interest. E.g. in topology one has the notion *open kernel* of a subset T of a topological space — this is the greatest open subset of T. Another example is the linear kernel, which was introduced in 2.4.1.

In many cases a contractive mapping $f : P \to P$ of a poset has the addional property that for every $T \subseteq P$ also $f(f(T)) = f(T)$ (Idempotency) holds and also $A \subseteq B \implies f(A) \subseteq f(B)$ (Isotonicity). If these conditions are satisfied f is called a *kernel function*, and this is the counterpart of a closure operator. In the following we require neither idempotency nor monotonicity from our contractive mappings!

2.2 Definition. Let $\mathfrak{X}$ be a set of sets, which contains with a set T also all subsets of T. In $\mathfrak{X}$ we have the order $\subseteq$. Let f be a contractive mapping on $\mathfrak{X}$. We define a *regressive chain* of length $r \in \mathbf{N}$ to be a set $\{f(X), f^2(X), \ldots, f^r(X)\}_>$ (so that $f(X) \supset \cdots \supset f^r(X)$), where X is an element of $\mathfrak{X}$, and where the iterates of f are defined in the usual manner: $f^1 := f$ and $f^{n+1}(X) := f(f^n(X))$.

A *fixed chain* (of f) is a chain $\mathfrak{Y} \subseteq \mathfrak{X}$ such that $f(X) = X$ holds for all $X \in \mathfrak{Y}$, and a *constant chain* (of f) is a chain $\mathfrak{C} \subseteq \mathfrak{X}$ for which all sets $f(C)$, $C \in \mathfrak{C}$, are equal.

The main theorem of this section states that for every $n \in \mathbf{N}$ there exists an $h(n) \in \mathbf{N}$, such that for every contractive function f in the power set $\mathfrak{P}(\{1, \ldots, h(n)\}$ there is a chain of size n which is regressive, fixed or constant. For its proof we need several preparations.

2.3 Definition. If A, B are finite subsets of $\mathbf{N}$ satisfying $A \subseteq B$ we define the *position* $p(A, B)$ of A in B as follows: If $A = \emptyset$ we put $p(A, B) := \emptyset$. If $A \neq \emptyset$, say $A = \{a_1, \ldots, a_k\}_<$, then there is a uniquely defined bijection φ of B onto the initial segment $\{1, \ldots, |B|\}$ of $\mathbf{N}$ which is $<$ - preserving. We put $x_i := \varphi(a_i)$ for $i = 1, \ldots, k$, and define $p(A, B) := \{x_1, \ldots, x_k\}$.

The following example illustrates this: Let $B := \{3, 4, 7, 9, 11, 14\}$ and $A := \{4, 9, 11\}$, then $p(A, B) = \{2, 4, 5\}$. And this $p(A, B)$ indicates that A contains the second, the fourth and the fifth number of B.

If A is a one-element subset of B, then its unique element is the $p(A, B)^{th}$ element of B.

The class of the contractive mappings has a subclass which is easier to handle:

2.4 Definition. Let $\mathfrak{C}$ be a non-empty set consisting of finite subsets of $\mathbf{N}$ such that $B \in \mathfrak{C}$ and $A \subseteq B$ implies $A \in \mathfrak{C}$. So the set $\{|B| \, | \, B \in \mathfrak{C}\}$ is an initial segment D of ω. Now a contractive mapping $f : \mathfrak{C} \to \mathfrak{C}$ is called *special* if, given any $t \in D$, for all $T \in \mathfrak{C}$ with $|T| = t$ the position of $f(T)$ in T is the same, hence only depending on t, so that here we can define the *position function* π of f on D by:

For $t \in D$ we put $\pi(t) := p(f(T), T)$, where T is an arbitrary set of $\mathfrak{E}$ with $|T| = t$. Of course, $|\pi(t)| = |f(T)|$ for every $T \in \mathfrak{E}$ with $|T| = t$.

We call f *n-special (for an $n \in \mathbf{N}$)*, if for the set $\mathfrak{A}$ of all n - element sets $\in \mathfrak{E}$ there holds: For each $A \in \mathfrak{A}$ the position $p(f(A), A)$ is the same set. So, a special contractive mapping f is n - special for all $n \in \mathbf{N}$, and conversely.

If e.g. f maps every non-empty set $E \in \mathfrak{E}$ onto its least element and $\emptyset$ onto $\emptyset$, f is special, and its corresponding position function π puts $\pi(0) = p(\emptyset, \emptyset) = \emptyset$ and $\pi(1) = \{1\}$.

In the following we perform a reduction process, by which we can reduce in some sense general contractive mappings to special ones. For this we need a form of Ramsey's theorem, which we now present.

2.5 Theorem. *Let n,p,q be natural numbers, where p and q are $\geq n$. Then there exists a natural number $r(p,q,n)$ (which only depends on p,q,n) with the following property:*

If S is a set with $|S| \geq r(p, q, n)$, and if $[S]^n = A \cup B$, then there exists a subset $P \subseteq S$ with $|P| = p$ and $[P]^n \subseteq A$, or there exists a subset $Q \subseteq S$ with $|Q| = q$ and $[Q]^n \subseteq B$.

Proof. For $n = 1$ and all $p, q \geq 1$ the statement is true. Here $r(p, q, 1) := p + q - 1$ fulfills the assertion: If $|S| \geq p + q - 1$ holds and $S = A \cup B$, then A must contain at least p elements or B at least q elements. Further we have:

(1) If $p = n$ or $q = n$ holds the assertion is fulfilled.

In the first case we put $r(n, q, n) := q$. If now $|S| \geq q$ and $[S]^n = A \cup B$ with $A \cap B = \emptyset$ holds, then there are n elements in S, whose set is in A, or otherwise all n-element subsets of S belong to B, and $[S]^n \subseteq B$ holds. Symmetrically we conclude the same in the second case.

Suppose now that $n \geq 2$ holds, and that the theorem is proved for $n - 1$ instead of n (and all p and q which are $\geq n - 1$). We prove the validity of the assertion for n by induction on the sum of p and q. Since p and q are $\geq n$, the least such sum is $2n$, and here the assertion holds by (1).

So we have to show, under the assumption that the theorem holds for $n - 1$: There exists a number $r(p, q, n)$ which satifies the statement, if it is already proved that a number $r(p', q', n)$ exists for the case where $p' + q' < p + q$.

Let now natural numbers p, q with $p > n < q$ be given, so that by induction hypothesis numbers $p_1 := r(p-1, q, n)$ and $q_1 := r(p, q-1, n)$ exist, which satisfy the assertion. We now verify it for the triple (p, q, n), and then our theorem is proved by induction. We show: There exists a number $r(p, q, n)$ which fulfills the statement, and for which holds:

(2) $r(p, q, n) \leq r(p_1, q_1, n - 1) + 1 := s$.

To this purpose let S be a set with at least s elements and $[S]^n = A \cup B$ and $A \cap B = \emptyset$. Let e be a fixed element of S, and $S' := S \setminus \{e\}$. We consider the decomposition $[S']^{n-1} = A' \cup B'$, where

A' is the set of those $(n-1)$ - element subsets T of S', for which $T \cup \{e\} \in A$, and

B' is the set of those $(n-1)$ - element subsets T of S', for which $T \cup \{e\} \in B$.

Because of (2) we have $|S'| \geq r(p_1, q_1, n - 1)$, and so, due to the induction hypothesis, there exist

1) a set P_1 of p_1 elements in S', such that $[P_1]^{n-1} \subseteq A'$, or

2) a set Q_1 of q_1 elements in S', such that $[Q_1]^{n-1} \subseteq B'$.

Suppose that 1) holds. Due to $p_1 = r(p-1, q, n)$ then there follows by considering $[P_1]^n = (A \cap [P_1]^n) \cup [B \cap [P_1]^n)$: There are q elements of P_1, such that all the n - element subsets of their set belong to B (and then we are done) or there exists a subset $X \subseteq P_1$ with $|X| = p - 1$ and $[X]^n \subseteq A$. In this case from $X \subseteq P_1$ and 1) we obtain

(*) $[X]^{n-1} \subseteq [P_1]^{n-1} \subseteq A'$.

Let $X_e := X \cup \{e\}$. Then $|X_e| = p$. It follows that $[X_e]^n \subseteq A$.

Indeed, let $N \subseteq X_e$ and $|N| = n$. For $e \notin N$ we have $N \subseteq X$ and $N \in [X]^n \subseteq A$ and thus $N \in A$.

For $e \in N$ we have $N = \{e\} \cup T$, where T is an $(n-1)$ - element subset of X and S'. So $T \in A'$ by (*), and by definition of A' now $T \cup \{e\} = N \in A$.

The case 2) is symmetric to case 1), and so (2) is proved, and with it the theorem.

The last theorem is easily extendable to the situation, where we have not only two, but finitely many classes:

2.6 Theorem. *Let $k \in \mathbf{N}$ and $n, p_1, \ldots, p_k$ be natural numbers with $p_\kappa \geq n$ for $\kappa = 1, \ldots, k$. Then there exists a natural number*

$r(p_1, \ldots, p_k, n) =: r$, *for which there holds:*

If S is a set with at least r elements and if $[S]^n = A_1 \cup \cdots \cup A_k$, then for at least one $\kappa \in \{1, \ldots, k\}$ there exists a set $P_\kappa \subseteq S$ with $|P_\kappa| = p_\kappa$ and $[P_\kappa]^n \subseteq A_\kappa$.

Proof. For $k = 2$ the assertion holds by 2.5. Let now k be an integer ≥ 2, for which the statement is already proved, so that a number $r(p_2, \ldots, p_{k+1}, n)$ already exists. We show: There exists a number

$$r(p_1, \ldots, p_{k+1}, n) \leq r(p_1, r(p_2, \ldots, p_{k+1}, n), n) =: s.$$

Let S be a set with $|S| = s$ and $[S]^n = A_1 \cup \cdots \cup A_{k+1}$, where the summands are pairwise disjoint. We consider the decomposition $[S]^n = A_1 \cup (A_2 \cup \cdots \cup A_{k+1})$ in only two classes. Here 2.5 can be applied, and there exists a set $P_1 \subseteq S$ with $|P_1| = p_1$ and $[P_1]^n \subseteq A_1$, or there exists a set $Q_1 \subseteq S$ with $|Q_1| = r(p_2, \ldots, p_{k+1}, n)$ and $[Q_1]^n \subseteq A_2 \cup \cdots \cup A_{k+1}$. In the last situation there exists an $i \in \{2, \ldots, k+1\}$ and a $P_i \subseteq Q_1$ with $|P_i| = p_i$ and $[P_i]^n \subseteq A_i$, and this proves our assertion by induction hypothesis.

From the last theorem we obtain immediately by putting all p_κ equal and $= p$:

2.7 Theorem. *Let p,k,n be natural numbers with $p \geq n$. Then there exists a number $R := R(p, k, n) \in \mathbf{N}$ such that the following holds: If S is a set with $|S| = R$, and if $[S]^n = A_1 \cup \cdots \cup A_k$, then there exists a set $P \subseteq S$ with $|P| = p$ and a $\kappa \in \{1, \ldots, k\}$ for which $[P]^n \subseteq A_\kappa$.*

Now we apply the idea which was used in the proof of [63], Satz 3.1 to obtain the following reduction lemma:

2.8 Lemma. *Given $n \in \mathbf{N}$, there exists a natural number $m := m(n)$, such that the following holds:*

If $M := \{1, \ldots, m\}$ and if $f : \mathfrak{P}(M) \to \mathfrak{P}(M)$ is a contractive mapping, then there exists a subset $S \subseteq M$ with $|S| \geq n$ such that the restriction of f to $\mathfrak{P}(S)$ is a special contractive mapping (in the sense of 2.4).

Proof. In this proof let $R(p, k, n)$ always be the least number satisfying 2.7. We put $m_0 := n$, and define recursively numbers $m_0, \ldots, m_n$ as follows: For $i = 1, \ldots, n$ we put:

(1_i) $m_i := R(m_{i-1}, 2^i, i)$,

so that $m_1 := R(m_0, 2^1, 1), \ldots, m_n := R(m_{n-1}, 2^n, n)$. Then $m := m_n$ satisfies our assertion. To see this we put $M =: M_n := \{1, \ldots, m_n\}$.

If T is an n - element subset of M, the set $f(T)$ has one of 2^n possible positions $p_1, \ldots, p_{2^n}$ in T, namely so many as there are subsets of T. Then we consider the representation $[M]^n = \mathfrak{A}_1 \cup \cdots \cup \mathfrak{A}_{2^n}$, where for $i \in \{1, \ldots, 2^n\}$ $\mathfrak{A}_i$ is the set of all those n - element subsets T of S, for which $f(T)$ has the position p_i in T. Due to (1_n) then there exists a subset $M_{n-1} \subseteq M_n$ of size m_{n-1} such that $[M_{n-1}]^n$ is a subset of one of the $\mathfrak{A}_i$, and then M_{n-1} is n - special.

Suppose that we have already defined sets $M_n \supseteq M_{n-1} \supseteq \cdots \supseteq M_i$ for an i with $0 < i < n$, so that $|M_j|$ is $= m_j$ and $(j+1)$ - special for $i \leq j < n$. Then, similar to the case $i = n$, and due to (1_i) there exists a set $M_{i-1} \subseteq M_i$ of cardinality m_{i-1}, which is i−special. Recursively we so obtain a sequence $M_n \supseteq M_{n-1} \supseteq \cdots \supseteq M_0$, where for each $i < n$ the set M_i is $(i+1)$ - special, and also j - special for all j with $i < j \leq n$. Finally M_0 is i - special for $i = 1, \ldots, n$. That means: f is a special contractive mapping over $M_0 =: S$. Now S has size $m_0 = n$ and so satisfies our assertion.

For the case of special mappings we now prove:

2.9 Theorem [66]. *Let $\mathfrak{E}$ be the set of all finite subsets of $\mathbf{N}$ and $f : \mathfrak{E} \to \mathfrak{E}$ a special contractive mapping. Then there is an infinite fixed chain, or for every $r \in \mathbf{N}$ there is a regressive chain of length r or a constant chain of length r.*

Proof. According to 2.4 the special mapping f is determined by its position function π. We distinguish two cases:

Case 1: There is an infinite subset $W \subseteq \mathbf{N}$ such that $|\pi(t)| = t$ for every $t \in W$. For the sets $T \in \mathfrak{E}$ with $|T| \in W$ we then have $f(T) = T$. Thus the sets $M_w := \{1, \ldots, w\}$, $w \in W$, form an infinite fixed chain.

Case 2. There exists a number $u \in \mathbf{N}$ such that $|\pi(t)| < t$ holds for all natural numbers $t \geq u$. We split Case 2 into two subcases:

Case 2a. $|\pi(t)| \to \infty$ for $t \to \infty$.

Now, given an $n_0 > u$ and $r \in \mathbf{N}$, we can successively construct a strictly ascending sequence of integers $n_0 < n_1 < \cdots < n_r$ such that
$$|\pi(t)| \geq n_0 \text{ for } t \geq n_1, |\pi(t)| \geq n_1 \text{ for } t \geq n_2, \ldots, |\pi(t)| \geq n_r \text{ for}$$
$t \geq n_{r-1}$.

If then a set $T \subseteq \mathbf{N}$ has n_r elements, it follows that $|f(T)| = |\pi(t)| \geq n_{r-1}, |f(f(T))| \geq n_{r-2}, \ldots, |f^{(r)}(T)| \geq n_0$. And now $\{f(T), f^{(2)}(T), \ldots, f^{(r)}(T)\}$ is a regressive chain of length r.

Case 2b. There exists a number $b \in \mathbf{N}$ and an infinite subset $U \subseteq \mathbf{N}$ such that $|\pi(t)| < b$ for all $t \in U$.

By the pigeonhole principle, then there also is an infinite set $X \subseteq \mathbf{N}$ such that for all $x \in X$ we have $|\pi(x)| = q$, where q is a fixed nonnegative integer $< b$. The case $q = 0$ is trivial: In this case for all $x \in X$ we have $\pi(x) = \emptyset$ and $f(M) = \emptyset$ for the $M \in \mathfrak{E}$ with $|M| = x$, and this yields an infinite constant chain. So we may assume $q > 0$.

To every $x \in X$ we ascribe a $(q+1)$ - vector $v(x) = (x_1, \ldots, x_{q+1})$ as follows. Let $\pi(x) = \{s_1, \ldots, s_q\}_<$. Then $v(x)$ is defined to be the vector of the lengths of the pieces into which $\pi(x)$ partitions the interval $[0, x]$:
$$v(x) := (x_1, \ldots, x_{q+1}) := (s_1, s_2 - s_1, s_3 - s_2, \ldots, s_q - s_{q-1}, x - s_q).$$
Hence we have $\pi(x) = (x_1, x_1 + x_2, \ldots, x_1 + \ldots + x_q)$ for $x \in X$.

We introduce a partial order in the set $\{v(x) | x \in X\}$ as follows: If $x', x'' \in X$ we put
$$v(x') \leq v(x'') \text{ iff } x'_i \leq x''_i \text{ for all } i \text{ - components}, i = 1, \ldots, q+1, \text{ of}$$
x', x''.

Now from Higman's theorem 6.3 there follows, since the set of all words over $\mathbf{N}$ is wqo:

(1) There is an infinite subset $Y \subseteq X$ such that for all $x', x'' \in Y$ with $x' \leq x''$ we have $v(x') \leq v(x'')$.

We can prove (1) also without referring to Higman's theorem, making use of an idea of Bernd Voigt:

For $i \in \{0, \ldots, q+1\}$ we define a set $A_i \subseteq [X]^2$ as follows: Let x', x'' be elements of X with $x' < x''$. We put $\{x', x''\} \in A_0$ iff $x'_\nu \leq x''_\nu$ for all $\nu = 1, \ldots, q+1$. If $\{x', x''\} \in A_0$ does not hold, there is a first index $i \in \{1, \ldots, q+1\}$ with $x'_i > x''_i$ and then we put $\{x', x''\} \in A_i$. The set $[X]^2$ is $= \cup\{A_i | i = 0, \ldots, q+1\}$. In a graph-theoretical formulation: The sets A_i, $i \in \{0, \ldots, q+1\}$, define an edge coloring of the complete graph, whose vertices are the numbers $x \in X$, with $q + 2$ colors. Now 6.4.8' entails that there exists an infinite subset $Y \subseteq \mathbf{N}$, such that $[Y]^2$ is a subset of an A_i with $i \in \{0, \ldots, q+1\}$. And this i can only be $i = 0$, since in $\mathbf{N}$ we have no infinite descending chains. So (1) is again proved.

Now, given $r \in \mathbf{N}$, we can construct a constant chain for f of length r. Let $y_1, \ldots, y_r$ be the first r elements of Y (in its natural order). We shall define recursively sets $M(y_r), M(y_{r-1}), \ldots, M(y_1)$, so that they have the same f - image. We abbreviate $y := y_r$ and $x := y_{r-1}$ and consider the position functions $\pi(x) = \{\xi_1, \ldots, \xi_q\}$ and $\pi(y) = \{\eta_1, \ldots, \eta_q\}$. Let $M(y)$ be an arbitrary set of y natural numbers. Then the set $f(M(y))$ contains the η_1^{th}, the $\eta_2^{th}, \ldots$, and the η_q^{th} element of $M(y)$. Now we define $M(x)$ as follows: $M(x)$ contains the ξ_1 last elements (in the order of $\mathbf{N}$) which are $\leq$ the η_1^{th} element of $M(y)$, and for every $i = 2, \ldots, q$ the $\xi_i - \xi_{i-1}$ last elements of $M(y)$, which are $\leq$ the η_i^{th} element of $M(y)$, and finally the $x - \xi_q$ last elements of $M(y)$. Then $M(x)$ has $\xi_1 + \sum \{\xi_i - \xi_{i-1} | i = 2, \ldots, q\} + x - \xi_q = x$ elements, and its ξ_i^{th} element is the same as the η_i^{th} element of $M(y)$. Since $\xi_1 \leq \eta_1$ and $\xi_i - \xi_{i-1} \leq \eta_i - \eta_{i-1}$ holds for $i = 2 \ldots, q$ and $x - \xi_q \leq y - \eta_q$ this construction is possible. Now $f(M(x))$ is $= f(M(y))$.

So we constructed to $M(y_r)$ a set $M(y_{r-1})$. In the same manner we construct to the latter set a set $M(y_{r-2})$ with $f(M(y_{r-2})) = f(M(y_{r-1}))$, and we continue this construction until we have reached a set $M(y_1)$ with the same f - value. Then the sets $M(y_1), \ldots, M(y_r)$ form a fixed chain of length r with respect to f.

2.10 Remark. Subsequently to the last theorem the question arises whether the statement can be sharpened so that there also exists an infinite chain of $\mathfrak{E}$, which is fixed, regressive or constant. A simple counter-example is given by the mapping $f : \mathfrak{E} \to \mathfrak{E}$, which maps every non-empty $T \in \mathfrak{E}$ onto the greatest element of T.

For special contractive mappings now there follows:

2.11 Theorem [66]. *Given $n \in \mathbf{N}$, there is a number $h^*(n) \in \mathbf{N}$ such that the following holds: If g is a special contractive mapping on $\mathfrak{P}(h^*(n))$, then there exists a chain of length n which is fixed, constant, or regressive.*

Proof. Assume that the statement is false. Then there exists a number $n^* \in \mathbf{N}$ such that for every $k \in \mathbf{N}$ there is a special contractive mapping f_k on $\mathfrak{P}(k)$ such that there is no chain of length n^* which is fixed, constant, or regressive for f_k.

We enumerate the set $\mathfrak{E}$ of all finite subsets of $\mathbf{N}$ in a sequence of type ω; $\mathfrak{E} = \{F_i | i \in \mathbf{N}\}$. The set $\{f_k(F_1) | k \in \mathbf{N}\}$ is finite, and so there is an infinite set of indices k, say N_1, such that all sets $f_k(F_1)$, $k \in N_1$, are equal. Further there exists an infinite subset $N_2 \subseteq N_1$ such that all sets $f_k(F_2)$, $k \in N_2$, are equal, and so on. Now we define a mapping $f : \mathfrak{E} \to \mathfrak{E}$ as follows: For $i \in \mathbf{N}$ let $f(F_i)$ be the common value $f_k(F_i)$ for $k \in N_i$. Then f is a special contractive mapping on $\mathfrak{E}$ since the mappings f_k have the same property. Further we see:

(1) There is no chain of length n^* which is fixed, constant, or regressive for f.

Otherwise such a chain would have the corresponding property also for one of the mappings f_k which gives a contradiction. So (1) would be valid. But it contradicts 2.9, and therefore the indirect assumption is false and the theorem proved.

Together with the reduction Lemma 2.8 the last theorem now immediately yields one of the main theorems of this section:

2.12 Theorem [66]. *Given $n \in \mathbf{N}$, there is a number $h(n) \in \mathbf{N}$ such that the following holds: If S is a set of $h(n)$ elements, and if $f : \mathfrak{P}(S) \to \mathfrak{P}(S)$ is a contractive mapping, then there exists an n - element chain of $\mathfrak{P}(S)$ which is fixed, constant, or regressive.*

2.13 Corollary. *If S is a set of $h(n)$ elements and if $f : \mathfrak{P}(S) \to \mathfrak{P}(S)$ is regressing, then there is an n - element chain of $\mathfrak{P}(S)$ which is constant or regressive.*

The Theorem 2.11 sharpens the following two theorems of Rado [146]:

2.14 Theorem [146]. *Given $n \in \mathbf{N}$, there is a positive integer $h_3(n)$ such that the following statement holds: If S is a set of $h_3(n)$ elements, and if $f(X)$, for every subset X of S, is a subset of X, then there always are subsets $X_0, \ldots, X_n$ of S such that $X_0 \subset \cdots \subset X_n$ and $f(X_0) \subseteq f(X_1) \subseteq \cdots \subseteq f(X_n)$.*

This theorem is a consequence of 2.12. For if there is an $n+1$- element chain of $\mathfrak{P}(S)$, which is fixed or constant, this is clear. And if there is a regressive chain $\{f^n(X), \ldots, f^1(X), f^0(X) = X\}$, then $X_\nu := f^{n-\nu}(X)$ for $\nu = 0, \ldots, n$ satisfies the assertion.

2.15 Theorem [146]. *Given positive integers n and j, there is a positive integer $h_2(n, j)$ such that the following holds: If S is a set of $h_2(n, j)$ elements, and if $f(X)$, for every subset X of S, is a subset of X having at most j elements, then there always are subsets $X_0, \ldots, X_n$ of S such that $X_0 \subset \cdots \subset X_n$ and $f(X_0) = \cdots = f(X_n)$.*

Proof. We take $h_2(n, j) := h(k)$ (of 2.12), where $k = \max\{j + 2, n\}$. Then by 2.12 there is a chain $X_0 \subset \cdots \subset X_{j+1}$, over which f is fixed, regressive or constant. The first two possibilities cannot occur, because in these cases $f(X_{j+1})$ had to contain at least $j + 1$ elements.

2.16 Remark. The proof of (2.11 and) 2.12 was non-constructive. A contradiction was deduced under the indirect assumption that no bound $h^*(n)$ (resp. $h(n)$) would exist. Also the proofs of 2.14 and 2.15 were non-constructive. Rado [146] had posed the question whether there exists a constructive proof. In [62] such a proof of 2.12, and then also of 2.14 and 2.15, was given.

The Theorem 2.12 has also a geometric interpretation. Its content can be reformulated in graph-theoretical concepts. To this purpose we first define:

2.17 Definition. For $n \in \mathbf{N}$ the *directed cubic graph* C_n has as vertices the $0, 1$ - sequences $(\varepsilon_1, \ldots, \varepsilon_n)$ ($\varepsilon_i \in \{0, 1\}$ for $i = 1, \ldots, n$), and its *directed* (or *oriented*) *edges* are the pairs (b, a), where for $a = (a_1, \ldots, a_n)$ and $b = (b_1, \ldots, b_n)$ we have $a_i \leq b_i$ for $i = 1, \ldots, n$, so that $a \leq b$ holds in the product order. The naming is explained by the fact that the vertices of C_n are the vertices of the unit n - cube $[0, 1]^n$ of the euclidean n - space $\mathbf{R}^n$.

There is a natural bijective mapping between C_n and the power set $\mathfrak{P}_n := \mathfrak{P}(\{1, \ldots, n\})$, namely that one which ascribes to the n - tuple $(\varepsilon_1, \ldots, \varepsilon_n)$ the subset T of those $i \in \{1, \ldots, n\}$ for which $\varepsilon_i = 1$ holds, in other words: T is that subset of $\{1, \ldots, n\}$, which has $(\varepsilon_1, \ldots, \varepsilon_n)$ as *characteristic function*.

This in mind one can reformulate the substance of 2.13 as follows:

2.18 Theorem. *For each $n \in \mathbf{N}$ there exists an $m = m(n) \in \mathbf{N}$ such that the following holds: If from each vertex $(\varepsilon_1, \ldots, \varepsilon_m)$ of C_m that is different from $(0, \ldots, 0)$ exactly one directed edge leads to a vertex which (in the product order) is strictly less than $(\varepsilon_1, \ldots, \varepsilon_m)$, then there exists a directed path of length n, i. e. a subset of vertices $v_1, \ldots, v_n$, where $\{v_i, v_{i+1}\}$ is a directed edge of the graph C_m for $i = 1, \ldots, n - 1$, or there is a vertex e, which is the common endpoint of n edges $\{a_i, e\}$, where the a_i, $i \in \{1, \ldots, n\}$ form a chain of $\mathfrak{P}_n$.*

Until here we had considered contractive mappings in power sets. Of course one can widen the reflections on more general posets. Klaus Leeb had posed the problem to find bounds such that any contractive mapping on a poset must have a monotone $(k + 1)$ - chain if the size of the poset exceeds those bounds. For posets of bounded width this problem was solved by G.W.Peck, P.Shor, W.T.Trotter, and D.B.West [136]. Before presenting their theorem we introduce some concepts which are used in their proof:

2.19 Definition. Let $(P, \leq)$ be a finite poset, and $g : P \to P$ a contractive mapping, also named *contraction* (so that $g(x) \leq x$ holds for all $x \in P$). A *monotone k-chain* is defined to be a chain $\{x_1, \ldots, x_k\}_<$ of $k \in \mathbf{N}$ elements for which $g(x_1) \leq \cdots \leq g(x_k)$.

If for $w, k \in \mathbf{N}$ a finite poset has width at most w and has a contraction with no monotone $(k + 1)$ - chain, it is called a (w, k) - *poset*. Let $f(w, k)$ be the smallest integer such that there is no (w, k) - poset with that many elements. In other words: Every contraction on every poset that has at least that many elements, but width at most w, has a monotone $(k + 1)$ - chain.

Then there holds:

2.20 Theorem [136]. $f(w, k) = (w + 1)^k$.

Proof. First we prove $f(w, k) \leq (w+1)^k$. Assume that P is a (w, k) - poset and that g is a contraction on it with no monotone $(k+1)$ - chain. We proceed by induction on k. If $k = 1$, and then P a $(w, 1)$ - poset, every element of P must be a minimal element. For, if an $x \in P$ would not be minimal we could choose a minimal element $m \leq g(x)$. Then $\{m, x\}$ would form a monotone 2 - chain since $m = g(m) \leq g(x)$ holds, a contradiction. Since the minimal elements of a poset are pairwise incomparable, now $|P| \leq w$ follows and thus $f(w, 1) \leq w + 1$.

Now assume $k > 1$, and that the assertion holds for all natural numbers $< k$. For each $x \in P$ let $G(x)$ be a longest monotone chain in P which has x as greatest element. Then x is said to have *effective height* $|G(x)| - 1$. So each element of P has an effective height $< k$. Let M be the set of elements with effective height $k - 1$. If M is deleted from P, then the restriction $g \restriction P \setminus M$ is a contraction on $P \setminus M$ having no monotone k - chain. Indeed, it is impossible that there exists an element $x \in P \setminus M$ with $g(x) \in M$. For such an element a longest monotone chain $G(g(x))$ could be prolonged by adjoining x as greatest element, thus producing a monotone $(k + 1)$ - chain in P. All elements of $P \setminus M$ have effective height $< k - 1$ in P, so that there are no more monotone $k-$ chains, and therefore $P \setminus M$ is a $(w, k - 1)$ - poset. So $P \setminus M$ has by definition a size $< f(w, k - 1)$, and this is $= (w + 1)^{k-1}$ by induction hypothesis. Thus $|P \setminus M| \leq (w + 1)^{k-1} - 1$ holds.

Next we find an upper bound for $|M|$. Let F be the set of fixed points of g in M.

(1) Two fixed points x, y of F cannot be comparable, and therefore

M contains at most w fixed points of g.

Indeed, if x, y would be comparable fixed-points of g in M, say $y < x$, then a monotone k - chain $G(y)$, having y as greatest element, could be prolonged to a monotone $(k+1)$ - chain by attaching x as a new greatest element. For the same reason there follows:

(2) There are no two comparable elements $x, y \in M$ which have the same g - image.

Finally no element $x \in M$ is mapped by g onto an element $y \neq x$ of M. For this would entail $x > y$, and again as before this would yield a monotone $(k + 1)$ - chain. Then by (2) at most w elements of $M \setminus F$ have the same g - image, which has to be in $P \setminus M$. This entails

$$|M \setminus F| \leq w \cdot |P \setminus M| \leq w \cdot ((w + 1)^{k-1} - 1), \text{ so that finally}$$
$$|P| = |P \setminus M| + |M| = |P \setminus M| + |F| + |M \setminus F|$$
$$\leq (w + 1)^{k-1} - 1 + w + w \cdot ((w + 1)^{k-1} - 1) = (w + 1)^k - 1, \text{ so that}$$

we have $f(w, k) \geq (w + 1)^k$.

In order to show that also $\leq$ holds we construct by induction on k a poset P_k having width w and size $(w + 1)^k - 1$, and a contraction g_k on it with no monotone $(k + 1)$ - chain; then P_k is a (w, k) - poset and $f(w, k) \geq (w + 1)^k$. Here $P_1 \subseteq \cdots \subseteq P_k$ will hold. Let P_1 be a single antichain of size w. This is a ranked poset where each element has rank 0. Let g_0 map each of these elements onto itself. Then g_0 has no monotone 2 - chain, and we are done for the case $k = 1$.

Suppose now that k is an integer > 1 and that for all $\kappa < k$ sets P_κ and mappings g_κ have already been defined which satisfy the required conditions. Then we add to P_{k-1} w disjoint chains to the top of P_{k-1}, each having $(w+1)^{k-1}$ elements, so that every new element of $P_k \setminus P_{k-1}$ lies above every element of P_{k-1}. The new elements form $(w + 1)^{k-1}$ ranks of w elements each. Now P_k has $(w+1)^{k-1} - 1 + (w+1)^{k-1} \cdot w = (w + 1)^k - 1$ elements and width w.

We put $g_k \upharpoonright P_{k-1} := g_{k-1}$. And let the minimal elements of $P_k \setminus P_{k-1}$ be fixed points of g_k. Let R be one of the $(w+1)^{k-1} - 1$ remaining ranks of $P_k \setminus P_{k-1}$. Then all of its elements are mapped by g_k onto a single element of P_{k-1}, and this so that elements of distinct R's have distinct g - images. This is possible since P_{k-1} has the same number $(w + 1)^{k-1} - 1$ of elements as the number of those ranks of $P_k \setminus P_{k-1}$, which don't contain minimal elements of this set. Hereby the uppermost w ranks of P_k map to the w minimal elements of P_{k-1}. The next highest w ranks of P_k map to the w elements of P_{k-1} at rank 1, and so on.

Now g_k has no monotone $(k+1)$ - chain. For by induction g_{k-1} on P_{k-1} has no monotone k - chain, and in addition it follows that no two elements of $P_k \setminus P_{k-1}$ can appear in a single monotone chain. For let $x > y$ hold for two elements x, y of $P_k \setminus P_{k-1}$.Then they are in the same maximal chain C of this set. If y is the minimal element of C, it is fixed, and x has a smaller g_k - image. If x and y have their g_k - images in the same rank of P_{k-1}, these are incomparable, and if they are in different ranks of P_{k-1}, then by construction $g_k(x) < g_k(y)$ holds.

9.3 Combinatorial properties of choice functions

In this section we consider the special case where such contractive mappings on sets of sets are investigated, which ascribe to each non-empty set a one-element subset of it, or what is essentially the same, a single element of it. These mappings are nothing else than choice functions. Before we consider a special case of 2.11 we introduce some concepts:

3.1 Definition. Let D be an initial segment of $\mathbf{N}$ and $f : D \to \mathbf{N}$ a function with $f(x) \leq x$ for all $x \in D$. The *graph* of $f : D \to \mathbf{N}$ is the set of all pairs $(x, f(x))$ with $x \in D$.

Then we define the concept *doubly monotonic f-chain of length l* in the following sense: It is a sequence $p_1, \ldots, p_l$ of l points of the graph of f,where $p_i = (x_i, f(x_i))$ holds for $i = 1, \ldots, l$ with $x_1 < \cdots < x_l$, and where the sequences $f(x_i)$ and $x_i - f(x_i)$ are monotonically increasing (the same: $\leq$ - preserving) with i. Of course, p_l is called the *endpoint* of this chain.

The *octant oct(p)* of a point $p = (p_1, p_2) \in \mathbf{R}^2$ is the set which contains all points $q = (q_1, q_2)$ of $\mathbf{R}^2$, for which $p_1 < q_1$, $p_2 \leq q_2$ and $0 \leq \frac{q_2 - p_2}{q_1 - p_1} \leq 1$ holds. See Figure 19.

310

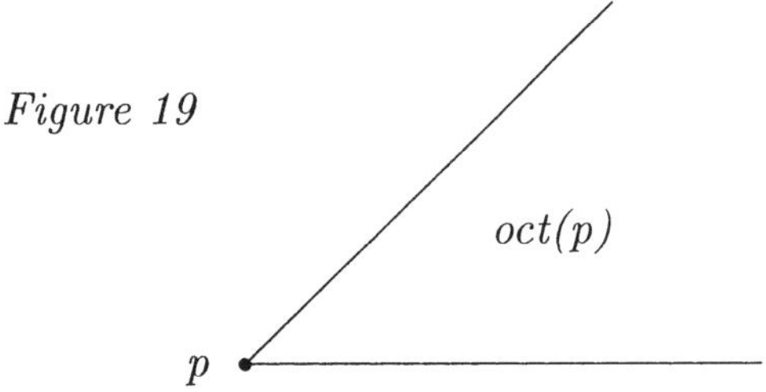

Figure 19

In this section we put $p \leq q$ for two points of $\mathbf{R}^2$, iff $q \in oct(p)$. This defines an order in $\mathbf{R}^2$, and relative to this the concepts chain and antichain.

3.2 Theorem [72]. *Let $n \in \mathbf{N}$. Then for each mapping $f : \{1, 2, \ldots, 2^n\} \to \mathbf{N}$ with $f(x) \leq x$ for $x \in \{1, \ldots, 2^n\}$ there exists a doubly monotonic f - chain of length $n + 1$. And this statement is tight.*

Proof. For $n = 0$ the assertion holds. Suppose now that the theorem is valid for a fixed $n \in \mathbf{N}$. Under this assumpton we first prove: (I) *If ν is an integer with $0 \leq \nu \leq 2^n - 1$ and $f_\nu : D_\nu := \{1, 2, \ldots, 2^n + \nu\} \to \mathbf{N}$ a function with $f_\nu(x) \leq x$ for $x \in D_\nu$, for which no doubly monotonic f_ν - chain of length $n + 2$ exists. Then there are $\nu + 1$ points $p_1, \ldots, p_{\nu+1}$ which are endpoints of a doubly monotonic f_ν - chain of length $n + 1$.* For $\nu = 0$, (I) holds by the induction hypothesis concerning n. Let now (I) be proved for a fixed ν and $f_{\nu+1} : D_{\nu+1} := \{1, 2, \ldots, 2^n + \nu + 1\} \to \mathbf{N}$ a function with $f_{\nu+1}(x) \leq x$ for $x \in D_{\nu+1}$, for which no doubly monotonic $f_{\nu+1}$ - chain of length $n + 2$ exists. Let $f_\nu := f_{\nu+1} \upharpoonright \{1, 2, \ldots, 2^n + \nu\}$. By induction hypothesis (on ν) there are at least $\nu + 1$ endpoints p_i of doubly monotonic f_ν - chains of length $n + 1$ (they are also $f_{\nu+1}$-chains). We take those points p_i of them which have the $\nu + 1$ lowest x - components and choose the arrangement of the $p_i := (x_i, y_i)$ so that $x_1 < \cdots < x_{\nu+1}$ holds. Since $f_{\nu+1}$ is supposed to have no doubly monotonic chain of length $n + 2$ it follows that each octant $oct(p_i)$ does not contain a p_j, $j \in \{1, \ldots, \nu + 1\}$. We define sets P and R (see Figure 20) as follows: P contains those points of the graph of $f_{\nu+1}$, that have an x - component $> x_{\nu+1}$ and a y - component $> y_{\nu+1}$. (These points are in the parallelotope P of Figure 20.) And R contains the points of the graph of $f_{\nu+1}$ that have an x - component $> x_{\nu+1}$ and a y - component $< y_{\nu+1}$. (These points are in the rectangle R of Figure 20.)

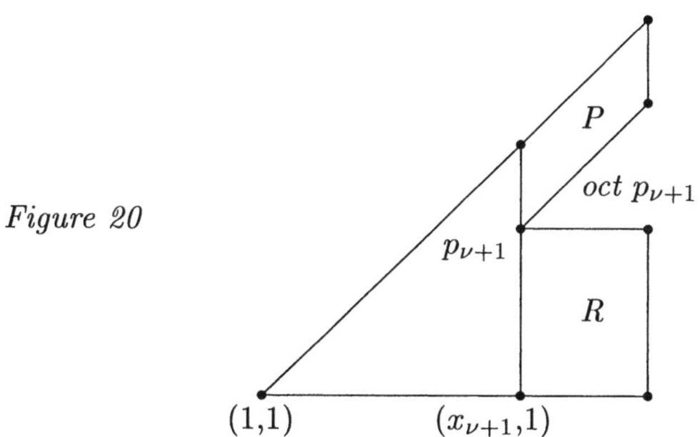

Figure 20

We define a function f^* by altering $f_{\nu+1}$ as follows: We eliminate $p_{\nu+1}$, and then we replace the points (a,b) of P by $g(a,b) := (a-1,b-1)$, and the points (a,b) of R by $g(a,b) := (a-1,b)$. The other points remain unchanged. So there arises a graph of a new function $f^* : D_\nu := \{1,\ldots,2^n+\nu\} \to \mathbf{N}$ which satisfies $f^*(x) \le x$ for $x \in D_\nu$. If there would exist a doubly monotonic f^*- chain of length $n+2$, then also there would exist a doubly monotonic $f_{\nu+1}$- chain of length $n+2$, which is excluded by our assumption. Now by induction hypothesis there exist $\nu + 1$ points which are endpoints of doubly monotonic f^*- chains of length $n + 1$. If $p_{\nu+2}$ is that point of them, which has the greatest x - component, then $g^{-1}(p_{\nu+2})$ is the endpoint of a doubly monotonic $f_{\nu+1}$- chain of length $n + 1$. So, together with $p_{\nu+1}$ we have $\nu + 2$ endpoints of doubly monotonic $f_{\nu+1}$- chains of length $n + 1$. And (I) is proved by induction. Let now $h : \{1,\ldots,2^{n+1}\} \to \mathbf{N}$ be a function with $h(x) \le x$ for all $x \in \{1,\ldots,2^{n+1}\}, \mu := 2^n - 1$ and $h_\mu : D_\mu := \{1,2,\ldots,2^n + \mu\} \to \mathbf{N}$ the restriction of h to D_μ. We assume that no doubly monotonic h_μ - chain of length $n + 2$ exists, otherwise we are done. Then, by (I), there exist $\mu + 1 = 2^n$ ends of doubly monotonic h_μ - chains of length $n + 1$. Let $Q = \{q_1,\ldots,q_{2^n}\}$ be their set, where the x - components x_i of the points q_i strictly increase with i. For $p \ne q$ in Q we have $p \notin oct(q)$ since otherwise p would prolong the chain ending in q. By induction hypothesis we have $x_1 \le 2^n$. Then $oct(q_1)$ intersects the segment S, which links the points $(2^{n+1},1)$ and $(2^{n+1},2^{n+1})$, in a segment of length $\ge 2^n$. Since no q_i is in $oct(q_j)$ for $i \ne j$ of $\{1,\ldots,2^n\}$ each $oct(q_i)$ intersects S in a segment which has a subsegment of length ≥ 1, which

312

is not contained in $oct(q_j)$ for $j \neq i$. So together $\cup\{oct(q_i)|i = 1,\ldots,2^n\}$ covers a subsegment of S of length $\geq 2^n + 2^n - 1 = 2^{n+1} - 1$, which means it covers all of S. See Figure 21.

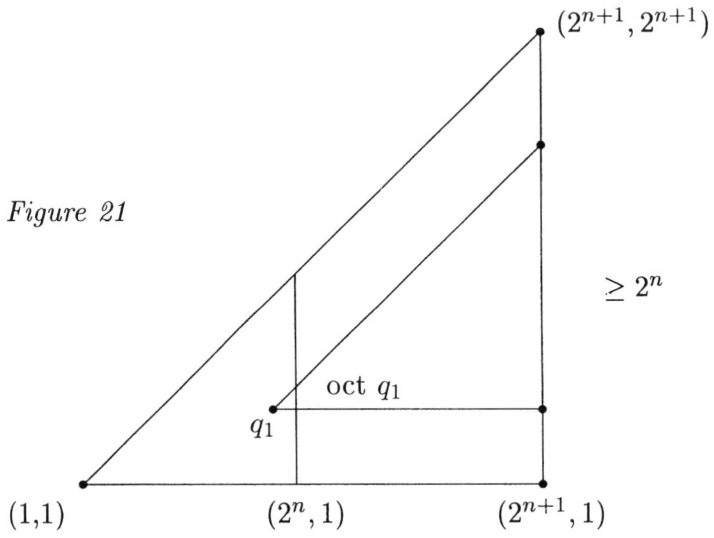

Figure 21

Now the point $(2^{n+1}, h(2^{n+1}))$ is in $oct(q)$ for some $q \in Q$. So it prolongs the chain ending in q and produces a doubly monotonic h - chain of length $n + 2$. And the first part of the theorem is proved by induction. Another proof of this is given in [113], page 497.

In order to see that the theorem is tight, we first prove:

3.3 Lemma. *Let $k \in \mathbf{N}$ and $f : D := \{1,\ldots,2k - 1\} \to \mathbf{N}$ a mapping satisfying $f(x) \leq x$ for $x \in D$. Let T be a k-element subset of D, over which f is strictly decreasing. Then T must be the set $\{k, k + 1,\ldots,2k - 1\}$, and we must have $f(k + \nu) = k - \nu$ for $\nu = 0,\ldots,k - 1$.*

Proof. For $k = 1$ this is trivial. So we assume $k > 1$. Let m be the least element of T. Then, due to $|T| = k$, we must have $m + k - 1 \leq 2k - 1$, which means $m \leq k$. Further there holds $f(m) \leq m \leq k$. And since the strictly decreasing set $\{f(x)|x \in D\}$ must have k elements, $f(m) \geq k$, and then also $f(m) = k$ holds. Now $m \geq f(m) = k$ entails $m = k$. Therefore T is the set of all integers of $[k, 2k - 1]$, and the rest easily follows. See Figure 22 (here $k = 5$).

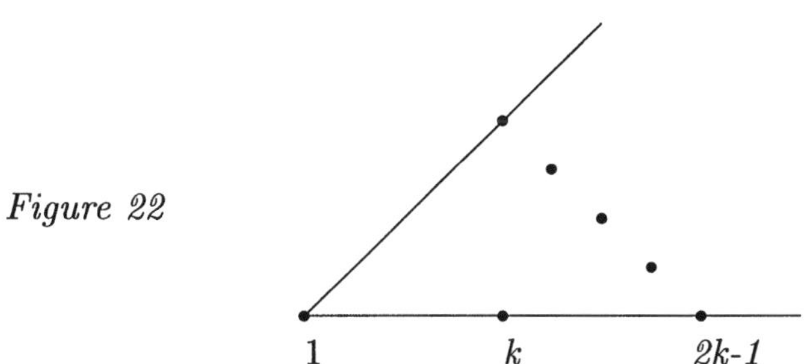

Figure 22

The next theorem proves that Theorem 3.2 is tight, and even something more:

3.4 Theorem [72]. *For each $n \in \mathbf{N}$ there is exactly one function $F_n : T_n := \{1, 2, \ldots, 2^n - 1\} \to \mathbf{N}$ with $F_n(x) \leq x$ for $x \in T_n$ such that there is no $(n+1)$-element subset of T_n over which F_n is monotonely increasing (the same: $\leq$ - preserving), not to mention $F_n(x)$ and $x - F_n(x)$.*

Proof. For $n \in \mathbf{N}$ we define F_n as follows: For each integer m *with* $0 \leq m < n$ and $\nu < 2^m$ we put $F_n(2^m + \nu) := 2^m - \nu$. See Figure 23:

So the graph of F_n is a union of n disjoint antichains of $\mathbf{R}^2$.

It can easily be verified that every subset of T_n, over which F_n is $\leq$ - preserving, has at most n elements, namely at most one of each segment $[2^\nu, 2^{\nu+1} - 1]$, $0 \leq \nu < n$.

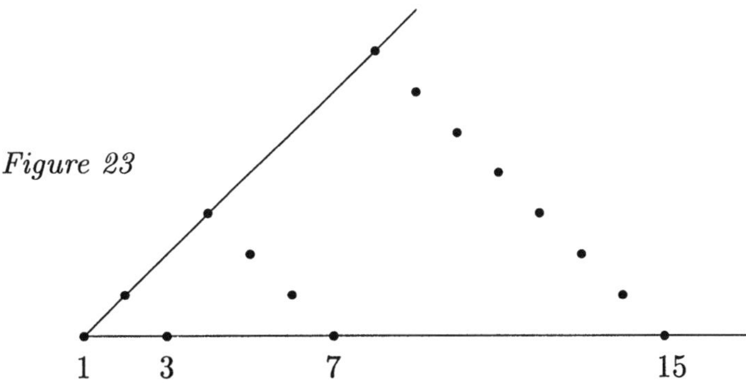

Figure 23

The theorem holds for $n = 1$. Suppose that it is valid for a fixed $n \in \mathbf{N}$, and let $f : T_{n+1} := \{1, \ldots, 2^{n+1} - 1\} \to \mathbf{N}$ be a mapping, satisfying $f(x) \leq x$ for $x \in T_{n+1}$, such that there is no $(n + 2)$ - element subset of T_{n+1} over which f is $\leq$ - preserving. A fortiori f is not doubly monotonic over the $(n + 2)$ - element subsets of T_{n+1}, and then we apply (I): We have $T_{n+1} = \{1, \ldots, 2^n + (2^n - 1)\}$, and so by (I) there exist 2^n points $p_i = (x_i, y_i)$, $i = 1, \ldots, 2^n$, which are endpoints of doubly monotonic f - chains of length $n + 1$. Let the numbering be so that $x_1 < \cdots < x_{2^n}$. Then we have $y_1 > \cdots > y_{2^n}$, for otherwise one of the chains ending in one of the p_i could be prolonged to a chain of length $n + 2$, with contradiction. Now it follows from 3.3 (with $k := 2^n$) that $f \restriction [k, 2k - 1]$ coincides with $F_n \restriction [k, 2k - 1]$. If $f \restriction [1, \ldots, k - 1]$ would be different from $F_n \restriction [1, \ldots, k - 1]$ there would, due to the induction hypothesis, exist an $(n + 1)$ - element subset of $[1, \ldots, k - 1]$, over which f is monotonically increasing, and this set could be enlarged by an element of $[k, k + 1]$ to an $(n + 2)$ - element subset over which f is still $\leq$ - preserving. This contradicts our assumption, and so the theorem is proved.

3.4$'$ Remark. A uniqueness theorem as in 3.4 no longer holds, if we take into consideration the monotonicity of f and $x - f(x)$. If e.g. $n = 2$, and if we define f by $f(1) = 1$, $f(2) = 1$, $f(3) = 3$, then f is defined over $\{1, 2, 2^2 - 1\}$ and different from F_2, but there is also no doubly monotonic f - chain of length 3. Theorem 3.3 can be brought into relation to Galton functions. We give an intuitive description of this. We consider a Galton-board (see Figure 24), with which the Gaussian bell curve $\exp(-x^2)$ can be generated .

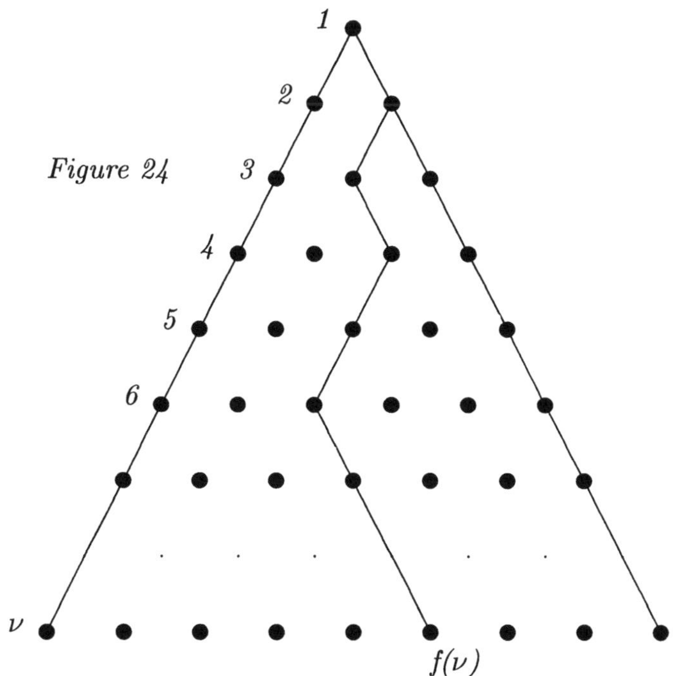

We identify the left border line s with an initial segment of $\mathbf{N}$. The way of a ball which runs through the board — which we call a *Galton-way* — can then be described by a function f which is defined as follows: In the νth line from above, the falling ball meets the $f(\nu)^{th}$ point (counting begins from left). Then we have $1 \leq f(\nu) \leq \nu$, and moreover

(2) $f(\nu + 1) = f(\nu)$ or $f(\nu + 1) = f(\nu) + 1$

for all numbers $\nu + 1$ of the range D of definition of f. Therefore f is monotonically increasing over D, and that also holds for the function $\nu - f(\nu)$. We call a function over an initial segment D of $\mathbf{N}$ and with values in $\mathbf{N}$ a *Galton-function*, if it satisfies (2) for all $\nu \in D$. Theorem 3.3 can then be reformulated as:

3.5 Theorem. *Let $f : D := \{1, \ldots, 2^n\} \to \mathbf{N}$ be a function satisfying $f(x) \leq x$ for $x \in D$. Then there exists a Galton-function over $\{1, \ldots, 2^n\}$, with which f coincides over at least $n+1$ elements of D.*

Proof. There is a subset $S \subseteq D$ of size $n + 1$, over which ν and $\nu - f(\nu)$ increase monotonically, and then one can easily construct a Galton-function g over D, which satisfies $g(\nu) = f(\nu)$ for $\nu \in S$.

Subsequently to 3.3 one can pose the question whether 3.3 has a related theorem for infinite ranges of definition. In this context there holds a theorem of Alexandroff/Urysohn [3]:

3.6 Theorem. *Let ω_α be a regular initial ordinal $> \omega_0$ and $f : \omega_\alpha \to \omega_\alpha$ a mapping which satisfies $f(x) < x$ for $0 < x < \omega_\alpha$ and $f(0) = 0$. Then there exists a subset of cardinality $\aleph_\alpha$ of ω_α, over which f is constant.*

Proof. In ω_α we introduce an order $\preceq$ (different from $\leq$) by $x \preceq y$ iff $x = y$ or $x = f^{(n)}(y)$ for some $n \in \mathbf{N}$. It is easily seen that this defines an order relation. Moreover each $y \in \omega_\alpha$ has only finitely many predecessors, namely the elements $f^{(n)}(y)$, $n \in \mathbf{N}$, which are $= 0$ for all n of a non-empty final segment of $\mathbf{N}$, since every strictly decreasing sequence of ordinals is finite. In particular this entails that $(\omega_\alpha, \preceq)$ is a tree T with 0 as least element, which forms the 0 - level L_0 of T. And each $y \in \omega_\alpha$ has finite height, namely the least $n \in \omega$ for which $f^{(n)}(y) = 0$.

If now the assertion would be false, every $x \in \omega_\alpha$ (and in particular 0) would have fewer than $\aleph_\alpha$ upper neighbors with respect to $\preceq$. For $n \in \omega$ let L_n be the n - level of the tree T. Now $|L_0| = 1$ and $|L_1| < \aleph_\alpha$ holds. If in general we have already proved $|L_n| < \aleph_\alpha$, then L_{n+1} is the set of all immediate successors relative to $\preceq$ of the elements of L_n, and hence the union of fewer than $\aleph_\alpha$ sets which all have cardinality $< \aleph_\alpha$. Thus, due to the regularity of $\aleph_\alpha$, also $|L_{n+1}|$ is $< \aleph_\alpha$. By induction we so obtain $|L_n| < \aleph_\alpha$ for all $n \in \mathbf{N}$, and therefore also $\cup\{L_n | n \in \omega\}$ would have cardinality $< \aleph_\alpha$, which is impossible, since this set is nothing else than ω_α.

If in 2.12 the function f is a choice function over $\mathfrak{P}(h(n))$, one can determine the least value $h(n)$ for which 2.12 holds. Namely the following theorem of Perry shows this. Other proofs of it were given by Lovasz ([114], p.500) and by Kleitman and Lewin [97]. We present here (in a slightly modified form) the proof of Kleitman/Lewin:

3.7 Theorem of Perry [140]. *Let k, n be natural numbers with $k \geq 2^n$, $S = \{1, \ldots, k\}$, and f a choice function defined on the non-empty subsets of S. Then there is a chain $C_1 \subset \cdots \subset C_{n+1}$ of subsets of S, such that all $f(C_\nu)$, $\nu = 1, \ldots, n + 1$, are equal. This statement is tight.*

Proof [97]. For $1 \leq i \leq k$ let s_i be the size of a maximal-sized chain of subsets of S whose f - image is i. First we prove:

Lemma. *If under the above assumptions $k = 2^n + j$ with $0 \leq j \leq 2^n$, then*

(1) $\sum_{i=1}^{k}(s_i - n) \geq 2j + 1$.

If the lemma is proved we are done, since (1) implies that at least one of the summands of the left side is > 0, so that the corresponding s_i is $\geq n + 1$.

Proof of the lemma. (1) can be put in the following form:

(1') $\sum_{i=1}^{k} s_i - k \cdot n \geq 2j + 1$.

The lemma is true for $n = 1$, $j = 0$. Indeed, k is ≥ 2, and therefore at least one of the s_i is ≥ 2, for there is at least one 2 - element subset T of S, and then $f(T)$ and $f(f(T))$ form a two-element chain. All others are ≥ 1. Thus $\sum \{s_i | i = 1, \ldots, k\} \geq k + 1$. Suppose now that the lemma is true for a fixed n and $j - 1$, where j is > 0. We prove that then it is also true for n, j. That means we have to verify (1).

Let S' be the set $\{1, \ldots, k - 1\}$, and s_i' the size of a maximal-sized chain of subsets of S' whose f - image is i. We have $k - 1 = 2^n + (j - 1)$, where $0 \leq j - 1 \leq 2^n$ and $k - 1 \in \mathbf{N}$. Then our induction hypothesis yields:

(2) $\sum_{i=1}^{k-1}(s_i' - n) = \sum_{i=1}^{k-1} s_i' - (k-1) \cdot n \geq 2 \cdot (j-1) + 1 = 2j - 1 > 0$.

Then for one summand we must have $s_{i_0}' - n > 0$, and thus

(3) $s_{i_0}' > n$.

Consider now $S'' := \{1, 2, \ldots, k\} \backslash \{i_0\}$, and let s_i'' be the size of a maximal-sized chain of subsets of S'' whose f - image is i. Then the induction hypothesis yields

(4) $\sum \{s_i'' | i \in S''\} - (k - 1) \cdot n \geq 2j - 1$.

For every $i = 1, \ldots, k$ we have $s_i' \leq s_i$ and $s_i'' \leq s_i$. Now we add i_0 to S'' and note that the set S itself increases the length of the maximal-sized chain with image $f(S)$. We obtain

(5) $\sum_{i=1}^{k} s_i - k \cdot n = 1 + \sum \{s_i'' | i \in S''\} + s_{i_0}' - (k - 1) \cdot n - n$,

by (3) and (4) the right-hand side is $\geq 1 + 2j - 1 + 1 = 2j + 1$. And so (1) is proved.

Summarizing we have obtained that the lemma holds for our fixed n and all $j \leq 2^n$. In particular it is valid for $j = 2^n$, in which case we have $k = 2j$. Here (1) becomes $\sum_{i=1}^{k} s_i - k \cdot n \geq k + 1$ and thus

(6) $\sum_{i=1}^{k} s_i - k \cdot (n+1) \geq 1$,

which is the lemma for $n+1$ and $j = 0$. So the lemma is proved by induction, and with it the main part of the theorem.

In order to see that the theorem is tight we consider the mapping $F_n : T_n \to \mathbf{N}$ of 3.4, where $T_n = \{1, 2, \ldots, 2^n - 1\}$. If M is an m - element subset of T_n with $m \geq 1$, we ascribe to M the $(F_n(m))^{th}$ member (in the usual order) of M. Then for each chain $C_1 \subset \cdots \subset C_t$ of the power set $\mathfrak{P}(T_n)$, over which this choice function is constant, we must have that the $F_n(|C_i|)$, $i = 1, \ldots, t$, form a monotonely increasing sequence. This has by 3.4 at most n elements, and so $t \leq n$ follows.

Some of the previous theorems can be interpreted as statements about subtrees of finite power sets. For there is a close connection between regressing mappings in finite power sets and subtrees of these. Referring to this we introduce some concepts:

3.8 Definition. Let S be a finite set and f a regressing mapping of $(\mathfrak{P}(S), \subseteq)$ into itself. Then we define an order relation $\preceq$ on $\mathfrak{P}(S)$ by:

For $A, B \subseteq S$ we put $A \preceq B \Longleftrightarrow$ There exists an $n \in \omega$ such that $A = f^n(B)$, where f^n is the nth iteration of f; in particular $f^0(B) = B$. It is immediately clear that $(\mathfrak{P}(S), \preceq)$ is a tree, the *tree defined by f*. The order $\preceq$ is a sub-order of $\subseteq$, and $\emptyset$ is its least element.

In analogy to a notion of graph theory we call a tree $(\mathfrak{P}(S), \leq_t)$ a *spanning tree* of the poset $(\mathfrak{P}(S), \subseteq)$, if $\leq_t$ is a subrelation of $\subseteq$ for which $\emptyset$ is the first element.

The correspondence between regressing mappings on $(\mathfrak{P}(S), \subseteq)$ and spanning trees of this set is now expressed by:

3.9 Theorem. *For a finite set S each regressing mapping on $(\mathfrak{P}(S), \subseteq)$ defines a spanning tree of this set, namely the tree defined in 3.8.*

If on the other hand $(\mathfrak{P}(S), \leq_t)$ is a spanning tree of $(\mathfrak{P}(S), \subseteq)$, then we define $f : \mathfrak{P}(S) \to \mathfrak{P}(S)$ by: $f(\emptyset) = \emptyset$, and for $\emptyset \neq T \subseteq S$ let $f(T)$ be the (uniquely defined) immediate predecessor of T with respect to $\leq_t$. Of course f is a regressing mapping on $(\mathfrak{P}(S), \subseteq)$.

Again in analogy to a graph-theoretical concept we define:

3.10 Definition. A tree $(T, \leq)$ is a *star*, if all elements of it which are different from its least element o are pairwise incomparable.

Theorems 2.13 and 3.7 can now be reformulated in the following forms:

3.11 Theorem. *For each $n \in \mathbf{N}$ there is an $m := m(n) \in \mathbf{N}$, for which there holds: Let S be a finite set with $|S| \geq m$ and $(\mathfrak{P}(S), \leq_t)$ a spanning tree of $(\mathfrak{P}(S), \subseteq)$. Then this tree has a chain of length n, or there exists a $T \subseteq S$, which has n upper neighbors in the tree, which form a chain in $(\mathfrak{P}(S), \subseteq)$.*

3.12 Theorem. *Let S be a set with 2^n elements, and $(\mathfrak{P}(S), \leq_t)$ a spanning tree of $(\mathfrak{P}(S), \subseteq)$, in which $\emptyset$ has the one-element subsets as upper neighbors, and where each (≥ 2) - element subset $T \subseteq S$ has as unique lower neighbor a one-element subset $\{x\}$ with $x \in T$, (so that the final segments T_x, which are generated by the one-element subsets $\{x\}$ of $(\mathfrak{P}(S), \leq_t)$, are stars), then there is an $(n + 1)$ - element chain of $(\mathfrak{P}(S), \subseteq)$, which is a subset of one of the stars T_x, which partition the set of non-empty subsets of $\mathfrak{P}(S)$.*

9.4 Combinatorial theorems on infinite power sets

Some order types τ have the property that each poset P with type τ has the property: However P is split into two parts A, B, at least one of them still has a subset of type τ. We call types τ which have (resp. don't have) this property *unsplittable* (resp. *splittable*). E.g. the linearly ordered set $\mathbf{Q}$ of rational numbers is unsplittable: If a subset $A \subseteq \mathbf{Q}$ has no subset of type $h_0 = \mathrm{tp}\mathbf{Q}$, A cannot be dense in $\mathbf{Q}$, and so there exists an interval $[a, b]$ with $a < b$, which is disjoint to A. But then $[a, b]$ is $\subseteq B := \mathbf{Q} \setminus A$, and B has a subset of type h_0.

Other examples of such unsplittable posets are the well-ordered sets whose order type is an indecomposable ordinal (see 4.8.8), e.g. the set $\mathbf{N}$ of natural numbers. Of course, the question arises how the set $\mathbf{R}$ behaves in this context. It turns out to be splittable.

In the discussion of the properties of infinite power sets the sets H_α of 4.3.3 and the sets C_α of 4.7.31 play an essential role. These subsets of $2((\omega_\alpha))$ are always considered as linearly ordered by the principle of first differences. Let us recall: C_α is dense, and H_α is strictly dense in C_α (4.7.32) . Further C_α has no gaps, and so it is a linearly ordered continuum in the sense of 7.1.5. The set $2((\omega_\alpha))$ is isomorphic to a

320

subset of C_α (4.7.33), but $2((\omega_\alpha))$, and thus also C_α, is not isomorphic to a subset of H_α (4.7.11)

In particular the set C_0 has no gaps and no first and no last element, and the set H_0, which has the order type h_0 of the set of rational numbers, is dense in C_0. Therefore C_0 is order-isomorphic to the set $\mathbf{R}$ of real numbers, and the sets C_α are the natural generalizations of it to higher cardinalities.

As a special case of 4.1.14 we have that the power set $(\mathfrak{P}(\omega_\alpha), \subseteq)$ is isomorphic to $(2((\omega_\alpha)), \leq_P)$, where $\leq_P$ is the product topology, and that there exists a $<$ - preserving mapping of $(\mathfrak{P}(\omega_\alpha), \subseteq)$ in $2((\omega_\alpha))$, if now $2((\omega_\alpha))$ is linearly ordered by first differences. This is one of the reasons why it is of interest to study the order properties of $2((\omega_\alpha))$ and its subset C_α. We introduce a concept for isotone functions:

4.1 Definition. Let $D \subseteq C_\alpha$ have at least two elements, and let $f : D \to C_\alpha$ be a $\leq$ - preserving function. Then $j \in D$ is said to be a *jump-spot* of f, if at least one of the following two statements holds:

a) j is not the first element of D and $\sup\{f(x)|x \in (D < j)\} < f(j)$.
Or

b) j is not the last element of D and $f(j) < \inf\{f(x)|x \in (D > j)\}$.

In analogy to the fact that a monotonically increasing function $f : \mathbf{R} \to \mathbf{R}$ has only countably many jump-spots, we here obtain for the cardinal $\mathfrak{k}_a$, which was defined in 4.3.5:

4.2 Theorem. *Let D and f be given as in 4.1. Then the set J of all jump-spots of $f : D \to C_\alpha$ has cardinality $\leq \mathfrak{k}_\alpha$.*

Proof. Each jump-spot x of f defines a segment S_x of C_α, which has more than one element and which contains no element of $f(D)$. Since H_α is dense in C_α we can choose an element $e_x \in S_x \cap H_\alpha$. The mapping which ascribes to each $x \in J$ the element e_x, is evidently $<$ - preserving, in particular injective, and thus J has a cardinality $\leq |H_\alpha| = \mathfrak{k}_\alpha$ (by 4.3.7).

We mention a very simple extension theorem:

4.3 Theorem. *Let M be a subset of C_α and $f_M : M \to C_\alpha$ a $\leq$ - preserving mapping. Let S be the least segment $\supseteq M$ of C_α. Then there exists a $\leq$ - preserving mapping $f : S \to C_\alpha$ with $f \restriction M = f_M$.*

Proof. We put $f \restriction M = f_M$. If $x \in S \backslash M$, there exists a greatest x containing segment S_x of S which is disjoint to M. Then we define

f over S_x to be constant $= c$, where c is any element which satisfies $c \geq f(t)$ for the $t \in M$, that are $<$ the elements of S_x, and $c \leq f(t)$ for the $t \in M$, which are $>$ the elements of S_x.

4.4 Theorem. *Let S be a subset of C_α with $|S| = 2^{\aleph_\alpha}$. The set F of all $\leq$ - preserving functions $f : S \to C_\alpha$ has cardinality $\leq 2^{\ell_\alpha}$. Since F contains all constant functions we have $|F| \geq |C_\alpha| = 2^{\aleph_\alpha}$. Using GCH this implies that F has cardinality $2^{\aleph_\alpha}$.*

If S contains a segment of C_α, which has more than one element, then $|F| = 2^{\ell_\alpha}$.

Proof. Let $f : S \to C_\alpha$ be a $\leq$ - preserving function, J the set of its jump-spots. If $a < b$ are elements of H_α and if the interval $[a, b]$ of C_α intersects S we choose an element of $[a, b] \cap S$. Let D be the set of all elements, which are so chosen. Then $|D| \leq \ell_\alpha$ holds since the set of all pairs $(a, b) \in H_\alpha \times H_\alpha$ has cardinality ℓ_α. Now f is uniquely defined by its values over the set $M := J \cup D$, whose cardinality is $\leq \ell_\alpha$ by 4.2. Indeed, if $t \in S \backslash M$, and if t is neither the first nor the last element of S, we have $s := \sup\{f(x) | x \in (S < t)\} \leq f(t) \leq \inf\{f(x) | x \in (S > t)\} =: j$, and since t is no jump-spot, we must have $s = j = f(t)$. If t is the first (resp. the last) element of S, $f(t)$ is $= j$ (resp. $= s$). It follows that $|F| \leq |C_\alpha^M| \leq |C_\alpha|^{\ell_\alpha} = (2^{\aleph_\alpha})^{\ell_\alpha} = 2^{\ell_\alpha}$.

Let now S contain a segment of C_α which has more than one element. Since H_α is dense in C_α the set S has a subset of type $h_\alpha = h_\alpha \cdot h_\alpha$ (by 4.4.5). Then S has a subset T which can be represented as the ordered sum $T = \sum\{T_i | i \in H_\alpha\}$, where each T_i has order type h_α, and where the $T_i, i \in H_\alpha$, are pairwise disjoint.

We wish to construct an injective mapping of the set of all subsets of H_α into the set F. To $H \subseteq H_\alpha$ we ascribe a mapping f_H by: $f_H(x) = x$ if $x \in \cup\{T_i | i \in H\}$, and over each T_i, where $i \in H_\alpha \backslash H$, we let f be constant $=$ an arbitrary fixed element of T_i. So every subset $H \subseteq H_\alpha$ defines a function $f_H \in F$, and different sets H define different functions f_H. So the set of $\leq$ - preserving functions over T has at least as many elements as the power set of H_α, which means $\geq 2^{\ell_\alpha}$ many. The rest follows with 4.3.

4.5 Lemma. *Let $T \subseteq C_\alpha$ have cardinality $|T| = 2^{\aleph_\alpha}$. Then the set $\mathfrak{U}$ of all subsets of C_α that have the same order type τ as T has cardinality $\leq 2^{\ell_\alpha}$.*

Proof. For each $U \in \mathfrak{U}$ there exists a $<$ - preserving surjective mapping $f_U : T \to U$. The mapping, which ascribes to each $U \in \mathfrak{U}$ the corresponding f_U, is injective, and since $\{f_U | U \in \mathfrak{U}\}$ has cardinality $\leq 2^{\mathfrak{k}_\alpha}$ by 4.4 also $\mathfrak{U}$ has a cardinality $\leq 2^{\mathfrak{k}_\alpha}$.

We can now prove the theorem that the GCH implies that all subsets of C_α of cardinality $2^{\aleph_\alpha}$ are splittable:

4.6 Theorem. *Using GCH there holds: Let $T \subseteq C_\alpha$ have cardinality $2^{\aleph_\alpha}$. Then we can partition T in two subsets A and B, such that T is neither embeddable in A nor in B. (In the terminology of 1.9.6 :* $\mathrm{tp}A$ *and* $\mathrm{tp}B$ *are* $< \mathrm{tp}\,T$.)*

Proof. The set $\mathfrak{U}$ of all subsets of T that have the same order type as T is by 4.5 of cardinality $\leq 2^{\mathfrak{k}_\alpha} = 2^{\aleph_\alpha}$ (due to GCH), and so we can find a well-ordering of their set in the form $\mathfrak{U} = \{T_\nu | \nu < \lambda\}$, where λ is an ordinal $\leq \omega(2^{\aleph_\alpha})$. First we choose two different elements a_0 and b_0 in T_0, then two different elements $a_1, b_1 \in T_1 \backslash \{a_0, b_0\}$. If in general we have a $\mu < \lambda$, for which elements a_ν, b_ν for $\nu < \mu$ are already defined, we choose two different elements $a_\mu, b_\mu \in T_\mu \backslash (\cup_{\nu < \mu} \{a_\nu, b_\nu\})$. This is possible since $\cup_{\nu < \mu} \{a_\nu, b_\nu\}$ has cardinality $< 2^{\aleph_\alpha}$. By transfinite induction we so obtain two disjoint subsets $A' = \{a_\nu | \nu < \lambda\}$ and $B = \{b_\nu | \nu < \lambda\}$ of T. We put $A := T \backslash B$. Then A and B satisfy our assertion, for T_ν contains b_ν, which is not in A, so that T_ν is no subset of A. Analogously T_ν is no subset of B, because it contains a_ν, which is not in B. So A and B have no subset of type $\mathrm{tp}\,T$.

Since for $\alpha = 0$ we have $\mathfrak{k}_0 = \aleph_0$ the last proof shows that without (!) the GCH the following can be proved:

4.7 Theorem. *Let $T \subseteq C_0$ (or $T \subseteq \mathbf{R}$) have cardinality $2^{\aleph_0}$. Then there exists a partition of T in two subsets A,B such that T is neither embeddable in A nor in B.*

Since from the sets $2((\omega_\alpha))$ and C_α each one is embeddable in the other (see 4.7.33), we obtain from 4.6 and 4.7:

4.7′ Theorem. *Every set $T \subseteq 2((\omega_0))$ of cardinality $2^{\aleph_0}$ can be partitioned in two subsets A and B, such that T is neither embeddable in A nor in B.*

And using GCH there holds: Every set $T \subseteq 2((\omega_\alpha))$ of cardinality $2^{\aleph_\alpha}$ can be partitioned in two subsets A,B, such that T is neither embeddable in A nor in B.

From 4.6 we can derive a similar decomposition theorem for power sets of sets of cardinality $\aleph_\alpha$, making use of characteristic sequences. First we establish a connection between the power set $\mathfrak{P}(\omega_\alpha)$ and the set C_α.

4.8 Theorem. *Using GCH there follows: Let S be a set of cardinality $\aleph_\alpha$. Then $\mathfrak{P}(S)$ has a subset $\mathfrak{C}_\alpha$, such that $(\mathfrak{C}_\alpha, \subseteq)$ has the order type λ_α of C_α.*

Since $2((\omega_\alpha))$, with the order by first differences, is embeddable in C_α, then also $\mathfrak{P}(S)$ has a subset which is isomorphic to $2((\omega_\alpha))$.

Proof. W.r.o.g. we can assume $S = \omega_\alpha$. Due to the GCH, H_α has cardinality $\mathfrak{k}_\alpha = \aleph_\alpha$, and therefore H_α is embeddable in the $\aleph_\alpha$ - universally ordered set $(\mathfrak{P}(\omega_\alpha), \subseteq)$. So there exists a subset $\mathfrak{H}_\alpha \subseteq \mathfrak{P}(\omega_\alpha)$, so that $(\mathfrak{H}_\alpha, \subseteq)$ has order type h_α.

The set of initial segments of H_α has a subset which is isomorphic to C_α, and thus also $\mathfrak{H}_\alpha$ has a set $\mathfrak{C}_\alpha \subseteq \mathfrak{P}(\omega_\alpha)$ of initial segments which is isomorphic to C_α.

4.9 Theorem. *Using GCH there holds: Let S be a set of cardinality $\aleph_\alpha$, and let τ be an order type of a linearly ordered set of cardinality $2^{\aleph_\alpha}$. Then $\mathfrak{P}(\omega_\alpha)$ can be partitioned in two sets $\mathfrak{A}$ and $\mathfrak{B}$, such that neither $\mathfrak{A}$ nor $\mathfrak{B}$ (both ordered by $\subseteq$) have a subset of type τ.*

Proof. It suffices to prove the theorem for the situation where $S = \omega_\alpha$ holds. In this case there exist by 4.7' two disjoint subsets A and B of $2((\omega_\alpha))$ with $2((\omega_\alpha)) = A \cup B$, such that neither A nor B has a subset of type τ. We consider the bijective $<$ - preserving mapping $f : 2((\omega_\alpha)) \to \mathfrak{P}(\omega_\alpha)$, which to every sequence $(x_\nu | \nu < \omega_\alpha) \in 2((\omega_\alpha))$ ascribes the set of those $\nu < \omega_\alpha$, for which $x_\nu = 1$ holds.

We put $\mathfrak{A} := f[A]$ and $\mathfrak{B} := f[B]$, so that $\mathfrak{P}(\omega_\alpha) = \mathfrak{A} \cup \mathfrak{B}$. Now $\mathfrak{A}$ has no subset of type τ, for in this case also A had to contain a subset of type τ. For $\mathfrak{B}$ we conclude analogously.

4.10 Theorem. *Using GCH there holds: Let S be a set of cardinality $\aleph_\alpha$. Then there is a partition of the power set $\mathfrak{P}(S)$ in two subsets $\mathfrak{A}$ and $\mathfrak{B}$, such that $(\mathfrak{P}(S), \subseteq)$ is neither embeddable in $(\mathfrak{A}, \subseteq)$ nor in $(\mathfrak{B}, \subseteq)$.*

Proof. Let τ be the type of the set $2((\omega_\alpha))$, which is ordered by first differences. Then we consider the representation $\mathfrak{P}(\omega_\alpha) = \mathfrak{A} \cup \mathfrak{B}$ of

324

4.9. Now $\mathfrak{P}(\omega_\alpha)$ has by 4.8 a subset of type tp2$((\omega_\alpha))$, but $\mathfrak{A}$ and $\mathfrak{B}$ do not.

Until here we had dealt with partitions of certain posets P in two subsets, and this mainly with respect to the question of whether these sets P are splittable. Now we consider also partitions of posets, in particular of H_α, C_α, U_α and power sets, in infinitely many subsets and investigate how sumptuous these subsets are.

4.11 Theorem [73]. *Let $\omega_\gamma := \mathrm{cf}(\omega_\alpha)$ and $\beta < \gamma$, and let $H_\alpha = \cup\{K_\nu | \nu < \omega_\beta\}$. Then one of the classes K_ν contains a subset of type h_α.*

Proof. There exists an interval $[a, b]$ of H_α, in which a class K_ν is dense. Suppose the contrary. Then there is an interval $[a_0, b_0]$ of H_α, which is disjoint to K_0. In general, let μ be an ordinal $< \omega_\beta$, such that for each $\nu < \mu$ an interval $[a_\nu, b_\nu]$ is defined, which is disjoint to K_ν, so that the $[a_\nu, b_\nu]$, $\nu < \mu$, form a decreasing sequence. Then it follows that $\{a_\nu | \nu < \mu\}$ and $\{b_\nu | \nu < \mu\}$ are not neighboring since H_α is an η_γ - set. So there exists an interval of H_α which is situated between these sets, and this has a subinterval $[a_\mu, b_\mu]$ which is disjoint to K_μ. By transfinite induction we so obtain a descending sequence $[a_\nu, b_\nu]$, $\nu < \omega_\beta$, of intervals of H_α. Their intersection is a non-empty segment S of H_α since $A := \{a_\nu | \nu < \omega_\beta\}$ and $B := \{b_\nu | \nu < \omega_\beta\}$ have a cardinality $\leq \aleph_\beta < \aleph_\gamma$ and are therefore not neighboring. But this yields a contradiction since S is disjoint to all K_ν and thus also to H_α.

Let now K_ν be a class which is dense in an interval $[a, b]$ of H_α. By 4.6.3' (a, b) has type h_α. Due to 4.4.5 we have $h_\alpha \cdot h_\alpha = h_\alpha$, and therefore $[a, b]$ has a subset M which has the form of an ordered sum $\sum\{S_i | i \in H_\alpha\}$, where the S_i are disjoint segments of M, which all have the order type h_α. Then K_ν has at least one element in each of the S_i and therefore a subset of type h_α.

As a corollary to 4.11 we list:

4.11'Theorem. *Let S be an η_α - set and $S = \cup\{K_\nu | \nu < \omega_\beta\}$, where $\beta < \alpha$ holds. Then one of the sets K_ν still contains an η_α - set.*

Proof. S has by 4.3.13 a subset of type h_α. If now ω_α is regular the assertion follows from 4.11. If ω_α is singular, then S is also an $\eta_{\alpha+1}$ - set by 3.3.5. Then S contains a set of type $h_{\alpha+1}$, and a fortiori the assertion follows with 4.11.

4.12 Remark. If in 4.11 we replace ω_β by ω_γ, the statement no longer remains valid. For if ω_α is regular, then, due to 4.3.12, H_α is the union of $\aleph_\alpha$ subsets which have no subsets of type ω_α (and ω_α^*), and a fortiori no subsets of type h_α.

If ω_α is singular and $\omega_\gamma := \mathrm{cf}(\omega_\alpha)$, say $\omega_\alpha = \sup\{\omega_{\alpha_\mu}|\mu < \omega_\gamma\}$, where all ω_{α_μ} are $< \omega_\alpha$,then from 4.4.3 we obtain a representaion $H_\alpha = \cup\{T_\mu|\mu < \omega_\gamma\}$, where T_μ has the type $h_{\alpha_\mu+1}$, so that each T_μ has no subset of type h_α.

Next we consider partitions of the sets C_α.

4.13 Theorem. *Let $C_\alpha = \cup\{K_\nu|\nu < \omega_\gamma\}$, where $\omega_\gamma = \mathrm{cf}(\omega_\alpha)$. Then one of the classes K_ν contains a subset of type h_α.*

Proof. One of the classes K_ν must be dense in an interval $[a, b]$ of C_α, where a and b are in H_α. Suppose the contrary. Then in the same way as in 4.11 we construct intervals $I_\nu := [a_\nu, b_\nu], \nu < \omega_\gamma$, which form a descending sequence, and where I_ν is disjoint to K_ν. Then the intersection $D := \cap\{I_\nu|\nu < \omega_\gamma\}$ is non-empty, since it contains $\sup\{a_\nu|\nu < \omega_\gamma\}$. But on the other hand D is disjoint to all $K_\nu, \nu < \omega_\gamma$, and thus also to C_α, a contradiction.

If in 4.13 ω_α is regular and thus $= \omega_\gamma$, ω_γ cannot be replaced by $\omega_{\alpha+1}$. For if we assume the GCH, then $|C_\alpha| = 2^{\aleph_\alpha} = \aleph_{\alpha+1}$ could be partitioned in its $\aleph_{\alpha+1}$ one-element subsets.

Concerning the cardinalities of the classes in 4.13 we see that one of the classes must have the same cardinality as C_α, even in a more general context:

4.14 Theorem. *If $C_\alpha = \cup\{K_\nu|\nu < \omega_\alpha\}$, then for at least one ν we have $|K_\nu| = 2^{\aleph_\alpha}(= |C_\alpha|)$. And further: At least one of the classes K_ν is not embeddable in H_α.*

Proof. We can assume w.r.o.g. that the K_ν are pairwise disjoint. If now $|K_\nu| < 2^{\aleph_\alpha}$ would hold for all $\nu < \omega_\gamma$, then, applying the theorem of König, we would have $|C_\alpha| = 2^{\aleph_\alpha} = \sum\{|K_\nu||\nu < \omega_\alpha\} < (2^{\aleph_\alpha})^{\aleph_\alpha} = 2^{\aleph_\alpha}$, with contradiction.

And if all $K_\nu, \nu < \omega_\alpha$, would be embeddable in H_α, then this would also hold for their union C_α by 4.4.6. But this contradicts 4.7.11.

For the proof of the main theorem of this section we still need some preparations.

4.15 Definition. Let α be an ordinal, $\omega_\gamma = \mathrm{cf}(\omega_\alpha)$, U a set with $|U| = \mathfrak{k}_\alpha$. For $\mu < \omega_\gamma$ we define S_μ to be the set of all sequences $(u_\nu | \nu \leq \mu)$ with $u_\nu \in U$ for $\nu \leq \mu$. So each sequence of S_μ has a last element u_μ. Then we put $S := \cup\{S_\mu | \mu < \omega_\gamma\}$.

If we have a partial order in U, we introduce a partial order in S by a variant of the principle of first differences: If $u = (u_\nu | \nu \leq \mu)$ and $u' = (u'_\nu | \nu \leq \mu')$ are different elements of S we put

$u < u' \iff$ there is an index, and then also a first index δ ($\leq \min\{\mu, \mu'\}$) for which $u_\delta \neq u'_\delta$ holds, and $u_\delta < u'_\delta$ (in U).

It can easily be verified that the so-defined relation $<$, augmented by the identity relation on S, is a partial order of S. In particular it follows that if an element $u \in U$ is a proper initial segment of an element $v \in U$, then u and v are incomparable in S.

We determine the cardinality of S:

4.16 Lemma. $|S| = \mathfrak{k}_\alpha$.

Proof. We have $|U| = \mathfrak{k}_\alpha = \sum\{2^{\aleph_\nu} | \nu < \alpha\} = \sup\{2^{\aleph_\nu} | \nu < \alpha\}$.

Case 1. ω_α is a regular initial ordinal. Then, due to the definition of $\mathfrak{k}_\alpha$, we can find a representation $U = \cup\{A_\nu | \nu < \omega_\alpha\}$ where $|A_\nu| = 2^{\aleph_\nu}$, and where the A_ν are pairwise disjoint. We put $M_\nu := \cup\{A_\iota | \iota \leq \nu\}$. Then $|M_\nu| = 2^{\aleph_\nu}$ for $\nu < \omega_\alpha$, and the sets M_ν, $\nu < \omega_\alpha$, form an ascending tower of subsets of U.

If $\mu < \omega_\alpha$ and if $(u_0, \ldots, u_\mu)$ is in S_μ, then each $u_\iota, \iota \leq \mu$, is in a set $M_{\nu(\iota)}$ with $\nu(\iota) < \omega_\alpha$. The set $\{\nu(\iota) | \iota \leq \mu\}$ cannot be cofinal in the regular ordinal ω_α, and so there is a $\nu < \omega_\alpha$ such that all $u_\iota, \iota \leq \mu$, are in M_ν. Then the sequence $(u_0, \ldots, u_\mu)$ is in $M_\nu^{\mu+1}$. This set has cardinality $(2^{\aleph_\nu})^{|\mu+1|} := k$. From $\mu < \omega_\alpha$ we conclude $|\mu + 1| = \aleph_\tau$ for an ordinal $\tau < \aleph_\alpha$. Now k is $= 2^{\aleph_\delta} \leq \mathfrak{k}_\alpha$, where $\delta = \max\{\nu, \tau\}$.

The set S is a subset of $\cup\{M_\nu^{\mu+1} | \nu < \alpha, \mu < \omega_\alpha\}$, which is the union of $|\alpha| \cdot |\omega_\alpha|$ sets of cardinality $\leq \mathfrak{k}_\alpha$. So its cardinality is also $\leq \mathfrak{k}_\alpha$.

Case 2. ω_α is singular. Here we find a representation $U = \cup\{M_\nu | \nu < \omega_\gamma\}$ with $|M_\nu| = 2^{\aleph_{\alpha_\nu}}$, where the $\alpha_\nu, \nu < \omega_\gamma$, form a strictly ascending sequence of ordinals $\leq \omega_\alpha$, whose supremum is ω_α.

If $\mu < \omega_\gamma$ and if $(u_0, \ldots, u_\mu)$ is in S_μ, then for each $\iota \leq \mu$ there is a $\nu(\iota) < \omega_\gamma$ such that $u_\iota \in M_{\nu(\iota)}$, and since $\{\nu(\iota) | \iota \leq \mu\}$ is not cofinal in ω_γ there is an index $\nu < \omega_\gamma$ such that all $u_0, \ldots, u_\mu$ are in M_ν. The set $M_\nu^{\mu+1}$ has cardinality $(2^{\aleph_{\alpha_\nu}})^{|\mu+1|}$, which is $\leq \mathfrak{k}_\alpha$ because of $|\mu+1| < \aleph_\gamma < \aleph_\alpha$. Finally $S \subseteq \cup\{M_\nu^{\mu+1} | \nu < \omega_\gamma, \mu < \omega_\gamma\}$ has cardinality $\leq \mathfrak{k}_\alpha \cdot \aleph_\gamma \cdot \aleph_\gamma = \mathfrak{k}_\alpha$ — and $|S| \geq \mathfrak{k}_\alpha$ is trivial.

Now we can prove the main theorem of this section:

4.17 Theorem [73]. *Let α be an ordinal, $\omega_\gamma := \mathrm{cf}(\omega_\alpha), U$ a set with $|U| = \mathfrak{k}_\alpha$ and $\mathfrak{P}(U) = \cup\{K_\nu | \nu < \omega_\gamma\}$. Then one of the sets K_ν contains a subset of type $\tau_\alpha := \mathrm{tp}(U_\alpha).$*

Proof. W.r.o.g. we can assume that U is the set U_α of 5.2.10. This has the advantage that we can use the additional order structure of U_α in the proof. We consider the set S of 4.15, which also has cardinality $\mathfrak{k}_\alpha$. Since the power set $\mathfrak{P}(U_\alpha)$ is $\mathfrak{k}_\alpha$ - universal, there exists an isomorphic mapping g of S onto a subset $g(S)$ of $\mathfrak{P}(U_\alpha)$. So each $g(s)$, $s \in S$, is a subset of U.

We define S^* to be the set of all sequences $(u_\nu | \nu < \lambda)$, where λ is an ordinal $< \omega_\gamma$ and $u_\nu \in U_\alpha$ for $\nu < \lambda$. So $S^* \supseteq S$, and S^* also contains the empty sequence $\mathfrak{o}$. If $\varphi = (u_\nu | \nu < \lambda) \in S^*$ and $x \in U$, we (of course) put $(\varphi, x) := (u_0, \ldots, u_\nu, \ldots (\nu < \lambda), x)$, and similar if x is before φ.

For $\mu < \omega_\gamma$ we denote the set of all sequences $\in S$, that have $(u_\nu | \nu \leq \mu)$ as initial segment, by $S(u_0, \ldots, u_\mu)$ or $S(u_\nu | \nu \leq \mu)$. This set is evidently a segment of S. For each $S(u_0, \ldots, u_\mu)$ of S we define:

$C(u_\nu | \nu \leq \mu)$ is the least segment of $\mathfrak{P}(U)$ which contains the image set $g[S(u_\nu | \nu \leq \mu)]$. It is the convex hull of $g[S(u_\nu | \nu \leq \mu)]$ in $\mathfrak{P}(U)$. First we prove:

(I) Let x and y be different elements of U, $\varphi \in S^*$,and let $X \in C(\varphi, x), Y \in C(\varphi, y)$. Then $x < y \implies X \subset Y$, and $x \parallel y \implies X \parallel Y$.

X is in the smallest segment of $\mathfrak{P}(U)$ which contains $g[S(\varphi, x)]$, and so there are sets $A, B \in g[S(\varphi, x)]$ with $A \subseteq X \subseteq B$. That means: There are two sequences $\varphi_1, \varphi_2 \in S$ with

(1) $g(\varphi, x, \varphi_1) \subseteq X \subseteq g(\varphi, x, \varphi_2)$.

For the same reason there are sequences ψ_1 and $\psi_2 \in S^*$ satisfying

(2) $g(\varphi, y, \psi_1) \subseteq Y \subseteq g(\varphi, y, \psi_2)$.

Let $x, y \in S$. Suppose now $x < y$. Then (1) and (2) yield $X \subseteq g(\varphi, x, \varphi_2) \subset g(\varphi, y, \psi_1) \subseteq Y$. Indeed, $x < y \implies (\varphi, x, \varphi_2) < (\varphi, y, \psi_1)$, and g is $<$ - preserving.

If $x \parallel y$ we conclude as follows: Suppose indirectly that X and Y are comparable, w.r.o.g. we can assume $X \subseteq Y$. Then we would have

$g(\varphi, x, \varphi_1) \subseteq X \subseteq Y \subseteq g(\varphi, y, \psi_2)$. But this is impossible because the sequences (φ, x, φ_1) and (φ, y, ψ_2) are incomparable and since g is an isomorphic mapping, and then $g(\varphi, x, \varphi_1)$ cannot be a subset of $g(\varphi, y, \psi_2)$. So (I) is proved.

If now K_0 has an element in common with each set $C(u_0)$, $u_0 \in U_\alpha$, then the set of these elements forms by (I) (in this case φ is the empty sequence) a subset of K_0 which is isomorphic to U_α, and we are done. So we can assume that there is an element $e_0 \in U_\alpha$, such that the set $C(e_0)$ is disjoint to K_0.

Suppose that for some $\mu < \omega_\gamma$ we have defined elements $e_\nu \in U_\alpha$ for $\nu < \mu$, such that each set $C(e_0, \ldots, e_\nu)$ is disjoint to K_ν (and, by the way, then also disjoint to all K_ρ with $\rho < \nu$), then we proceed as follows: We consider the set of all sequences $(e_0, \ldots, e_\nu, \ldots (\nu < \mu), u_\mu)$ with $u_\mu \in U_\alpha$ — which prolong the sequence $(e_\nu | \nu < \mu)$ by an element u_μ of U_α. If K_μ intersects each set $C(e_0, \ldots, e_\nu, \ldots (\nu < \mu), u)$, $u \in U_\alpha$, it follows by (I) that then K_μ has a subset of type τ_α, and we are done.

So we finally have to discuss the situation where the above construction could be performed over ω_γ steps. But this situation cannot happen: Suppose the contrary. Then we choose for each $\nu < \omega_\gamma$ elements $a_{\nu+1}$ and $b_{\nu+1}$ in U_α, which satisfy $a_{\nu+1} < e_{\nu+1} < b_{\nu+1}$ (in U_α). This is possible since U_α is an $\eta_{\gamma\gamma}$ - set. Now we define intervals $I_\nu := [A_\nu, B_\nu]$, $\nu < \omega_\gamma$, of $\mathfrak{P}(U_\alpha)$ as follows:

We choose an element $A_\nu \in g[S(e_0, \ldots, e_\nu, a_{\nu+1})] \subseteq C(e_0, \ldots, e_\nu, a_{\nu+1})$, and an element $B_\nu \in g[S(e_0, \ldots, e_\nu, b_{\nu+1})] \subseteq C(e_0, \ldots, e_\nu, b_{\nu+1})$. Then $A_\nu \subset B_\nu$ holds since each element of $S(e_0, \ldots, e_\nu, a_{\nu+1})$ is $<$ each element of $S(e_0, \ldots, e_\nu, b_{\nu+1})$, and g is $<$ - preserving. Now I_ν is disjoint to K_ν since K_ν does not intersect $C(e_0, \ldots, e_\nu)$, due to our indirect assumption, and $C(e_0, \ldots, e_\nu)$ contains A_ν and B_ν. By construction the I_ν form a decreasing sequence $I_0 \supseteq I_1 \supseteq \cdots \supseteq I_\nu \supseteq \cdots, \nu < \omega_\gamma$. In particular $A_0 \subseteq A_1 \subseteq \cdots \subseteq A_\nu \subseteq \cdots$, $\nu < \omega_\gamma$, holds. The intersection $\cap\{I_\nu | \nu < \omega_\gamma\}$ is disjoint to all $K_\nu, \nu < \omega_\gamma$, and then also to their union $\mathfrak{P}(U_\alpha)$. But, on the other hand, the set $A := \cup\{A_\nu | \nu < \omega_\gamma\}$ is contained in $\cap\{I_\nu, \nu < \omega_\nu\}$ and, of course, an element of $\mathfrak{P}(U_\alpha)$, which yields a contradiction.

The Theorem 4.11 has an analog for the partially (and $\aleph_\alpha$ - universally) ordered sets U_α of 5.2.10:

4.18 Theorem. *Let* $\omega_\gamma = \mathrm{cf}(\omega_\alpha)$, $\beta < \omega_\gamma$. *And let* $U_\alpha = \cup\{K_\nu | \nu < \omega_\beta\}$. *Then one of the sets* K_ν *contains an* $\eta_{\gamma\gamma}$ - *set.*

Proof. There exists an interval $[a, b]$ of U_α, in which one of the sets K_ν is dense. Suppose the contrary. Then there is an interval $[a_0, b_0]$ of U_α, which is disjoint to K_0. Let μ be an ordinal $< \omega_\beta$, such that for each $\nu < \mu$ already an interval $[a_\nu, b_\nu]$ of U_α has been constructed, so that $[a_\nu, b_\nu]$ is disjoint to K_ν, and so that the intervals $[a_\nu, b_\nu]$ form a decreasing sequence. Then the union of the sets $A := \{a_\nu | \nu < \mu\}$ and $B := \{b_\nu | \nu < \mu\}$ forms a chain C, and there exists a maximal chain $M \supseteq C$ of U_α. Since U_α is an $\eta_{\gamma\gamma}$ - set by 5.2.12, the set M is an η_γ - set by 5.2.4. Now A and B both have a cardinality $< \aleph_\gamma$, and therefore they cannot be neighboring in M. So there is an element of M between them. On the other hand the intersection $\cap\{[a_\nu, b_\nu] | \nu < \omega_\beta\}$ is by construction disjoint to all K_ν, and then also to their union U_α. This is a contradiction.

Let now K_ν be dense in the interval $[a, b]$ of U_α. By 5.2.12 resp. 5.2.8 U_α and the open interval (a, b) is an $\eta_{\gamma\gamma}$ - set, and by 5.2.7, then also $K_\nu \cap (a, b)$ is an $\eta_{\gamma\gamma}$ - set, which proves our assertion.

The analog to 4.11$'$ is:

4.19 Theorem. *Let S be an $\eta_{\alpha\alpha}$- set and $S = \cup\{K_\nu | \nu < \omega_\beta\}$, where $\beta < \alpha$ holds. Then one of the sets K_ν contains an $\eta_{\alpha\alpha}$- set.*

Proof. If the assertion would be false, we could again construct a decreasing sequence of intervals $[a_\nu, b_\nu]$, $\nu < \omega_\beta$, of S, such that $[a_\nu, b_\nu]$ is disjoint to K_ν for $\nu < \omega_\beta$. The intersection $\cap\{[a_\nu, b_\nu] | \nu < \omega_\beta\}$ would then be disjoint to $\cup\{K_\nu | \nu < \omega_\beta\}$, and thus to S. But the sets $\{a_\nu | \nu < \omega_\beta\}$ and $\{b_\nu | \nu < \omega_\beta\}$ both have cardinality $< \aleph_\alpha$, so that they cannot be neighboring in S, and therefore there exists an element of S between these sets.

Finally one can pose the question of whether it is possible to increase the number of classes in Theorems 4.13 and 4.17 while the conclusion remains valid. For a special situation this is possible, as the following theorem shows, which we mention without proof:

4.20 Theorem [69]**.** **a)** *If $C_\alpha = \cup\{K_\nu | \nu < \omega_\alpha\}$, then there is a $\nu < \omega_\alpha$, so that K_ν has a subset of type $\tau + \tau^*$ for every ordinal $\tau < \omega_{\alpha+1}$.*

b) *Let $|S| = \aleph_\alpha$, and $\mathfrak{P}(S) = \cup\{K_\nu | \nu < \omega_\alpha\}$. Then there is a $\nu < \omega_\alpha$, so that K_ν has a subset of type $\tau + \tau^*$ for every ordinal $\tau < \omega_{\alpha+1}$.*

If ω_α is regular, then these statements are surpassed by 4.13, resp. 4.17. For singular ω_α this is by no means the case.

Chapter 10
Comparison of order types

In 1.9.6 we introduced the $\leq$ - and the $<$ - relation for order types, which was defined by Fraïssé [44]. In the class of well-ordered sets we have already had a complete discussion since the order type of a well-ordered set is practically the same as an ordinal. (The exact definition of order type was given in 1.9.5.) In particular in the class of order types of well-ordered sets we have antisymmetry, so that this class is ordered by $\leq$. The situation changes radically if we consider the order types of more general linearly ordered sets, and even more of partially ordered sets. Here the relation $\leq$ is no longer antisymmetric, and so the class of all order types is only quasi-ordered by $\leq$.

A modest oversight over the class of order types is given by the $\aleph_\alpha$ - universally ordered sets. For the linearly ordered sets the sets H_α of 4.3.3 are $\aleph_\alpha$ - universal, and for partial orders the sets U_α of 5.2.10 are $\aleph_\alpha$ - universal. Here we recall 5.3.9', which states that the order of U_α can be extended to a linear order in U_α, with which U_α has the order type h_α of H_α.

10.1 Some general theorems on order types

1.1 Definition. If τ is an order type we define $|\tau|$ to be the cardinality of a realization of τ, and generally we ascribe to τ all those properties which all realizations of τ have, e.g. dense, countable, well-founded and so on. A first question arises: How many order types of a fixed cardinality exist ? We consider here only the infinite case. For this we obtain the following result, which is tight:

1.2 Theorem. *There are $2^{\aleph_\alpha}$ different types of linearly ordered sets of cardinality $\aleph_\alpha$.*

Proof. In this proof let 2 (resp.1) denote the order type of a 2-element (resp. 1-element) chain and η the type h_0 of the set $\mathbf{Q}$ of rationals with their usual order. Let $\mathfrak{C}$ be the set of all order types $\sum\{\tau_\nu | \nu < \omega_\alpha\}$, where $\tau_\nu = 2$ for even ordinals $\nu < \omega_\alpha$ and $\tau_\nu \in \{1, \eta\}$ for odd ordinals $\nu < \omega_\alpha$. The chains with such an order type evidently have cardinality $\aleph_\alpha$. And there holds:

332

(I) Two chains with such sums, which differ in at least one τ_ν, have different order types:

Let C be a chain of type $\sum\{\tau_\nu | \nu < \omega_\alpha\}$, and C' a chain of type $\sum\{\tau'_\nu | \nu < \omega_\alpha\}$, where τ_ν and τ'_ν are $= 2$ for even ν, resp. $\in \{1, \eta\}$ for odd ν. Suppose that there exists an isomorphic mapping $f : C \to C'$. Then f necessarily maps the first element of C onto the first element of C', further the second element of C onto the second of C'. If $\tau_1 = 1$, then also τ'_1 must be $= 1$, for otherwise τ'_1 would be $= \eta$. But this is impossible since C has a third element and C' would lack such an element. In general suppose that for all $\nu < \mu$, where $\mu < \omega_\alpha$ holds, we have already $\tau_\nu = \tau'_\nu$ and that for the segments T_ν of type τ_ν of C and the segments T'_ν of type τ'_ν of C' we have $f[T_\nu] = T'_\nu$, then also $\tau_\mu = \tau'_\mu$ must hold. For if we have $\tau_\mu = 1$, which implies that μ is odd, then the set $\cup\{T_\nu | \mu \leq \nu\}$ has a first element, and then this also must hold for $\cup\{T'_\nu | \mu \leq \nu\}$, and this entails $\tau'_\mu = 1$. If $\tau_\mu = \eta$, then for a similar reason also τ'_μ must be $= \eta$, and the isomorphic mapping f must map the initial segment of type $\tau_\mu = \eta$ of $\cup\{T_\nu | \mu \leq \nu\}$ onto the initial segment of type τ'_μ of $\cup\{T'_\nu | \mu \leq \nu\}$. By induction we so obtain that $\tau_\nu = \tau'_\nu$ for all $\nu < \omega_\alpha$, and therefore the chains C and C' have the same order type. And (I) is proved.

Now the cardinality of $\mathfrak{C}$ can easily be counted. A type of $\mathfrak{C}$ is by (I) completely determined by the set of the τ_ν, where the ν are odd ordinals $< \omega_\alpha$. For each such τ_ν we have two possibilities (to be $= 1$ or $= \eta$). This leads to $2^{\aleph_\alpha}$ order types in $\mathfrak{C}$.

It is easy to see that there cannot be more than $2^{\aleph_\alpha}$ order types (the types of the partially ordered sets included !) of cardinality $\aleph_\alpha$. For even the set of all relations over a set A of $\aleph_\alpha$ elements contains only $2^{|A \times A|} = 2^{\aleph_\alpha}$ elements.

We mention two simple examples:

1.3 Example. a) The type of a two-element antichain is incomparable with each type of a chain which contains at least two elements.

b) The type ω_0^* of the set of negative integers is incomparable with all types of well-ordered sets of type $\geq \omega_0$. For a well-ordered set has no infinite descending subset, and dually for inversely well-ordered sets.

In [13] Chajoth studied how the order type of a chain can alter if we change the position of elements in a linearly ordered set, resp. if we introduce a new element in a linearly ordered set. In the following his theorems are partially transferred onto the case of general posets, using the same methods:

1.4 Theorem. *One can change the order type of an arbitrary poset P by introducing a new element.*

Proof. If P has no first element, we can introduce a new element e and put $e < x$ for all $x \in P$. Then $P \cup \{e\}$ has a type different from that of P. In general, there exists the greatest well-ordered initial segment W of P, such that all elements of W are situated before all elements of $P \setminus W$. (If P has no first element we have $W = \emptyset$.) Let τ be the order type of W. If then we introduce a new element e and put $w < e < y$ for all $w \in W$ and $y \in P \setminus W$, then $P \cup \{e\}$ has another order type than P, since an isomorphic mapping of $P \cup \{e\}$ on P had to map the initial segment $W \cup \{e\}$ of $P \cup \{e\}$, which has type $\tau + 1$, onto an initial segment of P which also must have type $\tau + 1$. This contradicts the maximality of τ.

Contrary to the situation of the last theorem there exist many sets, in which one can omit single elements, and also greater subsets, without changing the order type of the set. This is e.g. the case with $\mathbf{N}$, $\mathbf{Z}$ and $\mathbf{Q}$, but not with $\mathbf{R}$, since the omission of a single element of $\mathbf{R}$ produces a gap, and $\mathbf{R}$ has no gaps.

1.5 Theorem. *Let S be a set of infinite cardinality m. The set of all those orders $\leq$ on S, for which $(S, \leq)$ retains its order type, if one removes finitely many elements of S, has the same cardinality as the set of all orders on S.*

Proof. For a poset A we define a poset A_ω as follows: For each $x \in A$ we choose a set $A(x)$ of the type ω of the natural numbers with x as first element, such that the sets $A(x)$, $x \in A$, are pairwise disjoint. We put $A_\omega := \cup\{A(x) | x \in A\}$ and order it by: Each set $A(x)$ is ordered as before, and for $a, b \in A_\omega$ with $a \in A(x)$, $b \in A(y)$, $x \neq y$ in A we put $a < b$ iff $x < y$. Roughly speaking: We enlarge each element $x \in A$ to a set of type ω, and order the new elements in the same way as the first elements of their ω - string.

(I) If now $(A, \leq_1)$ and $(B, \leq_2)$ are non-isomorphic posets, the types of A_ω and B_ω are also different. And conversely.

For suppose indirectly that there is an isomorphic mapping f from A_ω onto B_ω. If we have elements $x, y \in A \ (\subseteq A_\omega)$ with $x < y$, the images $f(x)$ and $f(y)$ cannot be in the same segment $B(z)$ of B_ω. For in this case there would be only finitely many elements between $f(x)$ and $f(y)$ whereas there are infinitely many elements of A_ω, in particular of $A(x)$, between x and y.

So, if $f(x) \in B(z)$, only elements of $A(x)$ can be mapped by f into $B(z)$, and analogously different elements of $A(x)$ cannot have images in different sets $B(z)$. And so f maps $A(x)$ bijectively onto $B(z)$. Herewith the first element x of $A(x)$ must be mapped onto the first element z of $B(z)$. Now $f \restriction A$ is an isomorphism of A into B. For the same reason $f^{-1} \restriction B$ is an isomorphism of B into A, and so finally $f \restriction A$ maps A isomorphically onto B. This proves (I).

If A_ω and B_ω have different order types, then A and B do also. For it is trivial that isomorphic posets A, B have isomorphic posets A_ω and B_ω.

Evidently the sets A_ω retain their order type if one removes finitely many elements of them, since this also holds for all sets of type ω. And since before we had established a bijection between the set T of order types of the posets $(S, \leq)$ and the set T^* of order types of the posets $(S_\omega, \leq)$, we have $|T^*| = |T|$, and we are done.

Finally we obtain a characterization of the finite linear order types:

1.6 Theorem. *The finite linear order types are characterized by the fact, that they don't alter, if in their realizations an element changes its position.*

Proof. Let τ be a linear order type with an infinite realization S. If S has no first element, we can take an arbitrary element of S and move it before all other elements of S. That changes the order type. Assume now that S has a first element, and let A be the greatest well-ordered initial segment of S. If this is the whole set S, then S has a type $\tau \geq \omega$. Then we take the first element of S and move it behind all other elements of S, so obtaining a set of type $\tau + 1$, and we are done. If $S \setminus A$ is non-empty, we choose an element $e \in S \setminus A$, place e behind A and before the elements of $(S \setminus A) \setminus \{e\}$ and obtain a linear order whose type differs from that of S, since now there is an initial segment of type $> \operatorname{tp}(A)$.

We supplement the definitions of section 4.1 by:

1.7 Definition. If α and β are order types, we call α an *initial segment* of β, if there exists an order type γ such that $\beta = \alpha + \gamma$. And symmetrically: α is a *final segment* of β, if there exists an order type γ with $\beta = \gamma + \alpha$.

We remark that if α, β are ordinals the equation $\beta = \alpha + \gamma$ determines γ uniquely, but $\beta = \gamma + \alpha$ does not. E.g. $\omega + \omega = (\omega + n) + \omega$ for every $n \in \mathbf{N}$.

If S is a linearly ordered set, A an initial segment and B the complementary final segment of S, we express this in the form $S = A + B$. An analogous definition applies to expressions $S = A + B + C + \cdots$, where we have finitely or denumerably many summands. Here A, B, C, ...are consecutive (and thus also disjoint) segments of S.

The next theorem was established by Lindenbaum. A proof of it was published by Sierpinski [160]:

1.8 Theorem (Lindenbaum) *Let α and β be linear order types, such that α is an initial segment of β, and β a final segment of α. Then $\alpha = \beta$.*

In another formulation: The equations

(1) $\alpha = \sigma + \beta$ and $\beta = \alpha + \rho$

imply $\alpha = \beta$.

Proof. From (1) we obtain $\beta = \sigma + \beta + \rho$. Starting from this equation we can choose disjoint linearly ordered sets B, S, B_1, R, where these sets have the order types $\beta, \sigma, \beta, \rho$ respectively such that

(2) $B = S + B_1 + R$.

There exists an isomorphism $f : B \to B_1$, and for the iterates f^n of f, $n \in \omega$, we prove:

(3) B has an initial segment $S + f[S] + \cdots + f^n[S] + f^{n+1}[B]$.

From (2) we obtain $B = S + f[B] + R = S + f[S + B_1 + R] + R = S + f[S] + f^2[B] + f[R] + R$, so that (3) holds for $n = 1$. Let now (3) be proved for a fixed $n \in \mathbf{N}$. Then B has an initial segment $S + f[S] + \cdots + f^n[S] + f^{n+1}[S + B_1 + R] = S + f[S] + \cdots + f^n[S] + f^{n+1}[S] + f^{n+2}[B] + f^{n+1}[R]$, and this proves (3) for $n + 1$, so that it follows by induction for all $n \in \mathbf{N}$.

Finally also the sum $S + f[S] + \cdots + f^n[S] + \cdots$ of denumerably many summands is an initial segment of B, so that there exists a final segment T of B with

(4) $B = S + f[S] + \cdots + f^n[S] + \cdots + T$.

Since S has the type σ, also all $f^n[S]$, $n \in \omega$, have the type σ, and so we obtain with $\tau := \mathrm{tp}(T)$ from (4) for the corresponding types

$\beta = \sigma + \sigma + \cdots + \tau$. On the other hand we have by (1) $\alpha = \sigma + \beta = \sigma + (\sigma + \sigma + \cdots) + \tau = \beta$.

10.2 Countable order types

A rough classification of the countable linear order types is given by the property of containing a subset of the type η of the rationals or not. Since η is the order type of an $\aleph_0$ - universally linearly ordered set, it is clear that all denumerable linearly ordered sets whose type is $\geq \eta$ are embeddable into each other. Moreover they are themselves also $\aleph_0$ - universal.

The linearly ordered sets that don't contain a subset of type η form the class of scattered linearly ordered sets. For this class we have Laver's theorems 8.6.7, which state that every set of scattered linear order types, that are pairwise incomparable, is finite. And there is no infinite strictly descending sequence of scattered linear order types.

An ascending sequence of type ω_1 of countable scattered order types is e.g. the set of all countable ordinals. And on the whole by 1.2 there exist $2^{\aleph_0}$ types of linearly ordered countable sets. From Dushnik/Miller [29] originates the following:

2.1 Theorem. *Every denumerably infinite linearly ordered set A contains a proper subset A' to which it is isomorphic.*

Proof. Let A be a denumerably infinite linearly ordered set. For $a, b \in A$ we put $a \sim b$ iff $[a, b]$ is finite. This yields an equivalence relation in A. The equivalence class of a is denoted by $C(a)$. Then we consider two cases:

Case 1. There exists an $a \in A$ such that $C(a)$ is infinite.

Then A is representable as ordered sum $A = A_1 + C(a) + A_3$ of three disjoint segments. The component $C(a)$ has one of the three types $\omega, \omega^*, \omega^* + \omega$. Suppose that its type is ω. Then we define a mapping $f : A \to A$ by $f(x) := x$ for $x \in A_1 \cup A_3$ and $f(x)$ as the immediate successor of x in $C(a)$. Then f is an isomorphic mapping of A onto a proper subset. In the cases where $C(a)$ has type ω^* or $\omega^* + \omega$ we again let the elements of $A_1 \cup A_3$ be fixed, and $C(a)$ can be mapped $<$ - preserving onto a proper subset.

Case 2. All congruence classes are finite.

Then we must have infinitely many congruence classes. Let F be the set of their first elements. If a and b are two different elements of F, w.r.o.g. with $a < b$, then there exists a $c \in F$ with $a < c < b$, otherwise also a and b would be congruent. Thus F is denumerable and dense. Its interval (a, b) has type η, and so A is isomorphic to a subset of (a, b), which is a proper subset of A.

2.2 Theorem [29]. *Let A be a denumerable linearly ordered set, for which each $<$ - preserving mapping $f : A \to A$ satisfies $f(a) \geq a$ for all $a \in A$. Then A is well-ordered.*

Proof. (I) A is scattered; it has no subset of type η.

Suppose the contrary, that $E \subseteq A$ is a subset of type η. Then choose an arbitrary element $b \in E$. The set $(E < b)$ has again type η, and so there exists a $<$ - preserving mapping f of A into $(E < b)$, in particular $f(b) < b$ holds, contradicting our assumption. And (I) is proved. Further we assume that A is infinite, since for finite A the statement is trivial.

Now we introduce in A the following equivalence relation: For $a, b \in A$ we put $a \sim b$ iff the closed interval with ends a, b is well-ordered. Evidently each equivalence class is a convex subset of A.

(II) Every equivalence class has a first element.

Suppose the contrary. Then there is an equivalence class C which has no first element. Let T be a coinitial inversely well-ordered subset $\{c_i | i < \omega\}$ of C of type ω^*, so that $i < j < \omega$ implies $c_i > c_j$ — such a set exists since A and C is infinite. And let τ_i be the order type of the interval $I_i := [c_{i+1}, c_i)$ which is well-ordered by construction, so that the τ_i are ordinals.

First we prove:

(III) There is an $m \in \omega$ such that a) the set $\{\tau_i | m \leq i < \omega\}$ has no greatest element, or b) it has a greatest element, and this occurs infinitely often in the sequence $\tau_i, i < \omega$.

Suppose the contrary. Then there exists a greatest element $\tau_{n(0)}$ in the sequence, and this occurs only finitely often. W.r.o.g. we can assume that $n(0)$ is the last index at which it occurs. By assumption then further there is an index $n(1) > n(0)$, such that $\tau_{n(1)}$ is the greatest element in $\{\tau_n | n \geq n(1)\}$, and we can assume that $n(1)$ is the last index n at which $\tau_{n(1)}$ occurs. This construction could be performed over ω steps and would yield a sequence $\tau_{n(0)} > \tau_{n(1)} > \cdots > \tau_{n(i)} > \cdots$, $i \in \omega$, of ordinals, which is strictly decreasing, and this is impossible. Thus (III) holds.

In the case a) of (III) we determine recursively numbers $n_i \in \omega$ for $i \in \omega$, which strictly increase with i, as follows: $n_0 := m$, and n_i is the first index $> i$, for which $\tau_{n_i} \geq \tau_i$ and $\tau_{n_i} > \tau_{n_j}$ for all $j < i$ holds. Then for each $i \in \omega$ there is a $<$ - preserving mapping $f_i : I_i := [c_{i+1}, c_i) \to I_{n_i}$ because of $\tau_i \leq \tau_{n_i}$. Let then f be the mapping which coincides with the f_i over I_i and which puts $f(x) = x$ for the $x \in (C \geq c_0)$ and for all $x \in A$, which are not in C. Now f is a $<$ - preserving mapping of A onto a subset of A, and there are elements x with $f(x) < x$. This contradicts our assumption, so that in case a) the statement (II) holds.

In case b) of (III) we have a sequence $m = n(0) < n(1) < \cdots < n(i) < \cdots$, $i < \omega$, satisfying $i < n(i)$ for $i < \omega$, for which all $\tau_{n(i)}, i < \omega$, are equal, and $= \max\{\tau_\nu | m \leq \nu < \omega\}$. Then we choose for each i with $m \leq i < \omega$ a $<$ - preserving mapping $f_i : [c_{i+1}, c_i) \to [c_{n(i+1)}, c_{n(i)})$ and define $f : A \to A$ by $f(x) := x$ for $x \in (C \geq c_{n(0)}) \cup (A \setminus C)$, and so that f coincides with all f_i over their corresponding range of definition. Then f is a $<$ - preserving mapping of A into A, and there are elements x with $f(x) < x$, contradicting our assumption. And therefore (II) holds. This implies that every equivalence class C is well-ordered, for each initial segment $[c, x]$, where c is the first and x an arbitrary element of C, is well-ordered.

After we have proved (II) it follows that there is only one equivalence class. Assume indirectly that there are at least two, C_1 and C_2. Their first elements a_1 and a_2 are not equivalent, so that the classes C_1, C_2 cannot be consecutive segments of A. There is at least one class between them, and its first element is between a_1 and a_2. Thus the set of first elements of the equivalence classes is dense and then also denumerable, and so it contains a subset of type η. But this contradicts (I). Now the unique equivalence class is nothing else but A, which is therefore well-ordered.

10.3 Uncountable order types

The Theorem 2.1 cannot be transferred on uncountable linearly ordered sets. This was proved by Dushnik/Miller [29]. They established a theorem of which we here mention a generalization on higher cardinalities, which follows with the same idea of proof:

3.1 Theorem. *Let C_α be the linearly ordered continuum of 4.7.31.*

Assuming $\mathfrak{k}_\alpha = \aleph_\alpha$, which follows from GCH, there holds: The continuum C_α contains a set E of power $c := 2^{\aleph_\alpha}$ ($= |C_\alpha|$), which is not isomorphic to any proper subset of itself.

Moreover the set E has only one automorphism, namely the identical mapping.

Proof. Let ω_c be the initial ordinal for the cardinality c. Let then $\mathfrak{F} := \{f_\nu | \nu < \omega_c\}$ be the set of all $<$ - preserving functions $f : C_\alpha \to C_\alpha$, which are different from the identity mapping id of C_α, which has only fixed elements. (For this it is sufficient that $f(x) \neq x$ holds for at least one $x \in C_\alpha$.) By 9.4.4 $\mathfrak{F}$ has cardinality $2^{\mathfrak{k}_\alpha} = c$, so that it is possible to arrange $\mathfrak{F}$ in the above manner.

For each $f \in \mathfrak{F}$ the set Fix f of fixed elements of f is not everywhere dense in C_α, that means: There is at least one open interval of C_α, which is disjoint to Fix f. For otherwise f would be the identity mapping. So there is a segment S of C_α with more than one element, which has no fixed element of f, and since $|S| = c$ holds, we have c elements p in S with $f(p) \neq p$.

We choose an element p_0 for which $f_0(p_0) =: q_0 \neq p_0$ holds. If μ is an ordinal $< \omega_c$, for which elements $p_\nu \neq q_\nu$ have already been defined for all $\nu < \mu$, we choose elements $p_\mu, q_\mu \in C_\alpha \setminus \cup_{\nu < \mu} \{p_\nu, q_\nu\}$ for which $f_\mu(p_\mu) = q_\mu \neq p_\nu$ holds. This is possible by construction, for we have:

Since f_μ is strictly increasing and $\neq$ id the set $A_\mu := \{p | f_\mu(p) \neq p\}$ has cardinality c. Then also $A'_\mu := A_\mu \setminus \cup_{\nu < \mu}\{p_\nu, q_\nu\}$ and $f_\mu[A'_\mu]$ have cardinality c. So there are c elements $p \in A'_\mu$ for which $f_\mu(p)$ is different from all p_ν and q_ν with $\nu < \mu$. Let p_μ be one of them. Then $f_\mu(p_\mu) =: q_\mu \neq p_\mu$ holds, and p_μ and q_μ are in $C_\alpha \setminus \cup_{\nu < \mu} \{p_\nu, q_\nu\}$. By transfinite induction, p_ν and q_ν are so constructed for all $\nu < \omega_c$.

If $C' = [a, b]$ is an interval of C_α, then there is a $<$ - preserving function, which is $=$ id on $C_\alpha \setminus C'$ and different from id on C', so that there is at least one element $x \in C'$ with $f(x) \neq x$. To see this choose two elements x, y in C_α with $a < x < y < b$ and a $<$ - preserving f which maps $[a, y]$ in $[a, x]$ — this can be done using 4.7.32 — and which leaves all other elements of C_α fixed. Since f is one of the $f_\nu \in \mathfrak{F}$ we obtain: There exists an index ν for which $p_\nu \in C'$. And this yields: $E := \cup\{p_\nu | \nu < \omega_c\}$ is dense in C_α.

Now E satisfies our assertion. Suppose the contrary, that $g : E \to E$ is a $<$ - preserving mapping. If g would be different from the identity mapping id_E of E, then by 9.4.3 there exists a $\leq$ - preserving mapping

340

$g^* : C_\alpha \to C_\alpha$ which extends g. Then g^* is also $<$ - preserving since E is dense in C_α, and so $g^* \in F$ is a f_ν. This yields a contradiction since $E \ni g(p_\nu) = g^*(p_\nu) = f_\nu(p_\nu) = q_\nu \notin E$.

If in 3.1 we take $\alpha = 0$, then C_0 is isomorphic to the set $\mathbf{R}$ of real numbers, and $\mathfrak{k}_0$ is $= \aleph_0$. Then there follows the original version of the theorem of Dushnik/Miller [29], and this now without using the GCH !

3.1′ Theorem. *The set of real numbers has a subset E of power $2^{\aleph_0}$ which is not isomorphic to any proper subset of itself. And this set has only one automorphism, the identical mapping.*

Next we prove several fundamental theorems of Sierpinski on order types of the subsets of $\mathbf{R}$. They can be generalized to higher cardinalities, practically without modifying his proofs.

3.2 Theorem [161]. *Let α be an ordinal and C_α the linearly ordered continuum of 4.7.31, $c = 2^{\aleph_\alpha}$ the cardinality of C_α, ω_c the initial ordinal of cardinality $|C_\alpha|$. Using $\mathfrak{k}_\alpha = \aleph_\alpha$ then there holds:*

a) There exists a transfinite sequence $\{\tau_\nu | \nu < \omega_c\}$ of linear order types $\tau_\nu \leq$ tp C_α, which is strictly decreasing: For $\mu < \nu < \omega_c$ we have $\tau_\mu > \tau_\nu$.

b) There exists a transfinite sequence $\{\tau_\nu | \nu < \omega_c\}$ of linear order types $\tau_\nu \leq$ tp C_α, which is strictly ascending with ν.

c) There exists a set of 2^c linear order types of cardinality c which are pairwise incomparable.

For $\alpha = 0$ these statements follow without the use of GCH since $\mathfrak{k}_0 = \aleph_0$ holds.

Proof. In the proof the method of the proof of 3.1 is again used. First we choose a well-ordering of the set F of all $<$- preserving functions $f : C_\alpha \to C_\alpha$ that are different from the identity mapping id of C_α. Since their set has cardinality c, we can arrange F as $F = \{f_\nu | \nu < \omega_c\}$, and, for technical reasons, we can do this in such a way that each element f of it is enumerated twice, at an even index 2β and at its successor index $2\beta + 1$.

Next we define elements x_ν and y_ν quite analogously to the proof of 3.1: Since f_0 is different from id we can choose an element $x_0 \in C_\alpha$ for which $f_0(x_0) = y_0 \neq x_0$.

For $\nu < \omega_c$ the set $\{x | f_\nu(x) = x\}$ is not dense in C_α. For otherwise f_ν would be the identity mapping. So there exists an open interval (a_ν, b_ν)

of C_α such that $f_\nu(x) \neq x$ for $x \in (a_\nu, b_\nu)$. And therefore we have $|\{x| f_\nu(x) \neq x\}| = c$.

Suppose that $\mu < \omega_c$ and that for all $\nu < \mu$ elements x_ν, y_ν have already been defined. Then, as in the proof of 3.1, there exist elements x_μ and y_μ with $f_\mu(x_\mu) = y_\mu$, which are different, and different from all x_ν, y_ν with $\nu < \mu$. By transfinite induction we so have defined elements x_ν, y_ν for all $\nu < \omega_c$.

If I is an open interval of C_α, then there exists, as in the proof of 3.1, a $<$ - preserving mapping $f : C_\alpha \to C_\alpha$, which satisfies $f(x) = x$ for $x \in C_\alpha \setminus I$, and $f(x) \neq x$ for at least one $x \in I$. This f is $= f_{2\beta}$ for some β, and we have by construction $f_{2\beta}(x_{2\beta})$ $(= y_{2\beta}) \neq x_{2\beta}$, so that $x_{2\beta} \in I$ holds. Thus we have obtained:

$H := \{x_{2\nu}|\nu < \omega_c\}$ is dense in C_α.

Then we put $Z := \{x_{2\nu+1}|\nu < \omega_c\}$. By construction $H \cap Z = \emptyset$. For $T \subseteq Z$ we put $E_T := T \cup H$.

The main idea of the proof consists in proving:

(I) Let A, B be subsets of Z with $A \setminus B \neq \emptyset$. Then E_A (with the induced order) cannot be isomorphic to a subset of E_B.

Suppose the contrary, that there is a $<$ - preserving mapping $g : E_A \to E_B$. The set H, and a fortiori its superset E_A, is dense in C_α. Now g can be extended to a $<$ - preserving function $f : C_\alpha \to C_\alpha$. This f is different from id. For otherwise we would have $g(x) = f(x) = x$ for $x \in E_A$ and thus $g[E_A] = E_A \subseteq E_B$ in contradiction to $\emptyset \neq A \setminus B =$ (since H is disjoint to A and B) $(A \cup H) \setminus (B \cup H) = E_A \setminus E_B$. So $f \in F$. And then there is an ordinal $\mu < \omega_c$ with $f = f_{2\mu}$.

We have $x_{2\mu} \in (H \subseteq) E_A$ and further $f_{2\mu}(x_{2\mu}) = y_{2\mu} \notin E_B = B \cup H$, for due to $B \subseteq Z$ only elements x_ν are in $B \cup H$. So it is impossible that $f_{2\mu}[E_A] \subseteq E_B$ holds. This contradicts $f_{2\mu}[E_A] = g[E_A] \subseteq E_B$, which follows from $f_{2\mu}(x) = f(x) = g(x)$ for $x \in E_A$, and (I) is proved.

For $\beta < \omega_c$ we put $Z_\beta := \{x_{2\nu+1}|\beta \leq \nu < \omega_c\}$, and obtain

$\beta < \gamma < \omega_c \implies (x_{2\beta+1} \in Z_\beta \setminus Z_\gamma$ and$) Z_\gamma \subset Z_\beta$.

If now for $\beta < \gamma < \omega_c$ we put in (I) $A = Z_\beta$ and $B = Z_\gamma$ then (I) yields: E_{Z_β} is not isomorphic to a subset of E_{Z_γ}. The order types of the sets $E_{Z_\beta}, \beta < \omega_c$, now form a set which satisfies a).

On the other hand we put $T_\beta := \{x_{2\nu+1}|\nu < \beta\}$. Then there follows:

$\beta < \gamma < \omega_c \implies x_{2\beta+1} \in T_\gamma \setminus T_\beta$ and thus $T_\beta \subset T_\gamma$. By (I) this entails $\text{tp } E_{T_\gamma} \not\leq \text{tp } E_{T_\beta}$ and further $\text{tp } E_{T_\gamma} \geq$ (and then also) $> \text{tp } E_{T_\beta}$, so that b) holds.

c) We have $|Z| = 2^{\aleph_\alpha}$. By 9.1.25 there is now an antichain $\mathfrak{A}$ of the power set $\mathfrak{P}(Z)$ of cardinality $2^{(2^{\aleph_\alpha})}$. So for $A, B \in \mathfrak{A}$ we have $A \setminus B \neq \emptyset \neq B \setminus A$, which by (I) entails tp $E_A \not\leq$ tp E_B and tp $E_B \not\leq$ tp E_A. And c) is proved.

The statement of 3.2,c) can partially be strengthened to the following:

3.3 Theorem [161]. *Suppose $\mathfrak{k}_\alpha = \aleph_\alpha$. Then there is a set of $2^{\aleph_\alpha}$ linearly ordered subsets of C_α of cardinality $2^{\aleph_\alpha}$, which pairwise differ in at most two points, so that their order types are pairwise incomparable.*

Proof. For $\beta < \omega_c$ we put $Q_\beta := \{x_{2\beta+1}\}$. Then the sets $E_{Q_\beta} = Q_\beta \cup H$, $\beta < \omega_c$, are incomparable by (I) and differ pairwise only in two elements.

3.4 Theorem [161]. *Suppose $\mathfrak{k}_\alpha = \aleph_\alpha$. Then there exist subsets H and H_1 of C_α of cardinality $2^{\aleph_\alpha}$, for which no order type exists which is strictly between them.*

Therefore the quasi-ordered set of all order types $\leq$ tp C_α is not dense.

Proof. We put (with the concepts of the proof of 3.2) $H_1 := H \cup \{x_1\}$. By definition of the T_β we have $T_0 = \emptyset$ and $T_1 = \{x_1\}$. These sets are $\subseteq Z$, and so there follows from (I): $H_1 = E_{T_1}$ is not isomorphic to a subset of $E_{T_0} = H$, in short tp $H_1 \not\leq$ tp H. Due to $H \subseteq H_1$ this implies tp $H <$ tp H_1. Now we can prove:

(II) There is no order type τ satisfying tp $H < \tau <$ tp H_1.

Suppose the contrary. Then there exists a subset $M \subseteq H_1$ with tp $M = \tau$ and tp $H <$ tp M. And there is a $<$ - preserving function $g : H \to M$. This can be extended to a $\leq$ - preserving function $f : C_\alpha \to C_\alpha$.

If we would have $f = \mathrm{id}_{C_\alpha}$ this would entail $g(x) = x$ for $x \in H$ and thus $g[H] = H$. But this yields $H \subseteq M \subseteq H_1 = H \cup \{x_1\}$, which can be strengthened to $H \subset M \subset H_1 = H \cup \{x_1\}$ because of tp $H <$ tp $M <$ tp H_1. This is a contradiction because there cannot be an element strictly between H and $H \cup \{x_1\}$. Therefore f is not $= \mathrm{id}_{C_\alpha}$ so that $f = f_{2\beta}$ holds for an ordinal $\beta < \omega(c)$. Due to $x_{2\beta} \in H$ we have $g(x_{2\beta}) = f(x_{2\beta}) = y_{2\beta} \notin H_1$, and a fortiori $g(x_{2\beta}) \notin M$ (which is $\subseteq H_1$). This contradicts $g[H] \subseteq M$, and so our statement is proved.

In several papers S.Ginsburg [56],[57],[58] studied related questions, among others the existence of order types ζ which satisfy $\sigma < \zeta < \tau$, if σ and τ are given linear order types. Further he investigates decompositions of linearly ordered sets into the union of disjoint subsets which have special properties, e.g. to be pairwise incomparable, or to be pairwise isomorphic.

10.4 Homogeneous posets

Usually a good oversight over the order-theoretical properties of a poset (and also other structured systems) can be obtained by considering its automorphisms, for they reflect its symmetry features. Some of the posets of practical importance, like **Z, Q, R,** are very "homogeneous": All their elements have the same structure-properties. In other posets, e.g. in the well-ordered ones, we have entirely different situations. We now investigate these phenomena in detail:

4.1 Definition. Let P be a poset. An *automorphism (or automorphic mapping)* of P is an isomorphic mapping $f : P \to P$ of P onto itself. The set of all automorphisms of P forms a group relative to the operation $\circ$ of concatenation, the so-called *automorphism group* of P. Its *unit element* is the identity mapping of P.

In 4.6.1 we defined a poset P to be *homogeneous*, if for each two elements $a, b \in P$ there is at least one automorphism of P which maps a onto b. If $|P| > 1$ and if for each two elements $a, b \in P$ there is one and only one automorphism which maps a onto b, then P is called *uniquely homogeneous.*

A poset P, which has only one automorphism, the identity mapping of course, is called *rigid.*

4.2 Example. The set **Z** of integers is uniquely homogeneous.

For if a,b are integers, we map every $x \in \mathbf{Z}$ onto $x + (b - a)$. This translation of **Z** maps a onto b and is an automorphism. It is uniquely determined since the immediate successor (resp. predecessor) of a must be mapped onto the immediate successor (resp. predecessor) of b and so on.

For the same reason the sets **Q** and **R** are homogeneous. But here not only translations effect automorphic mappings of **Q** (resp. **R**). As we have seen in 9.4.4 there are $2^{\aleph_0}$ strictly increasing mappings of $\mathbf{R}$ onto $\mathbf{R}$, and each of them is an automorphism of $\mathbf{R}$.

The fact that $\mathbf{Q}$ is homogeneous can be put into a more general context, namely there follows from 4.6.4:

4.3 Theorem. *If λ is an indecomposable ordinal the set T_λ of 4.5.7 is homogeneous. And as a special case of this the sets H_α of 4.3.3 are also homogeneous.*

Proof. By 4.6.4 the sets T_λ are homogeneous if λ is indecomposable. And by 4.5.10 the set $T(\omega_\alpha)$ is isomorphic to H_α.

For uniquely homogeneous linearly ordered sets we have a strong restriction for their possible order types. For these are subtypes of the type of $\mathbf{R}$:

4.4 Theorem. *Let $(L, \leq)$ be a linearly ordered set which is uniquely homogeneous. Then $(L, \leq)$ is embeddable in $\mathbf{R}$.*

Proof. For $|L| < 2$ the theorem is trivial. So we assume that L has at least two elements a, b. Then there is an automophism f of L with $f(a) = b \neq a$. W.r.o.g. we assume $a < f(a)$. Then the iterates of f generate a sequence
$$\cdots < f^{-n}(a) < \cdots < f^{-1}(a) < a < f(a) < f^2(a) < \cdots < f^n(a) < \cdots.$$

Let S be the smallest segment of L which contains all elements of this sequence. Then it follows rather immediately that f maps S into itself: For each element of S satisfies $f^z(a) \leq x \leq f^{z+1}(a)$ for some $z \in \mathbf{Z}$, and then also $f^{z+1}(a) \leq f(x) \leq f^{z+2}(a)$.

Now S turns out to be the whole set L. Otherwise we could construct an automorphism g of L by putting $g(x) := x$ for $x \in L \setminus S$ and $g(x) := f(x)$ for $x \in S$. Then g is different from the identity mapping of L, but maps all elements of $L \setminus S$ on themselves, contradicting the uniqueness.

So, if we look for uniquely homogeneous linearly ordered sets, we can find them only among sets which are order-isomorphic to subsets of $\mathbf{R}$. The question arises whether each such set is isomorphic to $\mathbf{Z}$. But Ohkuma has proved that there do exist uniquely homogeneous subsets of $\mathbf{R}$ which are dense in $\mathbf{R}$ and of cardinality $2^{\aleph_0}$. Before proving his theorem we mention:

4.5 Theorem. *A uniquely homogeneous linearly ordered set D, which is not isomorphic to $\mathbf{Z}$, must be dense.*

Proof. Suppose indirectly that a and b are neighboring elements of D, w.r.o.g. $a < b$. Then not only a has an immediate successor (namely b), but also every element $x \in D$ then has an immediate successor: If f is an automorphism of D which maps a onto x, it must map b on an immediate successor of x, and so on. Analogously each element $x \in D$ has an immediate predecessor. Now D has a segment S which is isomorphic to $\mathbf{Z}$, and then $D \setminus S$ must be empty. For otherwise we could map each element of $D \setminus S$ onto itself and each $x \in S$ onto its immediate successor, thus obtaining an isomorphic mapping f of D which is different from id_D. So $D = S$ holds , and D is isomorphic to $\mathbf{Z}$ with contradiction.

4.6 Theorem (Ohkuma [131]). *There exists a uniquely homogeneous linearly ordered set E which is not isomorphic to $\mathbf{Z}$.*

Proof. The construction of the set E will be performed in such a way that E is dense in $\mathbf{R}$ and a subgroup of the additive group $(\mathbf{R}, +)$, for which every automorphism of E is a translation, i. e. a function f, for which there exists a constant $d \in \mathbf{R}$ such that $f(x) = x + d$ for all $x \in E$. (And here, of course, d must be a difference $a - b$ with $a, b \in E$.) So it suffices to construct a subset E of $\mathbf{R}$, which satisfies the following two conditions:

(C_1) E is a subgroup of $\mathbf{R}$ (with $+$),

(C_2) All $<$ - preserving continuous functions $f : \mathbf{R} \to \mathbf{R}$, which are neither bounded from below nor from above (which thus form an automorphism of $\mathbf{R}$ since $f(\mathbf{R}) = \mathbf{R}$), and which are no translations of $\mathbf{R}$, don't generate an automorphism of E. That means: $f \restriction E$ is no automorphism of E.

Let ω_c be the initial ordinal $\omega(2^{\aleph_0})$ of $2^{\aleph_0}$. Then by 4.4 we can arrange the functions f of (C_2) in the form $\{f_\alpha | \alpha < \omega_c\}$.

The set E will be the union of a tower of subsets $E_\alpha \subseteq \mathbf{R}$, $\alpha < \omega_c$, which are constructed by induction, at the same time with a tower of sets $S_\alpha \subseteq \mathbf{R}$. Consider the following scheme:

1^0 E_α is an additive subgroup of $(\mathbf{R}, +)$.

2^0 $\beta < \alpha \implies E_\beta \subseteq E_\alpha$ and $S_\beta \subseteq S_\alpha$.

3^0 $E_\alpha \cap S_\alpha = \emptyset$.

4^0 $x \in E_\alpha$ and $s \in S_\alpha \implies \frac{x+s}{n} \in S_\alpha$ for all $n \in \mathbf{Z} \setminus \{0\}$.

5^0 For infinite α the cardinalities $|E_\alpha|$ and $|S_\alpha|$ are $\leq \max\{\aleph_0, |\alpha|\}$.

6^0 For $\beta < \alpha$ there is a $t_\beta \in E_\alpha$ such that $f_\beta(t_\beta) \in S_\alpha$ or $f_\beta^{-1}(t_\beta) \in S_\alpha$.

For $\alpha = 0$ let E_0 be the set $\mathbf{Q}$ of rational numbers. Then we choose an arbitrary irrational number z and put $S_0 := \{\frac{x+z}{n} | x \in E_0 \text{ and } n \in \mathbf{Z} \setminus \{0\}\}$. Now $E_0 \cap S_0 = \emptyset$. For otherwise there would be a $y \in E_0$, which is $= \frac{x+z}{n}$ for an $x \in E_0$ and an $n \in \mathbf{Z} \setminus \{0\}$. But then $ny - x = z$ would follow, and z would not be irrational. Thus 3) holds for $\alpha = 0$, also 4), for if $x \in E_0$ and $s = \frac{y+z}{n} \in S_0$ (with $y \in E_0, n \neq 0$ in $\mathbf{Z}$), then $\frac{x+s}{m}$ (with $m \in \mathbf{Z}\setminus\{0\}$) has the form $(nx + y + z)/nm$, which is an element of S_0. So for $\alpha = 0$ the conditions $1^0- 6^0$ are satisfied.

Suppose now that α is an ordinal $< \omega_c$, and that the properties $1^0 - 6^0$ are satisfied for all $\alpha' < \alpha$ (instead of α). Then we define E_α and S_α first for the case where α is a successor ordinal $\beta + 1$.

For $m \in \mathbf{Z}$ and $a \in E_\beta$ we put

$X_\beta(m, a) := \{x \in \mathbf{R}| f_\beta(x) = mx + a\}$, and

$X_\beta := \cup\{X_\beta(m,a)||m| \neq 1 \text{ in } \mathbf{Z}, a \in E_\beta\}$, $\quad Y_\beta := \cup\{X_\beta(1,a)|a \in E_\beta\}$, $\quad Z_\beta := \cup\{X_\beta(-1,a)|a \in E_\beta\}$, $\quad$ and

$W_\beta := \mathbf{R} \setminus (X_\beta \cup Y_\beta \cup Z_\beta \cup S_\beta)$, so that

(0) $\mathbf{R} = X_\beta \cup Y_\beta \cup Z_\beta \cup S_\beta \cup W_\beta$.

By 5^0 we have $|S_\beta| \leq |\beta|$ resp. $\leq \aleph_0$, so that in any case there holds:

(1) $|S_\beta| < 2^{\aleph_0}$.

The function g_β, which is defined by $g_\beta(x) := f_\beta(x) + x$ for $x \in \mathbf{R}$, is strictly increasing, and so the set $X_\beta(-1, a) = \{x \in \mathbf{R}|f_\beta(x) = -x + a\}$ contains exactly one element. This entails, due to $|E_\beta| \leq \max\{\aleph_0, |\beta|\} < 2^{\aleph_0}$:

(2) $|Z_\beta| \leq |E_\beta| < 2^{\aleph_0}$.

Next we prove for h_β, which is defined by $h_\beta(x) := f_\beta(x) - x$ for $x \in \mathbf{R}$:

(3) $h[Y_\beta] \subseteq E_\beta$.

Indeed, $x \in Y_\beta \implies x \in X_\beta(1, a)$ for some $a \in E_\beta \implies f_\beta(x) = x + a$, and $h_\beta(x) = f_\beta(x) - x = x + a - x = a \in E_\beta$.

h_β is continuous, and f_β is no translation (of type $x + const$). Thus h_β is not constant, and therefore the image set $H_\beta := h_\beta[\mathbf{R}]$ is a segment of $\mathbf{R}$ with more than one element, so that $|H_\beta| = 2^{\aleph_0}$.

(4) $|\mathbf{R} \setminus Y_\beta| = 2^{\aleph_0}$.

Due to $|E_\beta| < 2^{\aleph_0}$ and $|H_\beta| = 2^{\aleph_0}$ we have $|H_\beta \setminus E_\beta| = 2^{\aleph_0}$. This, together with $H_\beta \setminus E_\beta \subseteq h_\beta[\mathbf{R}] \setminus h_\beta[Y_\beta] \subseteq h_\beta[\mathbf{R} \setminus Y_\beta]$, entails that $h_\beta[\mathbf{R} \setminus Y_\beta]$ has a cardinality $\geq |H_\beta \setminus E_\beta| = 2^{\aleph_0}$, and this implies also (4) since $|\mathbf{R} \setminus Y_\beta| \geq |h_\beta[\mathbf{R} \setminus Y_\beta]|$.

(5) X_β or W_β has cardinality $2^{\aleph_0}$.

Since $|S_\beta|$ and $|Z_\beta|$ are $< 2^{\aleph_0}$ by (1) and (2), then (0) yields $\mathbf{R} \setminus Y_\beta \subseteq X_\beta \cup Z_\beta \cup S_\beta \cup W_\beta$ and by (4) $2^{\aleph_0} = |\mathbf{R} \setminus Y_\beta| = |X_\beta \cup W_\beta|$. This implies (5).

We can now perform the construction of s_β and t_β.

If $W_\beta \neq \emptyset$, we choose a $t_\beta \in W_\beta$. Then, by definition of W_β, we have $t_\beta \notin X_\beta \cup Y_\beta \cup Z_\beta$ and so $f_\beta(t_\beta) \neq mt_\beta + a$ for all $m \in \mathbf{Z}$ and all $a \in E_\beta$. Now we define $s_\beta := f_\beta(t_\beta)$.

If $W_\beta = \emptyset$ we put $F_\beta := \cup\{\frac{E_\beta}{n} | n \neq 0 \text{ in } \mathbf{Z}\}$. (Here $\frac{E_\beta}{n}$ is the set $\{\frac{x}{n} | x \in E_\beta\}$.) Evidently $|F_\beta| \leq |E_\beta|$ holds for infinite β and $|F_\beta| \leq \aleph_0$ for finite β. Next we prove:

(6) $X_\beta \cap (\mathbf{R} \setminus (S_\beta \cup F_\beta)) \neq \emptyset$.

Indeed, since in our case W_β is empty, we have $|X_\beta| = 2^{\aleph_0}$ by (5), and by (0):

(7) $\mathbf{R} \setminus Y_\beta \subseteq X_\beta \cup Z_\beta \cup S_\beta$.

Here the left side has cardinality $2^{\aleph_0}$ by (4), and because S_β and Z_β have cardinality $< 2^{\aleph_0}$ by (1) and (2), also X_β has cardinality $2^{\aleph_0}$. Since also $|F_\beta| \leq |E_\beta| \cdot \aleph_0 < 2^{\aleph_0}$ holds, (6) follows.

Due to (6) we can choose $s_\beta \in X_\beta \cap (\mathbf{R} \setminus (S_\beta \cup F_\beta))$, and we put $t_\beta := f_\beta(s_\beta)$.

(8) $s_\beta \neq mt_\beta + a$ for all $m \in \mathbf{Z}$ and all $a \in E_\beta$.

If we would have $s_\beta = mt_\beta + a$ for some $m \in \mathbf{Z}$, $a \in E_\beta$, then m must be $\neq 0$, since otherwise $s_\beta = a$ would be in E_β and F_β, contradicting the choice of s_β.

On the other hand from $s_\beta \in X_\beta$ we conclude: There exists an $n \in \mathbf{Z}$ and $b \in E_\beta$ with $t_\beta = f_\beta(s_\beta) = ns_\beta + b$, where $|n| \neq 1$. Then $s_\beta = m \cdot (ns_\beta + b) + a$ and $s_\beta \cdot (1 - mn) = mb + a$, so that $s_\beta = \frac{a+mb}{1-mn}$. ($m \cdot n$ is $\neq 1$ since $|n| \neq 1$.) But then s_β would be in F_β since a, b, mb are in the group E_β, and this contradicts the choice of $s_\beta \notin F_\beta$. So (8) is proved. Further we have:

(9) $t_\beta \notin S_\beta$.

348

In the first case (where $W_\beta \neq \emptyset$), we had $t_\beta \in W_\beta$, and $W_\beta \cap S_\beta = \emptyset$ by definition of W_β.

In the case where $W_\beta = \emptyset$, we had $t_\beta = f_\beta(s_\beta) = ns_\beta + b$ with $|n| \neq 1$, and $s_\beta \notin F_\beta$ due to the choice of s_β. If now $t_\beta \in S_\beta$ would hold, then $s_\beta = \frac{t_\beta - b}{n}$ had to be in S_β by 4^0, but we had $s_\beta \notin S_\beta$ in the case where $W_\beta = \emptyset$. And thus (9) holds.

We can resume: For $\beta = \alpha - 1$ we have defined elements t_β and s_β such that, together with the induction hypothesis, for all $\beta < \alpha$ we have by (8) and (9):

(D_1) $t_\beta \notin S_\beta$,

(D_2) $f_\beta(t_\beta) = s_\beta$ or $f_\beta^{-1}(t_\beta) = s_\beta$,

(D_3) for all $m \in \mathbf{Z}$ and all $a \in E_\beta$ there holds $s_\beta \neq mt_\beta + a$.

Now we can define the sets E_α and S_α :

$E_\alpha := \{mt_\beta + a \,|\, m \in \mathbf{Z}, a \in E_\beta \}$,

$S_{\alpha 1} := \{(s_\beta + mt_\beta + a)/n \,|\, m, n \in \mathbf{Z}, n \neq 0, a \in E_\beta\}$,

$S_{\alpha 2} := \{(s + mt_\beta)/n \,|\, m, n \in \mathbf{Z}, n \neq 0, s \in S_\beta\}$,

$S_\alpha := S_{\alpha 1} \cup S_{\alpha 2}$.

We now have to verify that E_α and S_α satisfy the conditions 1^0 – 6^0. Since E_β is a group by induction hypothesis, it follows immediately that also $(E_\alpha, +)$ is a group and that E_α contains E_β, so that 1^0 and 2^0 are fulfilled.

3^0 If $E_\alpha \cap S_\alpha$ would be $\neq \emptyset$, we would have $E_\alpha \cap S_{\alpha 1} \neq \emptyset$ or $E_\alpha \cap S_{\alpha 2} \neq \emptyset$. Suppose $E_\alpha \cap S_{\alpha 1} \neq \emptyset$. Then we can choose for each $x \in E_\alpha \cap S_{\alpha 1}$ elements $p, m, n \in \mathbf{Z}$ and $a, b \in E_\beta$ such that $x = mt_\beta + a = (s_\beta + pt_\beta + b)/n$. Then $mnt_\beta + na = s_\beta + pt_\beta + b$, so that $s_\beta = (mn - p) \cdot t_\beta + na - b$ follows with contradiction to (D_3).

Suppose now $E_\alpha \cap S_{\alpha 2} \neq \emptyset$. Then for each element $x \in E_\alpha \cap S_{\alpha 2}$ one can choose elements $p, m, n \in \mathbf{Z}$, $a \in E_\beta$, $s \in S_\beta$ such that $x = mt_\beta + a = (s + pt_\beta)/n$. Then $(mn - p) \cdot t_\beta = s - na$. Here $s - na$ is $\neq 0$ since $s \in S_\beta$ and $na \in E_\beta$, and because $E_\beta \cap S_\beta = \emptyset$ holds by 3^0. Then also on the left side $(mn - p)$ is $\neq 0$, and thus $t_\beta = (s - na)/(mn - p)$. Then by 4^0 $t_\beta \in S_\beta$, for we have $s \in S_\beta$ and $na \in E_\beta$. But this contradicts (D_1). Now 3^0 is satisfied for E_α, S_α.

4^0 For each $x \in E_\alpha$ there is an $m \in \mathbf{Z}$ and $a \in E_\beta$ such that $x = mt_\beta + a$. And for each $s \in S_\alpha$ there is $s \in S_{\alpha 1}$ or $s \in S_{\alpha 2}$, so that we have

$s = (s_\beta + pt_\beta + b)/n$ with $p, n \in \mathbf{Z}$, $b \in E_\beta$ or $s = (s' + pt_\beta)/n$ with $p, n \in \mathbf{Z}$, $s' \in S_\beta$.

In the first case for every integer $q \neq 0$ we have

$\frac{s+x}{q} = (s_\beta + pt_\beta + b)/qn + (nmt_\beta + an)/qn = (s_\beta + (p+mn) \cdot t_\beta + na + b)/qn \in S_{\alpha 1}$.

In the second case for every integer $q \neq 0$ there follows

$\frac{s+x}{q} = (s' + pt_\beta)/qn + (nmt_\beta + an)/qn = (s' + (mn+p) \cdot t_\beta + na)/nq$.

Here by 4) we have $s' + na \in S_\beta$, and then $\frac{s+x}{q}$ belongs to $S_{\alpha 2}$. So 4) is fulfilled for α.

5^0 From the definition of E_α there follows $|E_\alpha| \leq \aleph_0 \cdot |E_\beta| \leq \max\{\aleph_0, |E_\beta|\} \leq \max\{\aleph_0, |\beta|\}$. And $|S_\alpha| \leq \aleph_0 \cdot \aleph_0 \cdot |E_\beta| + \aleph_0 \cdot \aleph_0 \cdot |S_\beta| = \max\{\aleph_0, |E_\beta|, |S_\beta|\} \leq \max\{\aleph_0, |\beta|\}$, and so 5) is satisfied by α.

6^0 By the definition of E_α we have $t_\beta \in E_\alpha$, and by definition of $S_{\alpha 1}$ there follows $s_\beta \in S_{\alpha 1} \subseteq S_\alpha$. Further we had $f_\beta(t_\beta) =: s_\beta$ in the case, where $W_\beta \neq \emptyset$, and $f_\beta(s_\beta) =: t_\beta$ if $W_\beta = \emptyset$. Then 6) is satisfied for the immediate predecessor β of α, and by induction hypothesis then also for all ordinals $\beta < \alpha$.

We still have to supplement the definition of E_α and S_α for the case where α is a limit ordinal. Here we put

$E_\alpha := \cup\{E_\beta | \beta < \alpha\}$ and $S_\alpha := \cup\{S_\beta | \beta < \alpha\}$.

Condition 1^0 follows so: If a, b are in E_α, there is also a $\beta < \alpha$ with $a, b \in E_\beta$. This is a group which contains $a + b$ and their group-inverses, and so this also holds for E_α. Condition 2^0 is clearly satisfied. 3^0 If E_α and E_β would have an element in common, this would also be in some E_β with $\beta < \alpha$, with contradiction. 4^0 If $x \in E_\alpha$ and $s \in S_\alpha$, then by 2^0 there is a $\beta < \alpha$ such that $x \in E_\beta$ and $s \in S_\beta$, and 4^0 follows. 5^0 For each $\beta < \alpha$ we have that $|E_\beta|$ and $|S_\beta|$ are $\leq \max\{\aleph_0, |\beta|\}$. By 2^0 there follows $|E_\alpha| = \sup\{|E_\beta| | \beta < \alpha\} \leq \max\{\aleph_0, |\alpha|\}$ and $|S_\alpha| = \sup\{|S_\beta| | \beta < \alpha\} \leq \max\{\aleph_0, |\alpha|\}$, and 5^0 holds. Also Condition 6^0 follows immediately with 2^0.

Finally we can now define $E := \cup\{E_\alpha | \alpha < \omega_c\}$ and $S := \cup\{S_\alpha | \alpha < \omega_c\}$. These sets satisfy also the conditions $1^0 - 6^0$ (putting $E := E_{\omega_c}$ and $S := S_{\omega_c}$), and this is proved in the same way as before for limit ordinals.

In particular by 6^0 we have obtained the following: For every continuous function $f : \mathbf{R} \to \mathbf{R}$, which is $<$ - preserving and unbounded from below and from above, and which is not a translation, so that f is one of the f_β with $\beta < \omega_c$, there exists by 6^0 an element $t_\beta \in E$ for which a) $x := f_\beta(t_\beta) \in S$ or b) $x := f_\beta^{-1}(t_\beta) \in S$ holds. In the first case $f = f_\beta$ is no automorphism of E because $f_\beta(t_\beta)$ is in S, and thus not in E, which by 3^0 is disjoint to S. In the second case f_β maps x of S on the element $t_\beta \in E$, and then this has no element in E which is mapped on t_β by f_β.

So finally, if $a, b \in E$ and thus $a + (b - a) = b$, the translation T, which puts $T(x) := x + (b - a)$ maps a onto b. Of course, it is the only translation which does that. Since E is evidently not isomorphic to $\mathbf{Z}$, the theorem is proved.

Ohkuma had also determined the cardinality of the set of types of all uniquely homogeneous chains. We mention his theorem referring to this without proof:

4.7 Theorem [131]. *There is a set of cardinality 2^c (where $c = 2^{\aleph_0}$) of uniquely homogeneous chains, which are pairwise non-isomorphic.*

The theory of automorphisms of posets has meanwhile reached a great extent. It has, of course, an intensive contact to the theory of groups. One of its central intentions is to study the structure of the automorphism group of special classes of posets. Without going into the details we mention a few things which concern some characteristic features of this.

A relational structure $\mathfrak{A}$, e.g. a poset, is called k - *homogeneous* if each isomorphism between two k - element substructures of $\mathfrak{A}$ extends to an automorphism of $\mathfrak{A}$. In this context a theorem of Droste and MacPherson states:

4.8 Theorem [27]. *If a countable poset $(P, \leq)$ is 1 - and 4 - homogeneous, then it is k - homogeneous for each $k \in \mathbf{N}$. And there are infinitely many examples of certain countable posets showing that here the number 4 may not be replaced by 2 or 3.*

And Droste, MacPherson and Mekler proved:

4.9 Theorem [28]. *Assuming the set-theoretic assumption $\Diamond(\aleph_1)$, it is shown that for each $k \in \mathbf{N}$, there exist partially ordered sets of size $\aleph_1$ which embed each countable partial order and are k - homogeneous, but not $(k + 1)$ - homogeneous. This is impossible in the countable case for $k \geq 4$.*

We further mention a connection of order theory and group theory:

4.10 Definition. A group $(G, +)$, which is equipped with an order $\leq$ is called a *po-group*, if it satisfies: For each two elements $x, y \in G$ we have

(1) $x \leq y \Longrightarrow a + x + b \leq a + y + b$ for all $a, b \in G$.

In short this condidition can be formulated as: Every group translation is isotone.

A po-group, which is at the same time a lattice, is called a *lattice-ordered* group, in short an *l - group*.

This concept is of importance if we study the set of automorphisms of a linearly ordered set:

4.11 Definition. Let $(S, \leq)$ be a linearly ordered set, $\mathrm{Aut}(S)$ the set of its order-automorphisms. For $f, g \in \mathrm{Aut}(S)$ we put

$$f \leq g \Longleftrightarrow f(x) \leq g(x) \text{ for all } x \in S.$$

It is immediately clear that $\mathrm{Aut}(S)$ is partially ordered by $\leq$. But there holds more, namely:

4.12 Theorem. *Let S be a linearly ordered set. Then $(\mathrm{Aut}(S), \leq)$ is a lattice, and then an l-group.*

Proof. Let $f, g \in \mathrm{Aut}(S)$. Then we define a mapping $h : S \to S$ by $h(x) := \max\{f(x), g(x)\}$ for $x \in S$. We have to show that h is an automorphism of S. If $x, y \in S$ satisfy $x < y$, then $f(x) < f(y)$ and $g(x) < g(y)$ holds, and further $h(x) < h(y)$, so that h preserves $<$, and is therefore injective. We have to verify that it is also surjective. Let $y \in S$ be given. There are uniquely defined elements $x_1, x_2 \in S$ with $f(x_1) = y$ and $g(x_2) = y$. If $h(x_1) = f(x_1) = y$ holds, we are done. In the other case we have $h(x_1) = g(x_1) > f(x_1) = y$. Then there must hold $x_2 < x_1$ since g is strictly increasing. This implies $f(x_2) < (f(x_1) =) y$, and then $h(x_2) = g(x_2) = y$ follows. Of course, h is now also the least automorphism of S which is $\geq f$ and g.

In this context Holland proved the following fundamental theorem, which we mention without proof:

4.13 Theorem [87]. *Every l-group is isomorphic to a subgroup of the automorphism group of a linearly ordered set.*

And since every linearly ordered set can be embedded in an η_α - set, this statement can be supplemented by:

4.14 Theorem. *Every l-group is isomorphic to a subgroup of the automorphism group of a linearly ordered set of a normal type η_α ($= h_\alpha$).*

The automorphism groups of these sets have been studied in detail by E. Weinberg, see e.g. [177].

For a linearly ordered set S the group $\mathrm{Aut}(S)$ is usually not commutative. Consider e.g. $\mathbf{R}$ with its usual linear order. Let $f : \mathbf{R} \to \mathbf{R}$ be the mapping which puts $f(x) = x + 1$ for $x \in \mathbf{R}$, and $g : \mathbf{R} \to \mathbf{R}$ defined by $g(x) := \frac{x}{2}$ for $x \in \mathbf{R}$, then $g \circ f$ maps 1 onto 1, whereas $f \circ g$ maps 1 onto $1\frac{1}{2}$.

Chapter 11
Comparability graphs

In every poset $(P, \leq)$ the relation $\leq$ defines the corresponding comparability relation, which arises from $\leq$ by taking the symmetric hull of $\leq$. This concept effects a connection between order theory and graph theory. If we consider only the comparability relation of an order $\leq$, we lose information, but with this simplification we possibly can gain additional insight into the structure of the order.

How far-reaching this simplification can be is e.g. illustrated by the fact that all linear orderings on a set S produce the same comparability relation on S, namely the all-relation on S. In particular, an order $\leq$ on a set S generates the same comparability relation as its inverse order $\geq$.

11.1   General remarks

For the discussion of graph-theoretical properties of posets the directed graphs are of particular interest. They arise from the general graphs by equipping their edges with directions. For an edge $\{a, b\}$ of a graph there are two possibilities to give a direction to $\{a, b\}$, which means a linear order in $\{a, b\}$, one in which $a < b$ and one in which $b < a$ holds. In the first case we obtain the *oriented* or *directed edge* (a, b), which has a as *initial point* and b as *endpoint,* in the second conversely with (b, a).

In the illustration of a directed edge (a, b) we use a line segment with ends a and b and insert an arrow in it as follows: $a \;\longleftarrow\; b$. This shall give an association to a corresponding order theoretical meaning $a < b$.

We now define:

1.1 Definition. A *directed graph*, abbreviated *digraph*, is a pair $G = (V, D)$, where V is a set and D a subset of $(V \times V) \setminus \{(x, x) | x \in V\}$. The elements of V are again called *vertices* of G, and the elements of D *directed* (or *oriented*) *edges* of G. We emphasize that a digraph can have a directed edge (a, b) and at the same time the directed edge (b, a). In this case it contains, so to speak, the edge $\{a, b\}$ in both directions.

354

If we have a (non-oriented) graph $G = (V, E)$, and if we choose for each edge $\{a, b\}$ of E exactly one pair $f(\{a, b\}) \in \{(a, b), (b, a)\}$, we say that we have *oriented* G by f, and call $D := \{f(\{a, b\}) | \{a, b\} \in E\}$ the *orientation* of G (also of E), which is generated by f. We call $f(\{a, b\})$ the *orientation* of $\{a, b\}$, given by f.

If (V, D) is a directed graph, and if we put $E := \{\{a, b\} | (a, b) \text{ or } (b, a)$ are in $D\}$, then (V, E) is the *undirected graph corresponding* to (V, D).

We call a directed graph $G = (V, D)$ *transitively directed* (or *transitively oriented*) and D a *transitive orientation*, if $(a, b) \in D$ and $(b, c) \in D \implies (a, c) \in D$. This entails evidently that if we have a *directed path*, that means a finite sequence of directed edges $(a_1, a_2), (a_2, a_3), \ldots, (a_{n-1}, a_n)$, also (a_1, a_n) is in D.

1.2 Definition. A graph $G = (V, E)$ is called a *comparability graph*, if it is possible to find an order relation $\leq$ on V such that for each $\{a, b\} \in V \times V$ we have $\{a, b\} \in E \iff a < b$ or $b < a$.

A slightly different characterization of comparability graphs is given by:

1.3 Theorem. *A graph $G = (V, E)$ is a comparability graph iff it can be transitively oriented.*

Proof. Let $G = (V, E)$ be a comparability graph of an order relation $\leq$ on V. Then we orient each edge $\{a, b\}$ of E by the directed edge (a, b) if $a < b$, resp. by (b, a) if $b < a$ holds. This generates evidently a transitive orientation of G.

If we have a transitive orientation D of G, we put $a < b$ for all directed edges $(a, b) \in D$. Then the relation $<$, augmented by id_V, is an order relation on V, whose comparability graph is G.

Of course one is now interested to find more characterizations of the comparability graphs. First we mention some examples.

1.4 Example. Let $G = (V, E)$ be a graph with $V = \{v_1, \ldots, v_n\}$, where n is an odd number ≥ 5, and $E = \{\{v_i, v_{i+1}\} | i = 1, \ldots, n\}$, where $v_{n+1} = v_1$. Then G is no comparability graph.

Suppose the contrary. Then we could find an order $\leq$ in V which has G as comparability graph . W.r.o.g. let $v_1 < v_2$. Then v_2 cannot be $< v_3$ since this would imply $v_1 < v_3$, but $\{v_1, v_3\} \notin E$. And so $v_3 < v_2$ must hold. Successively we would obtain $v_i < v_{i+1}$ for the odd numbers $i \in \{1, \ldots, n\}$ and $v_i > v_{i+1}$ for the even numbers i. Since v_n

is comparable with v_1 we must have $v_n < v_1$ or $v_1 < v_n$. In the first case this would yield $v_n < (v_1 <) v_2$, in the second $v_1 (< v_n) < v_{n-1}$, which is impossible since v_n is not comparable with v_2, and v_1 not with v_{n-1}.

Contrary to the situation of 1.4 we have a positive result in the case where n is even:

1.5 Example. If the assumption of 1.4 is given with the difference that $n \geq 0$ is an even integer, then G is a comparability graph. And this has the form of the graph given in 7.3.6.

Another simple example can be constructed using bipartite graphs. We define them as follows:

1.6 Definition. A graph $G = (V, E)$ is said to be *bipartite* if there is a partition of V in two disjoint subsets A and B, such that each edge of E is of type $\{a, b\}$, where $a \in A$ and $b \in B$.

It follows trivially, that each bipartite graph is a comparability graph. For under the above assumptions we only have to ascribe to each edge $\{a, b\} \in E$ with $a \in A$, $b \in B$ the directed edge (a, b) (or to every edge $\{a, b\}$ the directed edge (b, a)) to obtain a transitive orientation of G. And these are usually not the only possibilities to reach this aim. But if we have in addition to our assumption that G is connected, we can prove more, namely:

1.7 Theorem. *Let $G = (V, E)$ be a connected bipartite graph. Then there is an order relation $\leq$ on V such that G is the comparability graph of $(V, \leq)$. And every order relation on V that generates the comparability graph G is $\leq$ or the reverse $\geq$ of $\leq$.*

Proof. As mentioned before there exists an order $\leq$ in V whose comparability relation generates G. Let A and B be subsets of V satisfying the condition of 1.6 and let w.r.o.g. $a \in A, b \in B$ be elements with $a < b$. Then for each $x \in B$ that is comparable with a, we must have $a < x$ since $x < a$ would entail $x < b$, contradicting the fact, that b and x are both in B. For a similar reason for all $y \in A$ that are comparable with b we must have $y < b$. The connectedness of G now yields that for every two comparable elements $x \in A$, $y \in B$ we have $x < y$. So the relation $a < b$ of the beginning determines the whole relation $<$. If before we would have had $b < a$ instead of $a < b$, then we would have obtained the reverse relation $\geq$, and the statement is proved.

The last theorem presents an example of a case where a comparability graph determines a generating poset "nearly uniquely", namely an order and its reverse order. Graphs with this property are called *pographs* ([6]). In general there are many orders on a set which generate the same comparability graph. To every chain in a poset there corresponds a *clique* in the comparability graph, that is a set of vertices which are pairwise linked. To the antichains of a poset there correspond *independent* subsets of the comparability graph — these are sets of pairwise unlinked vertices. Here the correspondence between the order and its comparability graph is more intensive.

Before we proceed we introduce two graph-theoretical concepts:

1.8 Definition. Let $G = (V, E)$ be a graph. A *subgraph* of G is a graph (V', E') where $V' \subseteq V$ and $E' \subseteq E$ holds.

For $V' \subseteq V$ we define the subgraph which is *induced* by V' as the pair (V', E'), where E' contains all edges $\{a, b\}$ of E, for which $a, b \in E'$.

For $E' \subseteq E$ we define the subgraph which is induced by E' as the subgraph which is induced by $\cup E'$. It contains all edges of E', but possibly more.

Concerning the behaviour of subgraphs (resp. induced subgraphs) of comparability graphs we have the following:

1.9 Remark. a) Every graph $G = (V, E)$ is a subgraph of a comparability graph, e.g. of the graph $(V, [V]^2)$, in which every two elements of V are linked, and which is the comparability graph of every linear order on V.

b) An induced subgraph of a comparability graph $G = (V, E)$ is also a comparability graph. For if $\leq$ is an order on V that generates the comparability graph G, and if $V' \subseteq V$ holds, the restriction $\leq \restriction V'$ has the subgraph of G, which is induced by V', as comparability graph.

c) A graph G is a comparability graph iff each connectivity component of G is a comparability graph.

The property to be a comparability graph transfers from the finite induced subgraphs of a graph to the whole graph. This can be proved using the compactness theorem of mathematical logic. Here we present a proof using Rado's Selection Lemma 8.1.6 (and therefore denote here the set of edges by I):

1.10 Theorem. *Let $G = (V, I)$ be a graph of which all finite induced subgraphs are comparability graphs. Then G is also a comparability graph.*

Proof. For each edge $i = \{a, b\} \in I$ let K_i be the set $\{(a, b), (b, a)\}$. We put $K := \cup\{K_i | i \in I\}$. For each finite $L \subseteq I$ the set L^* of edges which link elements of $U :=: \cup L$ is still a finite subset of I, and by assumption there exists a mapping $f_{L^*} : L^* \to K$, which generates a transitive orientation in the subgraph which is generated by U, and which therefore satisfies:

$\{a, b\}$ and $\{b, c\} \in L^*$, $f_{L^*}(\{a, b\}) = (a, b)$, $f_{L^*}(\{b, c\}) = (b, c) \implies$ $\{a, c\} \in L^*$ and $f_{L^*}(\{a, c\}) = (a, c)$. We put $f_L := f_{L^*} \upharpoonright L$.

According to the Rado Selection Lemma there is a function $f : I \to K$, such that for every finite $L \subseteq I$ there is a finite M with $L^* \subseteq M \subseteq I$ and $f(i) = f_M(i)$ for all $i \in L^*$. We define I^* to be the set of edges $i \in I$, where each i is oriented by $f(i)$.

Now the digraph (V, I^*) is transitively oriented: Let (a, b) and $(b, c) \in I^*$, and L^* the set of all edges of I whose both ends are in $\{a, b, c\}$. By the preceeding remarks there is a finite M with $L^* \subseteq M \subseteq I$ for which $f(i) = f_M(i)$ for all $i \in L^*$. Since f_M induces a transitive orientation in M, it follows: M contains (not only $\{a, b\}$ and $\{b, c\}$, but also) $\{a, c\}$, which of course is also in L^*, and then $f(\{a, c\}) = f_M(\{a, c\}) = (a, c)$, and thus (V, I^*) is transitively oriented by f and hence a comparability graph.

Theorems 1.10 and 1.3 now also imply:

1.11 Theorem. *If each finite induced subgraph of a graph G can be transitively oriented, then so also can G.*

11.2   A characterization of comparability graphs

In this section we present the characterizations of comparability graphs which were established by Ghouila-Houri [51] resp. Gilmore and Hoffman [54]. In [51] the folllowing concept was introduced:

2.1 Definition. Let $G = (V, E)$ be a graph which has been oriented by a function f, and let then D be the set of the oriented edges of E. Then G is called *quasi-transitively directed* if $(a, b) \in D$ and $(b, c) \in D \implies$ $\{a, c\} \in E$. And then D is said to be a *quasi-transitive orientation*.

358

Then it is trivial that a fortiori transitive implies quasi-transitive. On the other hand there holds the following theorem of Ghouila-Houri [51]:

2.2 Theorem. *Let $G = (V, E)$ be a finite graph, for which we have a quasi-transitive orientation Q. Then there also exists a transitive orientation T of G.*

Proof. We make the assumption that the assertion is true for $|V| < n$, where n is a fixed natural number, and consider a graph $G = (V, E)$ with $|V| = n$, which has a quasi-transitive orientation Q. First we remark:

(1) If three vertices a, b, c satisfy $(a, b) \in Q$, $(b, c) \in Q$ and $(c, a) \in Q$, each other vertex v of G is linked with 0,2 or 3 of the vertices a, b, c.

Indeed, if (w.r.o.g.) e.g. v is linked with b, we have two cases: $(v, b) \in Q$ or $(b, v) \in Q$ (see Figure 25). In the first case, due to $(b, c) \in Q$ we also have $\{v, c\} \in E$. In the second case (a, b) and $(b, v) \in Q$ implies $\{a, v\} \in E$ and (1) holds.

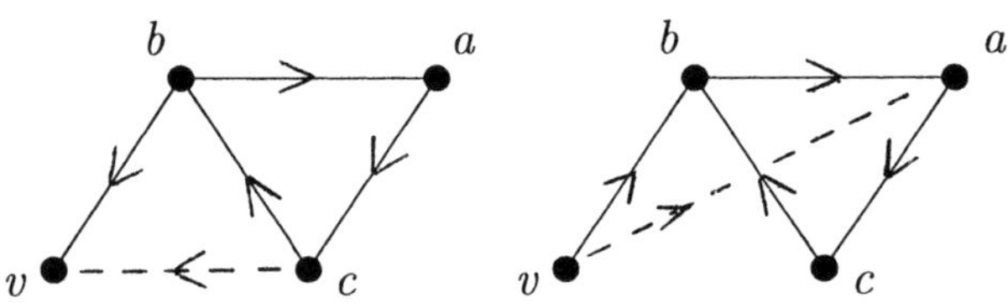

Figure 25

We now assume that Q is not transitive, otherwise we would be done. Then there exist three vertices x_1, x_2, x_3 satisfying

(2) $(x_1, x_2) \in Q$, $(x_2, x_3) \in Q$ and $(x_3, x_1) \in Q$.

By 1.9, c) we can w.r.o.g. assume that G is connected. Then, due to (1), we have to discuss two possible cases:

Case 1. Each vertex x of G which is different from x_1, x_2, x_3, and which is linked with one of these three, is linked with all of them.

Now we consider the vertex set $V' := V \setminus \{x_2, x_3\}$. If E' is the set of edges of G whose endpoints are in V', we consider the graph $G' := (V', E')$ — it is the subgraph of G, which is *induced* by V'. The orientation Q of G induces a quasi-transitive orientation Q' of G', and since G' has fewer than n vertices, it has by induction hypothesis a transitive orientation T'.

We extend T' to a transitive orientation T of G as follows: Let $x \in V'$ be a vertex which is linked with x_1, and then by assumption also with x_2, x_3. If $(x, x_1) \in T'$, we put the pairs (x, x_2) and (x, x_3) in T. If $(x_1, x) \in T'$, we put the pairs (x_2, x) and (x_3, x) in T. By (2) the vertices x_1, x_2, x_3 are pairwise linked in G, and the pairs (x_1, x_2), (x_2, x_3), and (x_1, x_3) and all elements of T' become elements of T. Then T is a transitive orientation of G, as can be verified using the following illustration in Figure 26, which considers the two possible cases $(x, x_i) \in T$ resp. $(x_i, x) \in T$ for $i = 1, 2, 3$.

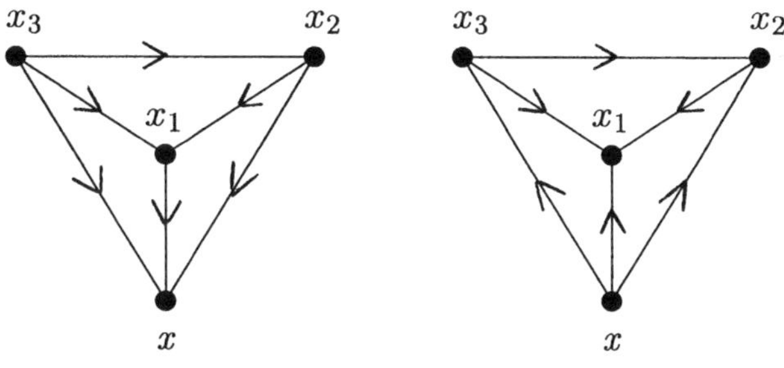

Figure 26

Case 2. There exists at least one vertex v of G which is different from x_i for all $i \in \{1, 2, 3\}$, and linked with two of these, but not with the third one. W.r.o.g., due to symmetry, we can assume that there is a vertex v different from these three which is linked with x_2 and x_3, but not with x_1.

Then we define A to be the set of all vertices of G that are linked with x_2 and x_3. So, among others, x_1 and $v \in A$. Let W denote the set of all two-element subsets of A that are not edges of G. Then (A, W) is a graph, and let C be the component of x_1 in this graph. Since v is not linked with x_1, v is in C, like x_1, and thus we have $|C| \geq 2$. Next we prove:

(3) Each element $c \in C$ satisfies $(c, x_2) \in Q$ and $(x_3, c) \in Q$.

If $(c, x_2) \notin Q$ would hold, we would have $(x_2, c) \in Q$. This, together with $(x_1, x_2) \in Q$ (which holds by (2)) entails $\{x_1, c\} \in E$ because of the quasi-transitivity of Q. But this contradicts the fact that x_1 and c are not linked in G.

And if $(x_3, c) \in Q$ would not hold, we would have $(c, x_3) \in Q$, and with $(x_3, x_1) \in Q$ (which holds by (2)) this would yield $\{c, x_1\} \in E$ with the same contradiction.

(4) Each vertex $x \in V$ that is not in C is linked with no or with all vertices of C.

Indeed, let $x \notin C$, $a, b \in C$, and let x be linked with a. We have to show that then also x is linked with b. First we see:

(5) x is linked with x_2 or x_3.

For if $(x, a) \in Q$, this together with $(a, x_2) \in Q$ — which holds by (3) — entails $\{x, x_2\} \in E$. And if $(a, x) \in Q$, then this with $(x_3, a) \in Q$ — which holds by (3) — , yields $\{x_3, x\} \in E$.

Now we can prove that also $\{x, b\} \in E$. Due to (5) it suffices to consider the following four cases:

a) Let $(x, x_2) \in Q$. This with $(x_2, x_3) \in Q$ entails $\{x, x_3\} \in E$.

b) Let $(x_2, x) \in Q$. This with $(b, x_2) \in Q$ (which holds by (3)) yields $\{b, x\} \in E$.

c) Let $(x, x_3) \in Q$. This with $(x_3, b) \in Q$ (which holds by (3)) entails $\{b, x\} \in E$.

d) Let $(x_3, x) \in Q$. This with $(x_2, x_3) \in Q$ yields $\{x_2, x\} \in E$.

So, in the cases b) and c) we are done. In the other two cases, a) and d), we have that x is linked with x_2 and x_3 and thus in A. If now x would not be linked with b, we would have $\{x, b\} \in W$, and because of $b \in C$ also x had to be in C, which contradicts the assumption. So (4) is proved.

Now we can construct T. We choose a transitive orientation T_C in the induced subgraph C which is possible because it has fewer than n vertices. And we choose a transitive orientation T_M in the induced subgraph of G which is generated by $M := (V \setminus C) \cup \{x_1\}$. This has fewer than n elements since C has at least two. Then we define T as the union of T_C and T_M and the set of all pairs (x, x'), where $\{x, x'\} \in E$ with $x \notin C$, $x' \in C$ and where $\{x, x'\}$ is oriented in the same way as $\{x, x_1\}$:

$$(x, x') \in T \iff (x, x_1) \in T, \text{ and of course, } (x', x) \in T \iff (x_1, x) \in T.$$

Together we so obtain a transitive orientation T. To see this we have to check:

(6) (a, b) and $(b, c) \in T \implies (a, c) \in T$.

This is trivial if a, b, c are all in C or all in M. For the rest we consider six cases: There are three cases where one of a, b, c is in $M \backslash \{x_1\}$ and the other two in C,   and three cases where one of a, b, c is in C and the other two in $M \backslash \{x_1\}$. So we have the six cases:

1) $a \in M \backslash \{x_1\}$, $b, c \in C$. 2) $b \in M \backslash \{x_1\}$, $a, c \in C$. 3) $c \in M \backslash \{x_1\}$, $a, b \in C$. 4) $a, b \in M \backslash \{x_1\}$, $c \in C$. 5) $a, c \in M \backslash \{x_1\}$, $b \in C$. 6) $b, c \in M \backslash \{x_1\}$, $a \in C$. See the following illustrations in Figure 27.

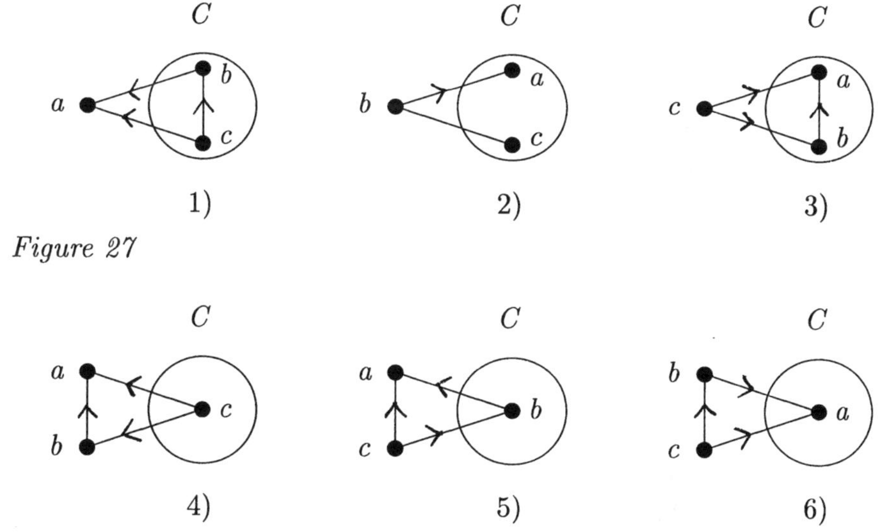

Figure 27

In the cases 1) and 3) we immediately see that (6) holds. Case 2) cannot happen. For here $(a, b) \in T$ would imply $(x_1, b) \in T$, and this $(c, b) \in T$, contradicting $(b, c) \in T$.

In case 4) we have $(b, c) \in T \implies (b, x_1) \in T$ .This with $(a, b) \in T$ and the fact that a, b, x_1 are in M and that T_M is transitive implies $(a, x_1) \in T$ and $(a, c) \in T$.

In case 5) we have $(a, b) \in T \implies (a, x_1) \in T$, and $(b, c) \in T \implies (x_1, c) \in T$, so that by transitivity of T_M we obtain $(a, c) \in T$.

In case 6) we have $(a, b) \in T \implies (x_1, b) \in T$, and $(b, c) \in T \implies (x_1, c) \in T$ since T_M is transitive. Then also $(a, c) \in T$. So T is a transitive orientation, and 2.2 is proved.

A bit later we shall see that the statement of 2.2 remains valid if the condition that G is finite is dropped.

2.3 Theorem of Ghouila-Houri [51]. *Let $G = (V, E)$ be a finite graph, and let G^* be the graph (V^*, E^*) which is defined as follows: V^* contains all pairs (a, b), for which $\{a, b\} \in E$. (Intuitively speaking: V^* contains all edges of E in both orientations. And note: These are in G^* the vertices, and not the edges!.)*

E^* is given by: Each vertex (a, b) of V^* is linked with (b, a) and with the vertices (b, c), for which $\{a, c\} \notin E$.

Then G has a transitive orientation iff the graph G^* is 2-chromatic.

Proof. a) Let D be a transitive orientation of G, then D is a subset of V^*.

(I) No two elements of D are linked in G^*.

For if elements $(a, b), (c, d)$ of D would be linked in G^*, we would have 1) $b = c$ and $\{a, d\} \notin E$, or 2) $d = a$ and $\{c, b\} \notin E$. Each of both is impossible since D is a transitive orientation. And (I) is proved.

Let $D_0 := \{(x, y)|(y, x) \in D\}$, so that D_0 contains the edges of D with "opposite direction" . Also we have by symmetry:

(II) No two elements of D_0 are linked in G^*.

Now we color all elements of D with the same color, say red, and all elements of D_0 with a second color, say green. Due to (I) and (II) this yields a 2-coloring of G^*.

b) Let now G^* be 2-chromatic, say V^* colored with red and green. So for each edge $\{a, b\} \in E$ exactly one of the oriented edges $(a, b), (b, a)$ is colored red, the other green. The red-colored elements of V^* now form a transitive orientation of G. For let (a, b) and (b, c) be red-colored elements of V^*. They cannot be linked, and therefore $\{a, c\}$ must be in G. Then the set of those elements of V^* which are colored in red is a quasi-transitive orientation of G. According to the previous theorem 2.2, then G also has a transitive orientation.

2.3′. Theorem. *The statements of 2.2 and 2.3 hold also, if the assumption that G is finite is dropped.*

Proof. 2.2: Let $G = (V, E)$ be a graph which has a quasi-transitive orientation. Then a fortiori each finite induced subgraph F of G has a quasi-transitive orientation, and by 2.2 then also a transitive orientation. According to 1.3 F is a comparability graph and by 1.10 also G, and thus G has a transitive orientation.

2.3: Let the assumptions of 2.3 be fulfilled, where the condition that G is finite is dropped. Let G have a transitive orientation. Then this also holds for each finite induced subgraph F of G, and the corresponding graph F^* is 2 - chromatic. Then by Theorem 2.5.5 of Erdös/De Bruijn also G^* is 2 - chromatic.

And if G^* is 2 - chromatic, the part b) of the proof of 2.3 also works for the general case, where G can be infinite. For before we had seen that quasi-transitive implies transitive in general.

Also Gilmore and Hoffman [54] established a characterization of the comparability graphs, which is essentially the same as that of Ghouila-Houri. We describe this in more detail. First we define, following [54]:

2.4 Definition. Let $G = (V, E)$ be a graph. A *cycle* in G is a sequence $a_1, \ldots, a_k$ (and we put $a_{k+i} := a_i$ for $i = 1, 2$) of $k \geq 3$ vertices of V such that $\{a_i, a_{i+1}\} \in E$ for $i = 1, \ldots, k$, for which the pairs (a_i, a_{i+1}), $i = 1, \ldots, k$, are pairwise different. (But it can happen that a pair (a_i, a_{i+1}) also appears in the reverse order as (a_{i+1}, a_i).) This cycle is *even* (resp *odd*), if k is even (resp. odd).

A *triangular chord* of the above cycle is an edge of G of type $\{a_i, a_{i+2}\}$ with $i \in \{1, \ldots, k\}$.

The criterion of Gilmore and Hoffman is now:

2.5 Theorem. [54]. *A graph $G = (V, E)$ is a comparability graph iff each odd cycle has a triangular chord.*

That this condition is necessary follows easily, and without making use of earlier theorems. For let $a_1, \ldots, a_k$ be an odd cycle (k odd) without a triangular chord. And suppose indirectly that we have an orientation of the edges of E which is to partially order the vertices of G. Then this must give any successive pair of edges of the cycle opposite orientations. Both must be directed towards or away from the common vertex of the pairs. In detail: If $\{a, b\}$ and $\{b, c\}$ are edges of the cycle while $\{a, c\}$ is not, then if (a, b) is the orientation which $\{a, b\}$ obtains, (c, b) must be the orientation given to $\{b, c\}$, for an orientation (b, c) would, due to the transitivity of order, entail the orientation (a, c), and thus $\{a, c\}$ also had to be in G, with contradiction. The situation, where (b, a) is the orientation given to $\{a, b\}$, is analogous. It is easily seen that only in an even cycle can all successive pairs of edges have opposite orientations. So the condition is necessary.

Suppose now that each odd cycle of G has at least one triangular chord. Then there follows:

(I) The graph $G^* = (V^*, E^*)$ of 2.3 has no odd cycle.

If in G^* we would have vertices $(a_1, a_2), (a_2, a_3), \ldots, (a_{n-1}, a_n)$ with $n \geq 3$ which form an odd cycle, then the sequence $a_1, \ldots, a_n$ forms an odd cycle of G. By assumption this has a triangular chord $\{a_i, a_{i+2}\}$. But then (a_i, a_{i+1}) and (a_{i+1}, a_{i+2}) wouldn't be linked in G^*. This yields a contradiction, and thus (I) holds.

We have obtained that all cycles of G^* are even. By the following simple lemma this implies that G^* is 2 - chromatic, and then by 2.3$'$ G has a transitive orientation and is thus a comparability graph.

2.6 Lemma. *Let G be a graph for which in all cycles $a_1, \ldots, a_n$, where the $a_i, i = 1, \ldots, n$, are pairwise different, n is even. Then G is 2 - chromatic.*

Proof. W.r.o.g. we can assume that G is connected. Let v be an arbitrary vertex of G. For $k \in \mathbf{N}$ let S_k be the (so-called k-*sphere* of v, that means the) set of all vertices x_k, for which $k+1$ is the length of a shortest sequence $v = x_0, \ldots, x_k$, where x_i is linked with x_{i+1} for $i = 0, \ldots, k - 1$. Put $S_0 := \{v\}$. Then we color all vertices of sets S_k with even k red and all other vertices green. This furnishes a 2-coloring of G. For if two vertices are linked, one must lie in a set S_k and the other one in S_{k-1} or S_{k+1}. To see this consider two vertices x, y of the same k-sphere of v and two shortest sequences $v = x_0, \ldots, x_k = x$ and $v = y_0, \ldots, y_k = y$ of the above type. Let m be the last index for which $x_l = y_l$ holds. Then x, y cannot be linked in G because otherwise the sequence $x_l, \ldots, x_k, y_k, \ldots, y_l$, diminished by the last element, would define a cycle of $2(k - l + 1) - 1$ vertices, which is impossible since this number is odd.

11.3 A characterization of the comparability graphs of trees

Next we study the comparability graphs of order-theoretical generalized trees, which are those posets P, in which for each $x \in P$ the set of its predecessors is linearly ordered. Wolk [180] gave a characterization of these graphs, and Jung [93] gave a shorter proof of this, which at the same time furnishes a complete overview of all orientations of a given comparability graph of this class. The central concept in Jung's proof is an equivalence relation in the set of vertices of a graph:

3.1 Definition. Let $G = (V, E)$ be a graph and $e = \{v, v'\}$ an edge of E. Then v is said to be a *V-root* of e, if there exists a $v'' \in V$ with $v'' \neq v'$, $\{v, v''\} \in E$ and $\{v', v''\} \notin E$.

A relation $<$ on V is called a *tree-orientation* of G, iff V with the relation $\leq$, that is $<$, augmented by the identity relation id_V, is a generalized tree.

$e = \{v, v'\}$ is called an n - *edge*, if exactly n of the vertices v, v' are V - roots of e. So $n \in \{0, 1, 2\}$.

3.2 Definition. Let $G = (V, E)$ be a graph. For $v, v' \in V$ we put $v \sim v' \iff v = v'$, or $\{v, v'\}$ is a 0 - edge of G.

3.3 Lemma. $\sim$ *is an equivalence relation in V.*

Proof. Only transitivity is non-trivial. Let $v_1 \sim v_2$ and $v_2 \sim v_3$. We assume that the vertices v_1, v_2, v_3 are different, otherwise the assertion is trivial. Then $\{v_1, v_2\}$ and $\{v_2, v_3\}$ are in E. Now v_2 is no V - root of the edge $\{v_1, v_2\}$ since this is a 0 - edge. And therefore v_1 must be linked with v_3, so that $\{v_1, v_3\} \in E$. We have to show that $\{v_1, v_3\}$ is also a 0 - edge. Suppose the contrary:

If w.r.o.g. v_1 is a V - root of $\{v_1, v_3\}$, then there is a $v_4 \neq v_3$ with $\{v_1, v_4\} \in E$ and $\{v_3, v_4\} \notin E$, and therefore we have $v_4 \neq v_2$. If $\{v_2, v_4\} \notin E$ would hold, v_1 would be a V - root of $\{v_1, v_2\}$, which is impossible since this is a 0 - edge. Also $\{v_2, v_4\} \in E$ would yield a contradiction since then v_2 would be a V - root of the 0 - edge $\{v_2, v_3\}$. So the lemma is proved.

From the definition it is clear that the vertices of an equivalence class are pairwise linked in G — they form a so-called *simplex* of G. And before we proceed to the main theorem we mention a property of the equivalence classes, although we don't need it for the proof. But it gives a better oversight. Namely there holds:

3.4 Lemma. *If C_1 and C_2 are two different equivalence classes, then either all elements of C_1 are linked with all elements of C_2, or each element of C_1 is unlinked with each element of C_2.*

Proof. Suppose that there is an edge $e = \{v_1, v_2\}$ with $v_1 \in C_1, v_2 \in C_2$. If x is an element of C_1 which is different from v_1, then x must also be linked with v_2, for otherwise v_1 would be a V - root of the 0 - edge

$\{v_1, x\}$. So all elements of C_1 must be linked with v_2. Symmetrically also all elements of C_2 must be linked with v_1. Then the rest easily follows.

3.5 Theorem [93]. *Let $G = (V, E)$ be a graph, and let $<$ be a tree-orientation in E. Then the following two properties hold:*

(i) *If C is an equivalence class of $\sim$, then the restriction $\leq \restriction C$ is a linear order on C.*

(ii) *If v is a V - root of $\{v, v'\} \in E$, we have $v < v'$.*

Proof. (i) All elements of C are by definition pairwise linked by 0 - edges. Therefore each two elements of C are comparable, so that C is linearly ordered.

(ii) If v is a V - root of $\{v, v'\}$, there exists a vertex $v'' \neq v'$ with $\{v, v''\} \in E$ and $\{v', v''\} \notin E$. Suppose that (ii) is false so that $v' < v$ holds. If now we would have $v < v''$, the transitivity of $<$ would yield $v' < v''$, so that $\{v', v''\} \in E$ would follow, with contradiction. In the case $v'' < v$ we would obtain that v'' and v' are comparable, since then they would be predecessors (in a tree) of the same v. Again $\{v', v''\} \in E$ would hold with contradiction.

(ii) yields that not both ends v, v' of an edge of E can be a V - root of $\{v, v'\}$ since this would imply $v < v' < v$. So we obtain the theorem:

3.6 Theorem. *A graph G which admits a tree- orientation has no 2 - edges.*

The main theorem now states that 3.5 can be inverted.

3.7 Theorem [93]. *Let $G = (V, E)$ be a graph without 2 - edges, and let $<$ be an orientation of G. Then $<$ is a tree-orientation iff (i) and (ii) are satisfied.*

Proof. It remains to prove that (i) and (ii) imply that $<$ is transitive and a tree-orientation, namely that every principal ideal $(v]$ of $(V, \leq)$ is linearly ordered. Let elements v_1, v_2, v_3 with $v_1 < v_2$ and $v_2 < v_3$ of V be given. Then $\{v_1, v_2\}$ and $\{v_2, v_3\}$ are edges of E. We have to prove $v_1 < v_3$ and make the indirect assumption that this does not hold.

v_2 is no V - root of $\{v_1, v_2\}$, for otherwise (ii) would imply $v_2 < v_1$ with contradiction. This implies that $\{v_1, v_3\} \in E$, and since we made the indirect assumption that $v_1 < v_3$ does not hold, we now have $v_3 < v_1$.

At least one of the edges $\{v_1, v_2\}, \{v_2, v_3\}, \{v_1, v_3\}$ is no 0 - edge, for otherwise we would have that v_1, v_2, v_3 are all in one equivalence class, and by (i) this class would be linearly ordered, and then $v_1 < v_3$ would follow, contradicting our indirect assumption.

Together with the indirect assumption we have $v_1 < v_2$, $v_2 < v_3$, $v_3 < v_1$.

Due to symmetry we can assume w.r.o.g. that $\{v_1, v_3\}$ is no 0 - edge. And so v_3 or v_1 is a V-root of $\{v_1, v_3\}$. But v_1 cannot be that, for otherwise we would by (ii) have $v_1 < v_3$, contradicting the indirect assumption. Thus v_3 is a V-root of $\{v_1, v_3\}$. So there further exists a vertex $v_4 \neq v_1$ with $\{v_1, v_4\} \notin E$ and $\{v_3, v_4\} \in E$. If $\{v_2, v_4\} \notin E$ holds, v_3 would be a V - root of $\{v_2, v_3\}$. This entails $v_3 < v_2$ by (ii), a contradiction. In the case $\{v_2, v_4\} \in E$ the vertex v_2 would be a V - root of $\{v_1, v_2\}$, and then $v_2 < v_1$, again with contradiction. So in every case our indirect assumption leads to a contradiction, and the transitivity of $<$ is proved.

Now we verify the tree-property: Let v, v', v'' be vertices with $v' < v$, $v'' < v$, and $v' \neq v''$. Then $\{v, v'\}$ and $\{v, v''\}$ are in E. If neither $v' < v''$ nor $v'' < v'$ would hold, then $\{v', v''\} \notin E$ follows, and v would be a V - root of $\{v, v'\}$, and by (ii) $v < v'$, contradicting $v' < v$. So all predecessors of a vertex v are comparable, and $(v]$ is linearly ordered.

3.8 Remark. By the above theorem all tree-orientations of a graph without 2 - edges are determined: First we orient all 1 - edges according to (ii). Then we choose an arbitrary linear order in each equivalence class. This induces an orientation of all 0 - edges.

In particular we have obtained the criterion of Wolk:

3.9 Theorem [180]. *A graph G has a tree-orientation iff G has no 2 - edges.*

For finite graphs one obtains from 3.7: A graph $G = (V, E)$ without 2 - edges has exactly $\Pi_{i \in I} |C_i|$! different tree-orientations. Here the $\{C_i, i \in I\}$ is the set of all equivalence classes of V.

Finally we mention without proof a theorem of Arditti/Jung [6]:

3.10 Theorem. *If $G = (V, E)$ is a comparability graph of two orders $\leq_1$ and $\leq_2$ on V, then the posets $(V, \leq_1)$ and $(V, \leq_2)$ have the same dimension.*

In view of this theorem it makes sense to define the *dimension* of a comparability graph $G = (V, E)$ as the dimension of an arbitrary poset $(V, \leq)$, which has G as comparability graph.

For the special case where G is a finite graph the statement of 3.10 was also obtained by Trotter/Moore/Sumner [171] and Gysin [61].

A penetrating analysis of finite comparability graphs is contained in Gallai's paper [47] of 1967, which characterizes these graphs in terms of the minimum list of graphs that are excluded as induced subgraphs.

We also refer to the survey article [95] "Comparability graphs" of D. Kelly.

References

[1] Abraham, U. and Bonnet, R.: Hausdorff's theorem for posets that satisfy the finite antichain property, Fund. Math. **159** (1999), 51- 69.

[2] Adhikari, S. D., Aspects of Combinatorics and Combinatorial Number Theory, CRC Press, 2002.

[3] Alexandroff, P. and Urysohn, P., Mémoire sur les espaces topologiques compacts, Verh. Nederl. Akad. Wetensch. Sect.I, **14** (1929), 1- 93.

[4] Alling, N. L., Foundations of Analysis over Surreal Number Fields, North-Holland-Amsterdam. New York. Oxford. Tokyo. 1987.

[5] Anderson, I., Combinatorics of Finite Sets, Clarendon Press, Oxford 1987.

[6] Arditti, J. C. and H. A. Jung, The dimension of finite and infinite comparability graphs, J. London Math. Soc. (2) **21** (1980), 31-38.

[7] Baker, K. A., Fishburn, P. C., and Roberts, F. S., Partial orders of dimension 2, Networks **2** (1971), 11-28.

[8] Banaschewski, B., Hüllensysteme und Erweiterung von Quasi-Ordnungen, Z. math. Logik Grundl. Math. **2** (1956), 117-130.

[9] Birkhoff, G., Lattice Theory 1948.

[10] Bogart, K., Maximal dimensional partially ordered sets. I, Discr. Math. **5** (1973), 21-32.

[11] Bonnet, R. and Pouzet, M., Extension et stratification d'ensembles dispersés. C. R. Acad. Sci. Paris **268** (A) (1969), 1512-1515.

[12] Carruth, P. W., Arithmetic of ordinals with applications to the theory of ordered abelian groups, Bull. Amer. Math. Soc. **48** (1942), 262-271.

[13] Chajoth, J., Beitrag zur Theorie der geordneten Mengen, Fund. Math. **16** (1930), 132-133.

[14] Conway, J. H., On numbers and games, Academic Press, London & New York (1976).

[15] Crawley, P. and Dean, R., Free lattices with infinite operations, Trans. Amer. Math. Soc. **92** (1959), 35-47.

[16] Cuesta Dutari, N., Algebra Ordinal, Rev. Acad. Ci. Madrid **48** (1954), 103-145.

[17] Cuesta Dutari, N., Matematica del Orden, Madrid 1958.

[18] Davey, B. A. and Priestley, H. A., Introduction to lattices and order, Cambridge University Press, Cambridge 2002.

[19] De Bruijn, N. G., Tengbergen, Ca.van E., and Kruyswijk, D., On the set of divisors of a number, Nieuw. Arch. Wisk. (2) **23** (1949-51), 191-193.

[20] De Bruijn, N.G. and Erdös, P., A colour problem for infinite graphs and a problem in the theory of relations, Nederl. Akad. Wetensch. Proc. Ser. A **54** (1951), 371-373.

[21] De Jongh, D. H. J. and Parikh, R., Well-partial orderings and hierarchies, Indag. Math. **39** (1977), 195-206.

[22] Devlin, K. J., and Johnsbraten, H., The Suslin Problem, 1974, Lecture Notes in Math. vol. 405, Springer Verlag (Berlin).

[23] Diestel, R., Graphentheorie, Springer 2000. And Graph Theory, Springer 1997, New York.

[24] Dilworth R .P., A decomposition theorem for partially ordered sets, Ann. Math. (2) **51** (1950), 161-166.

[25] Dilworth, R. P., Some combinatorial problems on partially ordered sets, in: Combinatorial Analysis, Proc. Symp. Pure and Appl. Math. X AMS (1960), 85-90.

[26] Dilworth, R.P., and Gleason, A.M., A generalized Cantor theorem, Proc. Amer. Math. Soc. **13** (1962), 704-705.

[27] Droste, M., and MacPherson, On k-homogeneous posets and graphs, J. Comb. Theory (Ser. A), **56** (1991), 1-15.

[28] Droste, M., MacPherson, D., and Mekler, A., Uncountable homogeneous partial orders, preprint 2001.

[29] Dushnik, B. and Miller, E., Concerning similarity transformations of linearly ordered sets, Bull.Amer. Math. Soc. **46** (1940), 322-326.

[30] Dushnik, B. and Miller, E., Partially ordered sets, Amer. J. Math. **63** (1941), 600-610.

[31] Dyer, E., A fixed point theorem, Proc. Amer. Math. Soc. **7** (1956), 662-672.

[32] Engel, K., and Gronau, H., Sperner Theory in Partially Ordered Sets, Teubner Verlagsges. Leipzig, 1985.

[33] Erdös, P., On a lemma of Littlewood and Offord, Bull. Amer. Math. Soc. **51** (1945), 898-902.

[34] Erdös, P. and Szekeres, G., A combinatorial problem in geometry. Comp. Math. **2** (1935), 463-470.

[35] Erdös, P., and Rado, R., A partition calculus in set theory, Bull. Amer. Math. Soc. **62** (1956), 427-489.

[36] Erdös, P., and Rado, R., A problem on ordered sets. J. London Math. Soc. **28** (1953), 426-438.

[37] Erdös, P. and Rado, R., unpublished manuscript, cited in [84].

[38] Erdös, P., Hajnal, A., Mate, A., Rado, R., Combinatorial set theory: Partition relations for cardinals, Akademiai Kiado, Budapest 1984.

[39] Erdös, P., Problem 4358, Amer. Math. Monthly **56** (1949), 480.

[40] Erdös, P., Some set-theoretical properties of graphs, Revista Universidad Nacional de Tucuman, Serie A vol.3 (1942), 363-367.

[41] Erdös, P., Ko, C., and Rado, R., Intersection theorems for systems of finite sets, Q. J. Math. Oxf. (2) **12** (1961), 313-320.

[42] Erné, M., Einführung in die Ordnungstheorie, B.I. Wissenschaftsverlag, Zürich 1982.

[43] Fishburn, P.C., Interval Orders and Interval Graphs, Wiley, New York.

[44] Fraissé, R., Sur la comparaison des types d'ordres, C. R. Acad. Sci. Paris **226** (1948), 1330-1331.

[45] Fraissé, R., Theory of Relations, Elsevier 2000, Amsterdam.

[46] Freese, R., An application of Dilworth's lattice of maximal antichains, Discr. Math. **7** (1974), 107-109.

[47] Gallai, T., Transitiv orientierbare Graphen, Acta Math. Acad. Sci. Hungar. **18** (1967), 25-66.

[48] Galvin, F. and Shelah, S., Some counterexamples in the partition calculus, J. Comb. Theory **15** (1973), 167-174.

[49] Galvin, F., A proof of Dilworth's chain decomposition theorem, Amer. Math. Monthly **101** (1994), 352-353.

[50] Galvin, F., On a partition theorem of Baumgartner and Hajnal, Colloquia Mathematica Societatis Janos Bolyai, **10** (1973).

[51] Ghouila-Houri, A., Caractérisation des graphes non orientés don't on peut orienter les arêtes de manière a obtenir le graphe d'une relation d'ordre, C. R. Acad. Sci. Paris **254** (1962),1370-1371.

[52] Gillman, L., Remarques sur les ensembles η_α , C. R. Acad. Sci. Paris **241** (1955), 12-13.

[53] Gillman, L., Some remarks on η_α -sets, Fund. Math. **43** (1956), 77-82.

[54] Gilmore, P. C. and Hoffman, A.J., A characterization of comparability graphs and of interval graphs, Canad. J. Math. **16** (1964), 539-548.

[55] Ginsburg, J., A note on the cardinality of infinite partially ordered sets, Pacific J. Math. **106** (1983), 265-270.

[56] Ginsburg, S., Further results on order types and decompositions of sets, Trans. Amer. Math. Soc. **77** (1954), 122-150.

[57] Ginsburg, S., Order types and similarity transformations, Trans. Amer. Math. Soc. **79** (1955), 341-361.

[58] Ginsburg, S., Some remarks on order types and decompositions of sets, Trans. Amer. Math. Soc. **74** (1953), 514-535.

[59] Gottschalk, W. H., Choice functions and Tychonoff's theorem, Proc. Amer. Math. Soc. **2** (1951), 172.

[60] Grätzer, G., General Lattice Theory, 2. edition, (2003), 684 pp., Birkhäuser, Boston-Basel-Berlin.

[61] Gysin, R., Dimension transitiv orientierbarer Graphen, Acta Math. Acad. Sci. Hungar. **29** (1977), 313-316.

[62] Harzheim, E., A constructive proof for a theorem on contractive mappings in power sets, Discr. Math. **45** (1983), 99-106.

[63] Harzheim, E. Ein kombinatorisches Problem über Auswahlfunktionen. Publ. Math. (Debrecen) **15** (1968), 19-22.

[64] Harzheim, E., Beiträge zur Theorie der Ordnungstypen, insbesondere der η_α -Mengen, Math. Ann. **154** (1964), 116-134.

[65] Harzheim, E., Blocksysteme auf einer Menge, in Theory of sets and topology (in honour of Felix Hausdorff) pp. 221-232, VEB Deutsch. Verlag Wissensch., Berlin 1972.

[66] Harzheim, E., Combinatorial theorems on contractive mappings in power sets, Discr. Math. **40** (1982), 193-201.

[67] Harzheim, E., Dualzerlegungen in totalgeordneten Mengen, Fund. Math. **53** (1963), 81-91.

[68] Harzheim, E., Ein Endlichkeitssatz über die Dimension teilweise geordneter Mengen, Math. Nachr. **46** (1970),183-188.

[69] Harzheim, E., Ein Satz der kombinatorischen Mengenlehre, Fund. Math. **61** (1968), 283-294.

[70] Harzheim, E., Einbettung totalgeordneter Mengen in lexikographische Produkte, Math. Ann. **170** (1967), 245-252.

[71] Harzheim, E., Einbettungssätze für totalgeordnete Mengen, Math. Ann. **158** (1965), 90-108.

[72] Harzheim, E., Eine kombinatorische Frage zahlentheoretischer Art, Publ. Math. (Debrecen) **14** (1967), 45-51.

[73] Harzheim, E., Kombinatorische Betrachtungen über die Struktur der Potenzmenge, Math. Nachr. **34** (1967), 123-141.

[74] Harzheim, E., Mehrfach wohlgeordnete Mengen und eine Verschärfung eines Satzes von Lindenbaum, Fund. Math. **53** (1964), 155-172.

[75] Harzheim, E., On the Dedekind-MacNeille closure of an ordered set and a theorem of Novák, Result. Math. **41** (2002),114-117.

[76] Harzheim, E., On topological properties of cartesian products of linearly ordered continua, Proceedings of the Conference on ordered sets and their application (Château de la Tourette, l'Arbrèsle, July 1982), 171-192.

[77] Harzheim, E., Remarks on Dilworth's decomposition theorem, Ars Comb. **16 B** (1983), 27-31.

[78] Harzheim, E., Some results on combinatorial set theory, in Combinatorial Structures and their Applications (Proc. Calgary Intern. Conf., Calgary, pp. 159-161. Gordon and Breach, New York (1970).

[79] Harzheim, E., Über universal geordnete Mengen, Math. Nachr. **36** (1968),195-213.

[80] Harzheim, E., Verallgemeinerung des Zerlegungssatzes von Jordan-Brouwer-Alexander auf Produkte lineargeordneter Kontinuen, Arch. Math. **46** (1986), 271-274.

[81] Hausdorff, F., Grundzüge der Mengenlehre, Chelsea Publishing Company, New York 1949.

[82] Hausdorff, F., Grundzüge einer Theorie der geordneten Mengen, Math. Ann. **65** (1908), 435-505.

[83] Hausdorff, F., Untersuchungen über Ordnungstypen I-V, Berichte der Sächs. Ges.d.Wiss. **58** (1906), 106-169, and **59** (1907).

[84] Higman, G., Ordering by divisibility in abstract algebras, Proc. London Math. Soc. **2** (1952), 326-336.

[85] Higman, G. and Neumann, B. H., Amer. Math. Monthly **59**, p. 257.

[86] Hiraguchi, T., On the dimension of orders, Sci. Rep. Kanazawa Univ. **4** (1955),1-20.

[87] Holland, C., The lattice-ordered group of automorphisms of an ordered set, Michigan Math. J. **10** (1963), 399-408.

[88] Hrbacek, K. and Jech, T., Introduction to Set Theory, Marcel Dekker. Inc. New York and Basel 1978.

[89] Jech, T., Nonprovability of Souslin's hypothesis, Comment. Math. Univ. Carolinae **8** (1967), 291-305.

[90] Jensen, R. B., Souslin's hypothesis is incompatible with V = L (Abstract), Notices Amer. Math. Soc. **15** (1968), 935.

[91] Jensen, R. B., Definable sets of minimal degree, In: Mathematical Logic and Foundations of Set Theory (Y. Bar-Hillel, ed.), pp. 122-128. North-Holland Publ. Co., Amsterdam, 1970.

[92] Jonsson, B., Universal relational systems, Math. Scand. **4** (1956), 193-208.

[93] Jung, H. A., Zu einem Satz von E. S. Wolk über die Vergleichbarkeitsgraphen von ordnungstheoretischen Bäumen, Fund. Math. **63** (1968), 217-219.

[94] Katona, G. O. H., A simple proof of the Erdös-Ko-Rado theorem, J. Comb. Theory B **13** (1972 b), 183-184.

[95] Kelly, D., Comparability graphs, in Graphs and Order, D. Reidel Publishing Company, Dordrecht/Boston/Lancaster, (ed. Ivan Rival) (1985), 3-40.

[96] Kinna, W. and Wagner, K., Über eine Abschwächung des Auswahlpostulates, Fund. Math. **42** (1955), 75-82.

[97] Kleitman, D. and Lewin, M., Amer. Math. Monthly **79** (1972), 152-154.

[98] Kleitman, D., On an extremal property of antichains in partial orders. In Combinatorics (eds. M. Hall and J. H. van Lint), Math. centre Tracts **55** (1974), 77-90. Mathematics Centre, Amsterdam.

[99] Kneebone, G. T. and Rotman, B., The Theory of Sets and Transfinite Numbers, Oldbourne, London 1966.

[100] Komm, H., On the dimenion of partially ordered sets. Amer. J. Math. **70** (1948), 507-520.

[101] Kruskal, J. B., Well-quasi-ordering and Rado's conjecture, unpublished.

[102] Kruskal, J. B., Well-quasi-ordering, the tree theorem, and Vazsonyi's conjecture, Trans. Amer. Math. Soc. **95** (1960), 210-225.

[103] Kühn, Daniela, On well-quasi-ordering infinite trees- Nash-Williams's theorem revisited, Math. Proc. Camb. Phil. Soc. **130** (2001), 401-408.

[104] Kurepa, G., Ensembles ordonnés et ramifiés, Thèse, Paris 1935; aussi Publ. Math. Univ. Belgrade, **IV**, 1935, 1-138.

[105] Kurepa, G., Ensembles ordonnés et leurs sous-ensembles bien ordonnés, C. R. Acad. Sci. Paris **242** (1956), 2202-2203.

[106] Kurepa, G., On the cardinal number of ordered sets and of symmetrical structures in

dependence on the cardinal number of its chains and antichains, Glasnik Mat.-Fiz. Astrnom. Drustvo, Mat. Fiz. Hrvatske Ser. II, **14** (1959), 183-203.

[107] Kurepa, G., On universal ramified sets. Glasnik Mat.-Fiz. Astr. **18** (1963), 17-26.

[108] Kurepa, G., Partitive sets and ordered chains, Rad. Jugoslav. Acad. Znan. Umjet. Odjel. Mat. Fiz. Techn. Nauke **6** (302), (1957), 197-235.

[109] Kurepa, G., Remark on recent two results of Dilworth and Gleason, Publications de l'institut Mathématique **4** (18), (1964), 107-108.

[110] Kurepa, G., Über das Auswahlaxiom, Math. Ann. **126** (1953), 381-384.

[111] Laver, R., On Fraissé's order type conjecture, Annals of Math. **93**, 89-111 (1971).

[112] Lindenbaum, A., Z. teorii uporzadkowania wielokrotnego (Sur la théorie de l'ordre multiple), Wiadomosci Matematyczne **37** (1934), 1-35.

[113] Löttgen, U. and Wagner, K., Über eine Verallgemeinerung des Jordanschen Kurvensatzes auf zweifach geordnete Mengen, Math. Ann. **139** (1959), 115-126.

[114] Lovasz, L., Combinatorial Problems and Exercises, Akademiai Kiado Budapest 1979.

[115] Lubell, D., A short proof of Sperner's theorem, J. Comb. Theory **1** (1966), 299.

[116] MacNeille, H.M., Partially ordered sets, Trans. Amer. Math. Soc. **42** (1937), 416-460.

[117] Meschalkin, L. D. , A generalization of Sperner's theorem on the number of subsets of a finite set, Theor. Probl. Appl. **8** (1963), 203-204.

[118] Mendelson, E., On a class of universal ordered sets, Proc. Amer. Math. Soc. **9** (1958), 712-713.

[119] Michael, E., A class of partially ordered sets, Amer. Math. Monthly **67** (1960), 448-449.

[120] Milner, E. C. and Sauer, N. On chains and antichains in well-founded partially ordered sets, J. London Math. Soc.(2), **24**, (1981), 15-33.

[121] Milner, E.C., Basic well-quasi-ordering and better-quasi-ordering theory, in Graphs and Order, Reidel Publishing Company, Dordrecht, Holland, (1985), 487-502.

[122] Nash-Williams, C. St. J. A., On well-quasi-ordering infinite trees, Proc. Cambridge. Philos. Soc. **61** (1965), 697-720.

[123] Nash-Williams, C. St. J. A., On well-quasi-ordering finite trees, Proc. Cambridge Philos. Soc. **59** (1963), 833-835.

[124] Nash-Williams, C. St. J. A., On well-quasi-ordering lower sets of finite trees. Proc.Cambridge Philos. Soc. **60** (1964), 369-384.

[125] Neumer, W., Über Mischsummen von Ordinalzahlen, Arch. Math. **5** (1954), 244-248.

[126] Novák, J., On partition of an ordered continuum, Fund. Math. **39** (1952), 53-64.

[127] Novák, V. Über eine Eigenschaft der Dedekind-MacNeilleschen Hülle, Math. Ann. **179** (1969), 337-342.

[128] Novotný, M., On representation of partially ordered sets by sequences of zeros and ones (cech). Casopis Pest. Mat. **78** (1953), 61-64.

[129] Novotný, M., Sur la représentation des ensembles ordonnés, Fund. Math. **39** (1952), 97-102.

[130] Oellrich, H. and Steffens, K., On Dilworth's decomposition theorem, Discr. Math. **15** (1976), 301-304.

[131] Ohkuma, T., Sur quelques ensembles ordonnés linéairement, Fund. Math. **43** (1956), 326-337.

[132] Ore, O. Theory of graphs, Amer. Math. Soc. Colloquium Publications 38, 1962.

[133] Padmavally, K., A remark on order types. Fund. Math. **42** (1955), 312-318.

[134] Padmavally, K., Generalization of rational numbers. Rev. Mat. Hisp. Amer. **12** (1952), 249-265.

[135] Padmavally, K., Generalization of the ordertype of rational numbers, Rev. Mat. Hisp. Amer. **14** (1954), 50-73.

[136] Peck, G. W., Shor, P., Trotter W. T., West, D. B., Regressions and monotone chains: A Ramsey-type extremal problem for partial orders, Combinatorica **4** (1) (1984), 117-119.

[137] Peirce C. S., On the algebra of logic, I-II. Amer. J. Math. **3** (1880), 15-57 and **7** (1884), 180-202.

[138] Perles, M. A., On Dilworth's theorem in the infinite case. Israel J. Math. **1** (1963),108-109.

[139] Perles, M. A., A proof of Dilworth's decomposition theorem for partially ordered sets, Israel J. Math. **1** (1963), 105-107.

[140] Perry, R.L., Representatives of subsets, Journal of Comb.Theory **3** (1967), 302-304.

[141] Pouzet, M., Sur les chaînes d'un ensemble partiellement bien ordonné, Publ. Dept. Math. (Lyon) **16** (1979), 21-26.

[142] Pouzet, M., Applications of well quasi-ordering and better quasi-ordering, in Graphs and Order, Reidel Publishing Company, Dordrecht, Holland (1985), 503-519.

[143] Pouzet, M., Généralisation d'une construction de Ben Dushnik-E.W.Miller, Comptes rendus **269A** (1969), 877-879.

[144] Pouzet, M., and Richard, D., (editors), Orders: Description and Roles, North-Holland, Mathematics Studies 99, Annals of Discrete Mathematics **23** (1984).

[145] Pretzel, O., Another proof of Dilworth's decomposition theorem, Discr. Math. **25** (1979), 91-92.

[146] Rado, R., A Theorem on Chains of Finite Sets, J. London Math. Soc. **42** (1967), 101-106.

[147] Rado, R., A Selection Lemma, Journ. Comb. Theory **10** (1971), 176-177.

[148] Rado, R., Axiomatic treatment of rank in infinite sets, Canad. J. Math. **1** (1949), 337-343.

[149] Rado, R., Partial well-ordering of sets of vectors, Mathematica **1** (1954), 89-95.

[150] Ramsey, F. P., On a problem of formal logic, Collected papers 82-111.

[151] Rival, I., ed., Algorithms and order, Kluwer, Dordrecht-Boston (1989).

[152] Rival, I., ed., Graphs and order, Dordrecht-Reidel, Boston (1984).

[153] Rival, I., ed., Ordered sets, Dordrecht-Reidel, Boston (1982).

[154] Rosenstein, J. G., Linear Orderings, 487 p., Academic Press, New York (1982).

[155] Schmidt, Diana, The relation between the height of a well founded partial ordering and the order types of its chains and antichains. Unpublished manuscript (1979).

[156] Schmidt, J., Zur Kennzeichnung der Dedekind-MacNeilleschen Hülle einer geordneten Menge. Arch. Math. **7** (1956), 241-249.

[157] Schröder, B. S. W., Ordered Sets, An Introduction, Birkhäuser, Boston, Basel, Berlin 2003.

[158] Schröder, E., Algebra der Logik, Vol.1, Teubner-Verlag, Leipzig 1890.

[159] Shen, A., and Vereshchagin, N. K., Basic Set Theory, Student Mathematical Library, Vol. 17 (2002).

[160] Sierpinski, W., Cardinal and Ordinal Numbers, Polska Akademia Nauk, Monografie Matematyczne, Warszawa 1958.

[161] Sierpinski, W., Sur les types d'ordre des ensembles linéaires, Fund. Math. **37** (1950), 253-264.

[162] Sierpinski, W., Sur un problème de la théorie des relations. Ann Scuolo Norm. Sup. Pisa **2** (1933), 285-287.

[163] Sierpinski, W., Sur un problème concernant les sous-ensembles croissants du continu, Fund. Math. **3** (1922), 109-112.

[164] Sierpinski,. W., Sur une propriété des ensembles ordonnés. Fund. Math. **36** (1949), 56-67.

[165] Solovay, R. M., and Tennenbaum, S., Iterated Cohen extensions and Souslin's problem. Ann. of Math. **94** (1971), 201-245.

[166] Sperner, E., Ein Satz über Untermengen einer endlichen Menge, Math. Zeitschr. **27** (1928), 544-548.

[167] Szpilrajn, J. Sur l'extension de l'ordre partiel, Fund. Math. **16** (1930), 386-389.

[168] Teichmüller, O., Braucht der Algebraiker das Auswahlaxiom ?, Deutsche Math. **4** (1939), 567-577.

[169] Tennenbaum, S., Souslin's problem, Proc. Nat. Acad. U.S.A. **59** (1968), 60-63.

[170] Trotter, W. T., Inequalities in dimension theory for posets, Proc. Amer. Math. Soc. **47** (1975), 311-316.

[171] Trotter, W. T., Moore, J. I., and Sumner, D.P., The dimension of a comparability graph, Proc. Amer. Math. Soc. **60** (1976), 35-38.

[172] Trotter, W.T., Partially Ordered Sets, Handbook of Combinatorics Vol I, Chapter 8 (1995), 433-480, Elsevier.

[173] Urysohn, P., Remarque sur ma note : Un théorème sur la puissance des ensembles ordonnés, Fund. Math. **6** (1924), 278.

[174] Urysohn, P., Un théorème sur la puissance des ensembles ordonnés. Fund. Math. **5** (1923), 14-19.

[175] Wagner, K., Verallgemeinerung des Brouwerschen Invarianzsatzes der Dimensionszahl mittels eines allgemeinen Stetigkeitsbegriffs von Abbildungen mehrfach geordneter Mengen, Math. Ann. **120** (1949), 514-532.

[176] Ward, M., A characterization of Dedekind structures, Bull. Amer. Math. Soc. **45** (1939), 448-451.

[177] Weinberg, E., Automorphism groups of minimal η_α-sets, in: Ordered Groups, Proceedings of the Boise State Conference, 1980, 71-79.

[178] Wolk, E. S., On decompositions of partially ordered sets. Proc. Amer. Math. Soc. **15** (1964), 197-199.

[179] Wolk, E. S., Partially well ordered sets and partial ordinals, Fund. Math. **60** (1967), 175-186.

[180] Wolk, E. S., The comparability graph of a tree, Proc. Amer. Math. Soc. **13** (1962), 789-795.

[181] Yamamoto, K., Logarithmic order of free distributive lattices, J. math. Soc. Japan **6** (1954), 343-353.

[182] Zorn, M., A remark on method in transfinite algebra, Bull. Amer. Math. Soc. **41** (1935), 667-670.

Index

List of symbols

$\mathbf{N}$, 1
$\mathbf{N}_0$, 1
ω, 1
$\mathbf{Z}$, 1
$\mathbf{Q}$, 1
$\mathbf{R}$, 1
$\mathbf{R}^n$, 1
$\emptyset$, 1
$\mathfrak{P}(\ )$, 1
$\subseteq$, 1
$\supseteq$, 1
$\subset$, 1
$\supset$, 1
$\circ$, 2
$\cup$, 3
$\cap$, 3
X, 4
$\times$, 4
Π, 4
AC, 5
On, 6
ω_ν, 8
$\aleph_\nu$, 8
GCH, 9
ZF, 9
$S \times S$, 11
R^c, 11
R^{-1}, 11
$I(S)$, 11
id_S, 11
$R \upharpoonright T$, 11
$R_1 \circ R_2$, 11
R^n, 12
TH, 12
$\leq$, 14
$<$, 14

$\nleq$, 14
$\nless$, 14
$>$, 14
$\geq$, 14
$\|$, 19
$\lhd$, 19
$\rhd$, 19
inf, 22
sup, 22
min, 22
max, 22
Mi, 23
Ma, 23
$h(a)$, 24
$h(P)$, 24
L_n, 24
$r(x)$, 27
$r_n(P)$, 28
$\mathfrak{I}(P)$, 29
$IS(T)$, 30
$FS(T)$, 30
$Conv\ T$, 30
$Fix(c)$, 30
$Inc(a)$, 31
$(P < a)$, 31
$(P \leq a)$, 31
$(P > a)$, 31
$(P \geq a)$, 31
$(P \| a)$, 31
$(a]$, 31
$[a)$, 31
$[a,b]$, 31
$(a,b]$, 31
$[a,b)$, 32
(a,b), 32
$A < B$, 31

$A > B$, 31
$A < z$, 31
$A > z$, 31
$\simeq$, 35
tp, 36
$\leq$ for types, 37
$<$ for types, 37
η, λ, 38
$L(T)$, 40
$U(T)$, 40
$C(T)$, 40
$DM(T)$, 40
type $(1,1)$, 47
type $(\infty,1)$, 47
type $(1,\infty)$, 47
type (∞,∞), 47
$LK(P)$, 55
$w(P)$, 57
$CCN(P)$, 57
$[S]^2$, 59
$\wedge$, 62
$\vee$, 62
$\omega(|M|)$, 71
$ord(M)$, 71
cf, 73
coin, 73
$\sum$, 85
X, 86
L, 87
$\mu((\tau))$, 88
$S((\tau))$, 88
H_α, 98
k_α, 99
h_α, 101
η_α, 101
H'_α, 105

8056027R0

Made in the USA
Lexington, KY
04 January 2011